T0202990

Lecture Notes in Computer Science   14423

The series Lecture Notes in Computer Science (LNCS), including its subseries Lecture Notes in Artificial Intelligence (LNAI) and Lecture Notes in Bioinformatics (LNBI), has established itself as a medium for the publication of new developments in computer science and information technology research, teaching, and education.

LNCS enjoys close cooperation with the computer science R & D community, the series counts many renowned academics among its volume editors and paper authors, and collaborates with prestigious societies. Its mission is to serve this international community by providing an invaluable service, mainly focused on the publication of conference and workshop proceedings and postproceedings. LNCS commenced publication in 1973.

Weili Wu · Guangmo Tong

Editors

Computing and Combinatorics

29th International Conference, COCOON 2023
Hawaii, HI, USA, December 15–17, 2023
Proceedings, Part II

 Springer

Editors
Weili Wu (ID)
University of Texas at Dallas
Richardson, TX, USA

Guangmo Tong (ID)
University of Delaware
Newark, DE, USA

ISSN 0302-9743   ISSN 1611-3349 (electronic)
Lecture Notes in Computer Science
ISBN 978-3-031-49192-4   ISBN 978-3-031-49193-1 (eBook)
https://doi.org/10.1007/978-3-031-49193-1

This Springer imprint is published by the registered company Springer Nature Switzerland AG
The registered company address is: Gewerbestrasse 11, 6330 Cham, Switzerland

Paper in this product is recyclable.

Preface

The papers in these proceedings, which consist of two volumes, were presented at the 29th International Computing and Combinatorics Conference (COCOON 2023), on December 15–17, 2023, in Honolulu, Hawaii, USA. The topics cover most aspects of theoretical computer science and combinatorics pertaining to computing.

In total 60 papers were selected from 146 submissions by an international program committee consisting of a large number of scholars from various countries and regions, distributed all over the world, including Asia, North America, Europe, and Australia. Each paper was evaluated by at least three reviewers. The decision was made based on those evaluations through a process containing a discussion period.

Authors of selected papers come from the following countries and regions: Australia, Canada, China (including Hong Kong, Macau, and Taiwan), Czechia, France, Germany, India, Israel, Japan, Sweden, and the USA. Many of these papers represent reports of continuing research, and it is expected that most of them will appear in a more polished and complete form in scientific journals.

We wish to thank all who have made this meeting possible and successful, the authors for submitting papers, the program committee members for their excellent work in reviewing papers, the sponsors, the local organizers, and Springer for their support and assistance. We are especially grateful to Lian Li and Xiaoming Sun, who lead the Steering committee, for making the ranking of COCOON go up significantly in recent years, and to Yi Zhu and Xiao Li, who made tremendous efforts on local arrangements and set-up.

December 2023

Weili Wu
Guangmo Tong

Organization

General Co-chairs

Peter Varman   Rice University, USA
Ding-Zhu Du   University of Texas at Dallas, USA

PC Co-chairs

Weili Wu   University of Texas at Dallas, USA
Guangmo Tong   University of Delaware, USA

Web Co-chairs

Xiao Li   University of Texas at Dallas, USA
Ke Su   University of Texas at Dallas, USA

Finance Co-chair

Jing Yuan   University of Texas at Dallas, USA

Registration Chair

Xiao Li   University of Texas at Dallas, USA

Local Chair

Yi Zhu   Hawaii Pacific University, USA

Program Committee Members

An Zhang	Hangzhou Dianzi University, China
Bhaskar Dasgupta	University of Illinois at Chicago, USA
Bo Li	Hong Kong Polytechnic University, China
Boting Yang	University of Regina, Canada
C. Pandu Rangan	Indian Institute of Technology Madras, India
Chee Yap	New York University, USA
Chia-Wei Lee	National Tatung University, Taiwan
Christos Zaroliagis	University of Patras, Greece
Chung-Shou Liao	National Tsing Hua University, Taiwan
Deshi Ye	Zhejiang University, China
Dominik Köppl	Tokyo Medical and Dental University, Japan
Eddie Cheng	Oakland University, USA
Gruia Calinescu	Illinois Institute of Technology, USA
Guohui Lin	University of Alberta, Canada
Haitao Wang	University of Utah, USA
Hans-Joachim Boeckenhauer	ETH Zurich, Switzerland
Ho-Lin Chen	National Taiwan University, Taiwan
Hsiang-Hsuan Liu	Utrecht University, The Netherlands
Jiangxiong Guo	Beijing Normal University at Zhuhai, China
Joong-Lyul Lee	University of North Carolina at Pembroke, USA
Jou-Ming Chang	National Taipei University of Business, Taiwan
Kai Jin	Sun Yat-sen University, China
Kunihiko Sadakane	University of Tokyo, Japan
Ling-Ju Hung	National Taipei University of Business, Taiwan
M. Sohel Rahman	Bangladesh University of Engineering and Technology, Bangladesh
Manki Min	Louisiana Tech University, USA
Micheal Khachay	Ural Federal University, Russia
Ovidiu Daescu	University of Texas at Dallas, USA
Pavel Skums	Georgia State University, USA
Peng Li	Chongqing University of Technology, China
Peng Zhang	Shandong University, China
Peter Rossmanith	RWTH Aachen University, Germany
Prudence Wong	University of Liverpool, UK
Qilong Feng	Central South University, China
Qiufen Ni	Guangdong University of Technology, China
Raffaele Giancarlo	University of Palermo, Italy
Ralf Klasing	CNRS and University of Bordeaux, France
Ryuhei Uehara	Japan Advanced Institute of Science and Technology, Japan

Contents – Part II

Algorithm in Networks

Contents – Part I

Combinatorics and Algorithms

Quantum Query Lower Bounds for Key Recovery Attacks on the Even-Mansour Cipher

Akinori Kawachi[✉] and Yuki Naito

Mie University, Tsu, Japan
kawachi@info.mie-u.ac.jp

Abstract. The Even-Mansour (EM) cipher is one of the famous constructions for a block cipher. Kuwakado and Morii demonstrated that a quantum adversary can recover its n-bit secret keys only with $\mathcal{O}(n)$ nonadaptive quantum queries. While the security of the EM cipher and its variants is well-understood for classical adversaries, very little is currently known of their quantum security. Towards a better understanding of the quantum security, or the limits of quantum adversaries for the EM cipher, we study the quantum query complexity for the key recovery of the EM cipher and prove every quantum algorithm requires $\Omega(n)$ quantum queries for the key recovery even if it is allowed to make adaptive queries. Therefore, the quantum attack of Kuwakado and Morii has the optimal query complexity up to a constant factor, and we cannot asymptotically improve it even with adaptive quantum queries.

Keywords: Quantum computing · Query complexity · Lower bounds · Symmetric-key cryptography

1 Introduction

Since the discovery of quantum algorithms for factorization and discrete logarithm problems by Shor [15], it has become widely known that many practical schemes based on public-key cryptography can be broken by quantum computers theoretically. Although the quantum computer that can be implemented with the current technology does not pose a threat to practical cryptographic schemes, it is essential to study the schemes that are secure enough against quantum computers that will be developed in the near future.

Much of the early work on quantum attacks focused on public-key cryptosystems, and only generic algorithms based on Grover's quantum search [9] were known to attack symmetric-key cryptosystems. However, recent studies have shown that more sophisticated quantum attacks are possible even against some symmetric-key cryptosystems. Kuwakado and Morii provided efficient quantum attacks against the well-known symmetric-key primitives such as the 3-round Feistel structure [12] and the Even-Mansour (EM) cipher [13] using Simon's quantum algorithm [16]. Following their celebrated results, several

© The Author(s), under exclusive license to Springer Nature Switzerland AG 2024
W. Wu and G. Tong (Eds.): COCOON 2023, LNCS 14423, pp. 3–16, 2024.
https://doi.org/10.1007/978-3-031-49193-1_1

papers revealed new quantum attacks against many symmetric-key constructions such as the work of Kaplan, Leurent, Leverrier, and Naya-Plasencia [10] that provided efficient quantum attacks on some of the most common block-cipher modes of operations for message authentication and authenticated encryption. The discovery of these quantum attacks against symmetric-key cryptosystems has led us to focus not only on analyses of the potential capabilities of quantum adversaries for public-key cryptography but also on those for symmetric-key cryptography.

In particular, the security of the EM cipher and its variants has been studied in many papers so far against classical and quantum adversaries. The EM cipher is a well-known construction for block ciphers and has a very simple structure to achieve the security of pseudorandom functions. For a random public permutation $\pi : \mathbb{Z}_2^n \to \mathbb{Z}_2^n$ and secret keys $k_1, k_2 \in \mathbb{Z}_2^n$, its encryption function is defined as $EM(x) := \pi(x + k_1) + k_2$.

The classical security of the EM cipher and its variants has been broadly studied. The original paper of Even and Mansour proved that classical adversaries require $\mathcal{O}(2^{n/2})$ queries to break the EM cipher [8]. Chen and Steinberger provided query lower bounds for generalizations of the EM cipher, called the iterated EM ciphers $iEM_t(x) := k_t + \pi_t(k_{t-1} + \pi_{t-1}(\cdots k_1 + \pi_1(k_0 + x) \cdots))$ [5]. They proved the tight query lower bound of $\Omega(2^{(t/(t+1))n})$ for attacking the variant that matches to query upper bounds of $\mathcal{O}(2^{(t/(t+1))n})$ by a generalization of Daemen's attack [7], which was pointed out by Bogdanov, Knudsen, Leander, Standaert, Steinberger, and Tischhauser [3]. Chen, Lambooij, and Mennink also studied the query bounds for security of "Sum of the EM ciphers" (SoEM), which are variants of the EM cipher [6]. For example, they proved that $\mathcal{O}(2^{n/2})$ queries are sufficient to classically attack $SoEM1(x) := \pi(x + k_1) + \pi(x + k_2) + k_1 + k_2$ for two independent keys k_1, k_2 and $SoEM21(x) := \pi_1(x + k_1) + \pi_2(x + k_1) + k_1$ for two independent permutations π_1, π_2, but $\Omega(2^{2n/3})$ queries are necessary to classically attack $SoEM22(x) := \pi_1(x + k_1) + \pi_2(x + k_2) + k_1 + k_2$ for independent keys k_1, k_2 and independent permutations π_1, π_2 beyond the birthday bound.

Also, quantum attacks on the EM cipher and its variants have been developed following the result of Kuwakado and Morii. Shinagawa and Iwata demonstrated quantum attacks on the variants of SoEM studied in [6] by extending the Kuwakado-Morii (KM) attack [14]. For example, they demonstrated that $SoEM1$ and $SoEM21$ can be broken only with $\mathcal{O}(n)$ quantum queries. Moreover, their new quantum algorithm that combines Simon's algorithm with Grover's algorithm can break $SoEM22$ with $\mathcal{O}(n2^{n/2})$ quantum queries, which is much lower than the classical query lower bound of $\Omega(2^{2n/3})$ [6]. Bonnetain, Hosoyamada, Naya-Plasencia, Sasaki and Schrottenloher constructed a new quantum algorithm that uses Simon's algorithm as a subroutine without quantum queries to oracles, and they succeeded in attacking the EM cipher with $\mathcal{O}(2^{n/3})$ classical queries, $\mathcal{O}(n^2)$ qubits, and offline quantum computation $\tilde{\mathcal{O}}(2^{n/3})$ [4].

On the other hand, little has been studied on the security of these schemes against quantum adversaries, or limits of capabilities of quantum adversaries, while the KM attack has been used to extend quantum attacks on other variants of the EM cipher. In many security proofs against quantum adversaries with oracle access, including the EM cipher and its variants, it is generally not possible to prove the security against quantum adversaries by conventional proof techniques used in the standard classical settings. This is because we need to assume that quantum adversaries have quantum access to cryptographic primitives. Indeed, many papers developed new techniques to show the limits of quantum adversaries against well-known symmetric-key cryptographic constructions (e.g., [17,18]).

The only example of the quantum security proof for the EM cipher, to the best of the authors' knowledge, is by Alagic, Bai, Katz, and Majenz [1]. They considered a natural post-quantum scenario that adversaries make classical queries to its encryption function EM, but can make quantum queries to the public permutation π. In this scenario, they demonstrated that it must hold either $q_\pi^2 q_{EM} = \Omega(2^n)$ or $q_\pi q_{EM}^2 = \Omega(2^n)$, where q_π (q_{EM}, respectively) is the number of queries to π (EM, respectively).

Therefore, it is important to understand better the limits of quantum adversaries for constructing quantumly secure variants of the EM cipher by studying the quantum query lower bounds for attacking the EM cipher.

In this paper, we investigate the limits of quantum adversaries against the original EM cipher to explore quantumly secure variants of the EM cipher. We prove lower bounds $\Omega(n)$ of quantum query complexity to recover n-bit secret keys of the EM cipher even if quantum adversaries are allowed to make adaptive queries. To the best of the authors' knowledge, this is the first result that provides new techniques for demonstrating the limits of adversaries against (variants of) the EM cipher with purely quantum queries. Our quantum query lower bound matches the upper bound $\mathcal{O}(n)$ of nonadaptive quantum queries provided by the KM attack up to a constant factor. This implies that their attack is optimal up to a constant factor in a setting of quantum query complexity, and thus, there is no asymptotically better quantum attack than the one based on Simon's algorithm even if it is allowed to make adaptive queries.

2   Overviews of Previous Results and Our Ideas

Since the structure of our proof is based on the optimality proof of (generalized) Simon's algorithm studied by Koiran, Nesme, and Portier [11], we briefly review Simon's algorithm and its optimality.

The problem solved by Simon's algorithm is commonly referred to as Simon's problem. The following is a generalized version of Simon's problem with any prime p. The oracle O hides some subgroup K of order $D = p^d$, where d is a non-negative integer.

Generalized Simon's (GS) Problem
Input: an oracle $O : \mathbb{Z}_p^n \to Y$ that is sampled uniformly at random from all the oracles that satisfy $x' = x + k \leftrightarrow O(x') = O(x)$ for some subgroup $K \leq \mathbb{Z}_p^n$ of order D;
Output: the generators of K.

The original Simon's problem corresponds to the case of $p = 2$ and $D = 2$. Then, $K = \{0, k\}$ for $k \in \mathbb{Z}_2^n \setminus \{0^n\}$. Simon's algorithm first makes $\mathcal{O}(n)$ nonadaptive queries $\sum_{x \in \mathbb{Z}_2^n} |x\rangle|0\rangle/\sqrt{2^n}$ to the oracle O and measures the second register. By the measurement, it obtains independent copies of the coset-uniform state $(|x_0\rangle + |x_0 + k\rangle)/\sqrt{2}$ for a random x_0 in the first register. Applying the quantum Fourier transform over $\mathbb{Z}_2^n$ (or, the Hadamard transform $H^{\otimes n}$) to them and measuring the resulting states, it obtains $\mathcal{O}(n)$ random linear constraints $\sum_{i<n} z_i \cdot k_i = 0$ with respect to the undetermined secret key $k = (k_0, \ldots, k_{n-1})$. From the constraints, it can identify k with constant probability.

The idea of the KM attack against the EM cipher is to construct the oracle of Simon's problem from the public permutation π and encryption function $EM(x) = \pi(x + k_1) + k_2$. In the KM attack, a quantum adversary is allowed to make quantum queries to π and EM in a quantum manner. Let $O(x) := EM(x) + \pi(x) = \pi(x + k_1) + \pi(x) + k_2$. The adversary applies Simon's algorithm to this function O. Since $O(x+k_1) = \pi(x+k_1)+\pi(x)+k_2 = O(x)$, The oracle O satisfies the direct part $x' = x + k_1 \rightarrow O(x') = O(x)$ and approximately satisfies the converse part with respect to random choices of π. Therefore, the KM attack succeeds in recovering k_1 using $\mathcal{O}(n)$ nonadaptive quantum queries to π and EM with constant probability by Simon's algorithm. It is obvious to recover k_2 from k_1 since $k_2 = EM(0) + \pi(k_1)$.

To prove the optimality of (generalized) Simon's algorithm, Koiran et al. studied quantum query lower bounds for its generalized decisional version. Let

$$F_D := \{O : \mathbb{Z}_p^n \rightarrow Y : \exists K \leq \mathbb{Z}_p^n, \forall x', \forall x \in \mathbb{Z}_p^n, \forall k \in K, x' = x + k \leftrightarrow O(x') = O(x)\},$$

where $D = |K| = p^d$ for a non-negative integer d.

Generalized Decisional Simon's (GDS) Problem
Input: an oracle $O : \mathbb{Z}_p^n \rightarrow Y$ that is sampled uniformly at random from F_p or F_1;
Output: "accept" if O is from F_p or "reject" if it is from F_1.

Note that F_1 is the set of all the $O : \mathbb{Z}_p^n \rightarrow Y$, The task of this problem is to distinguish between a function that hides some subgroup K of order p and a random function.

It is easy to see that if Simon's problem is solved with T queries, the GDS problem for $p = 2$ is also solved with the same T queries. Therefore, quantum query lower bounds of GDS problem for an arbitrary prime p directly lead to those for Simon's problem.

The argument of Koiran et al. [11] is based on the polynomial method [2] for the GDS problem. They analyzed the degree of the polynomial $Q(D)$ that represents the accepting probability for a random $O \in F_D$, where D is the order of the subgroup K that O hides. They showed an upper bound $\mathcal{O}(T)$ of $\deg(Q(D))$ for quantum algorithms with accepting probability $Q(D)$ and T queries to an oracle O that hides a subgroup K of order D, and further, a lower bound $\Omega(n)$ of $\deg(Q(D))$ for any polynomial $Q(D)$ that satisfies several conditions naturally posed on $Q(D)$, such as $Q(p) \geq 1 - \epsilon$, which corresponds to the case of F_p, $Q(1) \leq \epsilon$, which corresponds to the case of F_1, for a small constant ϵ, and $Q(p^i) \in [0, 1]$ for every $i \in \{0, 1, \ldots, n\}$.

Our goal, quantum query lower bounds for key recovery of the EM cipher, seems to be close to those for Simon's problem provided in [11]. However, there are actually technical gaps between these two problems. In the setting of the key recovery, a quantum adversary can make access to two oracles $EM(x)$ and $\pi(x)$ rather than a single oracle $O(x)$ in the setting of Simon's problem. The quantum query upper bound $\mathcal{O}(n)$ can be achieved by the KM attack that synchronously makes a (quantumly superposed) query x to $EM(x)$ and $\pi(x)$ and combines two answers to compute $O(x) = EM(x) + \pi(x)$. However, it would be possible to achieve better attacks by making different queries to two oracles in an adaptive manner.

We then provide a reduction of quantum query lower bounds in the standard query model to those in a special query model. In the special query model, which we refer to as a synchronized query model, any quantum adversary is posed to make a synchronized query to two oracles as done in the KM attack. If a quantum adversary A can recover the secret key with $T(n)$ queries to EM and π totally in the standard query model, we can easily modify A to another adversary A' that recovers it with $2T(n)$ queries in the synchronized query model.

In the synchronized query model, we can assume that a quantum algorithm has synchronized access to a oracle sequence $O : \mathbb{Z}_2^n \to (\mathbb{Z}_2^n)^2$, where $O(x) = (O_0(x), O_1(x))$ for a random permutation $O_0(x) = \pi(x)$ and the encryption function $O_1(x) = EM(x) = O_0(x + k_1) + k_2$. In our proof, we focus only on the inner key k_1 for simplification, which suffices to prove lower bounds since it is a special case when $k_2 = 0$. We define $O_1(x) = O_0(x + k_1)$. Then, our goal is to prove quantum query lower bounds for finding the inner key k_1 with synchronized queries to the oracle sequence $O(x) = (O_0(x), O_1(x)) = (O_0(x), O_0(x + k_1))$.

To apply the polynomial method as done in the proof of Koiran et al., we need to consider a generalized version of the oracle sequence $O(x) = (O_0(x), O_1(x))$ to represent the accepting probability as a polynomial in some single parameter. As a generalization, we consider an oracle sequence

$$
\begin{aligned}
O(x) &= (O_0(x), O_1(x), \ldots, O_{D-1}(x)) \\
&= (O_0(x + k_0), O_0(x + k_1), \ldots, O_0(x + k_{D-1}))
\end{aligned}
$$

of length $D = p^d$, where $K = \{k_0 = 0^n, k_1, \ldots, k_{D-1}\}$ is a subgroup in $\mathbb{Z}_p^n$ of the order D. We then analyze the accepting probability $Q(D)$ as a polynomial in D for a given oracle sequence O.

The major difference from the argument of Koiran et al. is an algebraic structure behind the oracles. In the cases of the GS and GDS problems, the subgroup is hidden in the single oracle. However, it is hidden in the correlation among D oracles in our setting. Recall that $x' = x + k$ for some $k \in K$ if and only if $O(x') = O(x)$ in Simon's problem. We need to reveal a similar algebraic structure to analyze of the degree of $Q(D)$.

Our idea is to characterize the order of oracles in the sequence O by the hidden subgroup K. Actually, we demonstrate that the definition of O is equivalent with the statement that $x' = x + k_i$ for $k_i \in K$ if and only if for $k_i \in K$ there exists some permutation σ_i over $\{0, \ldots, D-1\}$ it holds $O(x') = \sigma_i O(x)$.

Let us consider a small example $K = \{0^n, k_1, k_2, k_1 + k_2\} \leq \mathbb{Z}_2^n$ for $p = 2$ and $D = 4$, where $k_1 \neq k_2 \in \mathbb{Z}_2^n \setminus \{0^n\}$. The oracle sequence is defined as

$$O(x) = (O_0(x), O_0(x + k_1), O_0(x + k_2), O_0(x + k_1 + k_2))$$
$$= (O_{(0,0)}(x), O_{(0,1)}(x), O_{(1,0)}(x), O_{(1,1)}(x))$$

with some special indexing of the oracles. Then, we can see that

$$O(x + k_1) = (O_0(x + k_1), O_0(x), O_0(x + k_1 + k_2), O_0(x + k_2))$$
$$= (O_{(0,0)+(0,1)}(x), O_{(0,1)+(0,1)}(x), O_{(1,0)+(0,1)}(x), O_{(1,1)+(0,1)}(x))$$
$$= (O_{(0,1)}(x), O_{(0,0)}(x), O_{(1,1)}(x), O_{(1,0)}(x)).$$

Similarly, we have

$$O(x + k_2) = (O_{(1,0)}(x), O_{(1,1)}(x), O_{(0,0)}(x), O_{(0,1)}(x))$$
$$O(x + k_1 + k_2) = (O_{(1,1)}(x), O_{(1,0)}(x), O_{(0,1)}(x), O_{(0,0)}(x)).$$

Hence, every $k \in K$ corresponds to some permutation over the order of the oracles.

From the above characterization, we develop a variant of the argument of Koiran et al. based on the polynomial method with the analogous property of the oracle sequence that $O(x + k_i) = \sigma_i O(x)$ instead of the one of Simon's problem that $O(x + k) = O(x)$. As is obvious, the analogous property is different from that of Simon's problem, and hence, we need to fill this gap with other technical tricks in our proof.

3 Preliminaries

Before describing the main result, we briefly discuss the formal treatment of quantum query algorithms.

In the context of quantum query complexity, we usually assume the following framework for quantum query algorithms. A quantum algorithm A^O with a given oracle O has quantum memory of three registers $|x\rangle|y\rangle|z\rangle$, where the first one is the query register which stores a query to O, the second one is the answer register which stores an answer from O, and the third one is the working register which stores all the other than the query and answer registers. Let U_O be the oracle gate of $O : X \to Y$ that acts on the query and answer registers: $U_O|x\rangle|y\rangle = |x\rangle|O(x) \oplus y\rangle$ for every $x \in X$ and every $y \in Y$. A starts with the initial state $|0\rangle|0\rangle|0\rangle$, and applies an arbitrary unitary operator to all the three registers and then applies U_O to the two registers alternatively. Then, the A^O's final state is provided as $|\psi_T\rangle = U_T(U_O \otimes I)U_{T-1} \cdots U_1(U_O \otimes I)U_0|0\rangle|0\rangle|0\rangle$.

The A^O's output can be obtained by measuring a part of the final state in the computational basis. Note that this formulation allows A to make adaptive queries. In other words, A can make a query that depends on the answers to the previous queries.

In this paper, we need to deal with multiple oracles such as π and EM. We formulate the quantum query model with multiple oracles $O_0, O_1, \ldots, O_{N-1}$ by the model with a single oracle $O : \{0, 1, \ldots, N - 1\} \times X \to Y$ defined as $O(i, x) := O_i(x)$. In the framework for quantum query algorithms, this oracle can be implemented as $U_O |i, x\rangle |y\rangle |z\rangle = |i, x\rangle |O_i(x) \oplus y\rangle |z\rangle$ by extending the query register.

As described in Sect. 2, we also consider a special query model referred to as the synchronized query model. A quantum query algorithm A receives N answers $O_0(x), \ldots, O_{N-1}(x)$ simultaneously on a single query x at its oracle call in the synchronized query model. Formally, the oracle call can be implemented as $U_O |x\rangle |y_0, \ldots, y_{N-1}\rangle = |x\rangle |O_0(x) \oplus y_0, \ldots, O_{N-1}(x) \oplus y_{N-1}\rangle$. Similarly to the standard query model, A applies an arbitrary unitary operator to the registers, and then, the oracle operator U_O, with the all-zero initial state. We count the number of queries as the number of U_O used in the algorithm. We also regard the oracle O as a function $O : X \to Y^N$ by setting $O(x) := (O_0(x), \ldots, O_{N-1}(x))$ in this model.

As mentioned in Sect. 1, any quantum algorithm in the standard query model can be converted to the one in the synchronized query model from the following proposition. The proof is easily done by a standard reduction.

Proposition 1. *Let A be any quantum query algorithm with T queries in the standard query model. Then, there exists A' with $2T$ queries in the synchronized query model such that A''s output distribution is identical with A's one.*

From Proposition 1, if we obtain a query lower bound of T in the synchronized query model, we also obtain a query lower bound of $T/2$ in the standard query model. Thus, we focus on the synchronized query model in the remaining part of this paper.

We next discuss our target problem to prove the quantum query lower bounds for the key recovery of the EM cipher. As done in [11], we work on a decisional version of attacks against the EM cipher. In the key recovery problem for the EM cipher, we need to deal with multiple oracles such as π and EM, unlike the GDS problem. We are given two oracles $O_0 := \pi$ and $O_1 := EM$, where $O_0 : \mathbb{Z}_2^n \to \mathbb{Z}_2^n$ is a public permutation and $O_1(x) = \pi(x \oplus k_1) \oplus k_2$ for secret keys $k_1, k_2 \in \mathbb{Z}_2^n$. Then, the task is to recover k_1, k_2 via queries to O_0 and O_1. We focus on a special case $k_2 = 0^n$ of the key recovery problem since a lower bound for this special case implies that for the general case.

To apply the polynomial method similarly to [11], we consider a generalized version of the key recovery problem. One of the main technical contributions is a formalization of the generalized version, named generalized decisional inner-key only EM cipher (GDIKEM) problem, that is suitable for proving query lower bounds.

Note that query lower bounds of the key recovery problem in the standard query model can be obtained from the GDIKEM problem in the query synchronized model by Proposition 1. Therefore, we can suppose that a quantum query algorithm is provided an oracle sequence $O(x) = (O_0(x), \ldots, O_{N-1}(x))$ in the definition of the GDIKEM problem rather than a set of oracles $O_0, \ldots, O_{N-1}$ separately.

Before the definition of the GDIKEM problem, we consider a special index system $I = \{(i_0, \ldots, i_{d-1}) : i_0, \ldots, i_{d-1} \in \mathbb{Z}_p\}$ for the oracle sequences O. Let K be any subgroup of $\mathbb{Z}_p^n$ of order $D = p^d$. We fix the lexicographic first set $\{g_0^K, \ldots, g_{d-1}^K\}$ of generators for K. Then, any element $k_i \in K$ can be associated with $i \in I$ to satisfy $k_i := \sum_{j=0}^{d-1} i_j g_j^K$. Note that $k_i + k_{i'} = k_{i+i'}$ for $k_i, k_{i'} \in K$. For simplification, let 0 denote 0^d. We sometimes identify I with $\{0, 1, \ldots, D-1\}$ by the lexicographical order.

To formulate the GDIKEM problem, we define a set of oracle sequences of length D as $O(x) = (O_i(x))_{i \in I}$, where $O_i : \mathbb{Z}_p^n \to \mathbb{Z}_p^n$ is a permutation. Let

$$F_D := \left\{ O : \exists K \leq \mathbb{Z}_p^n \, (|K| = D), \forall x \in \mathbb{Z}_p^n, \forall i \in I, O_i(x) = O_0(x + k_i) \right\},$$

where $D = p^d$ for some d. For $O \in F_D$, we say that O hides a subgroup K.

Note that F_2 is a set of the oracles $O(x) = (O_0(x), O_1(x)) = (O_0(x + 0^n), O_0(x + k_1))$ for a subgroup $K = \{0^n, k_1\}$ in the case when $D = p = 2$, which corresponds to instances of the EM cipher only with an inner key k_1 and public random permutation O_0.

From the following reason, we can see that every $O \in F_D$ hides the unique subgroup K of order D. Assume that O hides two distinct subgroups K and K' of order D. For $k' \in K' \setminus K$, there exists some index i $O_i(x) = O_0(x + k')$. Then, some $k \in K$ is associated with the index i, and thus, $O_i(x) = O_0(x + k)$. Hence, $O_0(x + k') = O_0(x + k)$. However, since $x + k \neq x + k'$, O_0 cannot be a permutation. This is a contradiction. Therefore, a subgroup hidden by O is unique.

By analogy with the GDS problem, it would be natural to define the distinguishing task between oracle sequences from F_p and F_1. However, these oracle sequences from F_p and F_1 are of different output lengths. To align the lengths, we pad redundant oracles to them. We define a set $\hat{F}_{D,N}$ of oracle sequences of length N $\hat{O} = (O_0, \ldots, O_{N-1})$ such that $(O_0, \ldots, O_{D-1}) \in F_D$ and O_i is an arbitrary permutation over $\mathbb{Z}_p^n$ for $i \geq D$.

Now, we define the GDIKEM problem as follows.

GDIKEM Problem
Input: an oracle $\hat{O}$ that satisfies (i) $\hat{O} \in \hat{F}_{D,N}$ or (ii) $\hat{O} \in \hat{F}_{1,N}$.
Output: "accept" if (i) or "reject" if (ii).

F_1 contains all the permutations over $\mathbb{Z}_p^n$, and hence, $\hat{F}_{1,N}$ is the set of all the possible sequences permutations over $\mathbb{Z}_p^n$ of length N. On the other hand, F_2 contains pairs of the permutations $(O_0(x), O_0(x + k_1))$ for some subgroup $K = \{0^n, k_1\}$ of order 2 in the case when $D = p = 2$. Therefore, $\hat{F}_{1,N}$ and $\hat{F}_{2,N}$ correspond to the sets of accepting and rejecting instances of a decisional version (with redundant $N - 2$ padded oracles) of the attack against EM cipher, respectively.

In this paper, we show that every quantum algorithm A requires $\Omega(n)$ queries if $A^{\hat{O}}$ accepts for a randomly chosen oracle $\hat{O}$ in the case (i) with at most ϵ and for a randomly chosen oracle in the case (ii) with least $1 - \epsilon$, where ϵ is a fixed constant. If there exists a key-recovery quantum algorithm for permutations $O_0(x)$ and $O_1(x) = O_0(x+k_1)$ with some $k_1 \neq 0^n$, it also works for the GDIKEM problem. Thus, query lower bounds of the GDIKEM problem imply those of the key recovery.

4   Proof Sketch of Quantum Query Lower Bounds

We briefly sketch the proof of quantum query lower bounds for key recovery attacks against the EM cipher in this section. Most of technical details are omitted due to space limitations. See the full version to be published for the omitted details.

As used in the previous result of Koiran et al. [11], we characterize the acceptance probability of any quantum algorithm for the oracle O from a set of partial functions whose domain size by the number of queries using the polynomial method [2].

We say f extends s, which is also denoted by $f \supseteq s$, if $s(x) = f(x)$ for every $x \in \text{Dom}(s)$. For any function $f : X \to Y^N$ and any partial function $s : X \to Y^N$, we define

$$I_s(f) := \begin{cases} 1 & \text{if } f \text{ extends } s; \\ 0 & \text{otherwise.} \end{cases} = \prod_{\substack{x \in \text{Dom}(s), \\ s(x) = \bar{y}}} \Delta_{x,\bar{y}}(f),$$

where $\Delta_{x,\bar{y}}(f) = 1$ if $f(x) = \bar{y}$ and $\Delta_{x,\bar{y}}(f) = 0$ otherwise.

Similarly to [11], we can prove the following characterization (Theorem 1) of the acceptance probability with respect to $I_s(f)$ even in the synchronized query model. The proof follows from the same argument as the one of the standard polynomial method. (We omit its proof.)

Theorem 1. *Let A be any quantum algorithm with T queries in the synchronized query model. Then, there exists a set S of partial functions $s : X \to Y^N$ such that A accepts f with probability $P(f) := \sum_{s \in S} c_s I_s(f)$ for some real numbers c_s, where $|\text{Dom}(s)| \leq 2T$.*

As stated in Sect. 1, we focus on the degree of a polynomial that represents accepting probability of a quantum algorithm to prove the query lower bounds by the polynomial method.

In Sect. 3, we defined the GDIKEM problem to naturally fit some generalized decisional version of the attack against the EM cipher. From technical reasons, we focus on another equivalent formulation of the oracle set shown in the following lemma. (We omit its proof.)

Lemma 1. *Suppose that O hides a subgroup K. Then, we have*

$$F_D = \Big\{ O : \exists K \leq \mathbb{Z}_p^n \ (|K| = D)$$

$$\forall i \in I, \forall x, \forall x' \in \mathbb{Z}_p^n, \ x' = x + k_i \ (k_i \in K) \leftrightarrow O(x') = \sigma_i O(x) \Big\}.$$

From technical reasons, we define a subset $F_D^* := F_D \cap \{O : O_0 \in \Pi_K\}$ of the oracles. The set Π_K of permutations is defined as follows. Let K be the subgroup hidden by O. We consider the coset decomposition of $\mathbb{Z}_p^n$ for K: $\mathbb{Z}_p^n = \cup_{i<N/D}\{c_i + K\}$ for some fixed representatives, where $c_0 := 0^n$ and $N := |\mathbb{Z}_p^n| = p^n$. To construct Π_K, for every sequence $(a_0, \ldots, a_{(N/D)-1})$ of distinct N/D elements, we put a permutation π into Π_K such that $\pi(c_0) = a_0, \ldots, \pi(c_{(N/D)-1}) = a_{(N/D)-1}$ and the remaining values $\pi(x)$ for $x \notin c_0, \ldots, c_{(N/D)-1}$ are determined by the lexicographically first sequence of $N - (N/D)$ elements excluding $a_0, \ldots, a_{(N/D)-1}$ from $\mathbb{Z}_p^n$. Therefore, any permutation in Π_K is determined uniquely by specifying the values $\pi(c_0), \ldots, \pi(c_{(N/D)-1})$, and thus, $|\Pi_K| = p^n(p^n - 1) \cdots (p^n - (p^{n-d} - 1))$. We also define its padded version $\hat{F}_{D,N}^*$ by the same manner as $\hat{F}_{D,N}$.

We now provide a formal statement of our main theorem.

Theorem 2. *Let p be any prime, and let ϵ be any constant in $(0, 1/2)$. Suppose that A is any quantum algorithm with adaptive $T = T(n)$ quantum queries to a given oracle $\hat{O} : \mathbb{Z}_p^n \to (\mathbb{Z}_p^n)^N$, where $\hat{O}$ is sampled uniformly from (i) $\hat{F}_{p,N}^*$ or (ii) $\hat{F}_{1,N}^*$ for any fixed $N \geq p$. If $A^{\hat{O}}$ accepts with at least $1 - \epsilon$ in the case (i) and with at most ϵ in the case (ii), it holds that $T = \Omega(n)$.*

Immediately from Proposition 1 and Theorem 2, we obtain a quantum query lower bound of $\Omega(n)$ to recover secret keys in the EM cipher with constant success probability in the standard query model.

Proof of Theorem 2. We analyze the accepting probability that $A^{\hat{O}}$ accepts for an oracle $\hat{O} \in \hat{F}_{D,N}^*$. From Theorem 1, the accepting probability is

$$P(\hat{O}) = \sum_{\hat{O} \in \hat{F}_{D,N}^*} \sum_{\hat{s} \in \hat{S}} c_{\hat{s}} I_{\hat{s}}(\hat{O}) = \sum_{\hat{O} \in \hat{F}_{D,N}^*} \sum_{\hat{s} \in \hat{S}} c_{\hat{s}} \prod_{x \in \mathrm{Dom}\hat{S}, \bar{y} = \hat{s}(x)} \Delta_{x,\bar{y}}(\hat{O})$$

for some set $\hat{S}$ of partial functions. $\qquad\square$

We convert this multivariate polynomial $P(\hat{O})$ in $\{\Delta_{x,\bar{y}}(\hat{O})\}_{x,\bar{y}}$ into another univariate polynomial $Q(D)$ in D by averaging the redundant oracles, namely,

$$Q(D) := \frac{1}{|\hat{F}_{D,N}^*|} \sum_{\hat{O} \in \hat{F}_{D,N}^*} P(\hat{O}).$$

Recall that $\hat{O}$ is padded with $N - D$ redundant oracles to align the length of the oracle sequences. From the following lemma (Lemma 2), we can ignore such redundant oracles for the degree analysis of $Q(D)$. (We omit its proof.)

Lemma 2. *There exists a set of partial functions S such that for every $O \in F_D$ we have*

$$Q(D) = \frac{1}{|F_D^*|} \sum_{O \in F_D^*} \sum_{s \in S} c_s' I_s(O)$$

The following lemma shows $\deg(Q(D))$ is upper-bounded by the domain size of partial functions s.

Lemma 3. *Let A be any quantum algorithm with T queries in the synchronized query model. Then, we have $\deg(Q(D)) \leq \max_{s \in S} |\mathrm{Dom}(s)|$.*

By combining Theorem 1 and Lemma 3, the lower bound of T can be reduced to that of the degree of $Q(D)$. Koiran et al. provided the degree analysis in [11], which we apply in our proof.

Theorem 3 (Koiran et al. [11]). *Let $c > 0$ and $\xi > 1$ be constants and let P be a real polynomial with following properties: (i) $|P(\xi^i)| \leq 1$, for any integer $0 \leq i \leq n$, and (ii) $|dP(x_0)/dx| \geq c$, for some real number $1 \leq x_0 \leq \xi$. Then*

$$\deg(P) \geq \min \left\{ n/2, \left(\log_2 \left(\xi^{n+3} c \right) - 1 \right) / \left(\log_2 \left(\frac{\xi^3}{\xi - 1} \right) + 1 \right) \right\}.$$

Let A be any quantum algorithm solving GDIKEM problem for $|K| = p$ with bounded error probability ϵ and T queries in the synchronized query model. A^O rejects if $|K| = 1$ holds in GDIKEM problem, and A^O accepts if $|K| = p$. Then, $0 \leq |Q(p^i)| \leq 1$ ($0 \leq i \leq n$) and $Q(p) \geq 1 - \epsilon$ ($k \leq n$), $Q(1) \leq \epsilon$ holds from the property of A. Therefore, for the derivative of the polynomial Q, Q satisfies $|dQ(x_0)/dD| \geq \frac{1-2\epsilon}{p-1}$ for some x_0 ($1 \leq x_0 \leq p$) and $Q(p^i) \in [0,1]$ for any $i \in \{0, ..., n\}$. By applying Theorem 3 to the polynomial $P = 2Q - 1$, we obtain the following inequality

$$\deg(Q) \geq \min \left\{ n/2, \left(\log_2 \left(\frac{p^{n+3}}{p-1} (2 - 4\epsilon) \right) - 1 \right) / \left(\log_2 \left(\frac{p^3}{p-1} \right) + 1 \right) \right\} = \Omega(n).$$

Therefore, the remaining task for the proof of the lower bound is to show Lemma 3.

Proof of Lemma 3. From Lemma 2, we have

$$Q(D) = \frac{1}{|F_D^*|} \sum_{O \in F_D^*} \sum_{s \in S} c_s' I_s(O) = \sum_{s \in S} c_s' Q_s(O),$$

$$\text{where} \quad Q_s(D) := \frac{1}{|F_D^*|} \sum_{O \in F_D^*} I_s(O) = \Pr_{O \in F_D^*} [O \supseteq s].$$

It suffices to show that $\deg(Q_s(D)) \leq |\mathrm{Dom}(s)|$ for every $s \in S$. $\qquad\square$

We can assume that the identity 0^n is in $\mathrm{Dom}(s)$ for every partial function s by modifying a given algorithm A as follows. At the beginning, A makes the query

0^n with the initial state $|0^n\rangle|(0^n)^D\rangle|0^m\rangle$, stores $O(0^n)$ in the answer register, and swaps the answer register with a part of the working register. Afterwards, A applies the original operations to the zero-cleared registers except for the part that stores $O(0^n)$. Then, every $s \in S$ contains 0^n in its domain, and the modified algorithm keeps the original accepting probability and has the number $T + 1$ of queries if the original is T. Therefore, we can obtain a lower bound of T from the modified algorithm.

Let

$$A^i := \{a^{i,j} : \exists \ell \in I, s(a^{i,j}) = \sigma_\ell s(a^{i,1})\}$$

and

$$\mathrm{Dom}(s) := \left\{ \begin{array}{l} a^{1,1}, \ ..., \ a^{1,v_1} \in A^1 \\ a^{2,1}, \ ..., \ a^{2,v_2} \in A^2 \\ \quad\quad \vdots \\ a^{w,1}, \ ..., \ a^{k,v_w} \in A^w \end{array} \right\},$$

where $a^{1,1} := 0^n$.

By Lemma 1, we observe that $x' = x + k_\ell \ (k_\ell \in K) \leftrightarrow O(x') = \sigma_i O(x)$ for every $i \in I$ and every $O \in F_D^*$ that hides K. Since $O(a^{i,j}) = \sigma_\ell O(a^{i,1}) \leftrightarrow a^{i,j} = a^{i,1} + k_\ell \leftrightarrow a^{i,j} - a^{i,1} = 0^n + k_\ell \leftrightarrow O(a^{i,j} - a^{i,1}) = \sigma_\ell O(0^n)$, $O(a^{i,j}) = \sigma_\ell O(a^{i,1})$ if and only if $O(a^{i,j} - a^{i,1}) = \sigma_\ell O(0^n)$ for every i, j, every $\ell \in I$ and every $O \in F_D^*$.

Then, we modify s into another partial function $\tilde{s}$ by modifying s as follows. Let $s(a^{i,j}) = \sigma_\ell s(a^{i,1})$ for some $\ell \in I$. We set $\tilde{s}(a) := s(a)$ for every $a \in \mathrm{Dom}(s) \backslash \{a^{i,j}\}$. Since $0^n \in \mathrm{Dom}(s)$, we can also set $\tilde{s}(a^{i,j} - a^{i,1}) := \sigma_\ell s(0^n)$. Note that $\mathrm{Dom}(\tilde{s}) = (\mathrm{Dom}(s) \backslash \{a^{i,j}\}) \cup \{a^{i,j} - a^{i,1}\}$, and hence, $|\mathrm{Dom}(s)| = |\mathrm{Dom}(s')|$. From the modification, O extends s if and only if O extends $\tilde{s}$, and thus, we can analyze the probability that O extends $\tilde{s}$ instead of s.

From the above modification, we can suppose that $\mathrm{Dom}(s) = A^1 \cup A^2 \cup \cdots \cup A^w$ has the following form without loss of generality.

$$\mathrm{Dom}(s) = \left\{ \begin{array}{ll} a^{1,1}, \ ..., \ a^{1,v_1} \in A^1, \\ a^{2,1} \quad\quad\quad\ \in A^2, \\ \quad\quad \vdots \\ a^{w,1} \quad\quad\quad \in A^w \end{array} \right\},$$

where $a^{1,1} := 0^n$.

Let $K' := \langle A^1 \rangle$ and let $D' := |K'| = p^{d'}$ for some d'. For $O \in F_D^*$ that hides K, let

$$\mathcal{E}(O) \equiv \left[\bigwedge_{i=1}^{v_1} \exists \ell_i \in I : O(a^{1,i}) = \sigma_{\ell_i} O(0^n) \right].$$

We define

$$Q_s^R(D) = \Pr_{O \in F_D^*} [\mathcal{E}(O)], \quad Q_s^C(D) = \Pr_{O \in F_D^*} \left[O \supseteq s \ \middle| \ \mathcal{E}(O) \right].$$

Note that $Q_s(D) = Q_s^R(D) \cdot Q_s^C(D)$ since $\mathcal{E}(O)$ holds if $O \supseteq s$.

Since $\deg(Q_s(D)) = \deg(Q_s^R(D)) + \deg(Q_s^C(D))$, it suffices to estimate $\deg(Q_s^R(D))$ and $\deg(Q_s^C(D))$, which are given in Lemma 4. (We omit its proof.)

Lemma 4. *We have* $\deg(Q_s^R(D)) \leq v_1 - 1$ *and* $\deg(Q_s^C(D)) \leq w$.

5   Concluding Remarks

The oracle distribution (that is uniform over F_D^*) used for the quantum query lower bounds is artificially biased because of the condition "$O \in \Pi_K$" in the definition of F_D^*. This condition is crucial in the proof of Lemma 4 to show $\deg(Q_s^C(D)) \leq w$, although we omitted the technical details in this conference version due to space limitations. It is natural to use the uniform distribution over F_D to prove the average-case lower bounds, but the polynomial method fails because $Q_s^C(D)$ could be then exponential rather than polynomial. (See the full version for more details.) Hence, we need new proof techniques for quantum query lower bounds in the natural average case.

The obvious open problem is to prove the quantum security of classically secure variants of the EM cipher such as Iterated EM cipher [5] and SoEM [6], but there seem to be no approaches to them so far. The algebraic characterization of the oracle used in this paper could help to establish security proofs for quantum adversaries.

Acknowledgments. This work was supported by JSPS Grant-in-Aid for Scientific Research (A) Nos. 21H04879, 23H00468, (C) No. 21K11887, JSPS Grant-in-Aid for Challenging Research (Pioneering) No. 23K17455, and MEXT Quantum Leap Flagship Program (MEXT Q-LEAP) Grant Number JPMXS0120319794.

References

1. Alagic, G., Bai, C., Katz, J., Majenz, C.: Post-quantum security of the Even-Mansour cipher. In: Dunkelman, O., Dziembowski, S. (eds.) Advances in Cryptology – EUROCRYPT 2022. LNCS, vol. 13277, pp. 458–487. Springer, Cham (2022). https://doi.org/10.1007/978-3-031-07082-2_17
2. Beals, R., Buhrman, H., Cleve, R., Mosca, M.: Quantum lower bounds by polynomials. J. ACM **48**(4), 778–797 (2001)
3. Bogdanov, A., Knudsen, L.R., Leander, G., Standaert, F.-X., Steinberger, J., Tischhauser, E.: Key-alternating ciphers in a provable setting: encryption using a small number of public permutations. In: Pointcheval, D., Johansson, T. (eds.) EUROCRYPT 2012. LNCS, vol. 7237, pp. 45–62. Springer, Heidelberg (2012). https://doi.org/10.1007/978-3-642-29011-4_5
4. Bonnetain, X., Hosoyamada, A., Naya-Plasencia, M., Sasaki, Yu., Schrottenloher, A.: Quantum attacks without superposition queries: the offline Simon's algorithm. In: Galbraith, S.D., Moriai, S. (eds.) ASIACRYPT 2019. LNCS, vol. 11921, pp. 552–583. Springer, Cham (2019). https://doi.org/10.1007/978-3-030-34578-5_20

5. Chen, S., Steinberger, J.: Tight security bounds for key-alternating ciphers. In: Nguyen, P.Q., Oswald, E. (eds.) EUROCRYPT 2014. LNCS, vol. 8441, pp. 327–350. Springer, Heidelberg (2014). https://doi.org/10.1007/978-3-642-55220-5_19
6. Chen, Y.L., Lambooij, E., Mennink, B.: How to build pseudorandom functions from public random permutations. In: Boldyreva, A., Micciancio, D. (eds.) CRYPTO 2019. LNCS, vol. 11692, pp. 266–293. Springer, Cham (2019). https://doi.org/10.1007/978-3-030-26948-7_10
7. Daemen, J.: Limitations of the Even-Mansour construction. In: Imai, H., Rivest, R.L., Matsumoto, T. (eds.) ASIACRYPT 1991. LNCS, vol. 739, pp. 495–498. Springer, Heidelberg (1993). https://doi.org/10.1007/3-540-57332-1_46
8. Even, S., Mansour, Y.: A construction of a cipher from a single pseudorandom permutation. J. Cryptol. **10**(3), 151–162 (1997)
9. Grover, L.K.: A fast quantum mechanical algorithm for database search. In: Proceedings of the 28th ACM Symposium on Theory of Computing, pp. 212–218 (1996)
10. Kaplan, M., Leurent, G., Leverrier, A., Naya-Plasencia, M.: Breaking symmetric cryptosystems using quantum period finding. In: Robshaw, M., Katz, J. (eds.) CRYPTO 2016. LNCS, vol. 9815, pp. 207–237. Springer, Heidelberg (2016). https://doi.org/10.1007/978-3-662-53008-5_8
11. Koiran, P., Nesme, V., Portier, N.: The quantum query complexity of the abelian hidden subgroup problem. Theoret. Comput. Sci. **380**, 115–126 (2007)
12. Kuwakado, H., Morii, M.: Quantum distinguisher between the 3-round Feistel cipher and the random permutation. In: IEEE International Symposium on Information Theory, pp. 2682–2685. IEEE (2010)
13. Kuwakado, H., Morii, M.: Security on the quantum-type Even-Mansour cipher. In: Proceedings of the International Symposium on Information Theory and Its Applications, pp. 312–316 (2012)
14. Shinagawa, K., Iwata, T.: Quantum attacks on sum of Even-Mansour pseudorandom functions. Inf. Process. Lett. **173**, 106172 (2022)
15. Shor, P.W.: Polynomial-time algorithms for prime factorization and discrete logarithms on a quantum computer. SIAM J. Comput. **26**(5), 1484–1509 (1997)
16. Simon, D.R.: On the power of quantum computation. SIAM J. Comput. **26**(5), 1474–1483 (1997)
17. Zhandry, M.: How to construct quantum random functions. In: 53rd Annual IEEE Symposium on Foundations of Computer Science, FOCS 2012, pp. 679–687 (2012)
18. Zhandry, M.: How to record quantum queries, and applications to quantum indifferentiability. In: Boldyreva, A., Micciancio, D. (eds.) CRYPTO 2019. LNCS, vol. 11693, pp. 239–268. Springer, Cham (2019). https://doi.org/10.1007/978-3-030-26951-7_9

Extended Formulations via Decision Diagrams

Yuta Kurokawa[1], Ryotaro Mitsuboshi[1,2]($\boxtimes$) iD, Haruki Hamasaki[1,2] iD,
Kohei Hatano[1,2] iD, Eiji Takimoto[1] iD, and Holakou Rahmanian[3]

[1] Kyushu University, Fukuoka, Japan
ryotaro.mitsuboshi@inf.kyushu-u.ac.jp
[2] Riken AIP, Tokyo, Japan
[3] Amazon, Tokyo, Japan

Abstract. We propose a general algorithm of constructing an extended formulation for any given set of linear constraints with integer coefficients. Our algorithm consists of two phases: first construct a decision diagram (V, E) that somehow represents a given $m \times n$ constraint matrix, and then build an equivalent set of $|E|$ linear constraints over $n + |V|$ variables. That is, the size of the resultant extended formulation depends not explicitly on the number m of the original constraints, but on its decision diagram representation. Therefore, we may significantly reduce the computation time and space for optimization problems with integer constraint matrices by solving them under the extended formulations, especially when we obtain concise decision diagram representations for the matrices. We demonstrate the effectiveness of our extended formulations for mixed integer programming and the 1-norm regularized soft margin optimization tasks over synthetic and real datasets.
Eligible for best student paper.

Keywords: Extend formulation · Decision diagrams · Mixed integer programs

1 Introduction

Large-scale optimization tasks appear in many areas such as machine learning, operations research, and engineering. Time/memory-efficient optimization techniques are more in demand than ever. Various approaches have been proposed to efficiently solve optimization problems over huge data, e.g., stochastic gradient descent methods (e.g., [8]) and concurrent computing techniques using GPUs (e.g., [26]). Among them, we focus on the "computation on compressed data" approach, where we first compress the given data somehow and then employ an algorithm that works directly on the compressed data (i.e., without decompressing the data) to complete the task, in an attempt to reduce computation time and/or space. Algorithms on compressed data are mainly studied in string processing (e.g., [12,13,18,19,28]), enumeration of combinatorial objects (e.g., [21]), and combinatorial optimization (e.g., [2]). In particular, in the work on combinatorial optimization, they compress the set of feasible solutions that satisfy given

W. Wu and G. Tong (Eds.): COCOON 2023, LNCS 14423, pp. 17–28, 2024.
https://doi.org/10.1007/978-3-031-49193-1_2

constraints into a decision diagram so that minimizing a linear objective can be done by finding the shortest path in the decision diagram. Although we can find the optimal solution very efficiently when the size of the decision diagram is small, the method can only be applied to specific types of discrete optimization problems where the feasible solution set is finite, and the objective function is linear.

Whereas, we mainly consider a more general form of discrete/continuous optimization problems that include linear constraints with integer coefficients:

$$\min_{x \in X \subset \mathbb{R}^n} f(x) \quad \text{s.t.} \quad Ax \geq b \tag{1}$$

for some $A \in C^{m \times n}$ and $b \in C^m$, where X denotes the constraints other than $Ax \geq b$, and C is a finite subset of integers. This class of problems includes LP, QP, SDP, and MIP with linear constraints of integer coefficients. So our target problem is fairly general. Without loss of generality, we assume $m > n$, and we are particularly interested in the case where m is huge.

In this paper, we propose a pre-processing method that "rewrites" integer-valued linear constraints with equivalent but more concise ones. More precisely, we propose a general algorithm that, when given an integer-valued constraint matrix $(A, b) \in C^{m \times n} \times C^m$ of an optimization problem (1), produces a matrix $(A', b') \in C^{m' \times (n+n')} \times C^{m'}$ that represents its extended formulation, that is, it holds that

$$\exists s \in \mathbb{R}^{n'}, A' \begin{bmatrix} x \\ s \end{bmatrix} \geq b' \Leftrightarrow Ax \geq b$$

for some n' and m', with the hope that the size of (A', b') is much smaller than that of (A, b) even at the cost of adding n' extra variables. Using the extended formulation, we obtain an equivalent optimization problem to (1):

$$\min_{x \in X \subset \mathbb{R}^n, s \in \mathbb{R}^{n'}} f(x) \quad \text{s.t.} \quad A' \begin{bmatrix} x \\ s \end{bmatrix} \geq b'. \tag{2}$$

Then, we can apply any existing generic solvers, e.g., MIP/QP/LP solvers if f is linear or quadratic, to (2), combined with our pre-processing method, which may significantly reduce the computation time/space than applying them to the original problem (1).

To obtain a matrix (A', b'), we first construct a variant of a decision diagram called a Non-Deterministic Zero-Suppressed Decision Diagram (NZDD, for short) [11] that somehow represents the matrix (A, b). Observing that the constraint $Az \geq b$ can be restated in terms of the NZDD constructed as "every path length is lower bounded by 0" for an appropriate edge weighting, we establish the extended formulation $(A', b') \in C^{m' \times (n+n')} \times C^{m'}$ with $m' = |E|$ and $n' = |V|$, where V and E are the sets of vertices and edges of the NZDD, respectively. One of the advantages of the result is that the size of the resulting optimization problem depends only on the size of the NZDD and the number n of variables, but *not* on the number m of the constraints in the original problem. Therefore, if the matrix (A, b) is well compressed into a small NZDD, then we obtain an equivalent but concise optimization problem (2).

To clarify the differences between our work and previous work regarding optimization using decision diagrams, we summarize the characteristics of both results in Table 1. Notable differences are that (i) ours can treat optimization problems with any types of variables (discrete, or real), any types of objectives (including linear ones) but with integer coefficients on linear constraints, and (ii) ours uses decision diagrams for representing linear constraints while previous work uses them for representing feasible solutions of particular classes of problems. So, for particular classes of discrete optimization problems, the previous approach would work better with specific construction methods for decision diagrams. On the other hand, ours is suitable for continuous optimization problems or/and discrete optimization problems for which efficient construction methods for decision diagrams representing feasible solutions are not known. See the later section for more detailed descriptions of related work.

Table 1. Characteristics of previous work on optimization with decision diagrams (DDs) and ours.

	coeff. of lin. consts.	variables	objectives	DDs
Previous work ours	any type binary/integer	binary/integer any type	linear any type	feasible solutions lin. consts.

Then, to realize succinct extended formulations, we propose practical heuristics for constructing NZDDs, which is our third contribution. Since it is not known to construct an NZDD of small size, we first construct a ZDD of minimal size, where the ZDD is a restricted form of the NZDD representation. To this end, we use a ZDD compression software called zcomp [30]. Then, we give rewriting rules for NZDDs that reduce both the numbers of vertices and edges, and apply them to obtain NZDDs of smaller size of V and E. Although the rules may increase the size of NZDDs (i.e., the total number of edge labels), the rules seem to work effectively since reducing $|V|$ and $|E|$ is more important for our purpose.

Experimental results on synthetic and real data sets show that our algorithms improve time/space efficiency significantly, especially when (i) $m \gg n$, and (ii) the set C of integer coefficients is small, e.g., binary, where the datasets tend to have concise NZDD representations.

2 Related Work

Various computational tasks over compressed strings or texts are investigated in algorithms and data mining literature, including, e.g., pattern matching over strings and computing edit distances or q-grams [12,13,18,19,28]. The common assumption is that strings are compressed using the straight-line program, which is a class of context-free grammars generating only one string (e.g., LZ77 and +LZ78). As notable applications of string compression techniques to data mining and machine learning, Nishino et al. [25] and Tabei et al. [29] reduce the space complexity of matrix-based computations. So far, however, string compression-based approaches do not seem to be useful for representing linear constraints.

Decision diagrams are used in the enumeration of combinatorial objects, discrete optimization and so on. In short, a decision diagram is a directed acyclic graph with a root and a leaf, representing a subset family of some finite ground set Σ or, equivalently, a boolean function. Each root-to-leaf path represents a set in the set family. The Binary Decision Diagram (BDD) [4,16] and its variant, the Zero-Suppressed Binary Decision Diagram (ZDD) [16,20], are popular in the literature. These support various set operations (such as intersection and union) in efficient ways. Thanks to the DAG structure, linear optimization problems over combinatorial sets $X \subset \{0,1\}^n$ can be reduced to shortest/longest path problems over the diagrams representing X. This reduction is used to solve the exact optimization of NP-hard combinatorial problems (see, e.g., [2,3,5,14,24]) and enumeration tasks [21–23]. Among work on decision diagrams, the work of Fujita et al. [11] would be closest to ours. They propose a variant of ZDD called the Non-deterministic ZDD (NZDD) to represent labeled instances and show how to emulate the boosting algorithm AdaBoost* [27], a variant of AdaBoost [10] that maximizes the margin, over NZDDs. We follow their NZDD-based representation of the data. But our work is different from Fujita et al. in that, they propose specific algorithms running over NZDDs, whereas our work presents extended formulations based on NZDDs, which could be used with various algorithms.

The notion of extended formulation arises in combinatorial optimization (e.g., [7,32]). The idea is to re-formulate a combinatorial optimization with an equivalent different form, so that the size of the problem is reduced. For example, a typical NP-hard combinatorial optimization problem has an integer programming formulation of exponential size. Then a good extended formulation should have a smaller size than the exponential. Typical work on extended formulation focuses on some characterization of the problem to obtain succinct formulations (see, e.g., [9]). Our work is different from these in that we focus on the redundancy of the data and try to obtain succinct extended formulations for optimization problems described with data.

3   Preliminaries

The non-deterministic Zero-suppressed Decision Diagram (NZDD) [11] is a variant of the Zero-suppressed Decision Diagram(ZDD) [16,20], representing subsets of some finite ground set Σ. More formally, NZDD is defined as follows.

Definition 1 (NZDD). *An NZDD G is a tuple $G = (V, E, \Sigma, \Phi)$, where (V, E) is a directed acyclic graph (V and E are the sets of nodes and edges, respectively) with a single root with no-incoming edges and a leaf with no outgoing edges, Σ*

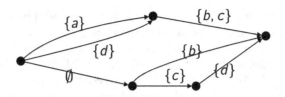

Fig. 1. An NZDD representing $\{\{a, b, c\}, \{b\}, \{b, c, d\}, \{c, d\}\}$.

is the ground set, and $\Phi : E \to 2^{\Sigma}$ is a function assigning each edge e a subset $\Phi(e)$ of Σ. More precisely, we allow (V, E) to be a multigraph, i.e., two nodes can be connected with more than one edge.

Furthermore, an NZDD G satisfies the following additional conditions. Let $\mathcal{P}_G$ be the set of paths in G starting from the root to the leaf, where each path $P \in \mathcal{P}_G$ is represented as a subset of E, and for any path $P \in \mathcal{P}_G$, we abuse the notation and let $\Phi(P) = \cup_{e \in P} \Phi(e)$.

1. For any path $P \in \mathcal{P}_G$ and any edges $e, e' \in P$, $\Phi(e) \cap \Phi(e') = \emptyset$. That is, for any path P, an element $a \in \Sigma$ appears at most once in P.
2. For any paths $P, P' \in \mathcal{P}_G$, $\Phi(P) \neq \Phi(P')$. Thus, each path P represents a different subset of Σ.

Then, an NZDD G naturally corresponds to a subset family of Σ. Formally, let $L(G) = \{\Phi(P) \mid P \in \mathcal{P}_G\}$. Figure 1 illustrates an NZDD representing a subset family $\{\{a, b, c\}, \{b\}, \{b, c, d\}, \{c, d\}\}$.

A ZDD [16,20] can be viewed as a special form of NZDD $G = (V, E, \Sigma, \Phi)$ satisfying the following properties: (i) For each edge $e \in E$, $\Phi(e) = \{a\}$ for some $a \in \Sigma$ or $\Phi(e) = \emptyset$. (ii) Each internal node has at most two outgoing edges. If there are two edges, one is labeled with $\{a\}$ for some $a \in \Sigma$ and the other is labeled with $\emptyset$. (iii) There is a total order over Σ such that, for any path $P \in \mathcal{P}_G$ and for any $e, e' \in P$ labeled with singletons $\{a\}$ and $\{a'\}$ respectively, if e is an ancestor of e', a precedes a' in the order.

We believe that constructing a minimal NZDD for a given subset family is NP-hard since closely related problems are NP-hard. For example, constructing a minimal ZDD (over all orderings of Σ) is known to be NP-hard [16], and construction of a minimal NFA which is equivalent to a given DFA is P-space hard [15]. On the other hand, there is a practical construction algorithm of ZDDs given a subset family and a fixed order over Σ using multi-key quicksort [30].

4   NZDDs for Linear Constraints with Binary Coefficients

In this section, we show an NZDD representation for linear constraints in problem (1) when linear constraints have $\{0, 1\}$-valued coefficients, that is, $C = \{0, 1\}$. We will discuss its extensions to integer coefficients in the later section. Let $\boldsymbol{a}_i \in \{0, 1\}^n$ be the vector corresponding to the i-th row of the matrix $A \in \{0,1\}^{m \times n}$ (for $i \in [m]$). For $x \in \{0,1\}^n$, let $\text{idx}(x) = \{j \in [n] \mid x_j \neq 0\}$, i.e., the set of indices of nonzero components of x. Then, we define $I = \{\text{idx}(c_i) \mid c_i = (a_i, b_i), i \in [m]\}$. Note that I is a subset family of $2^{[n+1]}$. Then we assume that we have some NZDD $G = (V, E, [n+1], \Phi)$ representing I, that is, $L(G) = I$. We will later show how to construct NZDDs.

The following theorem shows the equivalence between the original problem (1) and a problem described with the NZDD G.

Theorem 1. *Let $G = (V, E, [n+1], \Phi)$ be an NZDD such that $L(G) = I$. Then the following optimization problem is equivalent to problem (1):*

$$\min_{x \in X \subset \mathbb{R}^n, s \in \mathbb{R}^{|V|}} f(x) \tag{3}$$

$$s.t. \quad s_{e.u} + \sum_{j \in \Phi(e)} x'_j \geq s_{e.v}, \quad \forall e \in E,$$

$$s_{\text{root}} = 0, \quad s_{\text{leaf}} = 0,$$

$$x' = (x, -1),$$

where e.u and e.v are nodes that the edge e is directed from and to, respectively.

Before going through the proof, let us explain some intuition on problem (3). Intuitively, each linear constraint in (1) is encoded as a path from the root to the leaf in the NZDD G, and a new variable s_v for each node v represents a lower bound of the length of the shortest path from the root to v. The inequalities in (3) reflect the structure of the standard dynamic programming of Dijkstra, so that all inequalities are satisfied if and only if the length of all paths is larger than zero. In Fig. 2, we show an illustration of the extended formulation.

Proof. Let $x_\star$ and $(\hat{x}', \hat{s})$ be the optimal solutions of problems (1) and (3), respectively. It suffices to show that each optimal solution can construct a feasible solution of the other problem.

Let $\hat{x}$ be the vector consisting of the first n components of $\hat{x}'$. For each constraint $a_i^\top x \geq b_i$ ($i \in [m]$) in problem (1), there exists the corresponding path $P_i \in \mathcal{P}_G$. By repeatedly applying the first constraint in (3 along the path P_i, we have $\sum_{e \in P_i} \sum_{j \in \Phi(e)} \hat{z}'_j \geq \hat{s}_{\text{leaf}} = 0$. Further, since $\Phi(P_i)$ represents the set of indices of nonzero components of c_i, $\sum_{e \in P_i} \sum_{j \in \Phi(e)} \hat{z}'_j = c_i^\top \hat{x}' = a_i^\top \hat{x} - b_i$. By combining these inequalities, we have $a_i^\top \hat{x} - b_i \geq 0$. This implies that $\hat{x}$ is a feasible solution of (1) and thus $f(x_\star) \leq f(\hat{x})$.

Let $x'_\star = (x_\star, -1)$. Assuming a topological order on V (from the root to the leaf), we define $s_{\star,\text{root}} = s_{\star,\text{leaf}} = 0$ and $s_{\star,v} = \min_{e \in E, e.v=v} s_{\star,e.u} + \sum_{j \in \Phi(e)} z'_{\star,j}$ for each $v \in V \setminus \{\text{root}, \text{leaf}\}$. Then, we have, for each $e \in E$ s.t. $e.v \neq \text{leaf}$, $s_{\star,e.v} \leq s_{\star,e.u} + \sum_{j \in \Phi(e)} z'_{\star,j}$ by definition. Now, $\min_{e \in E, e.v=\text{leaf}} s_{\star,e.u} + \sum_{j \in \Phi(e)} z'_{\star,j}$ is achieved by a path $P \in \mathcal{P}_G$ corresponding to $\arg\min_{i \in [m]} a_i^\top x_\star - b_i$, which is ≥ 0 since $x_\star$ is feasible w.r.t. (1). Therefore, $s_{\star,e.v} \leq s_{\star,e.u} + \sum_{j \in \Phi(e)} z'_{\star,j}$ for $e \in E$ s.t. $e.v = \text{leaf}$ as well. Thus, $(x'_\star, s_\star)$ is a feasible solution of (3) and $f(\hat{x}) \leq f(x_\star)$.

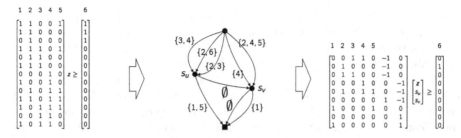

Fig. 2. An illustration of the extended formulation. Left: Original constraints as in (1). Middle: A NZDD representation of the left constraints. Right: The matrix form (3) of the middle diagram without constant terms. This example reduces the 13 constraints to 9 constraints by adding 2 variables.

Given the NZDD $G = (V, E)$, problem (3) contains $n + |V| - 2$ variables and $|E|$ linear constraints, where $|V| - 2$ variables are real. The most naive construction, where the resulting NZDD contains two nodes (root and leaf), and every root-to-leaf path corresponds to a constraint, has the same problem size as the original one. Thus, the problem size cannot be worse than the original one. In particular, if problem (1) is LP or IP, then problem (3) is LP, or MIP, respectively.

5   Extensions to Integer Coefficients

We briefly discuss how to extend our NZDD representation of linear constraints to the cases where coefficients of linear constraints belong to a finite set C of integers. There are two ways to do so.

Binary Encoding of Integers. We assume some encoding of integers in C with $O(\log |C|)$ bits. Then, each bit can be viewed as a binary-valued variable. Each integer coefficient can be also recovered with its binary representation. Under this attempt, the resulting extended formulation has $O(n \log |C| + |V|)$ variables and $O(|E|)$ linear constraints.

Extending Σ. Another attempt is to extend the domain Σ of an NZDD $G = (V, E, \Sigma, \Phi)$. The extended domain Σ' consists of all pairs of integers in C and elements in Σ. Again, integer coefficients are recovered through the new domain Σ'. The resulting extended formulation has $O(n|C| + |V|)$ variables and $O(|E|)$ linear constraints. While the size of the problem is larger than the binary encoding, its implementation is easy in practice and could be effective for C of small size.

6   Construction of NZDDs

We propose heuristics for constructing NZDDs given a subset family $S \subseteq 2^{\Sigma}$. We use the zcomp [30,31], developed by Toda, to compress the subset family

S to a ZDD. The zcomp is designed based on multikey quicksort [1] for sorting strings. The running time of the zcomp is $O(N \log^2 |S|)$, where N is an upper bound of the nodes of the output ZDD and $|S|$ is the sum of cardinalities of sets in S. Since $N \leq |S|$, the running time is almost linear in the input.

A naive application of the zcomp is, however, not very successful in our experiences. We observe that the zcomp often produces concise ZDDs compared to inputs. But, concise ZDDs do not always imply concise representations of linear constraints. More precisely, the output ZDDs of the zcomp often contains (i) nodes with one incoming edge or (ii) nodes with one outgoing edge. A node v of these types introduces a corresponding variable s_v and linear inequalities. Specifically, in the case of type (ii), we have $s_v \leq \sum_{j\Phi(e)} z'_j + s_{e.u}$ for each $e \in E$ s.t. $e.v = v$, and for its child node v' and edge e' between v and v', $s_{v'} \leq \sum_{j\in\Phi(e')} z'_j + s_v$. These inequalities are redundant since we can obtain equivalent inequalities by concatenating them: $s_{v'} \leq \sum_{j\in\Phi(e')} z'_j + \sum_{j\in\Phi(e)} z'_j + s_{e.u}$ for each $e \in E$ s.t. $e.v = v$, where s_v is removed.

Based on the observation above, we propose a simple reduction heuristics removing nodes of type (i) and (ii). More precisely, given an NZDD $G = (V, E)$, the heuristics outputs an NZDD $G' = (V', E')$ such that $L(G) = L(G')$ and G' does not contain nodes of type (i) or (ii). The heuristics can be implemented in $O(|V'| + |E'| + \sum_{e\in E'} |\Phi(e)|)$ time by going through nodes of the input NZDD G in the topological order from the leaf to the root and in the reverse order, respectively. The details of the heuristics is given in the full paper [17].

7   Experiments

We show preliminary experimental results on synthetic and real large data sets[1]. The tasks are, the mixed integer programming, and the 1-norm regularized soft margin optimization (see the full paper for details). Our experiments are conducted on a server with 2.60 GHz Intel Xeon Gold 6124 CPUs and 314 GB memory. We use Gurobi optimizer 9.01, a state-of-the-art commercial LP solver. To obtain NZDD representations of data sets, we apply the procedure described in the previous section. The details of preprocessing of data sets and NZDD representations are shown in the full paper.

7.1   Mixed Integer Programming on Synthetic Datasets

First, we apply our extended formulation (1) to mixed integer programming tasks over synthetic data sets. The problems are defined as the linear optimization with n variables and m linear constraints of the form $A\boldsymbol{x} \geq \boldsymbol{b}$, where (i) each row of A has k entries of 1 and others are 0s and nonzero entries are chosen randomly without repetition (ii) coefficients a_i of linear objective $\sum_{i=1}^{n} a_i x_i$ is chosen from 1,...,100 randomly, and (iii) first l variables take binary values in $\{0, 1\}$ and

---

[1] Codes are available at https://bitbucket.org/kohei_hatano/codes_extended_formula tion_nzdd/.

others take real values in $[0, 1]$. In our experiments, we fix $n = 25, k = 10$, $l = 12$ and $m \in \{4 \times 10^5, 8 \times 10^5 ..., 20 \times 10^5\}$. We apply the Gurobi optimizer directly to the problem denoted as mip and the solver with pre-processing the problem by our extended formulation (denoted as nzdd_mip, respectively. The results are summarized in Fig. 3. Our method consistently improves computation time for these datasets. This makes sense since it can be shown that when $m = O(n^k)$ there exists an NZDD of size $O(nk)$ representing the constraint matrix. In addition, the pre-processing time is within 2 s in all cases.

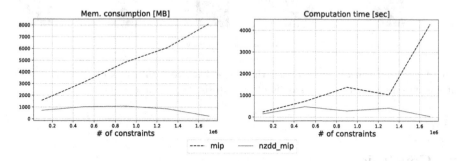

Fig. 3. The comparison for synthetic datasets of a MIP problem. The horizontal axis represents the number of constraints of the original problem.

7.2 1-norm Soft Margin Optimization on Real Data Sets

Next, we apply our methods on the task of the 1-norm soft margin optimization. This problem is a standard optimization problem in the machine learning literature, categorized as LP, for finding sparse linear classifiers given labeled instances. Details of this problem is shown in the full paper. We compare the following methods using a naive LP solver, (i) previous formulation (denoted as naive), (ii) our formulation over NZDD (denoted as nzdd_naive). Formulations of both (i) and (ii) is placed in the full paper. We measure its computation time (CPU time) and maximum memory consumption, respectively, and compare their averages over parameters. Further, we perform 5-fold cross validation to check the test error rates of our methods on real data sets. In fact, the test error rates are similar between methods for (i) and (ii). This means our extended formulation is comparable to the standard one in terms of generalization performance. Details of the cross validation is omitted and shown in the full paper.

We compare methods on some real data sets in the libsvm datasets [6] to see the effectiveness of our approach in practice. Generally, the datasets contain huge samples (m varies from 3×10^4 to 10^7) with a relatively small size of features (n varies from 20 to 10^5). The features of instances of each dataset is transformed into binary values. Results are summarized in Fig. 4. Note that these results exclude NZDD construction times since the compression takes around

1 s, except for the HIGGS dataset (around 13 s). Furthermore, the construction time of NZDDs can be neglected in the following reason: We often need to try multiple choices of the hyperparameters (ν in our case) and solve the optimization problem for each set of choices. But once we construct an NZDD, we can be re-use it for different values of hyperparameters without reconstructing NZDDs.

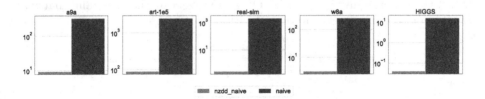

Fig. 4. Comparison of computation times (sec.) for real data sets of the soft margin optimization problem. The y-axis is plotted in the logarithmic scale.

8 Conclusion

We proposed a generic algorithm of constructing an NZDD-based extended formulation for any given set of linear constraints with integer constraints as well as specific algorithms for the 1-norm soft margin optimization and practical heuristics for constructing NZDDs. Our algorithms improve time/space efficiency on artificial and real datasets, especially when the datasets have concise NZDD representations.

Acknowledgements. We thank the reviewers for helpful comments. We also thank Mohammad Amin Mansouri for valuable discussions and initial development. This work was supported by JSPS KAKENHI Grant Numbers JP19H014174, JP19H04067, JP20H05967, and JP22H03649, respectively.

References

1. Bentley, J.L., Sedgewick, R.: Fast algorithms for sorting and searching strings. In: Proceedings of the Eighth Annual ACM-SIAM Symposium on Discrete Algorithms (SODA 1997), pp. 360–369 (1997)
2. Bergman, D., Cire, A.A., van Hoeve, W.J., Hooker, J.: Decision Diagrams for Optimization. Springer Cham (2016). https://doi.org/10.1007/978-3-319-42849-9
3. Bergman, D., van Hoeve, W.-J., Hooker, J.N.: Manipulating MDD relaxations for combinatorial optimization. In: Achterberg, T., Beck, J.C. (eds.) CPAIOR 2011. LNCS, vol. 6697, pp. 20–35. Springer, Heidelberg (2011). https://doi.org/10.1007/978-3-642-21311-3_5
4. Bryant, R.E.: Graph-based algorithms for Boolean function manipulation. IEEE Trans. Comput. **C-35**(8), 677–691 (1986)

5. Castro, M.P., Cire, A.A., Beck, J.C.: An MDD-based Lagrangian approach to the multicommodity pickup-and-delivery TSP. INFORMS J. Comput. **32**(2), 263–278 (2019). https://doi.org/10.1287/IJOC.2018.0881
6. Chang, C.C., Lin, C.J.: LIBSVM: a library for support vector machines. ACM Trans. Intell. Syst. Technol. **2**(27), 1–27 (2011)
7. Conforti, M., Cornuéjols, G., Zambelli, G.: Extended formulations in combinatorial optimization. 4OR **8**(1), 1–48 (2010)
8. Duchi, J., Hazan, E., Singer, Y.: Adaptive subgradient methods for online learning and stochastic optimization. J. Mach. Learn. Res. **12**(7), 2121–2159 (2011)
9. Fiorini, S., Huynh, T., Weltge, S.: Strengthening convex relaxations of 0/1-sets using Boolean formulas. Math. Program. **190**(1), 467–482 (2021)
10. Freund, Y., Schapire, R.E.: A decision-theoretic generalization of on-line learning and an application to boosting. J. Comput. Syst. Sci. **55**(1), 119–139 (1997)
11. Fujita, T., Hatano, K., Takimoto, E.: Boosting over non-deterministic ZDDs. Theoret. Comput. Sci. **806**, 81–89 (2020)
12. Goto, K., Bannai, H., Inenaga, S., Takeda, M.: Fast q-gram mining on SLP compressed strings. J. Discrete Algorithms **18**, 89–99 (2013)
13. Hermelin, D., Landau, G.M., Landau, S., Weimann, O.: A unified algorithm for accelerating edit-distance computation via text-compression. In: Proceedings of the 26th International Symposium on Theoretical Aspects of Computer Science (STACS 2009), LIPIcs, vol. 3, pp. 529–540 (2009)
14. Inoue, T., et al.: Distribution loss minimization with guaranteed error bound. IEEE Trans. Smart Grid **5**(1), 102–111 (2014)
15. Jiang, T., Ravikumar, B.: Minimal NFA problems are hard. SIAM J. Comput. **22**(6), 1117–1141 (1993)
16. Knuth, D.E.: The Art of Computer Programming, vol. 4A, Combinatorial Algorithms, Part 1. Addison-Wesley Professional, Upper Saddle River (2011)
17. Kurokawa, Y., Mitsuboshi, R., Hamasaki, H., Hatano, K., Takimoto, E., Rahmanian, H.: Extended formulations via decision diagrams (2022)
18. Lifshits, Y.: Processing compressed texts: a tractability border. In: Ma, B., Zhang, K. (eds.) CPM 2007. LNCS, vol. 4580, pp. 228–240. Springer, Heidelberg (2007). https://doi.org/10.1007/978-3-540-73437-6_24
19. Lohrey, M.: Algorithmics on SLP-compressed strings: a survey. Groups - Complexity - Cryptology **4**(2), 241–299 (2012)
20. Minato, S.I.: Zero-suppressed BDDs for set manipulation in combinatorial problems. In: Proceedings of the 30th International Design Automation Conference (DAC 1993), pp. 272–277 (1993)
21. Minato, S.I.: Power of enumeration - recent topics on BDD/ZDD-based techniques for discrete structure manipulation. IEICE Trans. Inf. Syst. **E100.D**(8), 1556–1562 (2017)
22. Minato, S.I., Uno, T.: Frequentness-transition queries for distinctive pattern mining from time-segmented databases. In: Proceedings of the 2010 SIAM International Conference on Data Mining (SDM), pp. 339–349 (2010)
23. Minato, S., Uno, T., Arimura, H.: LCM over ZBDDs: fast generation of very large-scale frequent itemsets using a compact graph-based representation. In: Washio, T., Suzuki, E., Ting, K.M., Inokuchi, A. (eds.) PAKDD 2008. LNCS (LNAI), vol. 5012, pp. 234–246. Springer, Heidelberg (2008). https://doi.org/10.1007/978-3-540-68125-0_22
24. Morrison, D.R., Sewell, E.C., Jacobson, S.H.: Solving the pricing problem in a branch-and-price algorithm for graph coloring using zero-suppressed binary deci-

sion diagrams. INFORMS J. Comput. **28**(1), 67–82 (2016). https://doi.org/10. 1287/IJOC.2015.0667

25. Nishino, M., Yasuda, N., Minato, S.I., Nagata, M.: Accelerating graph adjacency matrix multiplications with adjacency forest. In: Proceedings of the 2014 SIAM International Conference on Data Mining (SDM 2014), pp. 1073–1081 (2014)

26. Raina, R., Madhavan, A., Ng, A.Y.: Large-scale deep unsupervised learning using graphics processors. In: Proceedings of the 26th Annual International Conference on Machine Learning (ICML 2009), pp. 873–880 (2009)

27. Rätsch, G., Warmuth, M.K.: Efficient margin maximizing with boosting. J. Mach. Learn. Res. **6**, 2131–2152 (2005)

28. Rytter, W.: Grammar compression, LZ-encodings, and string algorithms with implicit input. In: Díaz, J., Karhumäki, J., Lepistö, A., Sannella, D. (eds.) ICALP 2004. LNCS, vol. 3142, pp. 15–27. Springer, Heidelberg (2004). https://doi.org/10. 1007/978-3-540-27836-8_5

29. Tabei, Y., Saigo, H., Yamanishi, Y., Puglisi, S.J.: Scalable partial least squares regression on grammar-compressed data matrices. In: Proceedings of the 22nd ACM SIGKDD International Conference on Knowledge Discovery and Data Mining (KDD 2016), pp. 1875–1884 (2016)

30. Toda, T.: Fast compression of large-scale hypergraphs for solving combinatorial problems. In: Fürnkranz, J., Hüllermeier, E., Higuchi, T. (eds.) DS 2013. LNCS (LNAI), vol. 8140, pp. 281–293. Springer, Heidelberg (2013). https://doi.org/10. 1007/978-3-642-40897-7_19

31. Toda, T.: ZCOMP: Fast Compression of Hypergraphs into ZDDs (2015). https:// www.sd.is.uec.ac.jp/toda/code/zcomp.html

32. Yannakakis, M.: Expressing combinatorial optimization problems by Linear Programs. J. Comput. Syst. Sci. **43**(3), 441–466 (1991). https://doi.org/10.1016/0022-0000(91)90024-Y

Greedy Gray Codes for Dyck Words and Ballot Sequences

Vincent Vajnovszki[1] and Dennis Wong[2(✉)]

[1] Université de Bourgogne, Dijon, France
vvajnov@u-bourgogne.fr
[2] Macao Polytechnic University, Macao, China
cwong@uoguelph.ca

Abstract. We present a simple greedy algorithm for generating Gray codes for Dyck words and fixed-weight Dyck prefixes. Successive strings in our listings differ from each other by a transposition, that is, two bit changes. Our Gray codes are both homogeneous and suffix partitioned. Furthermore, we use our greedy algorithm to produce the first known homogeneous 2-Gray code for ballot sequences, which are Dyck prefixes of all weights. Our work extends a previous result on combinations by Williams [Conference proceedings: Workshop on Algorithms and Data Structures (WADS), LNTCS 8037:525-536, 2013].

Keywords: Dyck word · lattice path · balanced parentheses · ballot sequence · homogeneous Gray code · greedy algorithm

1 Introduction

A *Dyck word* is a binary string with the same number of 1 s and 0 s such that any prefix contains at least as many 0 s as 1 s. Dyck words are in bijection with *balanced parentheses*, with an open bracket represented by a 0 and a close bracket represented by a 1 [4,7]. For example, all length six balanced parentheses are given by

$$((())), (()()), (())(), ()(()), ()()().$$

The Dyck words that correspond to the five balanced parentheses of length six are

$$000111, 001011, 001101, 010011, 010101.$$

Since the number of 0 s and 1 s of a Dyck word has to be the same, the length n of Dyck words has to be an even number. Dyck words can be used to encode lattice paths that end on their starting level and never pass below it.

A *ballot sequence* is a binary string of length n such that in any of its prefixes the number of 0s is greater than or equal to the number of 1s. As an example, the ten ballot sequences for length five are

$$00000, 00001, 00010, 00011, 01001, 00100, 00101, 00110, 01000, 01010.$$

W. Wu and G. Tong (Eds.): COCOON 2023, LNCS 14423, pp. 29–40, 2024.
https://doi.org/10.1007/978-3-031-49193-1_3

Such a length n sequence encodes a ballot counting scenario involving two candidates in which the number of votes collected by the first candidate is always greater than or equal to those collected by the second candidate throughout the count. Ballot sequences are also known as *Dyck prefixes*, which are prefixes of Dyck words. Ballot sequences and Dyck prefixes can also be used to encode lattice paths that end on the positive region and never pass below it.

The number of Dyck words is known as the *Catalan number*, and the number of ballot sequences is known as the *ballot number*. The enumeration sequences of Dyck words and ballot sequences are A000108 and A001405 in the Online Encyclopedia of Integer Sequences respectively [23]. The enumeration formulae for the number of Dyck words and the number of ballot sequences [2] of length n are given as follows:

- Catalan number: $\dfrac{1}{\frac{n}{2}+1}\dbinom{n}{\frac{n}{2}}$;
- Ballot number: $\dbinom{n}{\lfloor\frac{n}{2}\rfloor}$.

Dyck words and ballot sequences are well studied combinatorial objects that have a wide variety of applications. For example, Dyck words have been used to encode a wide variety of combinatorial objects including binary trees, balanced parentheses, lattice paths, and stack-sortable permutations [4,7,8,11,14,19,27,31]. Ballot sequences, on the other hand, have many applications ranging from constructing more sums than differences (MSTD) sets [33], generating n-node binary trees of different shapes [1,16], and enumerating random walks with various constraints [3,6,10,12,29]. For more applications of Dyck words and ballot sequences, see [9,13,20,24].

One of the most important aspects of combinatorial generation is to list the instances of a combinatorial object so that consecutive instances differ by a specified *closeness condition* involving a constant amount of change. Lists of this type are called *Gray codes*. This terminology is due to the eponymous *binary reflected Gray code* (BRGC) by Frank Gray, which orders the 2^n binary strings of length n so that consecutive strings differ in one bit. For example, when $n = 4$ the order is

$$0000, 1000, 1100, 0100, 0110, 1110, 1010, 0010,$$
$$0011, 1011, 1111, 0111, 0101, 1101, 1001, 0001.$$

The BRGC listing is a *1-Gray code* in which consecutive strings differ by one symbol change. In this paper, we are focusing on *transposition Gray code*, where consecutive strings differ by swapping the positions of two bits. A transposition Gray code is also a *2-Gray code*, where consecutive strings differ by at most two bit changes.

Several algorithms have been proposed to generate Dyck words. Proskurowski and Ruskey [15] devised a transposition Gray code for Dyck words. Later, efficient algorithms to generate such a listing were presented in [17,28]. Bultena and Ruskey [5], and later van Baronaigien [26] and Xiang et al. [32], developed algorithms to generate homogeneous transposition Gray codes for Dyck words. For example, the algorithm by Bultena and Ruskey generates the 42 Dyck words for $n = 10$ as follows:

0101010101, 0011010101, 0010110101, 0100110101, 0001110101, 0001101101,
0100101101, 0010101101, 0011001101, 0101001101, 0100011101, 0010011101,
0001011101, 0000111101, 0000111011, 0001011011, 0010011011, 0100011011,
0101001011, 0011001011, 0010101011, 0100101011, 0001101011, 0001110011,
0100110011, 0010110011, 0011010011, 0101010011, 0101000111, 0011000111,
0010100111, 0100100111, 0001100111, 0001010111, 0010010111, 0100010111,
0000110111, 0000101111, 0001001111, 0010001111, 0100001111, 0000011111.

The Gray code is said to be *homogeneous*, where the bits between the swapped 0 and 1 are all 0 s. Additionally, the Gray code is also a *suffix-partitioned* Gray code, where strings with the same suffix are contiguous. Vajnovszki and Walsh [25] discovered an even more restrictive Gray code that is *two-close*, where a 1 exchanges its position with an adjacent 0 or a 0 that is separated from it by a single 0. In contrast, Ruskey and Williams [18] provided a *shift Gray code* for Dyck words where consecutive strings differ by a prefix shift.

For ballot sequences, the problem of finding a Gray code for ballot sequences was first studied by Sabri and Vajnovszki [19]. Sabri and Vajnovszki proved that one definition of the reflected Gray code induces a 3-Gray code for k-ary ballot sequences, which is a generalization of ballot sequences that involves more than two candidates. Wong et al. [31] later provided an efficient algorithm to generate a 2-Gray code for ballot sequences. For example, the algorithm by Wong et al. generates the following cyclic 2-Gray code for ballot sequences for $n = 6$:

000111, 010011, 000011, 001011, 001001, 000001, 010001, 010101, 000101, 001101,
001100, 000100, 010100, 010000, 000000, 001000, 001010, 000010, 010010, 000110.

Another approach by Wong et al. to obtain a cyclic 2-Gray code for ballot sequences is by *filtering* the BRGC [31]. For more information about Gray codes induced by the BRGC, see [21] and [22]. However, these Gray codes for ballot sequences are not homogeneous. The greedy algorithm proposed in this paper can be used to generate the first known homogeneous 2-Gray code for ballot sequences.

2 Gray Codes for Dyck Words and Fixed-Weight Dyck Prefixes

In this section, we present a greedy algorithm to generate transposition Gray codes for fixed-weight Dyck prefixes and Dyck words.

In [30], Williams proposed a greedy algorithm to generate a transposition Gray code for combinations. The greedy algorithm by Williams can be summarized as follows:

Greedy Gray code algorithm for k-combinations: Starts with 1^k0^{n-k}. Greedily swap the leftmost possible 1 with the leftmost possible 0 before the next 1 and after the previous 1 (if there are any) such that the resulting string has not appeared before.

For example, the greedy algorithm generates the following 4-combinations for $n = 7$:

1111000, 1110100, 1101100, 1011100, 0111100, 0111010, 1011010,
1101010, 1110010, 1100110, 1010110, 0110110, 0101110, 1001110,
0011110, 0011101, 1001101, 0101101, 0110101, 1010101, 1100101,
1110001, 1101001, 1011001, 0111001, 0110011, 1010011, 1100011,
1001011, 0101011, 0011011, 0010111, 1000111, 0100111, 0001111.

We generalize the idea to fixed-weight Dyck prefixes and Dyck words. The *weight* of a binary string is the number of 1 s it contains. A *fixed-weight Dyck prefix* of weight k is a prefix of a Dyck word with its weight equal to k. Note that when $2k = n$, then the set of fixed-weight Dyck prefixes of weight k is equivalent to the set of Dyck words. The following simple greedy algorithm generates transposition Gray codes for fixed-weight Dyck prefixes and Dyck words of length n:

Greedy Gray code algorithm for fixed-weight Dyck prefixes: Starts with $(01)^k 0^{n-2k}$. Greedily swap the leftmost possible 1 with the leftmost possible 0 before the next 1 and after the previous 1 (if there are any) such that the resulting string is a Dyck prefix and has not appeared before.

Our Gray codes for fixed-weight Dyck prefixes and Dyck words are homogeneous and suffix-partitioned. Another way to understand the greedy algorithm is to greedily swap the leftmost possible 1 with the leftmost possible 0 in a homogeneous manner. As an example, the greedy algorithm generates the following Gray code for Dyck words for $n = 10$ (Dyck prefixes for $n = 10$ and $k = 5$):

0101010101, 0011010101, 0010110101, 0100110101, 0001110101, 0001101101,
0100101101, 0010101101, 0011001101, 0101001101, 0100011101, 0010011101,
0001011101, 0000111101, 0000111011, 0100011011, 0010011011, 0001011011,
0001101011, 0100101011, 0010101011, 0011001011, 0101001011, 0101010011,
0011010011, 0010110011, 0100110011, 0001110011, 0001100111, 0100100111,
0010100111, 0011000111, 0101000111, 0100010111, 0010010111, 0001010111,
0000110111, 0000101111, 0100001111, 0010001111, 0001001111, 0000011111.

Greedy Gray codes have been studied previously, with Williams [30] reinterpreting many classic Gray codes for binary strings, permutations, combinations, binary trees, and set partitions using a simple greedy algorithm. The algorithm presented in this paper can be considered as a novel addition to the family of greedy algorithms previously studied by Williams.

All strings considered in this paper are binary. Our algorithm uses a vector representation $S_1 S_2 \cdots S_k$ to represent a binary string with k ones, where each integer S_i corresponds to the position of the i-th one of the binary string. For example, the string $\alpha = 000110100011001$ can be represented by $S_1, S_2, S_3, S_4, S_5, S_6 = 4, 5, 7, 11, 12, 15$. We initialize the array $S_1, S_2, \ldots, S_k = 2, 4, \ldots, 2k$ for both Dyck words and fixed-weight Dyck prefixes. In addition, we set $S_0 = 0$ and $S_{k+1} = n + 1$. Pseudocode of the greedy algorithm to generate fixed-weight Dyck prefixes and Dyck words is given in Algorithm 1.

Algorithm 1. The greedy algorithm that generates a homogeneous transposition Gray code for fixed-weight Dyck prefixes and Dyck words.

1: **procedure** GREEDY-KDYCK-PREFIXES
2:     $S_1 S_2 \cdots S_k \leftarrow 2\,4 \cdots 2k$
3:     Print($S_1 S_2 \cdots S_k$)
4:     **for** i from 1 to k **do**
5:         **for** j from MAX($S_{i-1} + 1, i \times 2$) to $S_{i+1} - 1$ **do**
6:             **if** $S_1 S_2 \cdots S_{i-1}(j) S_{i+1} \cdots S_k$ has not appeared before **then**
7:                 $S_i \leftarrow j$
8:                 **go to** 4

Theorem 1. *The algorithm Greedy-kDyck-Prefixes generates a homogeneous transposition Gray code for fixed-weight Dyck prefixes that is suffix-partitioned for all n and k where $2k \leq n$.*

3   Proof of Theorem 1

In this section, we prove Theorem 1 for fixed-weight Dyck prefixes. The results also apply to Dyck words as the set of Dyck words is equivalent to the set of fixed-weight Dyck prefixes when $2k = n$. To this end, we begin by proving the following lemmas for fixed-weight Dyck prefixes.

Lemma 1. *The algorithm Greedy-kDyck-Prefixes terminates after visiting the Dyck prefix $0^{n-k}1^k$.*

Proof. Assume the algorithm terminates after visiting some string $b_1 b_2 \cdots b_n \neq 0^{n-k}1^k$. Since $b_1 b_2 \cdots b_n \neq 0^{n-k}1^k$, it must contain the suffix $10^i 1^j$ for some $n - k > i > 0$ and $k > j > 0$. It follows by the greedy algorithm that there exists a Dyck prefix of length n and weight k with the suffix $0^i 1^{j+1}$ in the listing since the algorithm terminates after visiting a string with the suffix $10^i 1^j$. If $j + 1 = k$, then clearly the only string with the suffix $0^i 1^k$ is $0^{n-k}1^k$. However, this string has a predecessor since it is not the initial string of the greedy algorithm. Moreover, by the greedy algorithm the predecessor of $0^{n-k}1^k$ is $0^t 10^{n-k-t}1^{k-1}$ for some $n - k > t > 0$, and all Dyck prefixes of length n and weight k with the suffix 01^{k-1} must have appeared before $0^{n-k}1^k$ in the listing. Therefore, the algorithm should terminate after visiting $0^{n-k}1^k$, a contradiction. Otherwise if $j + 1 < k$, then let α be the last string in the listing with the suffix $10^t 1^{j+1}$ for some $n - j - 2 > t > 0$. Since α appears before $b_1 b_2 \cdots b_n$ in the listing and $b_1 b_2 \cdots b_n$ has the suffix $10^i 1^j$, the algorithm must transpose the first 1 in the suffix 1^{j+1} of α with a 0 on the left to produce a later string with the suffix 01^j. It follows by the greedy algorithm that this is only possible if a string with the suffix $0^t 1^{j+2}$ appears before in the listing. Recursively applying the same argument implies that $0^{n-k}1^k$ exists in the listing, a contradiction since the algorithm would terminate after visiting $0^{n-k}1^k$ as discussed in the case of $j + 1 = k$. Therefore by proof by contradiction, the greedy algorithm terminates after visiting $0^{n-k}1^k$. □

Lemma 2. *If $0^i1^j0^t1\gamma$ is a length n Dyck prefix with weight k for some $i > 0$, $k > j > 0$, and $t > 0$, then the non-existence of $0^i1^j0^t1\gamma$ in the greedy listing implies the non-existence of $0^i1^{j-1}010^{t-1}1\gamma$ in the greedy listing.*

Proof. We prove the lemma by contrapositive. Suppose $\alpha = 0^i1^j0^t1\gamma$ is a Dyck prefix of weight k. Clearly $\beta = 0^i1^{j-1}010^{t-1}1\gamma$ is also a Dyck prefix of weight k and now consider the possible predecessor of β in our greedy listing. If the predecessor of β is of the form $0^{i-p}10^p1^{j-2}010^{t-1}1\gamma$ for some $p > 0$, then by the greedy algorithm, all Dyck prefixes of length n and weight k with the suffix $01^{j-2}010^{t-1}1\gamma$ should have appeared previously. The next string generated by the algorithm after β is thus α if α has not appeared before, or otherwise α must have appeared previously. In either case, α exists in the listing. Otherwise if the predecessor of β shares the same prefix 0^i1^{j-1} as β, then by the greedy algorithm, this is only possible if α appears before in the listing or α is the predecessor of β. Therefore, the string α exists if β exists, which completes the proof by contrapositive. □

We now prove Theorem 1 using the lemmas we proved in this section.

Theorem 1. *The algorithm Greedy-kDyck-Prefixes generates a homogeneous transposition Gray code for fixed-weight Dyck prefixes that is suffix-partitioned for all n and k where $2k \leq n$.*

Proof. Our algorithm permits only homogeneous transposition operations, and the listing is suffix-partitioned (as shown in Lemma 2). To demonstrate the Gray code property of our algorithm, we now prove it by contradiction.

Since the greedy algorithm ensures that there is no duplicated length n string in the greedy listing, it suffices to show that each Dyck prefix of length n and weight k appears in the listing.

Assume by contradiction that there exists a Dyck prefix $b_1b_2\cdots b_n \neq 0^{n-k}1^k$ that does not appear in the listing. Since $b_1b_2\cdots b_n \neq 0^{n-k}1^k$, the string $b_1b_2\cdots b_n$ contains the substring 10. Let $b_1b_2\cdots b_n = 0^i1^j0^t1\gamma$ for some $i > 0$, $k > j > 0$, and $t > 0$. Clearly, the string $0^i1^{j-1}010^{t-1}1\gamma$ is a Dyck prefix and by Lemma 2, the string $0^i1^{j-1}010^{t-1}1\gamma$ also does not exist in the greedy Dyck prefix listing. Repeatedly applying the same argument on $0^i1^{j-1}010^{t-1}1\gamma$ implies that the strings $0^{i+1}1^j0^{t-1}1\gamma$ and eventually $0^{n-k}1^k$ also do not exist in the listing, a contradiction to Lemma 1. □

4 Gray Codes for Ballot Sequences

In this section, we leverage Theorem 1 to construct the first known homogeneous 2-Gray code for ballot sequences. Our approach is to interleave strings from listings of homogeneous transposition Gray codes for fixed-weight Dyck prefixes, across all possible weight k, in order to create the homogeneous 2-Gray code for ballot sequences. To achieve this, we first prove the following lemma.

Lemma 3. *The string $b_1b_2\cdots b_{n-1}1$ is a Dyck prefix if and only if $b_1b_2\cdots b_{n-1}0$ is a Dyck prefix, provided that $2k < n - 1$.*

Proof. The forward direction is straightforward. For the backward direction, suppose that $2k < n - 1$ and that the string $b_1b_2 \cdots b_{n-1}0$ is a Dyck prefix. Since $2k < n - 1$, the prefix $b_1b_2 \cdots b_{n-1}$ has more 0 s than 1 s and thus both $b_1b_2 \cdots b_{n-1}1$ and $b_1b_2 \cdots b_{n-1}0$ are Dyck prefixes.   □

By Lemma 3, we can establish a one-to-one correspondence between Dyck prefixes $b_1b_2 \cdots b_{n-1}1$ of weight $k + 1$ and Dyck prefixes $b_1b_2 \cdots b_{n-1}0$ of weight k when $2k < n$. This correspondence enables us to construct a homogeneous 2-Gray code for ballot sequences.

The main idea of our algorithm is to utilize the same greedy strategy used for generating fixed-weight Dyck prefixes, with the addition of generating the correspondence to the generated Dyck prefix by Lemma 3. Specifically, whenever we produce a Dyck prefix $b_1b_2 \cdots b_n$ that terminates with a 1, we also generate its corresponding Dyck prefix $b_1b_2 \cdots b_{n-1}0$. Conversely, when we generate a Dyck prefix $b_1b_2 \cdots b_n$ that concludes with a 0 with $2k < n - 1$, we also generate its corresponding Dyck prefix $b_1b_2 \cdots b_{n-1}1$. Furthermore, if the application of the greedy strategy fails to produce a new string, we proceed to complement the last 1 in $b_1b_2 \cdots b_{n-1}$ and then update the value of $b_n = 1$. By making two relatively minor changes to the Algorithm 1, we can generate a homogeneous 2-Gray code for ballot sequences:

1. Before applying the greedy strategy to the current string $S_1S_2 \cdots S_k$, test whether $S_k = n$ or $S_k < n$ but with weight $k < \lfloor \frac{n}{2} \rfloor$. If $S_k = n$, then the algorithm generates its corresponding Dyck prefix $S_1S_2 \cdots S_{k-1}$. Similarly, if $S_k < n$ but with weight $k < \lfloor \frac{n}{2} \rfloor$, then the algorithm generates its corresponding Dyck prefix $S_1S_2 \cdots S_kn$;
2. After applying the greedy strategy to the current string $S_1S_2 \cdots S_k$ and it does not lead to the generation of any new string. If $S_k = n$, then the next string in the sequence is $S_1S_2 \cdots S_{k-2}n$. On the other hand, if $S_k < n$, then the following string in the sequence is $S_1S_2 \cdots S_{k-1}n$.

The algorithm starts with the initial string $(01)^k0^{n-2k}$ with $k = \lfloor \frac{n}{2} \rfloor$. Pseudocode of the algorithm to generate the Gray code for ballot sequences is given in Algorithm 2. As an example, the algorithm generates the following homogeneous 2-Gray code for ballot sequences for $n = 7$:

> 0101010, 0011010, 0010110, 0100110, 0001110, 0001101, 0001100,
> 0100100, 0100101, 0010101, 0010100, 0011000, 0011001, 0101001,
> 0101000, 0100010, 0100011, 0010011, 0010010, 0001010, 0001011,
> 0000111, 0000110, 0000101, 0000100, 0100000, 0100001, 0010001,
> 0010000, 0001000, 0001001, 0000011, 0000010, 0000001, 0000000.

Let α be a prefix of a Dyck word, and $\mathcal{G}(\alpha)$ be the list of strings obtained by applying Algorithm 1 with α as initial string. Clearly, for any such string α, $\mathcal{G}(\alpha)$ contains prefixes of Dyck words of the same length and same number of 1s as α, and in $\mathcal{G}(\alpha)$ there are no repeated strings.

Algorithm 2. The greedy algorithm that generates a homogeneous 2-Gray code for ballot sequences.

```
1: procedure GREEDY-BALLOT
2:     k = ⌊n/2⌋
3:     S₁S₂ ··· Sₖ ← 2 4 ··· 2k
4:     Print(S₁S₂ ··· Sₖ)
5:     if Sₖ = n then
6:         Sₖ ← n + 1
7:         k ← k − 1
8:         if S₁S₂ ··· Sᵢ₋₁(j)Sᵢ₊₁ ··· Sₖ has not appeared before then go to 4
9:         k ← k + 1
10:        Sₖ ← n
11:    else if k < ⌊n/2⌋ then
12:        Sₖ₊₁ ← n
13:        k ← k + 1
14:        if S₁S₂ ··· Sᵢ₋₁(j)Sᵢ₊₁ ··· Sₖ has not appeared before then go to 4
15:        k ← k − 1
16:        Sₖ₊₁ ← n + 1
17:    for i from 1 to k do
18:        for j from MAX(Sᵢ₋₁ + 1, i × 2) to Sᵢ₊₁ − 1 do
19:            if S₁S₂ ··· Sᵢ₋₁(j)Sᵢ₊₁ ··· Sₖ has not appeared before then
20:                Sᵢ ← j
21:                go to 4
22:    if Sₖ = n then
23:        Sₖ ← n + 1
24:        Sₖ₋₁ ← n
25:        k ← k − 1
26:        go to 4
27:    else if Sₖ = n − 1 then
28:        Sₖ ← n
29:        go to 4
```

Theorem 2. *The algorithm Greedy-Ballot generates a homogeneous 2-Gray code for ballot sequences for all n.*

Proof. The algorithm Greedy-Ballot starts with the string $(01)^{\lfloor n/2 \rfloor} 0^{n \bmod 2}$ with $k = \lfloor \frac{n}{2} \rfloor$. By Theorem 1, the algorithm generates all strings in $\mathcal{G}((01)^{\lfloor n/2 \rfloor} 0^{n \bmod 2})$ which contains all Dyck prefixes of weight $k = \lfloor \frac{n}{2} \rfloor$. Furthermore, according to Lemma 3 and lines 5–16 of the algorithm, the algorithm also generates all Dyck prefixes of weight $\lfloor \frac{n}{2} \rfloor - 1$ that end with a 0.

Since $\mathcal{G}((01)^{\lfloor n/2 \rfloor} 0^{n \bmod 2})$ ends with $0^{n - \lfloor n/2 \rfloor} 1^{\lfloor n/2 \rfloor}$, the algorithm generates all Dyck prefixes of weight $k = \lfloor \frac{n}{2} \rfloor$ and Dyck prefixes of weight $\lfloor \frac{n}{2} \rfloor - 1$ that end with a 0 until it reaches the string $0^{n - \lfloor n/2 \rfloor} 1^{\lfloor n/2 \rfloor}$ or $0^{n - \lfloor n/2 \rfloor} 1^{\lfloor n/2 \rfloor - 1} 0$. Then, as indicated in lines 22–29 of the algorithm, the next string generated by the algorithm is $0^{n - \lfloor n/2 \rfloor} 1^{\lfloor n/2 \rfloor - 2} 01$. Observe that $0^{n - \lfloor n/2 \rfloor} 1^{\lfloor n/2 \rfloor - 2} 01$ is generated in $\mathcal{G}((01)^{\lfloor n/2 \rfloor - 1} 0^2 0^{n \bmod 2})$ by Algorithm 1 after exhaustively generating all Dyck prefixes of weight $\lfloor \frac{n}{2} \rfloor - 1$ that end with a 0. Since all Dyck prefixes of weight $\lfloor \frac{n}{2} \rfloor - 1$ that end with a 0 have already been generated in our ballot sequence algorithm, the algorithm follows the same operations as $\mathcal{G}(0^{n-\lfloor\frac{n}{2}\rfloor}1^{\lfloor\frac{n}{2}\rfloor-2}01)$ and proceeds to generate all Dyck prefixes of weight $\lfloor\frac{n}{2}\rfloor-1$ that end with a 1. Therefore, all Dyck prefixes of weight $\lfloor\frac{n}{2}\rfloor-1$ are in the listing generated by the algorithm.

By repeatedly applying the same argument, the algorithm generates the fixed-weight Dyck prefixes with weight ranging from k to 0, which is the set of all ballot sequences of length n.

Moreover, since each listing in $\mathcal{G}$ is a homogeneous transposition Gray code and the operations in lines 5–16 and 22–29 of the algorithm only involve removing a 1 or swapping two nearby bits, the resulting sequence generated by the algorithm is a homogeneous 2-Gray code. □

5 Final Remarks

It is worth noting that an alternative homogeneous 2-Gray code for ballot sequences can be constructed by concatenating the homogeneous transposition Gray code listings of fixed-weight Dyck prefixes ranging from weight k to 0, and reversing the listings of fixed-weight Dyck prefixes with even (or odd) weights. For instance, let $\overline{\mathcal{G}}(\alpha)$ denote the reverse of the list of strings generated by applying Algorithm 1 with α as the initial string. A homogeneous 2-Gray code for ballot sequences for $n=7$ can be obtained by $\mathcal{G}(0101010) \cdot \overline{\mathcal{G}}(0101000) \cdot \mathcal{G}(0100000) \cdot \overline{\mathcal{G}}(0000000)$, which would result in the following listing:

0101010, 0011010, 0010110, 0100110, 0001110, 0001101, 0100101,
0010101, 0011001, 0101001, 0100011, 0010011, 0001011, 0000111,
0000011, 0001001, 0010001, 0100001, 0000101, 0000110, 0010010,
0100010, 0001010, 0001100, 0100100, 0010100, 0011000, 0101000,
0100000, 0010000, 0001000, 0000100, 0000010, 0000001, 0000000.

There is, however, no known simple algorithm to generate the reverse of the sequence generated by our algorithm for fixed-weight Dyck prefixes. This remains an open problem for future research.

Finally, efficient algorithms that generate the same Gray codes for Dyck words, fixed-weight Dyck prefixes and ballot sequences in constant amortized time per string were developed, and their details will be presented in the full version of the paper.

Acknowledgements. The research is supported by the Macao Polytechnic University research grant (Project code: RP/FCA-02/2022) and the National Research Foundation (NRF) grant funded by the Ministry of Science and ICT (MSIT), Korea (No. 2020R1F1A1A01070666).

A part of this work was done while the second author was visiting Kanazawa University in Japan. The second author would like to thank Hiroshi Fujisaki for his hospitality during his stay in Kanazawa.

Appendix: C Code to Generate Homogeneous 2-Gray Codes for k-Combinations, Dyck Words, Fixed-Weight Dyck Prefixes, and Ballot Sequences

```c
#include <stdio.h>
#include <stdlib.h>
#define INF 99999
#define MAX(a,b) (((a)>(b))?(a):(b))

int n, k, type, total = 0, s[INF], p[INF];

//--------------------------------------------------
int binToDec() {
    int i, j = 1, t = 0;
    for(i=1; i<=n; i++) if (s[j]==i) {t = t+(1<<(n-i)); j++;}
    return t;
}
//--------------------------------------------------
int greedy() {
    int i, j, t, r;

    if (type==4) {
        if (s[k]==n) {
            s[k] = n+1; k--;
            if (!p[binToDec()]) {p[binToDec()] = 1; return 1;}
            k++; s[k] = n;
        }
        else if (k<n/2) {
            s[k+1] = n; k++;
            if (!p[binToDec()]) {p[binToDec()] = 1; return 1;}
            k--; s[k+1] = n+1;
        }
    }

    for (i=1; i<=k; i++) {
        if (type==1) r = s[i-1]+1;
        else r = MAX(s[i-1]+1, i*2);

        for (j=r; j<s[i+1]; j++) {
            t = s[i]; s[i] = j;
            if (!p[binToDec()]) {p[binToDec()] = 1; return 1;}
            s[i] = t;
        }
    }

    if (type==4) {
        if (s[k]==n) {
            s[k] = n+1; s[k-1] = n; k--;
            p[binToDec()] = 1; return 1;
        }
        else if (s[k]==n-1) {
            s[k] = n;
            p[binToDec()] = 1; return 1;
        }
    }
    return 0;
}
//--------------------------------------------------
int main() {
    int i, j;

    printf(" =====================================\n");
    printf(" 1. Combinations\n");
    printf(" 2. Dyck words\n");
    printf(" 3. Prefix of Dyck words of weight k\n");
```

```
        printf(" 4. Ballot sequences\n");
        printf(" =======================================\n");

        printf(" Enter selection #: "); scanf("%d", &type);

        printf(" ENTER n: "); scanf("%d", &n);
        if (type!=2 && type!=4) {printf(" ENTER k: "); scanf("%d", &k);}
        else k = n/2;
        if (type==2 && n%2>0) {printf("n must be an even number. \n"); exit(0);}
        if (type==3 && k>n/2) {printf("k must be less than or equal to n/2. \n"); exit
            (0);}

        for (i=0; i<INF; i++) p[i] = 0;
        for (i=0; i<=k; i++) {if (type!=1) s[i] = i*2; else s[i] = i;}

        s[0] = 0; s[k+1] = n+1;
        p[binToDec()] = 1;

        do {
            j = 1;
            for (i=1; i<=n; i++) if (s[j]!=i) printf("0"); else {printf("1"); j++;}
            printf("\n"); total++;
        } while(greedy());
        printf("Total = %d\n", total);
}
```

References

1. Ahrabian, H., Nowzari-Dalini, A.: Generation of t-ary trees with ballot-sequences. Int. J. Comput. Math. **80**(10), 1243–1249 (2003)
2. Aigner, M.: Enumeration via ballot numbers. Discrete Math. **308**(12), 2544–2563 (2008)
3. Barton, D., Mallows, C.: Some aspects of the random sequence. Ann. Math. Stat. **36**(1), 236–260 (1965)
4. Benchekroun, S., Moszkowski, P.: A new bijection between ordered trees and legal bracketings. Eur. J. Combin. **17**(7), 605–611 (1996)
5. Bultena, B., Ruskey, F.: An Eades-McKay algorithm for well-formed parentheses strings. Inf. Process. Lett. **68**(5), 255–259 (1998)
6. Carlitz, L.: Sequences, paths, ballot numbers. Fibonacci Quart **10**(5), 531–549 (1972)
7. Deutsch, E.: A bijection on Dyck paths and its consequences. Discrete Math. **179**(1), 253–256 (1998)
8. Deutsch, E., Shapiro, L.: A bijection between ordered trees and 2-Motzkin paths and its many consequences. Discrete Math. **256**(3), 655–670 (2002)
9. Goulden, I., Jackson, D.: Combinatorial Enumeration. A Wiley-Interscience Publication, New York (1983)
10. Hackl, B., Heuberger, C., Prodinger, H., Wagner, S.: Analysis of bidirectional ballot sequences and random walks ending in their maximum. Ann. Comb. **20**(4), 775–797 (2016)
11. Labelle, J., Yeh, Y.N.: Generalized Dyck paths. Discrete Math. **82**(1), 1–6 (1990)
12. Lengyel, T.: Direct consequences of the basic ballot theorem. Stat. Probab. Lett. **81**(10), 1476–1481 (2011)
13. Mütze, T.: Combinatorial Gray codes - an updated survey. arXiv Preprint arXiv:2202.01280 (2022)
14. Panayotopoulos, A., Sapounakis, A.: On binary trees and Dyck paths. Math. Sci. Hum. **131**, 39–51 (1995)
15. Proskurowski, A., Ruskey, F.: Binary tree Gray codes. J. Algorithms **6**(2), 225–238 (1985)

16. Rotem, D., Varol, Y.: Generation of binary trees from ballot sequences. J. ACM **25**(3), 396–404 (1978)
17. Ruskey, F., Proskurowski, A.: Generating binary trees by transpositions. J. Algorithms **11**(1), 68–84 (1990)
18. Ruskey, F., Williams, A.: Generating balanced parentheses and binary trees by prefix shifts. In: Proceedings of the Fourteenth Symposium on Computing: The Australasian Theory, CATS 2008, Australia, vol. 77, pp. 107–115 (2008)
19. Sabri, A., Vajnovszki, V.: On the exhaustive generation of generalized ballot sequences in lexicographic and Gray code order. Pure Math. Appl. **28**(1), 109–119 (2019)
20. Savage, C.: A survey of combinatorial Gray codes. SIAM Rev. **4**, 605–629 (1997)
21. Sawada, J., Williams, A., Wong, D.: Inside the binary reflected gray code: flip-swap languages in 2-gray code order. In: Lecroq, T., Puzynina, S. (eds.) WORDS 2021. LNCS, vol. 12847, pp. 172–184. Springer, Cham (2021). https://doi.org/10.1007/978-3-030-85088-3_15
22. Sawada, J., Williams, A., Wong, D.: Flip-swap languages in binary reflected Gray code order. Theor. Comput. Sci. **933**, 138–148 (2022)
23. Sloane, N.: The on-line encyclopedia of integer sequences. https://oeis.org/. Sequence A000108 and A001405
24. Stanton, D., White, D.: Constructive Combinatorics. Springer, Heidelberg (2012)
25. Vajnovszki, V., Walsh, T.: A loop-free two-close Gray-code algorithm for listing k-ary Dyck words. J. Discrete Algorithms **4**(4), 633–648 (2006)
26. van Baronaigien, D.: A loopless gray-code algorithm for listing k-ary trees. J. Algorithms **35**(1), 100–107 (2000)
27. Viennot, G.: Theorié combinatoire des nombres d'Euler et de Genocchi. Séminaire de théorie des nombres. Publications Univ. Bordeaux I (1980)
28. Walsh, T.: Generation of well-formed parenthesis strings in constant worst-case time. J. Algorithms **29**(1), 165–173 (1998)
29. Wildon, M.: Knights, spies, games and ballot sequences. Discrete Math. **310**(21), 2974–2983 (2010)
30. Williams, A.: The greedy Gray code algorithm. In: Dehne, F., Solis-Oba, R., Sack, J.-R. (eds.) WADS 2013. LNCS, vol. 8037, pp. 525–536. Springer, Heidelberg (2013). https://doi.org/10.1007/978-3-642-40104-6_46
31. Wong, D., Calero, F., Sedhai, K.: Generating 2-Gray codes for ballot sequences in constant amortized time. Discrete Math. **346**(1), 113168 (2023)
32. Xiang, L., Ushijima, K., Tang, C.: Efficient loopless generation of Gray codes for k-ary trees. Inf. Process. Lett. **76**(4), 169–174 (2000)
33. Zhao, Y.: Constructing MSTD sets using bidirectional ballot sequences. J. Number Theory **130**(5), 1212–1220 (2010)

Efficiently-Verifiable Strong Uniquely Solvable Puzzles and Matrix Multiplication

Matthew Anderson[(⊠)] and Vu Le

Department of Computer Science, Union College, Schenectady, NY, USA
{andersm2,lev}@union.edu

Abstract. We advance the Cohn-Umans framework for developing fast matrix multiplication algorithms. We introduce, analyze, and search for a new subclass of strong uniquely solvable puzzles (SUSP), which we call *simplifiable SUSPs*. We show that these puzzles are efficiently verifiable, which remains an open question for general SUSPs. We also show that individual simplifiable SUSPs can achieve the same bounds on the matrix multiplication exponent ω that infinite families of SUSPs can. We construct, by computer search, larger SUSPs than known for small width. This, combined with our tighter analysis, strengthens the upper bound on ω from 2.66 to 2.505 obtainable via this computational approach, nearing the handcrafted constructions of Cohn-Umans.

Keywords: Matrix multiplication · Simplifiable strong uniquely solvable puzzle · Arithmetic complexity · 3D matching · Iterative local search

1 Introduction

Square matrix multiplication is a fundamental mathematical operation: Given $n \in \mathbb{N}$, a field $\mathbb{F}$, and matrices $A, B \in \mathbb{F}^{n \times n}$, compute the resulting matrix $C = AB$ where the entry $(i, k) \in [n]^2$ is $C_{i,k} = \sum_{j \in [n]} A_{i,j} B_{j,k}$. Early work by Strassen gave a recursive, divide-and-conquer algorithm for matrix multiplication that runs in time $O(n^{2.81})$ [16]. The situation steadily improved over the next two decades, culminating with the $O(n^{2.376})$ time Coppersmith-Winograd algorithm [10]. More recently, a series of refinements to the Coppersmith-Winograd algorithm has resulted in a state-of-the-art algorithm that runs in time $O(n^{2.37188})$ [2,12,13]. The question remains open: *What is the smallest ω for which there exists a matrix multiplication algorithm that runs in time $O(n^\omega)$?*

Instead of following the traditional approach of refinements to Coppersmith-Winograd, we pursue the framework developed by Cohn and Umans [8,9]. This framework connects the existence of efficient algorithms for matrix multiplication to the existence of combinatorial objects called *strong uniquely solvable puzzles* *(SUSP)*. An $(,sk)$-puzzle P is a subset of $\{1, 2, 3\}^k$ with cardinality $|P| = s$. The larger the *size* s of a strong uniquely solvable puzzle is for a fixed k, the

W. Wu and G. Tong (Eds.): COCOON 2023, LNCS 14423, pp. 41–54, 2024.
https://doi.org/10.1007/978-3-031-49193-1_4

more efficient a matrix multiplication algorithm is implied by the Cohn-Umans framework (see Lemma 1). Anderson et al. initiated a systematic computer-aided search for large puzzles that are SUSPs [4]. They developed algorithms that are sufficiently efficient in practice—using reductions to NP-hard problems, and sophisticated satisfiability and integer programming solvers—for verifying SUSPs and applied those algorithms to find large SUSPs of small width $k \le 12$.

There are several aspects of the work of Anderson et al. that warranted further study: (i) although the verification algorithm was shown to be experimentally effective, its worst-case performance was exponential time, (ii) the results used from [8] to imply efficient matrix multiplication algorithms were limited because they only found individual SUSPs of small width, rather than infinite families of SUSPs as in the constructions of [8], and (iii) they observed that for some pairs of SUSPs P_1, P_2, the Cartesian product $P_1 \times P_2$ was also an SUSP, but they did not provide a theoretical explanation as to why. These aspects limited the small-width SUSPs that were found in [4,5] to only achieve $\omega \le 2.66$.

Our Contributions. We make progress on the computer-aided search for large SUSPs and resolve the three limitations mentioned above by introducing a new class of SUSPs that we call *simplifiable SUSPs*.

In [4] they show that the problem of verifying whether a puzzle P is an SUSP reduces to determining whether a related tripartite hypergraph H_P has no nontrivial 3D matchings. In Sect. 3, we describe a polynomial-time simplification algorithm that takes a 3D hypergraph and attempts to simplify it to the trivial matching without changing the set of matchings the graph has. In this way, we define simplifiable SUSPs to be puzzles P whose 3D hypergraph H_P simplifies to the trivial matching. This gives a polynomial-time algorithm to generate a proof that P is an SUSP. In this way, simplifiable SUSPs are polynomial-time verifiable by definition, making them more feasible to search for.

Theorem 1. *Let P be an (s, k)-puzzle. There is an algorithm for determining whether P is a simplifiable SUSP. The algorithm runs in time $\mathrm{poly}(s, k)$.*

In Sect. 4, we show that simplifiable SUSPs have other interesting properties that make them a good candidate to search for when trying to improve bounds on ω. In particular, we show that simplifiable SUSPs are a natural generalization of *local SUSPs* from [8]. Local SUSPs are also efficiently verifiable, but since they are not densely encoded, they are difficult to search for. Relatedly, we show that simplifiable SUSPs are closed under Cartesian product, which is not the case for general SUSPs, and that this property allows a single simplifiable SUSP to generate an infinite family of SUSPs by taking all powers of the puzzle, and that simplifiable SUSPs can achieve any bound on ω that SUSPs can. The former allows the stronger infinite-family bound on ω of [8] to be applied, which strengthens the bounds on ω implied by individual simplifiable SUSPs.

Theorem 2. *Let $\epsilon > 0$, if there is a simplifiable (s, k)-SUSP P, then there is an algorithm for multiplying n-by-n matrices in time $O(n^{\omega+\epsilon})$ where $\omega \le \min_{m \in \mathbb{N}_{\ge 3}} 3 \cdot \frac{k \log m - \log s}{k \log(m-1)}$.*

Finally, in Sect. 5, we report finding new large simplifiable SUSPs of small width
that improve the bounds on ω from 2.66 to 2.505 via the computational Cohn-
Umans approach. The SUSPs we construct for small width are considerably
larger than those of the previous work [4,5,8], and imply stronger bounds on ω
for the same domain. Our results further the computational approach to developing efficient matrix multiplication algorithms using the Cohn-Umans framework
started by [4]. However, it is important to note that this computational approach
has yet to surpass the $\omega \leq 2.48$ bound implied by the infinite families of SUSPs
handcrafted in [8], or the state-of-the-art Coppersmith-Winograd refinements
with the record bound of $\omega \leq 2.37188$ [13].

Related Work. Some negative results are known for the Cohn-Umans framework that apply to our work as well. In particular, a series of articles [1,3,7,11]
showed that there exists an $\epsilon > 0$ such that this framework, as well as a variety
of other algorithmic approaches, cannot achieve $\omega \leq 2 + \epsilon$. This implies that our
approach cannot achieve the best potential result of $\tilde{O}(n^2)$, however, the authors
are unaware of a concrete value known for this ϵ.

We search for simplifiable SUSPs using a standard search technique called
iterative local search, c.f, e.g., [15]. Some comparison with our work can be drawn
to recent computational approach by Fawzi et al. who used reinforcement learning to generate low-rank representations of the matrix multiplication tensor [14],
producing algorithms with $\omega \leq 2.77$, which are weaker than our results.

2 Preliminaries

For a natural number $n \in \mathbb{N}$, we use $[n]$ to denote the set $\{1, 2, ..., n\}$. Sym_Q
denotes the symmetric group on the elements of a set Q.

Cohn et al. introduced the notion of puzzles and defined several useful sub-
classes [8]. For $s, k \in \mathbb{N}$, an $(,sk)$-puzzle is a subset $P \subseteq [3]^k$ with $|P| = s$. We
say that an (s, k)-puzzle has s rows and k columns. The columns are inherently
ordered and indexed by $[k]$. The rows are not inherently ordered, although it is
often convenient to assume that they are arbitrarily ordered and indexed by $[s]$.

Definition 1 (Strong Uniquely Solvable Puzzle (SUSP)). *An (s, k)-
puzzle P is strong uniquely solvable if $\forall \pi_1, \pi_2, \pi_3 \in Sym_P$, either (i) $\pi_1 = \pi_2 =
\pi_3$, or (ii) $\exists r \in P$ and $i \in [k]$ such that exactly two of the following conditions
are true: $(\pi_1(r))_i = 1, (\pi_2(r))_i = 2, (\pi_3(r))_i = 3$.*

For brevity, we call such puzzles *SUSPs*. Determining whether a puzzle is an
SUSP is in coNP. Anderson et al. studied this problem, devised a reduction
from this problem to a variant of the 3D perfect matching problem, and then
used it to develop a practical, but worst-case exponential time, algorithm [4].
Cohn et al. also introduced a subset of SUSPs, called *local SUSPs* that naturally
demonstrate that they are SUSPs. An (s, k)-puzzle P is *local strong uniquely
solvable* if for each $(u, v, w) \in P^3$ with u, v, w not all equal, there exists $c \in [k]$

such that $(u_c, v_c, w_c) \in \mathcal{L} = \{(1,2,1), (1,2,2), (1,1,3), (1,3,3), (2,2,3), (3,2,3)\}$. The task of determining whether a puzzle is a local SUSP can be done in time $O(s^3 \cdot k)$, by checking all triples of rows. Cohn et al. show that SUSPs can be converted to local SUSPs, albeit with a substantial increase in the parameters.

Proposition 1 ([8, Proposition 6.3]). *Let P be an (s, k)-SUSP, then there is a local $(s!, sk)$-SUSP P'. Moreover, SUSP capacity is achieved by local SUSPs.*

Note that the second consequence of this proposition is that any bound on ω that can be achieved by SUSPs can be achieved by local SUSPs.

From Matrix Multiplication to SUSPs. Using the concept of an SUSP, [9] showed how to define group algebras that allow matrix multiplication to be efficiently embedded into them. The existence of SUSPs implies upper bounds on the matrix multiplication exponent ω.

The *SUSP capacity* is defined as the largest constant C such that there exist SUSPs of size $(C - o(1))^k$ and width k for infinitely many values of k [8]. The constructions of Cohn et al. produce families $\mathcal{F}$ of $(s(k), k)$-SUSPs for infinitely many values of k. The key parameter that relates ω to the size of puzzles is the *capacity $C_{\mathcal{F}}$ of the family*, defined as the limit of $(s(k))^{\frac{1}{k}}$ as k goes to ∞. Cohn et al. showed the following bounds on ω as functions of capacity and SUSP size.

Lemma 1 ([8, Corollary 3.6]). *Let $\epsilon > 0$, (i) if there is a family $\mathcal{F}$ of SUSPs with capacity $C_{\mathcal{F}}$, then there is an algorithm for multiplying n-by-n matrices in time $O(n^{\omega+\epsilon})$ where $\omega \leq \min_{m \in \mathbb{N}_{\geq 3}} 3 \cdot \frac{\log m - \log C_{\mathcal{F}}}{\log(m-1)}$, and (ii) if there is an (s, k)-SUSP, there is an algorithm for multiplying n-by-n matrices in time $O(n^{\omega+\epsilon})$ where $\omega \leq \min_{m \in \mathbb{N}_{\geq 3}} 3 \cdot \frac{sk \log m - \log s!}{sk \log(m-1)}$.*

They also show that if the SUSP capacity is $C_{\max} = 3/2^{2/3}$, it immediately follows that $\omega = 2$. As mentioned in Sect. 1, subsequent work has shown that the SUSP capacity is strictly less than $C_{\max}$.

From SUSPs to 3D Matchings. Let G be a r-uniform hypergraph over r disjoint copies of a domain U. We only consider $r \in \{2, 3\}$ and use "2D graph" to refer to the case where $r = 2$ and "3D graph" to refer to the case where $r = 3$. We use the notation $V(G)$ to denote the vertex set of G and $E(G)$ to denote the edge set of G. We say that G has a *perfect matching* if there exists $M \subseteq E(G)$ such that $|M| = |U|$ and for all distinct pairs of edges $a, b \in M$, a and b are vertex disjoint, that is, $a_i \neq b_i, \forall i \in [r]$. Note that we only consider perfect matchings in this article, so often drop "perfect" for brevity. The *trivial matching* of G is the set $\{u^r \mid u \in U\}$. We call a matching M *nontrivial* if it is not the trivial matching of H_P. For two r-partite graphs G_1, G_2 over domains U_1 and U_2, respectively, we define their *tensor product* to be the r-partite graph $G_1 \times G_2$ over the Cartesian product of their domain sets $U_1 \times U_2$, and whose edges are the Cartesian product of their edge sets $E(G_1) \times E(G_2) = \{((u_1, u_2), (v_1, v_2)) \mid (u_1, v_1) \in E(G_1), (u_2, v_2) \in E(G_2)\}$.

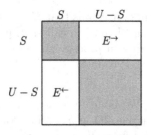

Fig. 1. Let G be 2D graph over the domain U. This diagram represents the partitioning of the adjacency matrix of G relative to a set $S \subseteq U$ which divides the adjacency matrix into four regions of edges, $S \times S$, $S \times (U - S)$, $(U - S) \times S$, $(U - S) \times (U - S)$. The edges in the gray regions survive the simplification to G' as in Lemma 3, while any edges in $E^{\rightarrow}$ or $E^{\leftarrow}$ are deleted from G.

Note that the adjacency matrix of the tensor product of two r-partite graphs is the Kronecker product of the two adjacency matrices of the graphs.

Anderson et al. show a reduction from checking whether an (s, k)-puzzle P is an SUSP to deciding whether there are no nontrivial perfect matchings in a related 3D graph H_P [4]. We briefly recall that construction. Define a function f to represent the inner condition of being an SUSP on triplets of rows $u, v, w \in P$ where $f(u, v, w) = 1$, if $\exists i \in [k]$ such that exactly two of the following hold: $u_i = 1, v_i = 2, w_i = 3$ and $f(u, v, w) = 0$, otherwise. Then, define H_P to be the 3D graph with domain P whose edges are $E(H_P) = \{(u, v, w) \mid f(u, v, w) = 0\}$. The trivial matching is a matching of H_P. We use the following result.

Lemma 2 ([5]). *P is an SUSP iff H_P has no nontrivial perfect matchings.*

3   Simplification and Efficiently-Verifiable SUSPs

The reduction from SUSP verification to the problem of 3D perfect matching, from Lemma 2, leads to a naïve worst-case $O(2^s \cdot \text{poly}(s, k))$-time algorithm for verification, without much hope for improvement as the latter problem is NP-complete. We overcome this obstacle by introducing a useful subset of SUSPs that are efficiently verifiable.

Let P be an (s, k)-puzzle and H_P be its corresponding 3D graph as in Lemma 2. If H_P has a nontrivial matching, the matching itself witnesses this fact. However, if H_P has only the trivial matching, there may not be a short witness of this fact. The subclass of SUSPs we develop naturally has short witnesses.

Our approach is based on the following insight about the 3D graph H_P: If H_P has a matching, the matching projects to three 2D matchings of the 2D faces of H_P. Moreover, if edges in one of the faces cannot be used for a matching of that face, none of the edges of H_P that project onto that edge can be used in a 3D matching of H_P. We iteratively apply this idea to efficiently *simplify* the 3D graph H_P, without changing the matchings it has, until it is reduced to a trivial matching or no further simplification can be made. If the 3D graph is reduced to the trivial matching, it means that H_P had no nontrivial matchings, and the puzzle P must be an SUSP. We call such puzzles *simplifiable SUSPs*. A by-product of this simplification process is a series of edges deletions of H_P, which provides a witness that P is an SUSP.

Simplifying 2D Graphs. We build up to simplifying 3D graphs and simplifiable SUSPs by first considering 2D graphs. The following lemma shows that certain edges can be removed from 2D graphs without eliminating matchings. See Fig. 1 for a visual representation of this lemma.

Lemma 3. *Let G be a 2D graph with domain U. Let $S \subseteq U$, $E^{\rightarrow} = S \times (U - S)$, and $E^{\leftarrow} = (U - S) \times S$. Let G' be a 2D graph with domain U and edges $E(G') = E(G) - E^{\rightarrow} - E^{\leftarrow}$. If $E^{\rightarrow} \cap E(G) = \emptyset$ or $E^{\leftarrow} \cap E(G) = \emptyset$, then G' has the same set of perfect matchings as G.*

Proof. Observe that since the edges of G' are a subset of the edges of G, G' cannot have a matching that G does not have. It remains to show that for each perfect matching M of G, M is also a perfect matching of G'.

Let $M \subseteq E(G)$ be a perfect matching of G. There are two cases to consider. Suppose $E^{\rightarrow} \cap E(G) = \emptyset$. Consider an edge $(u, v) \in M$. If $u \in S$, then $v \notin (U - S)$ since there are no edges in G that intersect with $S \times (U - S)$. Therefore, $v \in S$. Thus, for each $u \in S$, $(u, v) \in M$ and $v \in S$, so M matches S to S. If $u \in (U - S)$ and $(u, v) \in M$, then $v \notin S$ since for all $v \in S$ there already exists a one-to-one correspondence with $u' \in S$ where $(u', v) \in M$.

Thus, M must match S to S and match $U - S$ to $U - S$, that is, $M \subseteq (S \times S) \cup ((U - S) \times (U - S))$. Hence, M must be a perfect matching of G', because $M \cap (E^{\rightarrow} \cup E^{\leftarrow}) = \emptyset$ and therefore the edges in M are deleted. The case when $E^{\leftarrow} \cap E(G) = \emptyset$ is symmetric. □

Let $S \subseteq U$ be a subset of vertices in a 2D graph G with domain U for which the conditions of Lemma 3 are met. We say that S *induces a simplification* of G to G'. We now consider sequences of such simplifications. Let $G_0, G_1, \ldots, G_\ell$ be a sequence of 2D graphs with a common domain U and let $S_1, S_2, \ldots, S_\ell \subseteq U$ be sets such that S_i induces a simplification of G_{i-1} to G_i for $1 \leq i \leq \ell$. We say that G_0 *simplifies to* G_ℓ. The following is a corollary resulting from repeated application of Lemma 3 to the sets and 2D graphs in the above definition with a generalization to tensor products.

Corollary 1. *Let G, G' be 2D graphs over the domain U, and F be a 2D graph over the domain V. If G simplifies to G', then G and G' have the same set of perfect matchings and $G \times F$ simplifies to $G' \times F$.*

Proof (Sketch). Suppose G simplifies to G'. By definition, there exists $G_0, G_1, \ldots, G_\ell$ with $G = G_0$ and $G' = G_\ell$ and sets $S_1, S_2, \ldots, S_\ell$ for which S_i induces a simplification of G_{i-1} to G_i. Using Lemma 3, between G_{i-1} and G_i, one can show, by induction, that the set of perfect matchings for all G_i

are the same. Therefore, $G = G_0$ and $G' = G_\ell$ have the same set of perfect matchings.

One can argue that the sets $S_1 \times V, S_2 \times V, \ldots, S_\ell \times V$ induce the corresponding chain of simplifications $G \times F = G_0 \times F, G_1 \times F, \ldots, G_\ell \times F = G' \times F$. The argument for the individual simplification steps here proceeds analogously to the proof of Lemma 3 and shows the second half of the corollary.   □

Simplifying 3D Graphs. We lift the notion of simplification from 2D graphs to 3D graphs. Consider a 3D graph H with domain U. We construct three 2D graphs R_0, R_1, R_2, on the same domain U, which, respectively, correspond to projecting out the first, second, and third coordinates of H. For example, the edges of R_1 can be written $E(R_1) = \{(u, w) \mid \exists v \in U, (u, v, w) \in E(H)\}$. If H has a perfect matching, then it projects into a perfect matching for each of the R_f's. To see this, let M be a perfect matching of H, then following the projection, define $M_0 = \{(v, w) \mid \exists u \in U, (u, v, w) \in M\}$. By definition $M_0 \subseteq E(R_0)$. Because M is a perfect matching of H, $\{v \mid (u, v, w) \in M\} = \{w \mid (u, v, w) \in M\} = U$, and $|M| = |U|$, so M_0 is a perfect matching of R_0. The argument for R_1 and R_2 is analogous. Furthermore, one can argue that if a matching is nontrivial for H, then it is nontrivial for at least two of the R_f's.

We observe that simplifications induced on any of R_0, R_1, R_2, also induce a simplification of H. For brevity, the result below is stated only for R_0, but holds similarly for R_1 and R_2 using symmetric arguments.

Lemma 4. *Let H, R_0, U be defined as above. Let H' be the 3D graph over the domain U whose edges are $E(H') = E(H) - U \times ((S \times (U - S)) \cup ((U - S) \times S))$. If $S \subseteq U$ induces a simplification of R_0, then H' has the same set of perfect matchings that H does.*

Proof. Observe that since the edges of H' are a subset of the edges of H, H' cannot have a matching that H does not have. It remains to show that for each matching M of H, M is also a matching of H'.

Let M be a matching of H. Suppose, for the sake of contradiction, that M is not a matching of H'. There must exist an edge $(u, v, w) \in M$ that lies in the set of edges deleted in H'. Let M_0 be the projection of M into R_0, so that M_0 is a matching of R_0 and $(v, w) \in M_0$. By hypothesis and definition of H', $(v, w) \in (S \times (U - S)) \cup ((U - S) \times S)$. This is a contradiction to the fact that S simplifies R_0, because, by Lemma 3, $(S \times (U - S)) \cup ((U - S) \times S)$ does not intersect with any matchings of R_0.   □

When the conditions of Lemma 4 are met, we say that this set S *induces a simplification of H via R_0.* As before, we can lift the notion of simplification to a series of induced simplifications. Here it is more complex because changing H changes its projections. Let $S_1, S_2, \ldots, S_\ell \subseteq U$ and $f_1, f_2, \ldots, f_\ell \in \{0, 1, 2\}$. We define a series of tuples of graphs $(H_j, R_{0,j}, R_{1,j}, R_{2,j})$ with $0 \leq j \leq \ell$, where $H_0 = H$, $R_{0,0} = R_0, R_{1,0} = R_1, R_{2,0} = R_2$ and for $j > 0$, $R_{f_j,j}$ is the simplification of $R_{f_j,j-1}$ induced by S_j, H_j is the simplification of H_{j-1}

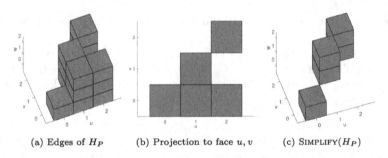

(a) Edges of H_P (b) Projection to face u, v (c) SIMPLIFY(H_P)

Fig. 2. Let $P = \{111, 321, 323\}$. On the left is a 3D grid representing the edges of H_P. In the middle is a projection of H_P onto the 2D face corresponding to u, v plane. Since $\{0\} \times \{1, 2\}$ contains no edges in this projection, edges of the form $\{1, 2\} \times \{0\} \times \{0, 1, 2\}$ can be simplified out of H_P. At the right is the result of applying this simplification plus one more simplification on the u, w face. The final instance is SIMPLIFY(H_P) and cannot be further simplified, showing that P is not a simplifiable SUSP.

induced by S_j via R_{f_j} and $R_{(f_j+1 \mod 3),j}$ and $R_{(f_j+2 \mod 3),j}$ are the result of reprojecting H_j. For brevity in describing this situation, we say that H *simplifies to H_ℓ*. As before, repeated application of Lemma 4 and Lemma 3 implies that H_ℓ has the same set of matchings as $H_0 = H$ does and results in the following 3D analog of Corollary 1.

Corollary 2. *Let H, H' be 3D graphs with the domain U and K be a 3D graph with domain V. If H simplifies to H', then H and H' have the same set of perfect matchings and $H \times K$ simplifies to $H' \times K$.*

Simplifiable SUSPs. We now apply the notion of simplification to help in checking whether an (s, k)-puzzle P is an SUSP. By Corollary 2, H_P has a nontrivial matching iff any simplification of H_P has a nontrivial matching. This suggests a way to construct a witness that P is an SUSP: If H_P simplifies to the trivial matching, then, by Corollary 2, H_P has no nontrivial matchings, and, by Lemma 2, P is an SUSP. The sequence of sets and their corresponding projection indexes are a witness that P is an SUSP. Moreover, if we exclude simplifications that do not change the 3D graph, the number of edges in the 3D graph—at most s^3—is a limit on the number of simplification steps that can occur.

Definition 2 (Simplifiable SUSP). *An (s, k)-puzzle P is a* simplifiable *SUSP if H_P simplifies to the trivial 3D perfect matching.*

By definition, simplifiable SUSP are SUSPs with short ($O(s^4)$ bit length) witnesses. To make this definition effective, we describe a polynomial-time algorithm that simplifies puzzles. In particular, the algorithm takes H_P; projects it onto its 2D faces, R_0, R_1, R_2; then, for each face, determines sets that induce maximal simplification of the faces; and, finally, applies those simplifications to H_P to form a new 3D graph H'_P. The algorithm repeats this until a fixed point is

---

Algorithm 1 : SIMPLIFY(H)

1: $f \leftarrow 0$; $sinceChange \leftarrow 0$; $R_0, R_1, R_2 \leftarrow$ PROJECT(H)
2: **while** $sinceChange < 3$ **do**
3:     $edgesToRemove \leftarrow$ CALCEDGESTOREMOVE(R_f)
4:     **for** $(u, v) \in edgesToRemove$ **do**
5:         **if** $f = 0$ **then** delete all edges $(*, u, v)$ from H
6:         **if** $f = 1$ **then** delete all edges $(u, *, v)$ from H
7:         **if** $f = 2$ **then** delete all edges $(u, v, *)$ from H
8:     **if** $edgesToRemove = \emptyset$ **then** $sinceChange \leftarrow sinceChange + 1$
9:     **else** $sinceChange \leftarrow 0$; $R_0, R_1, R_2 \leftarrow$ PROJECT(H)
10:     $f \leftarrow (f + 1) \bmod 3$
11: **return** H

---

reached. The resulting 3D graph is the fully simplified version of H_P. If that simplified graph is the trivial matching, this process witnesses that P is a (simplifiable) SUSP. An example of the results of this process are shown in Fig. 2. For completeness, this process is described in Algorithm 1.

In Algorithm 1, the subroutine PROJECT takes the 3D graph H and returns three 2D graphs R_0, R_1, R_2 that, respectively, correspond to projecting out the first, second, and third coordinates of G, as defined above. This subroutine can be naïvely implemented in $O(s^3)$ time.

The subroutine CALCEDGESTOREMOVE at Line 3 takes each of the 2D graphs corresponding to the faces and returns a list of edges that are not used in any maximum 2D matchings of that face. This subroutine can be implemented using the algorithm described in [17, Algorithm 2]. Their algorithm works by constructing the strongly connected components of the input 2D graph R_f, when R_f is viewed as a directed graph over P rather than a bipartite graph over $P \sqcup P$. The strongly connected components calculated by this algorithm inherently partition the vertex set $P = S_1 \cup S_2 \cup \ldots \cup S_\ell$.

Collapsing the 2D graph R_f down to its strongly connected components leaves us with a directed graph G_f with $V(G_f) = \{v_1, v_2, \ldots, v_\ell\}$ and $E(G_f) = \{(v_i, v_j) \mid \exists u \in S_i, w \in S_j$ such that $(u, v) \in E(R_f)\}$ with ℓ vertices v_j, one for each strongly connected component S_j. Furthermore, G_f must be an acyclic graph, otherwise the strongly connected components would have been larger. These strongly connected components are sets that induce the simplification of R_f. Let v_j be a vertex in G_f that has some incident edges but that has either no incoming or no outgoing edges. The latter property is sufficient to apply Lemma 3 and implies that S_j induces a simplification of R_f. Furthermore, this simplification corresponds to deleting all of the edges of v_j in G_f.

This process can be repeated until there are no more edges in G_f. Note that because G_f is acyclic, it will always be possible to find such a vertex v_j as long as there are edges remaining. This series of strongly connected components induces a complete simplification of R_f. This simplification is used to remove the corresponding edges in the 3D graph H in Lines 4–7 & 9. The 3D graph H is fully simplified when no edge can be removed from any of the three faces. By [17], the remaining edges in each of the projections R_f are "maximally matchable" in that they used in some perfect matching of R_f. Thus, once this happens, there can be no additional sets that can induce simplifications in any of the R_f that remove edges in R_f (or in H).

Since each edge of H, except for the diagonal, can be removed at most once, the algorithm must reach a fixed point within $3(|P|^3 - |P|)$ iterations of the main loop. The cost to update H and the projections in Lines 4–7 & 9 can be amortized, with careful bookkeeping, to cost $O(|P|^3)$ across the whole algorithm.

For 2D graphs whose domain is the (s, k)-puzzle P, the subroutine of [17] runs in $O(s^{2.5}/\sqrt{\log s})$ time. Combining the above analysis, the overall complexity of SIMPLIFY is $O(s^3 + s^3 \cdot s^{2.5}/\sqrt{\log s}) = O(s^{5.5}/\sqrt{\log s})$. The results of the above arguments can be summarized in the following lemma.

Lemma 5. *Let H be a 3D graph over P. In* $\mathrm{poly}(|P|)$ *time,* SIMPLIFY(H) *computes the complete simplification of H. Therefore, H has the same set of matchings as* SIMPLIFY(H).

By Definition 2, the 3D graph H_P associated with a simplifiable SUSP P simplifies to the trivial matching. Furthermore, by Lemma 5, SIMPLIFY(H_P) computes, in polynomial time, the complete simplification of H_P, preserving the matchings. These two facts imply a polynomial-time algorithm to determine whether a puzzle P is a simplifiable SUSP, proving Theorem 1.

Proof (Theorem 1). Perform the polynomial-time reduction from SUSP verification to 3D matching of [4] to produce the 3D graph H_P in time $\mathrm{poly}(s, k)$. Compute $H'_P = $ SIMPLIFY(H_P) in time $\mathrm{poly}(s)$. In time $O(s^3)$ verify and return whether or not H'_P is the trivial matching $\{(u, u, u) \mid u \in P\}$. The algorithm is correct by Lemma 2 and Lemma 5.   □

It is clear from the construction that simplifiable SUSPs are a subset of SUSPs, but they are also a generalize of the notion of local SUSPs.

Lemma 6. *Every local SUSP P is a simplifiable SUSP.*

Proof. By the definition, for every triple of rows $u, v, w \in P$, there is a column c such $(u_c, v_c, w_c) \in \mathcal{L}$. This implies, by the construction of H_P, that (u, v, w) is not an edge in H_P. This implies that H_P has no edges except where $u = v = w$. Therefore, H_P is the trivial matching and explicitly satisfies Definition 2 without any further simplification. We conclude that P is a simplifiable SUSP.   □

Intuitively, simplifiable SUSPs are an intermediate class between local SUSPs and SUSPs. The set containments are proper. There exist SUSPs that are not simplifiable, e.g., $P_1 = \{2233, 1232, 1123, 3311\}$, and simplifiable SUSPs that are not local, e.g., $P_2 = \{11, 23\}$. Simplifiable SUSPs have the efficient verification of local SUSPs, but the concise representation of general SUSPs. These properties make the prospect of searching for large simplifiable SUSPs more feasible.

4   Simplifiable SUSPs Generate Infinite Families

Lemma 1 gives bounds for the running time of matrix multiplication using infinite families of SUSPs and individual SUSPs. The first bound produces stronger results than the second. To see this, we define the *capacity* of an (s, k)-SUSP P to be $C_P = s^{\frac{1}{k}}$, this is analogous to the definition of capacity for families of SUSPs. Now, consider an SUSP P and an infinite family $\mathcal{F}$ with the same capacity $C_P = C_{\mathcal{F}}$. Lemma 1 gives a weaker upper bound on ω for the single puzzle than its does for the infinite family. For example, a $(14, 6)$-SUSP has capacity $14^{\frac{1}{6}}$ and the bound on ω using the dimensions of the puzzle is $\omega \leq 2.73$ and, although the first part of Lemma 1 does not apply, if we were to use the capacity of the puzzle instead of its dimensions, we get $\omega \leq 2.52$.

We show that simplifiable SUSPs can be turned into an infinite family of simplifiable SUSPs by taking Cartesian products (powers) of P with itself to product a family with the same capacity as P. This allows the first part of Lemma 1 to be applied, instead of the second, to produce a stronger bound on ω using the capacity of P.

Let P_1 be an (s_1, k_1)-puzzle and P_2 be an (s_2, k_2)-puzzle. We define the *product* of P_1 and P_2 to be the Cartesian product of their underlying sets: $P_1 \times P_2 = \{r_1 \circ r_2 \mid r_1 \in P_1, r_2 \in P_2\}$. Observe that $P_1 \times P_2$ is an $(s_1 \cdot s_2, k_1 + k_2)$-puzzle. Furthermore, if P is an (s, k)-puzzle, its m-th power is the Cartesian product of P with itself m times, P^m, and observe that this is an $(s^m, k \cdot m)$-puzzle. For a puzzle P, we can define the infinite family $\mathcal{F}_P = \{P^m \mid m \in \mathbb{N}\}$. Observe that $\mathcal{F}_P$ has capacity $(s^m)^{\frac{1}{k \cdot m}} = s^{\frac{1}{k}}$ matching the capacity of P.

We say an SUSP P *generates an infinite family of SUSPs*, if every puzzle in $\mathcal{F}_P$ is an SUSP. Unfortunately, the SUSP property is not generally preserved under Cartesian product or powering. For example, $P = \{2233, 1232, 1123, 3311\}$ is an SUSP, but $P \times P$ is not. A consequence of this is that not every SUSP generates an infinite family of SUSPs. Although SUSPs are generally not closed under powering, we show that simplifiable SUSPs are.

Lemma 7. *If P, Q are simplifiable SUSPs, $P \times Q$ is a simplifiable SUSP. Moreover, P generates an infinite family of simplifiable SUSPs.*

The proof is a direct consequence of Definition 2 and Corollary 2, and we defer it, for space, to the long version of this article [6]. Combining Lemma 7 with the first part of Lemma 1 we produce a tighter bound on ω from simplifiable SUSPs, which proves our main theorem (Theorem 2).

Although it is not the case that every SUSP generates an infinite family, there is evidence in both experimental results of [4,5] and some of the puzzle constructions of [8] that there are (non-local) SUSP of maximum size for their width that generate infinite families. For example, [8, Proposition 3.1] gives an infinite family with capacity $\sqrt{2}$ that is generated by the $(2, 2)$-SUSP $\{12, 33\}$.

Finally, using Lemma 6 and Proposition 1 we conclude that using simplifiable SUSPs does not inherently lead to weaker bounds on ω than SUSPs.

Lemma 8. *The SUSP capacity is achieved by SUSPs that are simplifiable.*

Table 1. Comparison of lower bounds for the maximum of size of width-k SUSPs and upper bounds on ω they imply. All the results in this work are simplifiable SUSPs.

	k	1	2	3	4	5	6	7	8	9	10	11	12
[8]	$s \geq$	1	2	3	4	4	10	10	16	36	36	36	136
	$\omega \leq$	3.00	2.88	2.85	2.85		2.80			2.74			2.70
[4]	$s \geq$	1	2	3	5	8	14	21	30	42	64	112	196
	$\omega \leq$	3.00	2.88	2.85	2.81	2.78	2.74	2.73	2.72	2.72	2.71	2.68	2.66
Us	$s \geq$	1	2	3	5	8	14	**23**	**35**	**52**	**78**	**128**	196
	$\omega \leq$	3.00	**2.67**	**2.65**	**2.59**	**2.57**	**2.52**	**2.505**	**2.52**	**2.53**	**2.53**	**2.52**	**2.52**

5   New Lowers Bounds on Maximum SUSP Size

The features of simplifiable SUSPs we proved in the previous sections make them well suited for discovery via computer search. We use iterative local search techniques to locate large simplifiable SUSPs with small width $k \leq 12$. For brevity, we defer discussion of our search algorithm and implementation to the long version of this article [6].[1]

We find simplifiable SUSPs that match or exceed the size of SUSPs found in prior work [4,8] for $k \leq 12$. We summarize our results in Table 1. Because the SUSPs we find are simplifiable, Theorem 2 implies that they produce stronger bounds on ω than the SUSPs of previous work that are analyzed using the weaker Lemma 1. This results in substantial improvements over prior work in this domain: decreasing the bound on ω by about 0.2. For $k \leq 5$, the sizes in [4] were shown to be maximum by exhaustive search, and our results match theirs. For $6 < k \leq 11$, we construct larger SUSPs than in the previous work. The long version of this article include examples of these maximal simplifable SUSPs [6].

The improvement in the bounds on ω appears to stall for $k \geq 8$. We do not believe that this reflects a real limit on the size of simplifiable SUSPs; rather, it represents a barrier for our search techniques and the large polynomial-time cost of running SIMPLIFY. Although our results improve substantially over [8] for $k \leq 12$, their construction achieves $\omega \leq 2.48$ as $k \to \infty$.

6   Conclusions

We propose and analyze simplifiable SUSPs, a new subclass of strong uniquely solvable puzzles. We prove that simplifiable SUSPs have nice properties: they are efficiently verifiable and generate infinite families of SUSP that lead to tighter bounds on ω. We report the existence of new large (simplifiable) SUSPs with width $7 \leq k \leq 11$ and strengthen the bound on ω that they imply compared to previous work. The SUSPs we have found through computer search are now close

---

[1] Implementations of our algorithms, along with a tool for verifying simplifiable SUSPs, are publicly available at https://bitbucket.org/paraphase/matmult-v2.

to producing the same bounds ($\omega \leq 2.505$) as those families of SUSP designed by human experts ($\omega \leq 2.48$).

New insights into the structure of (simplifiable) SUSPs or the search space seem necessary to progress. A primary bottleneck in the search is the run time of SIMPLIFY, which even if it quickly reaches a fixed point, the algorithm still spends $\Omega(s^3)$ time to construct an instance from a puzzle with $s \cdot k$ entries. We conjecture that if there are (s, k)-SUSPs, then there are simplifiable (s, k)-SUSPs, which is consistent with the SUSPs we found and report in Table 1.

Acknowledgments. The second author's work was funded in part by the Union College Summer Research Fellows Program. Both authors acknowledge contributions from other student researchers to various aspects of this research program. We thank our anonymous reviewers for their helpful suggestions.

References

1. Alman, J., Williams, V.V.: Further limitations of the known approaches for matrix multiplication. In: 9th Innovations in Theoretical Computer Science (ITCS). LIPIcs. Leibniz International Proceedings in Informatics, vol. 94, pp. Art. No. 25, 15. Schloss Dagstuhl. Leibniz-Zent. Inform., Wadern, Germany (2018). https://doi.org/10.4230/LIPIcs.ITCS.2018.25
2. Alman, J., Williams, V.V.: A Refined Laser Method and Faster Matrix Multiplication, pp. 522–539. SIAM (2020). https://doi.org/10.1137/1.9781611976465.32
3. Alon, N., Shpilka, A., Umans, C.: On sunflowers and matrix multiplication. Comput. Complex. **22**(2), 219–243 (2013). https://doi.org/10.1007/s00037-013-0060-1
4. Anderson, M., Ji, Z., Xu, A.Y.: Matrix multiplication: verifying strong uniquely solvable puzzles. In: Pulina, L., Seidl, M. (eds.) SAT 2020. LNCS, vol. 12178, pp. 464–480. Springer, Cham (2020). https://doi.org/10.1007/978-3-030-51825-7_32
5. Anderson, M., Ji, Z., Xu, A.Y.: Matrix multiplication: verifying strong uniquely solvable puzzles (2023). https://doi.org/10.48550/ARXIV.2301.00074
6. Anderson, M., Le, V.: Efficiently-verifiable strong uniquely solvable puzzles and matrix multiplication (2023). https://doi.org/10.48550/ARXIV.2307.06463
7. Blasiak, J., et al.: Which groups are amenable to proving exponent two for matrix multiplication? (2017). https://doi.org/10.48550/ARXIV.1712.02302
8. Cohn, H., Kleinberg, R., Szegedy, B., Umans, C.: Group-theoretic algorithms for matrix multiplication. In: 46th Annual IEEE Symposium on Foundations of Computer Science (FOCS), pp. 379–388 (2005). https://doi.org/10.1109/SFCS.2005.39
9. Cohn, H., Umans, C.: A group-theoretic approach to fast matrix multiplication. In: 44th Annual IEEE Symposium on Foundations of Computer Science (FOCS), pp. 438–449 (2003). https://doi.org/10.1109/SFCS.2003.1238217
10. Coppersmith, D., Winograd, S.: Matrix multiplication via arithmetic progressions. J. Symb. Comput. **9**(3), 251–280 (1990). https://doi.org/10.1016/S0747-7171(08)80013-2
11. Croot, E., Lev, V.F., Pach, P.P.: Progression-free sets in are exponentially small. Ann. Math. 331–337 (2017). https://doi.org/10.4007/annals.2017.185.1.7
12. Davie, A.M., Stothers, A.J.: Improved bound for complexity of matrix multiplication. Proc. R. Soc. Edinb. Sect. A: Math. **143**(2), 351–369 (2013)

13. Duan, R., Wu, H., Zhou, R.: Faster matrix multiplication via asymmetric hashing (2022). https://doi.org/10.48550/ARXIV.2210.10173
14. Fawzi, A., et al.: Discovering faster matrix multiplication algorithms with reinforcement learning. Nature **610**(7930), 47–53 (2022). https://doi.org/10.1038/s41586-022-05172-4
15. Russell, S., Norvig, P.: Artificial Intelligence: A Modern Approach. Pearson, London (2020)
16. Strassen, V.: Gaussian elimination is not optimal. Numer. Math. **13**(4), 354–356 (1969). https://doi.org/10.1007/BF02165411
17. Tassa, T.: Finding all maximally-matchable edges in a bipartite graph. Theoret. Comput. Sci. **423**, 50–58 (2012). https://doi.org/10.1016/j.tcs.2011.12.071

(min, +) Matrix and Vector Products for Inputs Decomposable into Few Monotone Subsequences

Andrzej Lingas[1](✉) and Mia Persson[2]

[1] Department of Computer Science, Lund University, 22100 Lund, Sweden
`Andrzej.Lingas@cs.lth.se`
[2] Department of Computer Science and Media Technology, Malmö University, 20506 Malmö, Sweden
`Mia.Persson@mau.se`

Abstract. We study the time complexity of computing the $(\min, +)$ matrix product of two $n \times n$ integer matrices in terms of n and the number of monotone subsequences the rows of the first matrix and the columns of the second matrix can be decomposed into. In particular, we show that if each row of the first matrix can be decomposed into at most m_1 monotone subsequences and each column of the second matrix can be decomposed into at most m_2 monotone subsequences such that all the subsequences are non-decreasing or all of them are non-increasing then the $(\min, +)$ product of the matrices can be computed in $O(m_1 m_2 n^{2.569})$ time. On the other hand, we observe that if all the rows of the first matrix are non-decreasing and all columns of the second matrix are non-increasing or *vice versa* then this case is as hard as the general one.

Similarly, we also study the time complexity of computing the $(\min, +)$ convolution of two n-dimensional integer vectors in terms of n and the number of monotone subsequences the two vectors can be decomposed into. We show that if the first vector can be decomposed into at most m_1 monotone subsequences and the second vector can be decomposed into at most m_2 subsequences such that all the subsequences of the first vector are non-decreasing and all the subsequences of the second vector are non-increasing or *vice versa* then their $(\min, +)$ convolution can be computed in $\tilde{O}(m_1 m_2 n^{1.5})$ time. On the other, the case when both vectors are non-decreasing or both of them are non-increasing is as hard as the general case.

1 Introduction

$(\min, +)$ *matrix product.* The $(\min, +)$ matrix product problem for two $n \times n$ integer matrices $A = (a_{i,j})$, $B = (b_{i,j})$ requires computing an $n \times n$ matrix $C = (c_{i,j})$ such that $c_{i,j} = \min\{a_{i,k} + b_{k,j} | 1 \leq k \leq n\}$. By the definition, this problem admits an $O(n^3)$-time algorithm. It is known to be equivalent to the fundamental all-pairs shortest-paths problem (APSP) [11]. If any of these two problems admits an $t(n)$-time algorithm then the other problem can be solved in

W. Wu and G. Tong (Eds.): COCOON 2023, LNCS 14423, pp. 55–68, 2024.
https://doi.org/10.1007/978-3-031-49193-1_5

$O(t(n))$ time [11]. Hence, the APSP hypothesis states that solving any of these two problems requires $n^{3-o(1)}$ time [22] and the current best algorithm for any of them runs in $\frac{n^3}{2^{\Theta(\sqrt{\log n})}}$ time [24].

The $(\min, +)$ matrix product as APSP has a large number of important applications. Because the prospects of deriving a substantially subcubic upper time bound for the general $(\min, +)$ matrix product are so vague, several authors studied the complexity of computing this product for restricted integer matrices. Already several decades ago, it was known that the $(\min, +)$ matrix product can be computed in $O(Mn^\omega)$ time, when the values of the entries in the input matrices are in the range $\{-M, ..., M\} \cup \{+\infty\}$ [2,26]. Here, ω stands for the smallest real number such that two $n \times n$ matrices can be multiplied using $O(n^{\omega+\epsilon})$ operations over the field of reals, for all $\epsilon > 0$ (i.e., the number of operations is $O(n^{\omega+o(1)})$ [1]). More recently, one succeeded to derive substantially subcubic upper time bounds when, e.g.,: one of the matrices has a small number of different entries in each row [25], the input matrices are of the so called bounded-difference (i.e., all pairs of horizontally and vertically adjacent entries differ by at most $O(1)$) [4], the input matrices are geometrically weighted [8], one of the matrices has a constant approximate rank [23], the entries of one of the matrices are of size $O(n)$ and its rows are non-decreasing [21], or just one of the matrices range over a constant number of integers [8].

Contributions on $(\min, +)$ *Matrix Product.* In this paper, we take a more general approach. We study the situation when each row of the first matrix A and each column of the second matrix B can be decomposed into a bounded number of monotone subsequences. When all the subsequences are non-decreasing or all of them are non-increasing, we obtain a substantially subcubic algorithm for the $(\min, +)$ matrix product already when the bound on the number of monotone subsequences of each row in A and each column in B is $O(n^{0.215})$. Namely, our algorithm runs in $O(m_a m_b n^{2.569})$ time, where m_a is an upper bound on the number of monotone subsequences of each row in A and m_b is an upper bound on the number of the monotone subsequences of each column in B. On the other hand, we observe that if all the rows of A are non-decreasing and all columns of B are non-increasing or *vice versa* then this case is as hard as the general case. When the entries in each row or column of one of the input matrices range over c different integers then it is sufficient that the columns or rows respectively of the other matrix can be decomposed into at most $n^{0.119}$ just monotone subsequences to subsume the upper time bound $O(cn^{2.688})$ [8] (see Fact 5) for the case without restrictions on the other matrix. Our results on $(\min, +)$ matrix product are summarized in Table 1.

$(\min, +)$ *Vector Convolution.* Our approach to the $(\min, +)$ matrix product is in fact similar to that to $(\min, +)$ convolution of two n-dimensional integer vectors taken by the authors in the prior paper [18]. The $(\min, +)$ convolution problem for two integer vectors $a = (a_0, ..., a_{n-1})$ and $b = (b_0, ..., b_{n-1})$ requires computing an $2n - 1$ dimensional vector $c = (c_0, ..., c_{2n-2})$ such that $c_k = \min\{a_\ell + b_{k-\ell} | \ell \in [\max\{k - n + 1, 0\}, \min\{k, n - 1\}]\}$ for $k = 0, ..., 2n - 2$. By the definition, the $(\min, +)$ vector convolution can be computed in $O(n^2)$

Table 1. Upper time bounds for computing the (min, +) matrix product of two $n \times n$ integer matrices A, B, where the rows of A and/or the columns of B admit decompositions into a bounded number of monotone subsequences (in particular non-decreasing or non-increasing) or the entries in each row of A or each column of B range over a constant number of integers.

matrix A/matrix B	c_b dif. values	m_b non-decr. subs.	m_b non-incr. subs.
c_a different values	$O(c_a c_b n^\omega)$	$O(c_a m_b n^{2.569})$	$O(c_a m_b n^{2.569})$
m_a non-decr. subs.	$O(m_a c_b n^{2.569})$	$O(m_a m_b n^{2.569})$	?
m_a non-incr. subs.	$O(m_a c_b n^{2.569})$	?	$O(m_a m_b n^{2.569})$
arbitrary	$O(c_b n^{2.688})$ [8]	?	?

time but again getting any substantially subquadratic upper time bound for this problem would be a breakthrough. The (min, +) vector convolution has also a large number of important applications ranging from stringology to knapsack problem [3,5,9,19].

Contributions on (min, +) *Convolution.* We correct the requirements on the monotonicity of the vector subsequences in the statement of Theorem 3.7 in [18] and provide a proof of the corrected theorem. It states that the (min, +) convolution of two n-dimensional integer vectors a and b, given with the decompositions of the sequences of their consecutive coordinates into m_a and m_b subsequences respectively, such that either all the subsequences of a are non-decreasing and all the subsequences of b are non-increasing or *vice versa*, can be computed in $\tilde{O}(m_a m_b n^{1.5})$ time. On the other hand, the case when both vectors are non-decreasing or both of them are non-increasing is as hard as the general case. Table 2 summarizes the updated results on (min, +) vector convolution (cf. [18]).

Table 2. Upper time bounds for computing the (min, +) convolution of two n-dimensional integer vectors either with coordinates having a bounded number of different values, or decompositions into a number of non-decreasing or non-increasing subsequences.

vector a/vector b	c_b dif. values	m_b non-decr. subs.	m_b non-incr. subs.
c_a different values	$\tilde{O}(c_a c_b n)$	$\tilde{O}(c_a n^{1.5})$	$\tilde{O}(c_a n^{1.5})$
m_a non-decr. subs.	$\tilde{O}(c_b n^{1.5})$	?	$\tilde{O}(m_a m_b n^{1.5})$
m_a non-incr. subs.	$\tilde{O}(c_b n^{1.5})$	$\tilde{O}(m_a m_b n^{1.5})$	?
arbitrary	$\tilde{O}(c_b n^{1.5})$	?	?

Techniques. Our algorithms for the (min, +) matrix product as well as those for the (min, +) vector convolution are mostly based on efficient reductions to collections of maximum or/and minimum witness problems for corresponding

Boolean matrix products or Boolean vector convolutions, respectively. One of our algorithms uses directly a method similar to that known for the extreme witnesses. For the definition of the extreme witness problems and facts on them see Preliminaries.

Paper Organization. The next section contains basic definitions and facts. Section 3 presents our results on $(\min, +)$ matrix product while Sect. 4 presents our results on $(\min, +)$ vector convolution.

2   Preliminaries

For two n-dimensional vectors $a = (a_0, ..., a_{n-1})$ and $b = (b_0, ..., b_{n-1})$ over a semi-ring $(\mathbb{U}, \oplus, \odot)$, their convolution over the semi-ring is a vector $c = (c_0, ..., c_{2n-2})$, where $c_i = \bigoplus_{l=\max\{i-n+1,0\}}^{\min\{i,n-1\}} a_l \odot b_{i-l}$ for $i = 0, ..., 2n - 2$. Similarly, for a $p \times q$ matrix A and a $q \times r$ matrix B over the semi-ring, their matrix product over the semi-ring is a $p \times r$ matrix $C = (c_{i,j})$ such that $c_{i,j} = \bigoplus_{m=1}^{q} a_{i,m} \odot b_{m,j}$ for $1 \leq i \leq p$ and $1 \leq j \leq r$. In particular, for the semi-rings $(\mathbb{Z}, +, \times)$, $(\mathbb{Z}, \min, +)$, $(\mathbb{Z}, \max, +)$, and $(\{0,1\}, \vee, \wedge)$, we obtain the arithmetic, $(\min, +)$, $(\max, +)$, and the Boolean convolutions or matrix products, respectively.

We shall use the unit-cost RAM computational model with computer word of length logarithmic in the maximum of the size of the input and the value of the largest input integer.

For a positive integer r, we shall denote the set of positive integers not greater than r by $[r]$.

A *witness* for a non-zero entry $c_{i,j}$ of the Boolean matrix product C of a Boolean $p \times q$ matrix A and a Boolean $q \times r$ matrix B is any index $k \in [q]$ such that $a_{i,k}$ and $b_{k,j}$ are equal to 1. Such a maximum index is the *maximum witness* for $c_{i,j}$ while such a minimum index is the the *minimum witness* for $c_{i,j}$. The *maximum witness problem* (*minimum witness problem*, respectively) is to report the maximum witness (minimum witness, respectively) for each non-zero entry of the Boolean matrix product of the two input matrices.

For positive real numbers p, q, s, $\omega(p,q,s)$ denotes the smallest real number such that an $n^p \times n^q$ matrix can be multiplied by $n^q \times n^s$ matrix using $O(n^{\omega(p,q,s)+\epsilon})$ operations over the field of reals, for all $\epsilon > 0$. For convenience, ω stands for $\omega(1,1,1)$.

Fact 1. *[10] The minimum witness problem and the maximum witness problem for the Boolean matrix product of two Boolean $n \times n$ matrices can be solved in $O(n^{2+\lambda})$ time, where λ satisfies the equation $\omega(1, \lambda, 1) = 1 + 2\lambda$. By currently best bounds on $\omega(1, \lambda, 1)$, $O(n^{2+\lambda}) = O(n^{2.569})$.*

The currently best bounds on $\omega(1, \lambda, 1)$ follow from a fact in [16] combined with the recent improved estimations on the parameters $\omega = \omega(1, 1, 1)$ and α, see [14,15]. They yield an $O(n^{2.569})$ upper bound on the running time of the algorithm for minimum and maximum witnesses in [10] (originally, $O(n^{2.575})$).

The following fact is well known (cf. [12]).

Fact 2. *Let p and q be two n-dimensional integer vectors. The arithmetic convolution of p and q can be computed in $\tilde{O}(n)$ time. Hence, also the Boolean convolution of two n-dimensional vectors can be computed in $\tilde{O}(n)$ time.*

Let $c = (c_0, ..., c_{2n-2})$ be the Boolean convolution of two n-dimensional Boolean vectors a and b. A *witness* of $c_i = 1$ is any $l \in [\max\{i - n + 1, 0\}, \min\{i, n - 1\}]$ such that $a_l \wedge b_{i-l} = 1$. A *minimum witness* (or *maximum witness*) of $c_i = 1$ is the smallest (or, the largest, respectively) witness of c_i. The *minimum witness problem*, or *maximum witness problem* for the Boolean convolution of two n-dimensional Boolean vectors is to determine the minimum witnesses or the maximum witnesses, respectively, for all non-zero entries of the Boolean convolution of the vectors.

Fact 3. *(Theorem 3.2 in [18]). The minimum witness problem (maximum witness problem, respectively) for the Boolean convolution of two n-dimensional vectors can be solved in $\tilde{O}(n^{1.5})$ time.*

For a sequence s of integers, we shall denote the minimum number of monotone subsequences into which s can be decomposed by $mon(s)$.

Fact 4. *[13, 20]. A sequence s of n integers can be decomposed into $O(mon(s) \log n)$ monotone subsequences in $O(n^{1.5} \log n)$ time.*

Fact 5. *(Theorem 3.2 in [8]). Let A and B be two $n \times n$ integer matrices, where the entries of one of the matrices range over at most c different integers. The (min, +) matrix product of A and B can be computed in $O(cn^{2.688})$ time.*

3   (min,+) Matrix Product

Consider two $n \times n$ integer matrices A and B. If we are given decompositions of the rows of A and the columns of B into monotone subsequences such that either all the subsequences are non-decreasing or all of them are non-increasing then we can use the algorithm depicted in Fig. 1 in order to compute the (min, +) matrix product of A and B. First, for all $i, j \in [n]$, for each subsequence a_i^o of the i-th row of A and each subsequence b_j^r of the j-th column of B, we compute the Boolean vectors $char(a_i^o)$ and $char(b_j^r)$ indicating with ones the entries of the row and column covered by a_i^o or b_j^r, respectively. Next, we form the Boolean matrices A^o whose rows are the vectors $char(a_i^o)$ and the Boolean matrices B^r whose columns are the vectors $char(b_j^r)$. Then, depending if the subsequences are non-decreasing or non-increasing, for each pair of matrices A^o, B^r, we compute either the minimum witnesses of the Boolean matrix product of A^o and B^r or the maximum witnesses of this Boolean product, respectively. We use the extreme witnesses to update the current entries of the computed (min, +) matrix product of A and B. The correctness of the reduction to extreme witnesses for the Boolean matrix product of A^o and B^r in the algorithm is implied by the following observation.

Input: two $n \times n$ integer matrices $A = (a_{i,j})$ and $B = (b_{i,j})$, for each $i \in [n]$, a decomposition of the i-th row of A into m_a subsequences a_i^o, and for each $j \in [n]$, a decomposition of the j-th column of B into m_b subsequences b_j^r, such that either all the subsequences are non-decreasing or all of them are non-increasing.

Output: the $(\min, +)$ matrix product $C = (c_{i,j})$ of A and B.

1: **for** each $o \in [m_a]$ **do**
2:   **for** each $i \in [n]$ **do**
3:     form a Boolean vector $char(a_i^o)$ with n coordinates indicating with ones the entries of the i-th row of A covered by a_i^o
4:   **end for**
5:   form a Boolean matrix A^o, where for $i \in [n]$, $char(a_i^o)$ is the i-th row
6: **end for**
7: **for** each $r \in [m_b]$ **do**
8:   **for** each $j \in [n]$ **do**
9:     form a Boolean vector $char(b_j^r)$ with n coordinates indicating with ones the entries of the j-th column of B covered by b_j^r
10:   **end for**
11:   form a Boolean matrix B^r, where for $j \in [n]$, $char(b_j^r)$ is the j-th column
12: **end for**
13: initialize the $C = (c_{i,j})$ matrix by setting each $c_{i,j}$ to $+\infty$
14: **for** each pair A^o and B^r **do**
15:   **if** the subsequences are non-decreasing **then** compute the minimum witnesses $(wit(d_{i,j}))$ for the Boolean matrix product $(d_{i,j})$ of A^o and B^r
16:   **if** the subsequences are non-increasing **then** compute the maximum witnesses $(wit(d_{i,j}))$ for the Boolean matrix product $(d_{i,j})$ of A^o and B^r
17:   **for** $i = 1$ **to** n **do**
18:     **for** $j = 1$ **to** n **do**
19:       **if** $d_{i,j} \neq 0$ **then** $c_{i,j} \leftarrow \min\{a_{i,wit(d_{i,j})} + b_{wit(d_{i,j}),j}, c_{i,j}\}$
20:     **end for**
21:   **end for**
22: **end for**
23: $C \leftarrow (c_{i,j})$
24: **return** C

Fig. 1. An algorithm for computing the $(\min, +)$ product of two $n \times n$ integer matrices A and B given with decompositions of all rows of A into m_a subsequences and decompositions of all columns of B into m_b subsequences such that either all the subsequences are non-decreasing or all the subsequences are non-increasing.

Remark 1. Let $A = (a_{i,j})$ and $B = (b_{i,j})$ be two $n \times n$ integer matrices. Next, let a' be a subsequence of the sequence of entries in an i-th row of A and let b' a subsequence of the sequence of entries in an j-th column of B. If a' and b' are non-decreasing then if the set $\{a_{i,k} + b_{k,j} | k \in [n] \wedge a_{i,k} \in a' \wedge b_{k,j} \in b'\}$ is not empty then the minimum sum in the set is achieved by the pair minimizing the index k. Analogously, if a' and b' are non-increasing then the minimum sum is achieved by the pair maximizing the index k.

Theorem 6. *Let $A = (a_{i,j})$ and $B = (b_{i,j})$ be two $n \times n$ integer matrices. Suppose that for each $i, j \in [n]$, there is given a decomposition of the i-th row of A into at most m_a subsequences and a decomposition of the j-th column of B into* *at most m_b subsequences, where either all the subsequences are non-decreasing or all of them are non-increasing. Then, the $(\min, +)$ product of A and B can be computed in $O(m_a m_b n^{2.569})$ time.*

Proof. By Remark 1, the following condition holds: (*) if $d_{i,j} \neq 0$ in line 19 of the algorithm in Fig. 1 then $\min\{a_{i,k} + b_{k,j} | k \in [n] \wedge char(a_i^o)_k = 1_i \wedge char(b_j^r)_k = 1\}$ is equal to the first argument of the minimum in this line, i.e., $a_{i,wit(d_{i,j})} + b_{wit(d_{i,j}),j}$. Hence, none of the entries of the output matrix C has a lower value than the corresponding entry of the $(\min, +)$ matrix product of A and B. Conversely, if the i, j entry of the $(\min, +)$ matrix of A and B equals $a_{i,k} + b_{k,j}$ then there exist o, r such that $a_{i,k} \in a_i^o$ and $b_{k,j} \in b_j^r$, i.e., more precisely $char(a_i^o)_k = 1$ and $char(b_j^r)_k = 1$. Hence, again by (*) and line 19 in the algorithm, the $c_{i,j}$ entry in the output matrix has value not larger than the corresponding i, j entry of the $(\min, +)$ matrix product of A and B.

The time complexity of the algorithm is dominated by the $m_a m_b$ computations of minimum or maximum witnesses of the Boolean product of two Boolean $n \times n$ matrices. Thus, by Fact 1 the algorithm runs in $O(m_a m_b n^{2.569})$ time. □

Example 1. We shall assume the notation from our first algorithm. Suppose that two input integer matrices A and B have size 6×6 and that each row of A can be decomposed into at most 3 non-decreasing subsequences while each column of B can be decomposed into at most 2 non-decreasing subsequences. Suppose in particular that the fourth row a_4 of A is $(1, 7, 3, 9, 8, 4)$ while the fifth column b_5 of B is $(5, 11, 2, 7, 13, 10)$. Then, it is easy to see that in the $(\min, +)$ product $(c_{i,j})$ of A and B, $c_{4,5} = 5$ holds. Note that a_4 can be decomposed into three following non-decreasing subsequences $a_4^1 = (1, \ , 3, \ , \ , 4)$, $a_4^2 = (\ , 7, \ , \ , 8, \ )$, $a_4^3 = (\ , \ , \ , 9, \ , \ )$ while b_5 can be decomposed into two non-decreasing subsequences $b_5^1 = (5, 11, \ , \ , 13, \ )$ and $b_5^2 = (\ , \ , 2, 7, \ , 10)$. Their characteristic Boolean vectors are $char(a_4^1) = (1, 0, 1, 0, 0, 1)$, $char(a_4^2) = (0, 1, 0, 0, 1, 0)$, $char(a_4^3) = (0, 0, 0, 1, 0, 0)$, and $char(b_5^1) = (1, 1, 0, 0, 1, 0)$, $char(b_5^2) = (0, 0, 1, 1, 0, 1)$, respectively. For $o \in [3]$ and $r \in [2]$, the inner Boolean product of the vectors $char(a_4^o)$ and $char(b_5^r)$ yields the $d_{4,5}$ entry of the Boolean matrix product of the Boolean matrices A^o and B^r in the algorithm. The minimum witness of the entry $d_{4,5}$ is 1 for $o = 1$, $r = 1$, 3 for $o = 1$, $r = 2$, 2 for $o = 2$, $r = 1$, and 4 for $o = 3$, $r = 2$, respectively. For the other combinations of o and r, it is undefined. Hence, $c_{4,5}$ is computed as the minimum of $1 + 5, 3 + 2, 7 + 11, 9 + 7$ which is 5 as required.

We shall call a sequence of integers *uniform* if all its elements have the same value.

A uniform subsequence of a matrix row or column covering all entries in the row or column having the same fixed value is both non-increasing and non-decreasing. If the entries in the row or column can have at most c different values then the row or column can be easily decomposed into at most c uniform subsequences. Hence, if the entries in rows or columns of one of the input matrices range over relatively few different integers then it is sufficient to decompose the rows or columns of the other matrix into relatively few monotone subsequences in order to obtain relatively efficient algorithm for the $(\min, +)$ matrix product. The aforementioned subsequences do not have to be simultaneously non-decreasing or non-increasing as the counterpart subsequences in the first matrix are uniform and hence are both non-decreasing and non-increasing.

Remark 2. Let $A = (a_{i,j})$ and $B = (b_{i,j})$ be two $n \times n$ integer matrices. Next, let a' be a subsequence of the sequence of entries in an i-th row of A and let b' a uniform subsequence of the sequence of entries in an j-th column of B. If a' is non-decreasing and the set $\{a_{i,k} + b_{k,j} | k \in [n] \land a_{i,k} \in a' \land b_{k,j} \in b'\}$ is not empty then the minimum sum in the set is achieved by the pair minimizing the index k. Analogously, if a' is non-increasing then the minimum sum is achieved by the pair maximizing the index k.

Remark 2 yields our second algorithm for the $(\min, +)$ product described in the following theorem. For the algorithm and the proof of the theorem the reader is referred to the full version [17].

Theorem 7. *Let* $A = (a_{i,j})$ *and* $B = (b_{i,j})$ *be two* $n \times n$ *integer matrices. Suppose that at least one of the two following conditions holds:*

1. *the entries in each column of B range over at most c integers and for each $i \in [n]$, there is given a decomposition of the i-th row of A into at most m monotone subsequences;*
2. *the entries in each row of A range over at most c integers and for each $j \in [n]$, there is given a decomposition of the j-th column of B into at most m monotone subsequences,*

Then, the $(\min, +)$ product of A and B can be computed in $O(mcn^{2.569})$ time.

Clearly, if the entries in each row of A range over at most c_a different integers and the entries in each column of B range over at most c_b different integers then the $(\min, +)$ product can be computed in $O(c_a c_b n^\omega)$ time by reduction to $c_a c_b$ Boolean matrix products.

Recall that for an integer sequence s, $mon(s)$ denotes the minimum number of monotone subsequences into which s can be decomposed. By Fact 4, we obtain immediately the following corollary from Theorem 7.

Corollary 1. *Let A and B be two $n \times n$ integer matrices. Let m_1 be the maximum of $mon(d)$ over all sequences d formed by consecutive entries in the rows of A and let m_2 be the maximum of $mon(d)$ over all sequences d formed by consecutive entries in the columns of B. If the entries in each row of A range over at most c_a different integers then the $(\min, +)$ matrix product of A and B can be computed in $O(m_2 c_a n^{2.569} \log n)$ time. Similarly, if the entries in each column of B range over at most c_b different integers than the product can be computed in $O(m_1 c_b n^{2.569} \log n)$ time.*

Note that when m_1 or m_2 does not exceed $n^{0.119}$ then the upper bound of Corollary 1 subsumes that of Fact 5.

Finally, we demonstrate that the case when the rows of the first matrix are non-decreasing and the columns of the second matrix are non-increasing or *vice versa* is as hard as the general case.

Theorem 8. *The problem of computing the* (min, +) *matrix product of two* $n \times n$ *integer matrices* $A = (a_{i,j})$ *and* $B = (b_{i,j})$, *where for* $i \in [n]$, *the rows* $a_{i,1}, ..., a_{i,n}$ *of* A *are non-decreasing and the columns* $b_{1,j}, ..., b_{n,j}$ *of* B *are non-increasing or vice versa is equally hard as computing the product for arbitrary* $n \times n$ *integer matrices.*

Proof. Let M be the maximum absolute value of an entry in the matrices A, B. Transform the matrix A to a matrix A' by setting $a'_{i,k} = a_{i,k} + 2kM$ for i, $k \in [n]$. Observe that each row in A' is non-decreasing. Similarly, define the matrix B' by setting $b'_{k,j} = b_{k,j} - 2kM$ for j, $k \in [n]$. Similarly observe that each column of B' is non-increasing. Now, consider the (min, +) matrix products $C = (c_{i,j})$ of A, B and $C' = (c'_{i,j})$ of A', B'. For i, $j \in [n]$, we have

$$c'_{i,j} = \min\{(a_{i,k} + 2kM) + (b_{k,j} - 2kM) | k \in [n]\} = c_{i,j}.$$

The proof for the case where the rows of the first matrix are non-increasing and the columns of the second matrix are non-decreasing is symmetric. □

We summarize our results on the (min, +) matrix product in Table 1.

4 (min, +) Convolution

If we are given decompositions of the two input n-dimensional vectors a and b into monotone subsequences such that either all the subsequences of a are non-decreasing and all the subsequences of b are non-increasing or *vice versa* then we can use the algorithm depicted in Fig. 2 in order to compute the (min, +) convolution of a and b. First, for each subsequence a^i of a and each subsequence b^j of b, we compute the Boolean vectors $char(a^i)$ and $char(b^j)$ indicating with ones the coordinates of a or b covered by a^i or b^j, respectively. Next, depending if the subsequences are non-decreasing and non-increasing, respectively, or *vice versa*, for each pair of such subsequences a^i and b^j, we compute the minimum witnesses of the Boolean convolution of $char(a^i)$ and $char(b^j)$ or the maximum witnesses of this Boolean convolution, respectively. We use the extreme witnesses to update the current coordinates of the computed (min, +) convolution. The correctness of the reduction to extreme witnesses of the Boolean convolution of $char(a^i)$ and $char(b^j)$ in the algorithm is implied by the following observation.

Remark 3. Let $a = (a_0, ..., a_{n-1})$ and $b = (b_0, ..., b_{n-1})$ be two n-dimensional integer vectors. Next, let a' be a subsequence of $a_0, ..., a_{n-1}$ and let b' be a subsequence of $b_0, ..., b_{n-1}$. For each $k \in \{0, ..., 2n-2\}$, if a' is non-decreasing and b' non-increasing then if the set $\{a_\ell + b_{k-\ell} | a_\ell \in a' \wedge b_{k-\ell} \in b'\}$ is not empty then the minimum sum in the set is achieved by a pair minimizing the index ℓ (thus maximizing $k - \ell$). Analogously, if a' is non-increasing and b' is non-decreasing then the minimum sum is achieved by a pair maximizing the index ℓ (thus, minimizing the index $k - \ell$).

Hence, we obtain the following theorem, correcting Theorem 3.7 in [18].

Input: two n-dimensional vectors $a = (a_0, ..., a_{n-1})$ and $b = (b_0, ..., b_{n-1})$ with integer coordinates and their decompositions into m_a and m_b subsequences a^i and b^j respectively such that either all the subsequences a^i are non-decreasing and all the subsequences b^j are non-increasing or *vice versa*.

Output: the $(\min, +)$ convolution $c = (c_0, ..., c_{2n-2})$ of a and b.

1: **for** each a^i **do**
2:     form a Boolean vector $char(a^i)$ with n coordinates indicating with ones the coordinates of a covered by a^i
3: **end for**
4: **for** each b^j **do**
5:     form a Boolean vector $char(b^j)$ with n coordinates indicating with ones the coordinates of b covered by b^j
6: **end for**
7: initialize the vector $c = (c_0, ..., c_{2n-2})$ by setting all its coordinates to $+\infty$
8: **for** each pair a^i, b^j **do**
9:     **if** the subsequence a^i is non-decreasing **then** compute the minimum witnesses $wit(d_0), ..., wit(d_{2n-2})$ of the Boolean convolution $(d_0, ..., d_{2n-2})$ of $char(a^i)$ and $char(b^j)$
10:     **if if** the subsequence a^i is non-increasing **then** compute the maximum witnesses $wit(d_0), ..., wit(d_{2n-2})$ of the Boolean convolution $(d_0, ..., d_{2n-2})$ of $char(a^i)$ and $char(b^j)$
11:     **for** $k = 0$ **to** $2n - 2$ **do**
12:         **if** $d_k \neq 0$ **then** $c_k \leftarrow \min\{a_{wit(d_k)} + b_{k-wit(d_k)}, c_k\}$
13:     **end for**
14: **end for**
15: $c \leftarrow (c_0, ..., c_{2n-2})$
16: **return** c

Fig. 2. An algorithm for computing the $(\min, +)$ convolution c of two n-dimensional integer vectors a and b given with their decompositions into m_a and m_b subsequences respectively such that either all the subsequences of a are non-decreasing and all the subsequences of b are non-increasing or *vice versa*.

Theorem 9. *Let a and b be two n-dimensional integer vectors given with the decompositions of the sequences of their consecutive coordinates into m_a and m_b monotone subsequences respectively such that either all the subsequences of a are non-decreasing and all the subsequences of b are non-increasing or* vice versa. *The algorithm depicted in Fig. 2 computes the $(\min, +)$ convolution of a and b in $\tilde{O}(m_a m_b n^{1.5})$ time.*

Proof. The proof of the correctness of the algorithm depicted in Fig. 2 is analogous to that of the correctness of the algorithm depicted in Fig. 1. In particular, we obtain the following implication from Remark 3: (***) if $d_k \neq 0$ in line 12 of the algorithm in Fig. 2 then $\min\{a_\ell + b_{k-\ell} | char(a^i)_\ell = 1 \wedge char(b^j)_{k-\ell} = 1\}$ is equal to the first argument of the minimum in this line, i.e., $a_{wit(d_k)} + b_{k-wit(d_k)}$. Hence, none of the coordinates of the output vector has a lower value than the corresponding coordinate of the $(\min, +)$ convolution of a and b. Conversely, if the k coordinate of the $(\min, +)$ convolution of a and b equals $a_\ell + b_{k-\ell}$ then there exists i, j such that $char(a^i)_\ell = 1$ and $char(b^j)_{k-\ell} = 1$. Hence, again by

(***) and line 12 in the algorithm, the c_k coordinate in the output vector has value not larger than the corresponding coordinate of the (min, +) convolution of a and b.

The time complexity analysis of the algorithm in Fig. 2 is also similar to that of the algorithm in Fig. 1. It is dominated by the $m_a m_b$ runs of the $\tilde{O}(n^{1.5})$-time algorithm for the extreme witnesses of the Boolean convolution of two n-dimensional Boolean vectors given in Fact 3.   □

Example 2. We shall assume the notation from the first algorithm in this section. Suppose that $a = (a_0, ..., a_5) = (1, 7, 3, 9, 8, 4)$ and $b = (b_0, ..., b_5) = (13, 7, 11, 5, 10, 12)$. Then, it is easy to see that in the (min, +) vector convolution $(c_0, ..., c_{10})$ of a and b, in particular $c_4 = \min\{a_0 + b_4, a_1 + b_3, a_2 + b_2, a_3 + b_1, a_4 + b_0\} = 11$ holds. Similarly as in Example 1, a can be decomposed into three non-decreasing subsequences $a^1 = (1, , 3, , , 4)$, $a^2 = (, 7, , , 8,)$, $a^3 = (, , , 9, ,)$. On the other hand, b can be decomposed into two non-increasing subsequences $b^1 = (13, , 11, , 10,)$ and $b^2 = (, 7, , 5, , 2)$. Their corresponding characteristic Boolean vectors are $char(a^1) = (1, 0, 1, 0, 0, 1)$, $char(a^2) = (0, 1, 0, 0, 1, 0)$, $char(a^3) = (0, 0, 0, 1, 0, 0)$, and $char(b^1) = (1, 0, 1, 0, 1, 0)$, $char(b^2) = (0, 1, 0, 1, 0, 1)$, respectively. The minimum witness of the entry d_4 in the Boolean vector convolution $(d_0, ..., d_{10})$ of $char(b^i)$ and $char(b^j)$ is 0 for $i = 1$, $j = 1$, 4 for $i = 2$, $j = 1$, 1 for $i = 2$, $j = 2$, and 3 for $i = 3$, $j = 2$, respectively. For the other combinations of $i \in [3]$ and $j \in [2]$, it is undefined. Hence, c_4 is computed as the minimum of $1 + 10, 8 + 13, 7 + 5, 9 + 7$ which is 11 as required.

When the consecutive coordinates of the two input n-dimensional integer vectors are simultaneously non-decreasing or non-increasing the problem of computing the their (min, +) convolution appears to be as hard as in the general case [6].

Fact 10. *[6] The problem of computing the (min, +) convolution of two integer vectors $a = (a_0, ..., a_{n-1})$ and $b = (b_0, ..., b_{n-1},)$, where the sequences $a_0, ..., a_{n-1}$ and $b_0, ..., b_{n-1}$ are both non-decreasing or both non-increasing, is equally hard as computing the convolution for arbitrary n-dimensional integer vectors.*

Proof. Let M be the maximum absolute value of a coordinate in the a, b vectors. Transform the vectors a, b into vectors a', b' by setting $a_i' = a_i + 2iM$ and $b_i' = b_i + 2iM$ for $i = 0, ..., n - 1$. Observe that both sequences $a_0', ..., a_{n-1}'$ and $b_0', ..., b_{n-1}'$ are non-decreasing. Consider the (min, +) convolutions $c = (c_0, ..., c_{2n-2})$ of the vectors a, b and $c' = (c_0', ..., c_{2n-2}')$ of the vectors a', b'. For $k = 0, ..., 2n - 2$, we have

$$c_k' = \min\{(a_\ell + 2\ell M) + (b_{k-\ell} + 2(k - \ell)M) | \ell \in [\max\{k - n + 1, 0\}, \min\{k, n - 1\}]\}$$

$$= c_k + 2kM.$$

Analogously, we can reduce the problem of computing the convolution of two arbitrary n-dimensional integer vectors to that where both input vectors form non-increasing sequences by using the transformation $a_i'' = a_i - 2iM$ and $b_i'' = b_i - 2iM$ for $i = 0, ..., n - 1$.   □

When the entries of one of the input n-dimensional integer vectors range over a relatively few distinct integers then following the general idea of the proof of Lemma 2.1 in [7], we can proceed as follows. First, we can decompose the sequence of consecutive coordinates of the aforementioned vector into a relatively few uniform subsequences. Then, we can sort the coordinates of the other vector and divide the sorted sequence into interval groups of almost equal size. Next, we can run Boolean vector convolution on pairs composed of characteristic Boolean vectors covering with ones a group of the other vector and a uniform subsequence of the first vector, respectively. Further, using the results of the Boolean convolutions, for a fixed uniform subsequence, for $k = 0, ..., 2n - 2$, we can find the group with the smallest index containing an element whose mate belongs to the uniform subsequence. By brute-force search in the group, we can find a smallest element having a mate in the uniform subsequence in order to update the computed k coordinate of the (min, +) convolution. The algorithm is described in the full version [17]. Its correctness is implied by the following observation based on the sorted order of the groups of the other vector and the uniformity of considered subsequences of the first vector.

Remark 4. Let $a = (a_0, ..., a_{n-1})$ and $b = (b_0, ..., b_{n-1})$ be two n-dimensional integer vectors. Next, let $a_0, ..., a_{n-1}$ be divided into subsequences g^i such that no element in g^i is greater that any element in g^{i+1} for $i = 1, 2, ...$ Suppose that b' is a uniform subsequence of $b_0, ..., b_{n-1}$. Then, for $k = 0, ..., 2n - 2$, $\min\{a_q + b_{k-q} | b_{k-q} \in b'\}$ is equal to $\min\{a_q + b_{k-q} | a_q \in g^m \wedge b_{k-q} \in b'\}$, where m is the minimum i such that there is $a_q \in g^i$ for which $b_{k-q} \in b'$.

We obtain the following generalization of Lemma 2.1 in [7], for its proof see the full version [17].

Theorem 11. *Let a and b be two n-dimensional integer vectors. Suppose that the entries of a or the entries of b range over at most h distinct integers. The (min, +) convolution of a and b can be computed in $\tilde{O}(hn^{1.5})$ time.*

Because of the correction of Theorem 3.7 from [18] and Theorem 11, several entries in Table 1 in [18] need to be updated. Table 2 presents the updated version of the table.

Acknowledgments. Thanks go to Alejandro Cassis for pointing the flaw in the statement of Theorem 3.7 in [18], providing Fact 10, and suggesting Theorem 11. This research was partially supported by Swedish Research Council grant 2018-04001.

References

1. Alman, J., Vassilevska Williams, V.: A refined laser method and faster matrix multiplication. In: Proceedings of SODA, pp. 522–539 (2021)

2. Alon, N., Galil, Z., Margalit, D.: On the exponent of the all pairs shortest path problem. J. Comput. Syst. Sci. **54**(2), 255–262 (1997)
3. Bringmann, K., Cassis, A.: Faster knapsack algorithms via bounded monotone min-plus-convolution. In: Proceedings of ICALP, pp. 31:1–31:21 (2022)
4. Bringmann, K., Grandoni, F., Saha, B., Vassilevska Williams, V.: Truly subcubic algorithms for language edit distance and RNA folding via fast bounded-difference min-plus product. SIAM J. Comput. **48**(2), 481–512 (2019)
5. Bremner, D., et al.: Necklaces, convolutions and X+Y. Algorithmica **69**, 294–314 (2014)
6. Cassis, A.: Personal communication (2023)
7. Chan, T.M., He, Q.: More on change-making and related problems. In: Proceedings of ESA 2020, Article No. 29, pp. 29:1–29:14 (2020)
8. Chan, T.M.: More algorithms for all-pairs shortest paths in weighted graphs. SIAM J. Comput. **39**(5), 2025–2089 (2010)
9. Cygan, M., Mucha, M., Wegrzycki, K., Wlodarczyk, M.: On problems equivalent to (min, +)-convolution. ACM Trans. Algorithms **15**(1), 14:1–14:25 (2019)
10. Czumaj, A., Kowaluk, M., Lingas, A.: Faster algorithms for finding lowest common ancestors in directed acyclic graphs. Theor. Comput. Sci. **380**(1–2), 37–46 (2007)
11. Fischer, M.J., Meyer, A.R.: Boolean matrix multiplication and transitive closure. In: Proceedings of 12th Symposium on Switching and Automata Theory, pp. 129–131 (1971)
12. Fisher, M.J., Paterson, M.S.: String-matching and other products. In: Proceedings of 7th SIAM-AMS Complexity of Computation, pp. 113–125 (1974)
13. Fomin, F.V., Kratsch, D., Novelli, J.: Approximating minimum cocolorings. Inf. Process. Lett. **84**, 285–290 (2002)
14. Le Gall, F.: Powers of tensors and fast matrix multiplication. In: Proceedings of 39th International Symposium on Symbolic and Algebraic Computation, pp. 296–303 (2014)
15. Le Gall, F., Urrutia, F.: Improved rectangular matrix multiplication using powers of the coppersmith-winograd tensor. In: Proceedings of SODA 2018, pp. 1029–1046 (2018)
16. Huang, X., Pan, V.Y.: Fast rectangular matrix multiplications and applications. J. Complex. **14**, 257–299 (1998)
17. Lingas, A., Persson, M.: (min, +) Matrix and Vector Products for Inputs Decomposable into Few Monotone Subsequences. CoRR abs/2309.01136 (2023)
18. Lingas, A., Persson, M.: extreme witnesses and their applications. Algorithmica **80**(12), 3943–3957 (2018). (Prel. version in Proc. COCOA 2015)
19. Muthukrishnan, S.: New results and open problems related to non-standard stringology. In: Galil, Z., Ukkonen, E. (eds.) CPM 1995. LNCS, vol. 937, pp. 298–317. Springer, Heidelberg (1995). https://doi.org/10.1007/3-540-60044-2_50
20. Yang, B., Chen, J., Lu, E., Zheng, S.Q.: A comparative study of efficient algorithms for partitioning a sequence into monotone subsequences. In: Cai, J.-Y., Cooper, S.B., Zhu, H. (eds.) TAMC 2007. LNCS, vol. 4484, pp. 46–57. Springer, Heidelberg (2007). https://doi.org/10.1007/978-3-540-72504-6_4
21. Gu, Y., Polak, A., Williams, V.V., Xu, Y.: Faster monotone min-plus product, range mode, and single source replacement paths. In: Proceedings of ICALP 2021, pp. 75:1–75:20 (2021)
22. Williams, V.V.: On some fine-grained questions in algorithms and complexity. In: Proceedings of the ICM, vol. 3, pp. 3431–3472. World Scientific (2018)
23. Williams, V.V., Xu, Y.: Truly Subcubic Min-Plus Product for Less Structured Matrices, with Applications. CoRR abs/1910.04911 (2019)

24. Williams, R.: Faster all-pairs shortest paths via circuit complexity. In: Proceedings of 26th STOC, pp. 664–673. ACM (2014)
25. Yuster, R.: Efficient algorithms on sets of permutations, dominance, and real-weighted APSP. In: Proceedings of 20th SODA, pp. 950–957 (2009)
26. Yuval, G.: An algorithm for finding all shortest paths using $N^{2.81}$ infinite-precision multiplication. Inf. Process. Lett. **11**(3), 155–156 (1976)

A Sub-quadratic Time Algorithm for Computing the Beacon Kernel of Simple Polygons

Binay Bhattacharya, Amirhossein Mozafari$^{(\boxtimes)}$, and Thomas C. Shermer

School of Computing Science, Simon Fraser University, Burnaby, Canada
{binay,amozafar,shermer}@sfu.ca

Abstract. In 2011, Biro et al. [4] initiated the concept of *beacon attraction trajectory* motivated by routing messages in sensor network systems. Let P be a polygonal region such that there is a point particle at each point in P. When we activate a *beacon* at a point $b \in P$, each particle in P greedily moves toward b. For a point $p \in P$, if the particle at p reaches b, we say b *attracts* p. We call a point $b \in P$ a *beacon kernel point* of P if a beacon at b attracts all points in P. The *beacon kernel* of P is defined as the set of all beacon kernel points of P. In 2013 [3] Biro presented a naive quadratic time algorithm to compute the beacon kernel of polygonal domains and showed that this bound is tight. But, obtaining a sub-quadratic time algorithm for computing the beacon kernel of simple polygons remained open. In this paper, we answer to this open problem by presenting an $O(n^{1.5} \log^2 n)$ time algorithm for computing the beacon kernel of simple polygons.

1 Introduction

Studying the behaviour of point particles in a polygonal region under the influence of an attraction actuator, called *beacon*, is an active area in computational geometry due to its applications in several computer science branches such as robot motion planning and network systems [1,10,11]. The problem first appeared in the context of sensor network systems in early 2000s [10]. Consider a network of sensors in a polygonal region P that gather information and send it to a destination point (base) b in the region. Each sensor has a range such that it only can pass a message to the sensors within its range (we call these sensors *neighbor sensors*). Greedy routing protocol is widely used in such circumstances as each sensor only needs to know the location of itself, the base and its neighbor sensors. Specifically, each sensor passes its message to the neighbor sensor that is closest to b (if all the neighbor sensors are farther to b than itself the sensor does not pass its message). Two main problems can arise here. The first problem is determining the sensors that can successfully send their messages to the base using the above greedy protocol and the other is determining the locations for a base such that all sensors can successfully send their information to the base. If we assume that P is uniformly filled with sensors and the range of each sensor

© The Author(s), under exclusive license to Springer Nature Switzerland AG 2024
W. Wu and G. Tong (Eds.): COCOON 2023, LNCS 14423, pp. 69–81, 2024.
https://doi.org/10.1007/978-3-031-49193-1_6

is infinitely small, then for each pair of points (p, b) in P, we can assign a path inside P that indicates the trajectory of a message that a sensor at p tries to send to b. Note that p can successfully send its message to b if and only if this path ends up at b.

In general, we use the *beacon/particle* terminology to model such problems. In this terminology, beacons and particles are *pointwise* objects which means at each time they can reside in only one point of P. We assume that initially there is a particle at each point in P and beacon is an attraction actuator that exerts magnetic pull on the particles. Without any confusion, when we say *a point p in P*, based on the context, we either refer to the location p in the polygon or the particle that initially resides at p. When we activate a beacon at a point $b \in P$, the particles inside P *greedily* move toward b (we assume that the particles do not interact with each other). Specifically, consider the particle initially resides at a point $p \in P$. At each time, the particle moves in a direction that maximizes the reduction of its distance from b while it remains inside the polygon. If the particle reaches b or such a direction does not exist, the particle stands still. In the first case, we say b *attracts* p. The attraction region of b denoted by $A(b)$ is defined as the set of points in P that can get attracted by b. For a point $p \in P$, the inverse attraction region of p is $\{b \in P : p \in A(b)\}$. $b \in P$ is called a *(beacon) kernel point* of P if it attracts all points in P $(A(b) = P)$. The *(beacon) kernel* of P is defined as the set of all kernel points of P and is denoted by $Ker(P)$. Figure 1, depicts an example of $Ker(P)$ which as we can see, it may not form a connected region.

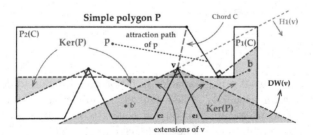

Fig. 1. $Ker(P)$ is shown as the green regions. $DW(v)$ is showed in gray. C is a chord that divides P into two sub-polygons $P_1(C)$ and $P_2(C)$. (Color figure online)

1.1   Previous Works

The concept of beacon attraction trajectory was first introduced by Biro et al. in 2011 [3,5] as a framework to address problems involving greedy routing toward a destination point in polygonal regions. A polygonal region is called a simple polygon when it has no hole and its boundary does not intersect itself. They gave an $O(n)$ (resp. $O(nh)$) time algorithm for computing the attraction region of a point $b \in P$ when P is a simple polygon with n vertices (resp. a polygonal domain with n vertices and h holes). They also proposed an $O(n^2)$ time algorithm for computing the inverse attraction region of a given point p in polygonal domains. In addition, they proposed a naive $O(n^2)$ time algorithm for computing the beacon kernel of polygonal domains. They showed that for polygonal domains, the kernel might have quadratic number of vertices but for simple polygons, this bound is linear [3]. Since then, obtaining a sub-quadratic time algorithm for computing the bacon kernel of simple polygons remained as an open problem. In 2015, Kouhestani et al. presented an $O(n \log n)$ (resp. $O(n)$) time algorithm for computing the inverse attraction region of a point in monotone polygons (resp. polygonal terrains) [13]. In 2018, Kostitsyna et al. obtained an $O(n \log n)$ time algorithm for computing the inverse attraction region of a point in simple polygons and showed that this bound is optimal [12]. In 2019, Bae et al. [2] studied rectilinear polygons and showed that $\lfloor \frac{n}{6} \rfloor$ (resp. $\lfloor \frac{n+4}{8} \rfloor$) beacons are always sufficient and sometimes necessary to attract any point in simple (resp. monotone) rectilinear polygons. They also showed that the beacon kernel of rectilinear polygons can be computed in linear time. In [6] Bose and Shermer introduced the concept of *attraction-convex* polygons. A polygon P is called attraction-convex if any point $b \in P$ can attract any point $p \in P$ ($P = Ker(P)$). They provided a linear time algorithm to detect whether a simple polygon is attraction-convex.

Despite recent studies on various problems regarding beacon based trajectory of points in polygonal regions, it is still unknown whether the beacon kernel of simple polygons can be computed in sub-quadratic time. Let us call this problem the *beacon kernel problem (BKP)*. In this paper, we address this problem by providing the first sub-quadratic time algorithm to solve the BKP.

2 Preliminaries

Let us assume that P is a given simple polygon with n vertices and (p, b) is a pair of points in P where we have a beacon at b. Also, let $\vec{p_b}$ be the unit vector from p toward b. We call a unit vector $\vec{z}$ a *greedy direction* of p with respect to b if:

1. $\vec{z}$ (located at p) points toward the inside of P (including its boundary).
2. $\vec{z} \cdot \vec{p_b} > 0$ (inner product of two vectors).
3. $\vec{z} \cdot \vec{p_b}$ is maximum over all unit vectors satisfying the first two conditions.

If a point $d \in P$ does not have a greedy direction with respect to b, we call d a *dead point* of P with respect to b. Note that if p is a vertex of P, p might have more than one greedy direction with respect to b. We call a path τ in P an *attraction path* of p with respect to b if it starts at p and at each point $x \in \tau$, the path continues along a greedy direction of x with respect to b. We say b *attracts* p if there is an attraction path of p with respect to b that ends at b.

Observation 1. *If b attracts p, the attraction path of p (with respect to b) that ends at b is unique.*

This follows from the fact that if two attraction paths from p ends at b, P must have a *hole* enclosed by the paths that contradicts the simplicity of P. Henceforth, if b attracts p, we consider the attraction path of p (with respect to b) as the one that ends at b. In Fig. 1, the attraction path from p with respect to b is depicted by the red dashed line segments.

Let τ be a path in P starting from a point p. For two points x_1 and x_2 in τ, let us denote the distance between x_1 and x_2 along τ by $d_\tau(x_1, x_2)$. We say $x_1 < x_2$ if $d_\tau(p, x_1) < d_\tau(p, x_2)$. We say τ is a *distance decreasing* path with respect to b if and only if for any two points $\{x_1, x_2\} \in \tau$ if $x_1 < x_2$ then $d(x_1, b) > d(x_2, b)$ where $d(x_1, b)$ (similarly $d(x_2, b)$) is the Euclidean distance between x_1 and b. Based on the definition of an attraction path, we have the following observation:

Observation 2. *Any attraction path of p with respect to b is distance decreasing.*

Therefore, if there is no distance decreasing path from p to b, then b can not attract p. A *chord* C of P is defined as a line segment such that its endpoints lie on ∂P (the boundary of P) and its interior completely lies in the interior of P. C divides P into two sub-polygons. We denote these sub-polygons by $P_1(C)$ and $P_2(C)$ (see Fig. 1). For a reflex vertex v with incident edges e_1 and e_2, consider the two half-planes $H_1(v)$ and $H_2(v)$ induced by the lines perpendicular to $sup(e_1)$ (the supporting line of e_1) and $sup(e_2)$ at v containing e_1 and e_2 respectively. The *dead wedge* of v is defined as the interior of the cone induced by $H_1(v) \cap H_2(v)$ and is denoted by $DW(v)$. Also, we define the *perpendicular extensions (for short extensions)* of v as the two half-lines from v perpendicular to e_1 and e_2 respectively enclosing $DW(v)$. See Fig. 1 for an example of $H_1(v)$, $DW(v)$ and the extensions of v.

Observation 3 *[3]. If a beacon lies on $DW(v)$, it can not be a kernel point.*

For example, in Fig. 1, the point $b' \in DW(v)$ can not be a kernel point because it can not attract the points in the interior of e_1. Indeed, the above observation not only gives us a necessary condition for a point to be a kernel point but also gives us a sufficient condition:

Theorem 1. *A point $b \in P$ is a kernel point if and only if it is not contained in the dead wedge of any reflex vertex of P.*

The proof of the above theorem can be found in [3]. In order to simplify our algorithm, we assume that all given points are in general position by which we mean no three points are collinear. First, we consider the *discrete beacon kernel problem (DBKP)* and provide a sub-quadratic time algorithm to solve it. Next, we use our algorithm for solving the DBKP to obtain a sub-quadratic time algorithm for the BKP.

3 The Discrete Beacon Kernel Problem

Let P be a given simple polygon with n vertices and $\mathcal{X}$ be a given set on m points in P. In the DBKP, we want to find out which points from $\mathcal{X}$ are kernel points for P. The idea for solving the DBKP is applying Theorem 1 to the points in $\mathcal{X}$ to see which points *survive* from all reflex vertices of P (not lying in the dead wedge of any reflex vertex). For a reflex vertex v, we say x is *eliminated* by v if x lies in $DW(v)$. In order to compute the points in $\mathcal{X}$ that are eliminated by at least one reflex vertex, we use the following theorem from Matoušek [14]:

Theorem 2. *For a given set of m points in $\mathbb{R}^2$, one can preprocess it in $O(m^{1+\delta})$ time and $O(m)$ space for simplex range searching such that each query can be answered in $O(m^{0.5} \log m)$ time*[1].

Here, $\delta > 0$ is an arbitrary fixed number. The idea behind Theorem 2 is recursively building a *simplicial partition* (partitioning of the plane into a set of simplices each simplex contains a suitable number of points) to form a partition tree $\mathcal{T}$ with $O(\log m)$ height. When a query simplex Q is given, we start from the root of $\mathcal{T}$ and in each internal node of the tree, we proceed into the children whose corresponding regions *crosses* Q (has non-empty intersection with both inside and outside of Q). Therefore, according to the above theorem, by spending $O(m^{1+\delta})$ time for preprocessing, we can specify the points in $\mathcal{X}$ eliminated by a given reflex vertex in $O(m^{0.5} \log m)$ time described as an union of simplices containing the eliminated points (we do not need to report the points here). We can modify this data structure to detect the points lying in $\{\cup DW(v) : v \text{ is a reflex vertex of P}\}$ as follows: after preprocessing and building the partition tree, we query each $DW(v)$ where v is a reflex vertex. But, when a simplex is completely lies in $DW(v)$, we *mark* it as eliminated. After querying all dead wedges, for each point $x \in \mathcal{X}$, we perform a point location to find the *leaf simplex* (a simplex corresponding to a leaf-node of $\mathcal{T}$) containing it. Next, we traverse $\mathcal{T}$ from the corresponding leaf-node to the root and if we see any marked simplex, we report x as eliminated. Based on Theorem 1, the points of $\mathcal{X}$ that do not get eliminated are the kernel points. Because the height of $\mathcal{T}$ is $O(\log m)$, this step can be done in $O(m \log m)$ time.

Theorem 3. *The discrete beacon kernel problem can be solved in $O(m \log m + nm^{0.5} \log m)$ time.*

4 The Beacon Kernel Problem

Our general framework to solve the BKP is as follows: we first introduce a data structure for P called the *split decomposition tree (SDT)* similar to the polygon-cutting decomposition of Chazelle [7]. Next, we use the SDT to build a set of *candidate points* $\mathcal{K}$ for the vertices of $Ker(P)$ and store them in an appropriate

---

[1] In the original paper, this time complexity is $O(m^{0.5} 2^{O(\log^* m)})$. For the sake of simplicity, here we use $O(m^{0.5} \log m)$.

partial order. Finally, we run our algorithm for solving the DBKP on $\mathcal{K}$ to obtain a set of kernel point $\mathcal{K}^*$. We show that the points of $\mathcal{K}^*$ are indeed the vertices of $Ker(P)$. Using $\mathcal{K}^*$ and the partial order on it, we can construct $Ker(P)$.

4.1   The Split Decomposition Tree of P

We start by computing a triangulation Δ of P and its dual graph $\mathbb{T}_\Delta$. The triangulation of P can be done in linear time using Chazelle's polygon triangulation algorithm [8]. $\mathbb{T}_\Delta$ is a graph for which there is a node corresponding to each triangle in Δ and two nodes in $\mathbb{T}_\Delta$ are connected by an edge if their corresponding triangles share an edge. Note that because P is simple, $\mathbb{T}_\Delta$ is a tree (this is because we can always embed $\mathbb{T}_\Delta$ in P such that each node of $\mathbb{T}_\Delta$ lies in its corresponding triangle in Δ). We can see that each subtree $T \subseteq \mathbb{T}_\Delta$ corresponds to a connected region in P which is obtained by the union of the triangles corresponding to the nodes in T. The *centroid* of $\mathbb{T}_\Delta$ is defined as a node for which by removing it (and its incident edges) from $\mathbb{T}_\Delta$, the maximum size of each connected component is minimum. Therefore, the size of each connected component of the remaining graph is at most $\lceil |\mathbb{T}_\Delta|/2 \rceil$. Note that because the degree of each node of $\mathbb{T}_\Delta$ is at most three, by removing a node, we would have at most three connected components. In order to avoid confusion in our algorithm, we always assume that the trees are rooted and if there are multiple choices for selecting a centroid, we choose the one closest to the root. Based on this assumption, the centroid of $\mathbb{T}_\Delta$ (and each of its subtrees) is unique and can be computed in linear time [15]. Suppose that $|\mathbb{T}_\Delta| > 2$ ($|\mathbb{T}_\Delta|$ is the number of nodes in $\mathbb{T}_\Delta$) and c is the centroid of $\mathbb{T}_\Delta$. Also, suppose that c has degree three (resp. two) and $\{T_1, T_2, T_3\}$ (resp. $\{T_1, T_2\}$) are the subtrees of $\mathbb{T}_\Delta$ emanating from removing c from $\mathbb{T}_\Delta$ such that T_1 has the greatest size. We say two subtrees $\{S_1, S_2\}$ are obtained from *splitting* $\mathbb{T}_\Delta$ over c if $S_1 = T_1$ and S_2 is the join of T_2 and T_3 by c (resp. $S_2 = T_2 \cup \{c\}$). Based on the definition of centroid, we have $|S_1| \geq |\mathbb{T}_\Delta|/3$ and $|S_2| \leq 2|\mathbb{T}_\Delta|/3$.

Next, we build a data structure called the *split decomposition tree* for $\mathbb{T}_\Delta$ denoted by $SDT(\mathbb{T}_\Delta)$ which is a binary tree such that each of its nodes corresponds to a subtree of $\mathbb{T}_\Delta$. We build $SDT(\mathbb{T}_\Delta)$ recursively by splitting the subtrees of $\mathbb{T}_\Delta$ over their centroids starting from $\mathbb{T}_\Delta$. For each node $\omega \in SDT(\mathbb{T}_\Delta)$, let us denote the sub-polygon of P corresponding to its subtree by $P(\omega)$. If ω is an internal node with two children ω_1 and ω_2, we call the chord separating $P(\omega_1)$ and $P(\omega_2)$, *the chord corresponding to* ω and we store it in ω.

Proposition 1. $SDT(\mathbb{T}_\Delta)$ *can be constructed in* $O(n \log n)$ *time.*

This is because the height of $SDT(\mathbb{T}_\Delta)$ is $O(\log n)$ and the subtrees in each level of $SDT(\mathbb{T}_\Delta)$ are disjoint. Let $\mathcal{C}$ be the set of all chords corresponding to the internal nodes of $SDT(\mathbb{T}_\Delta)$. Note that because $SDT(\mathbb{T}_\Delta)$ has linear number of nodes, the number of chords in $\mathcal{C}$ is also linear. For a node $\omega \in SDT(\mathbb{T}_\Delta)$, we say ω *contains* p if $p \in P(\omega)$. Let $L(v)$ be the set of leaf-nodes of $SDT(\mathbb{T}_\Delta)$ containing a vertex v, then we have:

$$\sum_{v \text{ is a vertex}} \sum_{P' \in \text{sub-polygons of } L(v)} |P'| = O(n) \tag{1}$$

where $|P'|$ is the number of vertices in P'. This is because we have $O(n)$ leaf-node sub-polygons in P and each of these sub-polygons contains $O(1)$ number of vertices. We say a chord C separates a point p from a vertex v if $p \in P_1(C)$ and $v \in P_2(C)$ or vice versa. Now, for a vertex v, let $\mathcal{C}_v \subseteq \mathcal{C}$ be the set of chords corresponding to the nodes in $SDT(\mathbb{T}_\Delta)$ on the path(s) from the root to the leafs in $L(v)$. Here, the construction of $SDT(\mathbb{T}_\Delta)$ implies the following proposition:

Proposition 2. *For any point p and any vertex v of P (except possibly an $O(1)$ number of vertices), a chord in $\mathcal{C}_v$ separates p from v.*

Note that based on our definition, if p lies on a chord, the chord separates it from all vertices in the polygon.

4.2   Computing Candidate Kernel Points

According to Theorem 1, each vertex of $Ker(P)$ (other than a vertex of P) is either the intersection of two extensions or the intersection of an extension and an edge of P. Using this property, we build a set $\mathcal{K}$ of candidate points such that if κ is a vertex of $Ker(P)$ then $\kappa \in \mathcal{K}$. In order to build $\mathcal{K}$, we consider each pair (w, E_w) where w is a vertex of P and E_w is an extension (if w is a reflex vertex) or an edge incident to w. We compute a set $\mathcal{K}(E_w)$ of candidate points on E_w such that:

$$\mathcal{K} = \bigcup_{w \text{ is a vertex of } P} \{\mathcal{K}(E_w) : \ E_w \text{ is an extension or an edge incident to } w\}$$

Note that if κ is a vertex of $Ker(P)$ on E_w such that $\kappa \notin \mathcal{K}(E_w)$, then κ should be the intersection of E_w and E_v for some vertex v and $\kappa \in \mathcal{K}(E_v)$. Also, we store the points in $\mathcal{K}(E_w)$ in sorted order based on their distances to w. Suppose that a pair (w, E_w) is given and we are going to compute $\mathcal{K}(E_w)$. In order to do that, we provide an algorithm to compute the candidates on E_w *with respect to a given chord C* denoted by $\mathcal{K}_C(E_w)$. Then, we show that how we can use this algorithm to build $\mathcal{K}(E_w)$. In the following, we discuss this approach in detail.

Let us assume that C is a given chord with two induced sub-polygons $P_1(C)$ and $P_2(C)$ such that $w \in P_2(C)$ (the case $w \in P_1(C)$ is symmetrical). We build $\mathcal{K}_C(E_w)$ as a set of points with constant size such that if $\kappa \in P_2(C)$ is a vertex of $Ker(P)$ obtained by the intersection of E_w and an extension of a reflex vertex in $P_1(C)$, then $\kappa \in \mathcal{K}_C(E_w)$. In order to build $\mathcal{K}_C(E_w)$, we introduce two types of partitions. One is with respect to the given chord C and we call it the *C-partition*. The other is with respect to a reflex vertex $v \in P_1(C)$ which is called the *v-partition*. In the following, we define these two types of partitions.

The C-Partition for $P_2(C)$: Suppose that C is horizontal such that $P_1(C)$ lies below C (in the neighborhood of C). We first compute all chords of $P_2(C)$ induced by $sup(C)$. Next, we modify the triangulation Δ on $P_2(C)$ such that each of these new chords becomes an edge of the triangulation. We denote this new triangulation for $P_2(C)$ by $\Delta_2(C)$. According to [16], $\Delta_2(C)$ can be computed in $O(|P_2(C)|)$ time. Let $\mathbb{T}_2(C)$ be the dual graph of $\Delta_2(C)$ rooted at a node ρ corresponding to the triangle incident to C. For a node $\nu \in \mathbb{T}_2(C)$, we say ν crosses $sup(C)$ from left (resp. right), if the sequence of triangles corresponding to the path from ρ to ν crosses (by the first time) $sup(C)$ from the left (resp. right) side of C. If the path does not cross $sup(C)$ we say ν does not cross $sup(C)$. The C-partition of $P_2(C)$ is defined as the following three regions (See Fig. 2(a)):

1. $P_2^{up}(C)$ is the union of the triangles corresponding to the nodes that do not cross $sup(C)$.

2. $P_2^{left}(C)$ (resp. $P_2^{right}(C)$) is the union of triangles corresponding to the nodes that cross $sup(C)$ from the left (resp. right) side of C.

The v-Partition of P: Suppose that v is a reflex vertex such that $\{e_1, e_2\}$ are its incident edges. Also, let C_{e_1} and C_{e_2} be the two chords from v along $sup(e_1)$ and $sub(e_2)$. These chords divides P into three sub-polygons. Let us denote the sub-polygon containing e_1 (resp. e_2) by $S(e_1)$ (resp. $S(e_2)$). Also, we call the sub-polygon containing neither of e_1 and e_2, the sub-polygon in front of v denoted by $S(v)$ (see Fig. 2(b)).

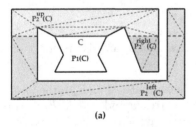
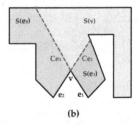

(a)   (b)

Fig. 2. (a) The C-partition of $P_2(C)$ (b) the v-partition of P.

We recall that the half-plane containing e_1 (resp. e_2) with the boundary passing v and perpendicular to e_1 (resp. e_2) is denoted by $H_1(v)$ (resp. $H_2(v)$). Let us denote the half-lines from v containing C_{e_1} and C_{e_2} by $\bar{C}_{e_1}$ and $\bar{C}_{e_2}$ respectively. Also, for a point b and an edge e, we denote the perpendicular projection of b on $sup(e)$ by $h_b(e)$.

Lemma 1. *For any reflex vertex $v \in P_1(C)$ with incident edges e_1 and e_2, we have:*

1. *If b is a point in the interior of $H_1(v)$ (resp. $H_2(v)$), then there is no attraction path from a point in the interior of $e_1 \cap [v, h_b(e_1)]$ (resp. $e_2 \cap [v, h_b(e_2)]$) to b.*
2. *For a point $b \in S(v)$, if $b \in DW(v)$, then there is a reflex vertex $v' \neq v$ such that $b \in DW(v')$.*

Proof. 1) Suppose that b is a beacon in $H_1(v)$ and t is a point in the interior of $e_1 \cap [v, h_b(e_1)]$ (the case for e_2 is similar). Note that t exists because $b \in H_1(v)$ as in Fig. 3 (a). Also, suppose that there is an attraction path π_t from t to b that passes $\bar{C}_{e_1}$ at a point x. But in this case, $d(b,t) < d(b,x)$ which means π_t is not a distance decreasing path with respect to b that contradicts Observation 2.

2) Suppose that $b \in DW(v)$ and B is the visible portion of ∂P from b (as in Fig. 3 (b)). Note that the points on B can directly get attracted by b. Let us assume that the sequence of triangles corresponding to the path in $\mathbb{T}_\Delta$ between the two nodes containing v and b enters $DW(v)$ by passing the extension perpendicular to e_2 as in Fig. 3 (b) (the other case is similar). Let $\gamma = vb'$ (b' is an endpoint of B) be the portion of ∂P from v to B starting with e_1. Because $b \in S(v)$, γ intersects $\bar{C}_{e_2}$ after leaving v. This implies that γ can not be distance decreasing with respect to b (because of the right angle between $\bar{C}_{e_2}$ and the extension of v perpendicular to e_2). Therefore, there must be the last (from b') vertex $v' \in \gamma$ such that $v'b'$ is distance decreasing but this happens only when $b \in DW(v')$. □

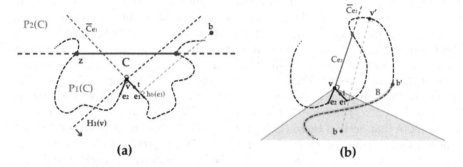

(a)   **(b)**

Fig. 3. In this figure, the dashed black curves represent any combination of vertices. In (a), b can not attract t by a path passing $\bar{C}_{e_1}$ and in (b), b lies in the dead wedge of v' because the points of ∂P around v' go into different directions by activating a beacon at b.

Note that in the above proof $d_\gamma(v', b') < d_\gamma(v, b')$. Suppose that $\mathcal{R}$ is a region of the C-partition of $P_2(C)$. Here, we show that each reflex vertex $v \in P_1(C)$ can generate at most one half-plane $H_\mathcal{R}(v)$ called the *eliminating half-plane of v* on $\mathcal{R}$ such that for any point $b \in \mathcal{R}$, if $b \in H_\mathcal{R}(v)$, then b can not be a kernel point. Furthermore, if $b \notin H_\mathcal{R}(v)$, b can not be eliminated by v. Based on the part 1 of Lemma 1, if $P_2(C) \subseteq S(e_1)$ (resp. $P_2(C) \subseteq S(e_2)$), then $H_2(v)$ (resp. $H_1(v)$) is the eliminating half-plane of v on each region of the C-partition of $P_2(C)$. On the other hand, based on the part 2 of the lemma, if $P_2(C)$ contained in $S(v)$, then v does not need to generate any eliminating half-plane on $P_2(C)$ (because all points in $P_2(C) \cap DW(v)$ also gets eliminated by another reflex vertices along ∂P). Therefore, we only need to care about the case where C intersects C_{e_1} or C_{e_2} (or both).

Observation 4. *If C intersects C_{e_1} or C_{e_2}, then at most one extension of v can intersect $sup(C)$.*

The reason is that the extensions of v make right angles with C_{e_1} and C_{e_2}. For $P_2^{up}(C)$, if none of the extensions of v intersect $sup(C)$, $P_2^{up}(C) \cap DW(v) = \emptyset$ and so, v does not generate an eliminating half-plane for $P_2^{up}(C)$. Otherwise, if the extension of v perpendicular to e_1 (resp. for e_2) intersects C, $H_1(v)$ (resp. $H_2(v)$) would be the eliminating half-plane for $P_2^{up}(C)$. Note that in this case, Observation 4 implies that the intersection of $P_2^{up}(C)$ and this eliminating half-plane lies in $DW(v)$ and therefore, the eliminating half-plane correctly eliminates the points of $P_2^{up}(C)$ in it.

For $P_2^{left}(C)$ ($P_2^{right}(C)$ is symmetrical), suppose that z is the left vertex of C. If z lies in the sub-polygon in front of v, then $P_2^{left}(C)$ is a subset of $S(v)$ and therefore, v does not need to generate any eliminating half-plane on $P_2^{left}(C)$ (part 2 of Lemma 1). Now, suppose that $z \in S(e_2)$ (the case $z \in S(e_1)$ is symmetrical). Then, if $b \in H_1(v) \cap P_2^{left}(C)$, any path from t (a point in the interior of e_1) to b should pass $\bar{C}_{e_1}$ (see Fig. 3 (a)). Therefore, by the part 1 of Lemma 1, $H_1(v)$ would be the eliminating half-plane for $P_2^{left}(C)$. Using the algorithm of Chazelle et al. in [9], by spending $O(n \log n)$ time for preprocessing P, we can compute the emanating chords of each reflex vertex in $O(\log n)$ time. Therefore, we can compute all eliminating half-planes of the reflex vertices in $P_1(C)$ on the regions of the C-partition in $O(|P_1(C)| \log n)$ time. Finally, for each of these regions, we compute the union of the eliminating half-planes. Let us denote the polygonal chains corresponding to the boundaries of these unions in $P_2^{up}(C)$, $P_2^{left}(C)$ and $P_2^{right}(C)$ by $\mathcal{W}_2^{up}(C)$, $\mathcal{W}_2^{left}(C)$ and $\mathcal{W}_2^{right}(C)$ respectively (note that each of these chains are convex). Using the divide-and-conquer schema, these chains can be computed in $O(|P_1(C)| \log |P_1(C)|)$ time. Now, having a pair (w, E_w), we set $\mathcal{K}_C(E_w)$ as the intersections (if any) of E_w with $\mathcal{W}_2^{up}(C)$, $\mathcal{W}_2^{left}(C)$ and $\mathcal{W}_2^{right}(C)$ (which implies $|\mathcal{K}_C(E_w)| \leq 3$). Because of the convexity of the chains, $\mathcal{K}_C(E_w)$ can be computed in $O(\log |P_1(C)|)$ time. The above discussion gives us the following proposition:

Proposition 3. *We can compute $\mathcal{K}_C(E_w)$ and $\mathcal{K}_C(E_v)$ for all vertices $w \in P_2(C)$ and $v \in P_1(C)$ in $O(n \log n)$ time.*

Instead of the entire polygon P, we can work with a sub-polygon $P(\omega) \subseteq P$ where $\omega \in SDT(\mathbb{T})$ is an internal node with two children ω_1 and ω_2. Let C' be the chord stored in ω (C' divides $P(\omega)$ into $P(\omega_1)$ and $P(\omega_2)$). In this case, if we assume that w is a vertex in $P(\omega_2)$, we denote the set of candidate points induced by the reflex vertices of $P(\omega_1)$ on E_w by $\mathcal{K}_{C'}^{P(\omega)}(E_w)$ (the case $w \in P_1(C)$ is symmetrical). Algorithm 1 generates $\mathcal{K}(E_w)$ (the set of all candidate points on E_w) sorted based on their distances to w. In addition, for each point in $\mathcal{K}(E_w)$, we store the pair of vertices (one of them is w) creating it in the point. Finally, we set $\mathcal{K}$ as the union of all $\mathcal{K}(E_w)$ where w is a vertex of P and E_w is an extension or an edge incident to w.

Algorithm 1 BUILD_CANDIDATE (E_w)

1: Let $\mathcal{K}(E_w) = \emptyset$.
2: Let Π_w be the set of root-leaf paths in $SDT(\mathbb{T}_\Delta)$ to the leafs containing w.
3: **for** each π in Π_w **do**
4:     **for** each internal node $\omega \in \pi$ with corresponding chord C' **do**
5:         Let ω_1 and ω_2 be the children of ω such that $w \in P(\omega_2)$.
6:         Add $\mathcal{K}_{C'}^{P(\omega)}(E_w)$ to $\mathcal{K}(E_w)$.
7:     **end for**
8:     Let ℓ be the leaf-node of π.
9:     Add the intersections of E_w with the extensions of the reflex vertices in $P(\ell)$ to $\mathcal{K}(E_w)$.
10: **end for**
11: Sort the points in $\mathcal{K}(E_w)$ based on their distances to w.
12: **return** $\mathcal{K}(E_w)$.

Theorem 4. *If κ is a vertex of $Ker(P)$ and not a vertex of P then $\kappa \in \mathcal{K}$.*

Proof. Suppose that κ is a vertex of $Ker(P)$ such that it is not a vertex of P. Therefore, κ should be the intersection of an extension E_v incident to a reflex vertex v and an extension (or an edge) E_w incident to a vertex w. We need to prove that $\kappa \in \mathcal{K}(E_w) \cup \mathcal{K}(E_v)$. Suppose that $\omega_{split} \in SDT(\mathbb{T}_\Delta)$ is the node where the two paths from the root to the leafs containing w and v split (if such a node does not exist, w and v lie on the same leaf-node sub-polygon and we catch such candidates in the line 9 of Algorithm 1). Let ω_1 and ω_2 be the children of ω_{split} such that $w \in P(\omega_2)$. Suppose that $\kappa \in P(\omega_2)$ (note that if E_w is an incident edge of w, then κ always lie in $P(\omega_2)$ but if $\kappa \in P(\omega_1)$, we catch it while processing v). Then $\kappa \in \mathcal{K}_{C'}^{P(\omega)}(E_w)$ and thus, based on line 6 of Algorithm 1, $\kappa \in \mathcal{K}(E_w)$ and so, $\kappa \in \mathcal{K}$. $\square$

Proposition 4. *The set of candidate points $\mathcal{K}$ can be computed in $O(n \log^2 n)$ time.*

Proof. In each level of $SDT(\mathbb{T}_\Delta)$, the corresponding sub-polygons of the nodes are internally disjoint. For each internal node $\omega \in SDT(\mathbb{T}_\Delta)$ with two children ω_1 and ω_2, we need to compute $\mathcal{K}_{C'}^{P(\omega)}(E_w)$ and $\mathcal{K}_{C'}^{P(\omega)}(E_v)$ for all vertices $v \in P(\omega_1)$, $w \in P(\omega_2)$ and their extensions and incident edges. This costs $O(|P(\omega)| \log |P(\omega)|)$ and so $O(n \log n)$ for all nodes in a level of $SDT(\mathbb{T}_\Delta)$. Because $SDT(\mathbb{T}_\Delta)$ has $O(\log n)$ levels, the total time complexity of computing $\mathcal{K}$ would be $O(n \log^2 n)$. In addition, the total number of vertices in the regions corresponding to the leafs of $SDT(\mathbb{T}_\Delta)$ is linear (here, a vertex may count more than once due to its presence in more than one such regions). For each of such vertices, its corresponding root-leaf path in Algorithm 1 generates $O(\log n)$ candidate points. Therefore, we have $O(n \log n)$ candidate points in $\mathcal{K}$ and the total cost of sortings is $O(n \log^2 n)$. $\square$

4.3   Building the Beacon Kernel

After computing $\mathcal{K}$, we first run the DBKP algorithm on it to find a subset $\mathcal{K}^*$ which are the beacon kernel points in $\mathcal{K}$. Note that for each node $\omega \in SDT(\mathbb{T}_\Delta)$, we have $O(|P(\omega)|)$ candidates in $\mathcal{K}$ and thus, $|\mathcal{K}| = O(n \log n)$. Based on Theorem 3, $\mathcal{K}^*$ can be computed in $O(n^{1.5} \log^2 n)$ time. Having $\mathcal{K}^*$, we pick a point $\kappa \in \mathcal{K}^*$. We know that κ is the intersection of two extensions or an extension and an edge of P. Specifically, let κ be the intersection of E_v and E_w for two vertices v and w. For each of E_v and E_w, one side of κ can not be in $Ker(P)$ because it is either lies in the dead wedge of v or w or outside of P (depending on whether E_v or E_w is an extension or an edge). This implies that κ is a vertex of $Ker(P)$. Now, the candidate points on E_v and E_w are sorted along on E_v and E_w. So, by traversing these candidate points from κ along E_v and E_w in the direction that does not get eliminated, we can get the vertices of $Ker(P)$ incident to κ. By repeating this process, we build all the edges of the connected component of $Ker(P)$ containing κ and therefore the component itself. We remove all vertices of the component from $\mathcal{K}^*$ and pick a new $k \in \mathcal{K}^*$ (we can sort $\mathcal{K}^*$ once to facilitate this operation) and repeat the above process until all components of $Ker(P)$ are built.

Theorem 5. *Given a simple polygon P with n vertices, $Ker(P)$ can be computed in $O(n^{1.5} \log^2 n)$ time.*

References

1. Al-Karaki, J.N., Kamal, A.E.: Routing techniques in wireless sensor networks: a survey. IEEE Wirel. Commun. **11**(6), 6–28 (2004)
2. Bae, S.W., Shin, C.S., Vigneron, A.: Tight bounds for beacon-based coverage in simple rectilinear polygons. Computat. Geom. **80**, 40–52 (2019)
3. Biro, M.: Beacon-based routing and guarding. PhD thesis, State University of New York at Stony Brook (2013)
4. Biro, M., Gao, J., Iwerks, J., Kostitsyna, I., Mitchell, J.S.: Beacon-based routing and coverage. In: 21st Fall Workshop on Computational Geometry (2011)
5. Biro, M., Iwerks, J., Kostitsyna, I., Mitchell, J.S.B.: Beacon-based algorithms for geometric routing. In: Dehne, F., Solis-Oba, R., Sack, J.-R. (eds.) WADS 2013. LNCS, vol. 8037, pp. 158–169. Springer, Heidelberg (2013). https://doi.org/10.1007/978-3-642-40104-6_14
6. Bose, P., Shermer, T.C.: Attraction-convexity and normal visibility. Comput. Geom. **96**, 101748 (2021)
7. Chazelle, B.: A theorem on polygon cutting with applications. In: 23rd Annual Symposium on Foundations of Computer Science, pp. 339–349. IEEE (1982)
8. Chazelle, B.: Triangulating a simple polygon in linear time. Discret. Comput. Geom. **6**(3), 485–524 (1991)
9. Chazelle, B., Guibas, L.J.: Visibility and intersectin problems in plane geometry. In: Proceedings of the First Annual Symposium on Computational Geometry, pp. 135–146 (1985)
10. Karp, B., Kung, H.-T.: GPSR: greedy perimeter stateless routing for wireless networks. In Proceedings of the 6th Annual International Conference on Mobile Computing and Networking, pp. 243–254 (2000)

11. Kim, Y.-D., Yang, Y.-M., Kang, W.-S., Kim, D.-K.: On the design of beacon based wireless sensor network for agricultural emergency monitoring systems. Comput. Stand. Interfaces **36**(2), 288–299 (2014)
12. Kostitsyna, I., Kouhestani, B., Langerman, S., Rappaport, D.: An optimal algorithm to compute the inverse beacon attraction region. arXiv preprint arXiv:1803.05946 (2018)
13. Kouhestani, B., Rappaport, D., Salomaa, K.: On the inverse beacon attraction region of a point. In: CCCG (2015)
14. Matoušek, J.: Efficient partition trees. In: Proceedings of the Seventh Annual Symposium on Computational Geometry, pp. 1–9 (1991)
15. Megiddo, N.: Linear-time algorithms for linear programming in R^3 and related problems. SIAM J. Comput. **12**(4), 759–776 (1983)
16. Sojka, E.: A simple and efficient algorithm for sorting the intersection points between a Jordan curve and a line. In Fifth International Conference in Central Europe in Computer Graphics and Visualisation, pp. 524–533 (1997)

An Approach to Agent Path Planning Under Temporal Logic Constraints

Chaofeng Yu, Nan Zhang$^{(\boxtimes)}$, Zhenhua Duan, and Cong Tian

Institute of Computing Theory and Technology, and ISN Laboratory,
Xidian University, Xi'an 710071, China
nanzhang@xidian.edu.cn, {zhhduan,ctian}@mail.xidian.edu.cn

Abstract. The capability of path planning is a necessity for an agent to accomplish tasks autonomously. Traditional path planning methods fail to complete tasks that are constrained by temporal properties, such as conditional reachability, safety, and liveness. Our work presents an integrated approach that combines reinforcement learning (RL) with multi-objective optimization to address path planning problems with the consideration of temporal logic constraints. The main contributions of this paper are as follows. (1) We propose an algorithm LCAP^2 to design extra rewards and accelerate training by tackling a multi-objective optimization problem. The experimental results show that the method effectively accelerates the convergence of the path lengths traversed during the agent's training. (2) We provide a convergence theorem based on the fixed-point theory and contraction mapping theorem.

Keywords: Path planning · Reinforcement learning ·
Pareto-dominance · Unified temporal logic · Fixed-point theory

1 Introduction

Path planning [9] plays a crucial role in various domains, including autonomous driving, robot navigation, and unmanned aerial vehicle (UAV) navigation [1,18]. It contributes significantly to enhancing efficiency, reducing costs, and optimizing resource utilization [12]. In the context of robot navigation, path planning ensures collision-free movement and efficient task execution. Furthermore, it is desirable for the agent to accomplish the task with minimal time, minimum energy, and minimum jerk [9]. Given that intelligent systems are prone to errors, the path-planning policy of agents necessitates a certain level of fault tolerance. This implies that they should be capable of re-planning an optimal sub-path from an erroneous position when they deviate from the originally planned trajectory.

This research is supported by National Natural Science Foundation of China under Grant Nos. 62272359 and 62172322; Natural Science Basic Research Program of Shaanxi Province under Grant Nos. 2023JC-XJ-13 and 2022JM-367.

W. Wu and G. Tong (Eds.): COCOON 2023, LNCS 14423, pp. 82–93, 2024.
https://doi.org/10.1007/978-3-031-49193-1_7

Traditional path planning methods, such as the graph search-based A* algorithm [19], exhibit inefficiency in path search and require complete environmental information. Sampling-based methods often converge slowly. Intelligent optimization techniques, including ant colony algorithms and genetic algorithms [13], tend to encounter challenges in escaping local optima.

In contrast, reinforcement learning (RL) [14] provides a powerful approach to tackling sequential decision problems by improving policies through continuous interactions with the environment, guided by reward signals. RL-based path planning methods have gained significant traction in various applications. However, challenges remain, particularly in terms of training stability and robustness. Concurrently, the path planning task assigned to agents are often intricate and multi-phased. Within a designated area, agents may be required to visit multiple locations sequentially. Tasks with temporal properties can be described by temporal logic formulas [17]. Hasanbeig et al. [10,11] propose an RL algorithm to synthesize policies that satisfy linear time properties for Markov Decision Processes (MDPs) [3]. Comparing to dynamic programming, the number of iterations is reduced by one order of magnitude.

However, the approach does not address the challenge of inadequate stability in the training process of RL, nor does it effectively mitigate the duration of the training process. Therefore, we introduce $LCAP^2$, an algorithm that utilize multi-objective optimization to evaluate different positions in the destination and provide extra rewards during the RL procedure. Experiments show that it effectively shortens the average path length.

The structure of this paper is outlined as follows: Sect. 2 provides an introduction to the path planning problem and presents fundamental concepts. Moreover, it introduces essential notions such as Pareto dominance. Section 3 introduces the proposed $LCAP^2$ algorithm and outlines the process of reward design. In Sect. 4, we empirically demonstrate the efficacy of the proposed algorithm. In Sect. 5, we give a convergence theorem. Finally, we explore potential extensions of the ideas presented in this paper.

2   Preliminaries

Multi-phase tasks require agents to reach several target areas in a specific order and complete the sub-tasks. When the agent reaches the target area, it receives a positive reward signal. On the contrary, when it reaches an unsafe area, it will receive a negative punishment signal.

Markov Decision Process (MDP). An MDP is defined as a six tuple $\mathfrak{M} = (S, A, s_0, P, \mathcal{AP}, L)$ over a finite set of states S, where A is a finite set of actions; $s_0 \in S$ is the initial state; $P : S \times A \times S \to [0,1]$ specifies transition probabilities, $P(s, a, s')$ is the probability of transitioning from s to s' with action a; AP is a finite set of atomic propositions, and $L : S \to 2^{\mathcal{AP}}$ is a labeling function.

An MDP $\mathfrak{M}$ describes the interactions between the agent and environment. At state $s \in S$, given the policy function $\pi : S \times A \to [0,1]$, agent makes the action $a = \arg\max_{a' \in A} \pi(s, a')$, and the agent is assigned a reward according to the reward function $R : S \times A \times S \to \mathbb{R}$. State-action value function $Q : S \times A \to \mathbb{R}$, $Q(s, a)$ denotes the expected discounted reward that the agent can get after performing action a at state s. $Q(s, a)$ is usually assigned a fixed value before training. It will be updated by back propagation of the reward signal:

$$\begin{cases} Q(s, a) \leftarrow (1 - \alpha)Q(s, a) + \alpha[R(s, a, s') + \gamma \max_{a' \in A} Q(s', a')] \\ Q(s'', a'') \leftarrow Q(s'', a'') \end{cases} \quad (1)$$

where α is the learning rate, γ is the discount factor, s' is the next state after performing action a in state s.

We describe temporal properties through Unified Temporal Logic (UTL) with the infinite model, then construct a product MDP using original MDP $\mathfrak{M}$ and the Büchi Automaton (BA) $\mathcal{B}$ converted from UTL formula. The agent's trajectory is trained to satisfy UTL [20] properties by learning to synthesize a policy through product MDP. UTL combines all characteristics of traditional Linear Temporal Logic (LTL) and Propositional Projection Temporal Logic (PPTL) [4]. Therefore, UTL can be used to describe full regular and omega-regular properties, which are often encountered in the field of formal verification. A UTL property can be characterized by an automaton. We converts UTL properties into BA. After that, a product MDP based on MDP and BA is constructed. Then, the state-action value function is iteratively calculated by Q-learning [16]. For UTL syntax and semantics, please refer to [20].

Büchi Automaton (BA). A BA is a five-tuple, $\mathcal{B} = (\mathcal{Q}, \Sigma, \Delta, q_0, \mathcal{F})$, where $\mathcal{Q}$ denotes a finite set of states; $\Sigma = 2^{\mathcal{AP}}$ is a finite alphabet; $\Delta : \mathcal{Q} \times \Sigma \to 2^{\mathcal{Q}}$ is a transition function; $q_0 \in \mathcal{Q}$ is initial state; $\mathcal{F}$ is the set of accepting conditions.

Let Σ^ω be the set of all infinite words over Σ, an infinite word $\omega \in \Sigma^\omega$ can be accepted by a BA if and only if there exists an infinite run $\theta \in \mathcal{Q}^\omega$ starting from q_0, where $\theta[i + 1] \in \Delta(\theta[i], \omega[i]), i \geq 0$ and $inf(\theta) \cap F \neq \emptyset$ ($inf(\theta)$ is the set of states that are visited infinitely often in the sequence θ). The accepted language of the BA $\mathcal{B}$ is the set of all infinite words accepted by the BA $\mathcal{B}$.

The task's property is described in UTL, which is a complete and sound system. If the property is exclusively described in LTL, the LTL3BA tools [2] is called. If the property is described in UTL or PPTL, PPTL2LNFG is called to transform the formula into labeled normal form graph (LNFG) [5,6]. Then an LNFG can be transformed to a GBA, and it can further be transformed to a BA [7].

Multi-objective optimization [15] allows us to consider multiple objectives simultaneously and find a set of optimal solutions known as the Pareto frontier. By exploring the trade-offs among different objectives, multi-objective optimization offers a more comprehensive understanding of the problem space. The definition of a multi-objective optimization problem is as follows:

$$\begin{cases} min & \boldsymbol{y} = \boldsymbol{F}(\boldsymbol{x}) = (f_1(\boldsymbol{x}), f_2(\boldsymbol{x}), \dots, f_m(\boldsymbol{x}))^T. \\ s.t. & g_i(\boldsymbol{x}) \le 0, \quad i = 1, 2, \dots, q. \\ & h_j(\boldsymbol{x}) = 0, \quad j = 1, 2, \dots, p. \end{cases} \quad (2)$$

where $\boldsymbol{x} = (x_1, \dots, x_n) \in \boldsymbol{X} \subset \mathbb{R}^n$ denotes n-dimensional decision vector, $\boldsymbol{X}$ denotes n-dimensional decision space, $\boldsymbol{y} = (y_1, \dots, y_m) \in Y \subset \mathbb{R}^n$ is m-dimensional target vector. $g_i(\boldsymbol{x}) \le 0, i = 1, 2, \dots, q$, represent q inequality constraints, and $h_j(\boldsymbol{x}) = 0, j = 1, 2, \dots, p$, represent p equation constraints.

Pareto Dominance. Suppose $\boldsymbol{x}_A, \boldsymbol{x}_B$ are two feasible solutions to the multi-objective optimization problem defined above, then $\boldsymbol{x}_A$ is said to be Pareto dominant compared to $\boldsymbol{x}_B$ if and only if: $\forall i \in \{1, 2, \dots, m\}.f_i(\boldsymbol{x}_A) \le f_i(\boldsymbol{x}_B) \wedge \exists j \in \{1, 2, \dots, m\}.f_j(\boldsymbol{x}_A) < f_j(\boldsymbol{x}_B)$, which denoted as $\boldsymbol{x}_A \succ \boldsymbol{x}_B$, and called as $\boldsymbol{x}_A$ dominate $\boldsymbol{x}_B$.

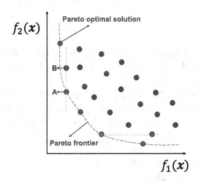

Fig. 1. Illustrative Example of Pareto Optimality in Objective Space

A feasible solution $\boldsymbol{x}^\star$ is called a Pareto optimal solution if and only if there is no feasible solution $\boldsymbol{x}$, such that $\boldsymbol{x} \succ \boldsymbol{x}^\star$. The Pareto optimal solution set $\boldsymbol{P}^\star$ is the set of Pareto optimal solutions. Pareto frontier $\boldsymbol{PF}^\star$ is the surface formed by the target vector corresponding to the optimal solution in $\boldsymbol{P}^\star$. Figure 1 shows the Pareto frontier and Pareto optimal solutions in a minimization problem with two objective functions $(f_1(\boldsymbol{x}), f_2(\boldsymbol{x}))$. It is evident that none of the alternatives can claim dominance over solution A. That is, A is a Pareto optimal solution.

3 LCAP²

In this section, we present an innovative algorithm, named LCAP², designed specifically for tackling the intricate challenges of **L**ogically-**C**onstrained **A**gent **P**ath **P**lanning (LCAP²) problems within the realms of RL. The following are the details of the algorithm:

Algorithm 1 LCAP^2(G, $\mathcal{P}$, $episode_{max}$, $episode_{threshold}$, $iteration_{max}$, α, γ, ε)

Input: gridworld environment $G = (w, h, start, end, A_G, T, R_G)$, UTL Constraint $\mathcal{P}$,
$episode_{max}$, $episode_{threshold}$, $iteration_{max}$, α, γ, ε
   //G is an instantiated object that defines width, height of the gridworld and
   the start and end positions; A_G denotes the action space; T shows the type of a
   position; R_G is the reward; $\mathcal{P}$ is a constraint defined by UTL formula; $episode_{max}$
   is the upper limit of the episodes in a training session; $episode_{threshold}$ determines
   the number of episode to start adding extra rewards; α is the learning rate; γ is the
   discount factor; ε is the probability of taking a random action in $\varepsilon-$ greedy policy

Output: Q-function
1: Transform $\mathcal{P}$ into the corresponding BA $\mathcal{B} = (Q_{\mathcal{B}}, \Sigma, \Delta, q_0, \mathcal{F})$
2: Modeling the gridworld G into an MDP $\mathfrak{M} = (S, A, s_0, P, AP, L)$
3: Combine $\mathcal{B}$ and $\mathfrak{M}$ to construct the product MDP $\mathfrak{M}_{\mathcal{B}} = (S^*, A, s_0^*, P^*, AP^*, L^*)$
4: **Variable:** $Q = dict = \{\}$, $episode = iteration = flag = R_{flag} = r' = q = q' = 0$
5: **Variable:** $\overline{hm} = \overline{vm} = \triangle z_i = 0 (i = 1, 2, 3, 4)$, $s^* = s^{*\prime} = []$, $label = a^* = $""
6: **while** $episode <= episode_{max}$ **do** //value iteration process
7:      $episode + +$
8:      **if** $flag == 0$ and $episode >= episode_{threshold}$ **then** //set the time to start
        adding extra rewards
9:          $flag = 1$
10:     **end if**
11:     $s^* = s_0.append(q_0)$ //reset the current state to the initial state of $\mathfrak{M}_{\mathcal{B}}$
12:     $\triangle z_i = 0$, $i = 1, 2, 3, 4$, $R_{flag} = 0$, $r' = 0$ //reset the values
13:     **while** $iteration < iteration_{max}$ **do**
14:         $Q[str(s^*)] = \{$"up":0, "down":0, "left":0, "right":0, "stay":0$\}$
15:         $a^* = max(Q[str(s^*)])$, $key = Q[str(s^*)].get$ with the probability of $1 - \varepsilon$
16:         **if** $a^* == $ "up" **then** //determine $\triangle z_i (i = 1, 2, 3, 4)$ and next state s'
17:             $\triangle z_1 + +$, $s' = [s^*[0] - 1, s^*[1]]$
18:         **else if** $a^* == $ "down" **then**
19:             $\triangle z_2 + +$, $s' = [s^*[0] + 1, s^*[1]]$
20:         **else if** $a^* == $ "left" **then**
21:             $\triangle z_3 + +$, $s' = [s^*[0], s^*[1] - 1]$
22:         **else if** $a^* == $ " right" **then**
23:             $\triangle z_4 + +$, $s' = [s^*[0], s^*[1] + 1]$
24:         **else if** $a^* == $ " stay" **then**
25:             $\triangle z_4 + +$, $s' = [s^*[0], s^*[1]]$
26:         **end if**
27:         $label = L(s')$, $q' = \Delta(q, label)$ //q' is the next state of $\mathcal{B}$
28:         $s^{*\prime} = s'.append(q')$ //$s^{*\prime}$ is the next state of s^*
29:         $Q[str(s^{*\prime})] = \{$"up":0, "down":0, "left":0, "right":0, "stay":0$\}$
30:         **if** $s^{*\prime}[2] \in \mathcal{F}$ **then** //reach terminal state
31:             $R_{flag} = 1$ //reward for reaching terminal state
32:             Compute the target vector $[\overline{hm}, \overline{vm}]$ corresponds to $s^{*\prime}$ through Eq. (4)
33:             **if** $flag$ **then** //start to add an additional reward
34:                 **if** not $dict$ **then**
35:                     $dict[s^{*\prime}] = [\overline{hm}, \overline{vm}]$ //add "$s^{*\prime} : [\overline{hm}, \overline{vm}]$" to $dict$
36:                 **end if**
37:                 **for** key in $dict$ **do**
38:                     **if** $\overline{hm} <= dict[key][0]$ and $\overline{vm} <= dict[key][1]$ and $(\overline{hm} + \overline{vm}) <$
        $sum(dict[key])$ **then** //this solution is dominated
39:                         $*$

40:                    del $dict[key]$ //remove the dominated solution
41:                    $dict[s^{*\prime}] = [\overline{hm}, \overline{vm}]$ //add the new Pareto optimal solution
42:                **end if**
43:                **if** $\overline{hm} > dict[key][0]$ or $\overline{vm} > dict[key][1]$ **then** // not a Pareto
        optimal solution
44:                    break
45:                **end if**
46:            **end for**
47:            **if** $s^{*\prime}$ in $dict$ **then** //$s^{*\prime}$ corresponds to a Pareto optimal solution
48:                $r' = sum(min(list(dict.values()), axis = 0))/(\overline{hm} + \overline{vm})$
49:            **end if**
50:        **end if**
51:        Break
52:    **end if**
53:    $Q(s^*, a^*) = (1 - \alpha)Q(s^*, a^*) + \alpha[R_{flag} + r' + \gamma max(Q[str(s^{*\prime})].values())]$
54:    $s^* = s^{*\prime}$ //step forward
55:    $iteration + +$
56:    **end while**
57: **end while**

In Algorithm 1, we first transform UTL constraints and the reinforcement learning task into a Büchi automaton and an MDP. These are then combined to form a product MDP. The main part of the algorithm (lines 6 to 57) involves value iteration. Lines 13 to 56 explain how agents execute actions within trajectories, covering state transitions, action selection, Q-value initialization, and updates. Once the endpoint is reached, lines 33 to 50 detail the reward design using multi-objective optimization. This includes computing and updating Pareto optimal solutions and an additional reward calculation. Overall, Algorithm 1 provides a concise framework that leverages reinforcement learning, temporal logic, and Pareto optimal solutions for effective path planning.

3.1 Property Extraction for Path Planning in Gridworld

In RL, the gridworld problem serves as a prevalent and illustrative environment model for investigating and exploring the capabilities of various RL algorithms. Gridworld, an abstract representation consisting of a two-dimensional grid comprised of squares, is employed to symbolize discrete states or spatial locations. The agent possesses the capacity to execute diverse actions, encompassing movements in vertical and horizontal directions, thereby eliciting corresponding rewards depending upon the action performed and the agent's present location.

Our focus initially revolves around comprehending the task requirements inherent to the gridworld scenario. By leveraging the label function, we transform the demand encapsulated within the UTL formula into an associated BA. Given an MDP $\mathfrak{M} = (S, A, s_0, P, \mathcal{AP}, L)$ and an BA $\mathcal{B} = (\mathcal{Q}, \Sigma, \Delta, q_0, \mathcal{F})$, where $\Sigma = 2^{\mathcal{AP}}$, a **product MDP** is defined as a tuple $\mathfrak{M} \otimes \mathcal{B} = \mathfrak{M}_\mathcal{B} = (S^*, A, s_0^*, P^*, \mathcal{AP}^*, L^*)$, where $S^* = S \times \mathcal{Q}, s_0^* = (s_0, q_0), \mathcal{AP}^* = \mathcal{Q}, L^* : S \times \mathcal{Q} \rightarrow 2^\mathcal{Q}$, such that $L^*(s, q) = q, P^* : S^* \times A \times S^* \rightarrow [0, 1]$ is the transition probability function such that $(s_i \xrightarrow{a} s_j) \wedge (q_i \xrightarrow{L(s_j)} q_j) \Rightarrow P^*((s_i, q_i), a, (s_j, q_j)) = P(s_i, a, s_j)$. Over the states of the product MDP we also define accepting condition $\mathcal{F}^*$ such that $s^* = (s, q) \in \mathcal{F}^*$.

3.2   Formulating Multi-Objective Optimization Problem

During the training process, the agent learns an optimal path and concurrently strives to identify the optimal goal. These two elements possess a mutually reinforcing relationship, where advancements in one aspect can effectively facilitate progress in the other. As illustrated in Fig. 2, the scenario arises where the agent accomplishes the last sub-task but deviates towards a sub-optimal goal. Therefore, considering the path planning task's objective of reaching a predefined region, our approach revolves around optimizing the refined target region. By leveraging this optimized region, we design an additional reward signal.

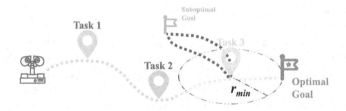

Fig. 2. The robot is encouraged to explore during the early stages of training. So it is possible for the task to conclude with a sub-optimal goal.

In the subsequent step, we establish two generic optimization sub-objectives that are intrinsically linked to the path, namely, the optimization of consumption hm along the horizontal direction and consumption vm along the vertical direction. By delineating these generic sub-objectives, we aim to capture the essence of optimizing energy expenditure and resource allocation within the multi-objective optimization framework, thereby paving the way for a comprehensive exploration of trade-offs and potential synergies in the pursuit of optimal solutions.

$$
\begin{cases}
min & \boldsymbol{y} = \boldsymbol{F}(\boldsymbol{x}) = (hm(\boldsymbol{x}), vm(\boldsymbol{x}))^T. \\
s.t. & x_{min} \leq x \leq x_{max}, \\
& y_{min} \leq y \leq y_{max}, \\
& 0 \leq \triangle z_i, i = 1, 2, 3, 4.
\end{cases}
\tag{3}
$$

Let $S_x = \{x | x_{min} \leq x \leq x_{max}; y_{min} \leq y \leq y_{max}; 0 \leq \triangle z_i, i = 1, 2, 3, 4\}$ be the feasible set of the above multi-objective optimization problem. $x \in S_x$ is the feasible solution, $\boldsymbol{x}_j = (x_j, y_j, \triangle z_{1,j}, \triangle z_{2,j}, \triangle z_{3,j}, \triangle z_{4,j}) \in S_x \subset \mathbb{R}^6$, represents the jth sample, i.e. the training situation when the agent reaches area t for the jth time, $j = 1, 2, \ldots, n$. x_j, y_j represent the location in gridworld, $\triangle z_{1,j}, \triangle z_{2,j}, \triangle z_{3,j}, \triangle z_{4,j}$ represent the distance traveled by the jth learning sample in four directions. $\overline{hm_n(\boldsymbol{x})} = \frac{1}{n}\sum_{j=1}^{n} hm(\boldsymbol{x}_j) = \frac{1}{n}\sum_{j=1}^{n}\sum_{i=1}^{4}\eta_i\triangle z_{i,j} + b_j$ denotes the average consumption of the first n learning sessions in the left and right directions, $\overline{vm_n(\boldsymbol{x})} = \frac{1}{n}\sum_{j=1}^{n} vm(\boldsymbol{x}_j) = \frac{1}{n}\sum_{j=1}^{n}\sum_{i=1}^{4}\zeta_i\triangle z_{i,j} + a_j$ denotes the average consumption of the first n learning sessions in the up and down directions. $\eta_i, \zeta_i(i = 1, 2, 3, 4)$ are coefficients, and a_j, b_j denotes the static consumption incurred by the agent while waiting in place.

3.3 Reward Design Method

We update the average consumption in an incremental manner:

$$
\overline{hm_k(\boldsymbol{x})} = \frac{1}{k}[(k-1) \times \overline{hm_{k-1}(\boldsymbol{x})} + \sum_{i=1}^{4}(\eta_i\triangle z_{i,k} + b_k)]
$$

$$
= \overline{hm_{k-1}(\boldsymbol{x})} + \frac{1}{k}(\sum_{i=1}^{4}(\eta_i\triangle z_{i,k} + b_k) - \overline{hm_{k-1}(\boldsymbol{x})})
$$

(4)

In fact, it is not meaningful to calculate the average consumption from the beginning. Therefore, we set the $episode_{threshold}$. When $episode >$ $episode_{threshold}$, we start to calculate the average consumption incrementally. In the process of updating the target vector, we maintain a dictionary where the key is the coordinate of the Pareto optimal solution and the corresponding value is the target vector. When an agent reaches a new target point, it compares the target vector with all the values in the dictionary. When it is confirmed that the target point reached corresponds to the Pareto optimal solution, an appropriate reward signal is given to each path that reaches the target area:

$$
r' = \frac{\min_{\boldsymbol{x}\in\widehat{S_x}}\left(\overline{hm(\boldsymbol{x})} + \overline{vm(\boldsymbol{x})}\right)}{\overline{hm(\boldsymbol{x}^\star)} + \overline{vm(\boldsymbol{x}^\star)}}
$$

(5)

where the numerator represents the lowest average consumption sum in the set $\widehat{S_x} \subset S_x$, which are feasible solutions obtained. And the denominator is the average consumption sum of the optimal solutions visited in the current episode. The Q value is updated according to the following formulas:

$$
R'(s, a, s') = \begin{cases} r' & \textit{if } s' \textit{ is the state corresponding to } \boldsymbol{x}^\star \\ 0 & \textit{otherwise} \end{cases}
$$

(6)

$$
Q(s, a) \leftarrow (1-\alpha)Q(s, a) + \alpha[R(s, a, s') + R'(s, a, s') + \gamma\arg\max_{a'\in A} Q(s', a')]
$$

(7)

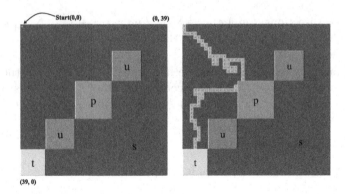

Fig. 3. Illustrations of agent path planning in a 2-D space.

4   Experiments

As shown in Fig. 3, we tested the effectiveness of the LCAP[2] method on gridworld G [8]. The task $s \rightarrow p \rightarrow t$ *while avoiding the unsafe area* u illustrated in Fig. 3 can be described by the UTL formula [10]:

$$\Diamond(p \wedge \Diamond t) \wedge \Box(t \rightarrow \Box t) \wedge (u \rightarrow \Box u) \tag{8}$$

The underlying rationale behind Eq. (8) is that the agent needs to, at a certain point in the future, sequentially visit the state represented by p and then reach the state represented by t, ensuring $\Diamond(p \wedge \Diamond t)$. Furthermore, the agent should stay there $\Box(t \rightarrow \Box t)$ while avoiding unsafe areas $\Diamond(u \rightarrow \Box u)$. We can build the BA $\mathcal{B}$ associated with (8) as in Fig. 4a, and the product MDP is obtained from BA $\mathcal{B}$ and MDP $\mathfrak{M}$ according to the construction rule of the product MDP.

For example, the initial state of the product MDP $\mathfrak{M}_\mathcal{B}$ is $s_0^* = (s_0, q_0) = (0, 0, q_0)$, where $s_0 = (0, 0)$(in the gridworld) is the initial state of the MDP $\mathfrak{M}$, and q_0 is the initial state of BA. If the agent steps "down", then we have $(0, 0) \xrightarrow{down} (1, 0)$ and $q_0 \xrightarrow{L((1,0))} q_0$, so the transition probability is corresponding to the original one $P^*((0, 0, q_0), down, (1, 0, q_0)) = P((0, 0), down, (1, 0))$.

Suppose the initial episode of the agent is being executed, and at the state $s_{cur}^* = (15, 14, q_0)$, the agent selects the action "left". The corresponding next state in the $\mathfrak{M}$ is $(15, 15)$, and its associated atomic propositions, is obtained through the label function $L^*(s_{cur}^*) = p$. Subsequently, BA transitions from state q_0 to q_1 based on p. By combining the $\mathfrak{M}$ state and the $\mathcal{B}$ state, we obtain the corresponding state $s_n^* = (15, 15, q_1)$ in $\mathfrak{M}_\mathcal{B}$. At this point, since the agent has not yet received any reward signal, the state-action value functions for both states are initialized to 0. As the agent transitions from the current state s_{cur}^* to a new state s_n^*, the incurred cost of moving left is captured as the leftward movement consumption($\triangle z_3+ = 1$). Upon reaching the target area, the target vector can be calculated using Eq. (4). Consider the scenario where the agent is still at state s_{cur}^*, and the Q-values for states s_{cur}^* and s_n^* are presented in following table:

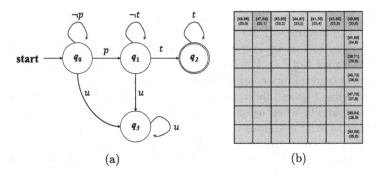

(a)   (b)

Fig. 4. (a) BA $\mathcal{B}$ for (8). (b) Illustration of the target vectors in target area.

a					
s^*	Up	Down	Left	Right	Stay
s^*_{cur}	0.03	0.44	0.12	0.52	0.32
s^*_n	0.00	0.67	0.80	0.14	0.28

Notably, the Q-value associated with the "right" action is the highest, imply-ing that the agent is most likely to choose this action. If the agent opts for the "right" action, its Q-value will be updated according to the following equation:

$$Q[str(s^*_{cur})]["right"] = (1 - \alpha)Q[str(s^*_{cur})]["right"] + \alpha[0 \\ + \gamma \max_{a' \in A} Q[str(s^*_n)][a] = 0.488 \tag{9}$$

The decision vector is $\boldsymbol{x} = (x, y, \triangle z_1, \triangle z_2, \triangle z_3, \triangle z_4)$. We set the coefficients a_j, b_j to 0. And since the location(x, y) of the target region in this problem does not affect the optimization objective, the $(hm(\boldsymbol{x}), vm(\boldsymbol{x})) = (1/2 \times (\triangle z_3 + \triangle z_4), 1/2 \times (\triangle z_1 + \triangle z_2))$. As depicted in Fig. 4b, consider the target vector $[43, 65]$ associated with each terminal state $(33, 5)$. If the agent reaches the state $(33, 4)$, we proceed to evaluate whether the target vector corresponding to this state represents a Pareto optimal solution. If the target vector satisfies the Pareto

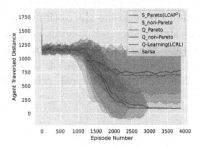

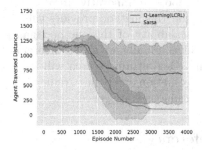

Fig. 5. Experimental results

optimality conditions, an additional reward, denoted as $r' = (38+64)/(43+65) = 0.94$, is given according to Eq. (5).

We examine the average distance traversed by the agent under various training configurations. Notably, Fig. 5 illustrates that the agent performs better and achieves faster convergence compared to the Q-learning algorithm in this specific problem domain. Figure 6 indicates that the incorporation of LCAP^2 enhances the training stability when using the same learning algorithm. This observation suggests that LCAP^2 mitigates the issue of the agent wandering within a region to some extent, thereby reducing sampling complexity and expediting the training process. These outcomes support the effectiveness of our proposed method.

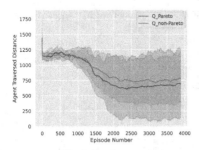 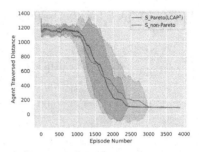

Fig. 6. Apply LCAP^2 on Q-learning and *Sarsa*

5   Convergence Theorem

Theorem 1. *For an MDP* $\mathfrak{M} = (S, A, s_0, P, \mathcal{AP}, L)$*, let* $\boldsymbol{X} : S \times A \to \mathbb{R}$ *be the state-action value function,* $\widetilde{\mathfrak{S}}$ *be the set of state-action value functions, the operator* T *is defined on* $\widetilde{\mathfrak{S}}$: $T(\boldsymbol{X}) = (1 - \alpha)\boldsymbol{X} + \alpha(0 + \gamma \cdot \boldsymbol{X}')$*, then operator* T *converges and has a fixed-point.* T *converges at the only fixed-point, and the rate of convergence is* $\frac{1-\alpha}{1-\alpha\gamma}$.

6   Conclusion

We propose an innovative approach called LCAP^2. The simulation results suggest that this method provides valuable assistance in learning the optimal path. The scalability of this method has practical significance for solving complex problems. Finally, we give a theorem demonstrating the convergence of the algorithm. Our future research will focus on using methods such as value function approximation and policy gradient to study RL problems with logical constraints in continuous state spaces.

References

1. Aggarwal, S., Kumar, N.: Path planning techniques for unmanned aerial vehicles: a review, solutions, and challenges. Comput. Commun. **149**, 270–299 (2020)

2. Babiak, T., Křetínský, M., Řehák, V., Strejček, J.: LTL to Büchi automata translation: fast and more deterministic. In: Flanagan, C., König, B. (eds.) TACAS 2012. LNCS, vol. 7214, pp. 95–109. Springer, Heidelberg (2012). https://doi.org/10.1007/978-3-642-28756-5_8
3. Baier, C., Katoen, J.P.: Principles of Model Checking. MIT Press, Cambridge (2008)
4. Duan, Z.: Temporal logic and temporal logic programming. Science Press (2005)
5. Duan, Z., Tian, C., Yang, M., He, J.: Bounded model checking for propositional projection temporal logic. In: Du, D.-Z., Zhang, G. (eds.) COCOON 2013. LNCS, vol. 7936, pp. 591–602. Springer, Heidelberg (2013). https://doi.org/10.1007/978-3-642-38768-5_52
6. Duan, Z., Tian, C., Zhang, L.: A decision procedure for propositional projection temporal logic with infinite models. Acta Informatica 45(1), 43–78 (2008)
7. Duan, Z., Tian, C., Zhang, N.: A canonical form based decision procedure and model checking approach for propositional projection temporal logic. Theor. Comput. Sci. 609, 544–560 (2016)
8. Gao, Q.: Deep Reinforcement Learning with Temporal Logic Specifications. Ph.D. thesis, Duke University (2018)
9. Gasparetto, A., Boscariol, P., Lanzutti, A., Vidoni, R.: Path planning and trajectory planning algorithms: a general overview. Motion Oper. Plan. Robot. Syst. Background Pract. Approach. 29, 3–27 (2015)
10. Hasanbeig, M., Abate, A., Kroening, D.: Logically-constrained reinforcement learning. arXiv preprint arXiv:1801.08099 (2018)
11. Hasanbeig, M., Jeppu, N.Y., Abate, A., Melham, T., Kroening, D.: Deepsynth: automata synthesis for automatic task segmentation in deep reinforcement learning. In: Proceedings of the AAAI Conference on Artificial Intelligence, vol. 35, pp. 7647–7656 (2021)
12. Hayat, S., Yanmaz, E., Brown, T.X., Bettstetter, C.: Multi-objective UAV path planning for search and rescue. In: 2017 IEEE International Conference on Robotics and Automation (ICRA), pp. 5569–5574. IEEE (2017)
13. Masehian, E., Sedighizadeh, D.: Multi-objective robot motion planning using a particle swarm optimization model. J. Zhejiang Univ. Sci. C 11, 607–619 (2010)
14. Sutton, R.S., Barto, A.G.: Reinforcement Learning: An Introduction. MIT Press, Cambridge (2018)
15. Van Moffaert, K., Nowé, A.: Multi-objective reinforcement learning using sets of pareto dominating policies. J. Mach. Learn. Res. 15(1), 3483–3512 (2014)
16. Watkins, C.J., Dayan, P.: Q-learning. Mach. Learn. 8, 279–292 (1992)
17. Xu, Z., Topcu, U.: Transfer of temporal logic formulas in reinforcement learning. In: IJCAI: Proceedings of the Conference, vol. 28, p. 4010. NIH Public Access (2019)
18. Yijing, Z., Zheng, Z., Xiaoyi, Z., Yang, L.: Q learning algorithm based UAV path learning and obstacle avoidance approach. In: 2017 36th Chinese Control Conference (CCC), pp. 3397–3402. IEEE (2017)
19. Yu, J., Hou, J., Chen, G.: Improved safety-first a-star algorithm for autonomous vehicles. In: 2020 5th International Conference on Advanced Robotics and Mechatronics (ICARM), pp. 706–710. IEEE (2020)
20. Zhang, N., Yu, C., Duan, Z., Tian, C.: A proof system for unified temporal logic. Theor. Comput. Sci. 949, 113702 (2023)

The Heterogeneous Rooted Tree Cover Problem

Pengxiang Pan[1], Junran Lichen[2], Ping Yang[1], and Jianping Li[1(✉)]

[1] School of Mathematics and Statistics, Yunnan University,
East Outer Ring South Road, University Town,
Kunming 650504, People's Republic of China
{pengxiang,jianping}@ynu.edu.cn
[2] School of Mathematics and Physics, Beijing University of Chemical Technology,
No.15, North Third Ring East Road, Chaoyang District,
Beijing 100029, People's Republic of China
J.R.Lichen@buct.edu.cn

Abstract. We consider the heterogeneous rooted tree cover (HRTC) problem. Concretely, given an undirected complete graph $G = (V, E)$ with a root $r \in V$, an edge-weight function $w : E \to R^+$ satisfying the triangle inequality, a vertex-weight function $f : V \backslash \{r\} \to R_0^+$, and k construction teams having nonuniform construction speeds $\lambda_1, \lambda_2, \ldots, \lambda_k$, we are asked to find k trees for these k construction teams to cover all vertices in V, each tree starting at the same root r, *i.e.*, k trees having a sole common vertex called root r, the objective is to minimize the maximum completion time, where the completion time of each team is the total construction weight of its related tree divided by its construction speed.

In this paper, we first design a $58.3286(1 + \delta)$-approximation algorithm to solve the HRTC problem in time $O(n^3(1 + \frac{1}{\delta}) + \log(w(E) + f(V \backslash \{r\})))$ for any $\delta > 0$. In addition, we present a $\max\{2\rho, 2 + \rho - \frac{2}{k}\}$-approximation algorithm for resolving the HRTC problem in time $O(n^2)$, where ρ is the ratio between the maximum and minimum speed of these k teams.

Keywords: Rooted tree cover · Nonuniform speeds · Approximation algorithms · Complexity of algorithms

1 Introduction

The subgraph cover problems, including the cycle cover problem and the tree cover problem, form a much-studied family of combinatorial optimization problems. These problems have wide range of practical applications, such as routings of multi-vehicles [2,7,12], nurse station location [4] and data gathering in wireless

This paper is supported by the National Natural Science Foundation of China [Nos. 12361066, 12101593]. Junran Lichen is also supported by Fundamental Research Funds for the Central Universities [No.buctrc202219], and Jianping Li is also supported by Project of Yunling Scholars Training of Yunnan Province [No. K264202011820].

W. Wu and G. Tong (Eds.): COCOON 2023, LNCS 14423, pp. 94–105, 2024.
https://doi.org/10.1007/978-3-031-49193-1_8

sensor networks [13,15]. In some applications, the minimization of the latest service completion time at the service locations is more relevant. As a result, there is a growing body of literature on cover problems under the min-max objective.

Considering the vehicles to have nonuniform speeds in routing planning, Gørtz et al. [8] in 2016 proposed the heterogeneous traveling salesman problem, which we refer to as the heterogeneous rooted cycle cover (HRCC) problem. In the HRCC problem, given a complete graph $G = (V, E)$ equipped with an edge-weight function $w : E \rightarrow R^+$ that satisfies the triangle inequality, a root $r \in V$ and k vehicles with nonuniform speeds $\lambda_1, \lambda_2, \ldots, \lambda_k$, we are asked to find k cycles for these k vehicles to cover all vertices in V, each vehicle starting at r, i.e., these k cycles having a sole common vertex r, the objective is to minimize the maximum completion time, where the completion time of a vehicle is the total weight of its related cycle divided by its speed. For the HRCC problem, Gørtz et al. [8] in 2016 presented a constant factor approximation algorithm.

The rooted cycle cover (RCC) problem [6], also called the k-traveling salesman problem, which is an important special version of the HRCC problem, where $\lambda_i = 1$ for each $i \in \{1, 2, \ldots, k\}$. Employing a splitting strategy, Frederickson et al. [6] in 1978 gave a $(\frac{5}{2} - \frac{1}{k})$-approximation algorithm to solve the RCC problem. For the version $k = 1$ of the RCC problem, i.e., the metric traveling salesman problem, Christofides [3] provided a famous 3/2-approximation algorithm by using an algorithm for solving the Euler tour problem.

In addition, many researches focus on the tree cover problems of graphs, in which vertices are all covered by a set of k trees. Taking the service handling times of vertices into consideration, Nagamochi [10] proposed the rooted tree cover (RTC) problem, which is modelled as follows. Given a complete graph $G = (V, E)$ equipped with an edge-weight function $w : E \rightarrow R^+$ satisfying the triangle inequality, a vertex-weight function $f : V \backslash \{r\} \rightarrow R_0^+$, a root $r \in V$ and k construction teams, it is asked to find k trees for these k teams to cover all vertices in V, each tree starting at the same root r, the objective is to minimize the maximum total weight among these k trees, where the total weight of a tree is the summation of edge weights and vertex weights in that tree, equivalently, the objective is to minimize the maximum completion time, where the completion time of a construction team is the total construction weight of that tree divided by its speed for the case that speeds of these k teams are all same one.

The RTC problem is NP-hard [1] even for the case $k = 2$ and $f(\cdot) \equiv 0$. Many research papers have been focused on the development of constant factor approximation algorithms to resolve the RTC problem. Using a tree partition technique, Nagamochi [10] in 2005 presented a $(3 - \frac{2}{k+1})$-approximation algorithm to resolve the RTC problem. Xu and Wen [14] in 2010 gave a lower bound of 10/9 for the RTC problem. Moreover, the other relevant results can be found in [5,9,11,16].

In practice, the construction efficiencies or construction speeds of multiple construction teams are often different similar to the vehicle speeds of the HRCC problem. Motivated by the observation and the RTC problem, we address the heterogeneous rooted tree cover (HRTC) problem. Concretely, given an undirected complete graph $G = (V, E; w, f)$ with a root $r \in V$, an edge-weight function $w : E \to R^+$ satisfying the triangle inequality, a vertex-weight function $f : V \backslash \{r\} \to R_0^+$, and k construction teams having nonuniform construction speeds $\lambda_1, \lambda_2, \ldots, \lambda_k$, we are asked to find k trees $\mathcal{T} = \{T_i \mid i = 1, 2, \ldots, k\}$ for these k construction teams to cover all vertices in V, each starting at the same root r, i.e., k trees having a sole common vertex called root r, the objective is to minimize the maximum completion time, where the completion time of each team is the total construction weight of its related tree divided by its construction speed. In a formulaic way, the min-max objective is written as $\min_{\mathcal{T}} \max \{\frac{w(T_i) + f(T_i)}{\lambda_i} \mid i = 1, 2, \ldots, k\}$.

As far as what we have known, the HRTC problem has not been considered in the literature. The aforementioned HRCC problem (without vertex weights) has been studied in [8], but we cannot directly use the algorithms for the HRCC problem to solve the HRTC problem, because the HRTC problem includes the vertex weights. However, modifying the technique in [8], we intend to design first approximation algorithm with constant approximation ratio to solve the HRTC problem. In addition, we shall present second approximation algorithm with lower time complexity to resolve the HRTC problem.

The remainder of this paper is organized as follows. In Sect. 2, we present some terminologies and fundamental lemmas to state descriptions of approximation algorithms; In Sect. 3, we design first constant factor approximation algorithm to solve the HRTC problem; In Sect. 4, we design second approximation algorithm with lower time complexity to resolve the HRTC problem; In Sect. 5, we provide our conclusion and further research.

2   Terminologies and Fundamental Lemmas

All graphs considered in the paper are assumed to be finite, undirected and loopless. Given a graph $G = (V, E)$, to contract a vertex subset $V' \subseteq V$ is to replace these vertices by a single vertex incident to all the edges which were incident in G to any vertex in V'. The resulting graph is denoted by G/V' with vertex set $V \cup \{v'\} \backslash V'$ and edge set $E \cup \{uv' \mid uv \in E, u \in V \backslash V', v \in V'\} \backslash E(G[V'])$, where v' is viewed as a new vertex obtained by contracting the vertex subset V'. For a vertex set V and a set $\mathcal{T} = \{T_i \mid i = 1, 2, \ldots, k\}$ of trees (or cycles), if $V \subseteq \bigcup_{i=1}^k V(T_i)$, we say that $\mathcal{T}$ covers V.

For any two sets X_1 and X_2, $X_1 + X_2$ is a multiset obtained by adding all elements in $X_1 \cap X_2$ to $X_1 \cup X_2$. Especially, for any two graphs $G = (V, E)$ and $G' = (V', E')$, denote $G \cup G' = (V \cup V', E \cup E')$ and $G + G' = (V \cup V', E + E')$, respectively.

In designing a constant factor approximation algorithm for the HRTC problem, we need the following definition, which is obtained by slightly modifying the definition in [8].

Definition 1. *Given an undirected graph $G = (V, E; w, f)$ with two constants $M > 0$ and $\varepsilon > 0$, where $w : E \to R^+$ is an edge-weight function and $f : V \backslash \{r\} \to R_0^+$ is a vertex-weight function, let $\mathcal{F}_i$ be a set of trees in G starting at the same vertex r for each integer $i \geq 0$. Then the collection $\{\mathcal{F}_i\}_{i \geq 0} = \bigcup_{i \geq 0} \mathcal{F}_i$ is referred to as $(\alpha, \beta)_{M, \varepsilon}$-assignable, if it has the following properties*

(1) $w(F) + f(F) \leq \alpha \cdot (1 + \varepsilon)^i M$, for each tree $F \in \mathcal{F}_i$, $i \geq 0$;

(2) $\sum_{j \geq i}(w(\mathcal{F}_i) + f(\mathcal{F}_i)) \leq \beta M \cdot \Lambda((1 + \varepsilon)^{i-1})$ for each $i \geq 0$, where $\Lambda((1 + \varepsilon)^{i-1})$ is the sum of speeds that at least $(1 + \varepsilon)^{i-1}$.

For Definition 1, we can regard the sum of edge weights and vertex weights of each tree as a whole weight, and using the ASSIGN algorithm in Gørtz et al. [8], we can obtain the following important result.

Lemma 1. [8] Given an $(\alpha, \beta)_{M,\varepsilon}$-assignable collection $\{\mathcal{F}_i\}_{i \geq 0}$ of trees starting at r, we can use the ASSIGN algorithm to assign all trees in $\{\mathcal{F}_i\}_{i \geq 0}$ to k construction teams in time $O(n^3 + \log(w(E) + f(V\backslash\{r\})))$, satisfying that the completion time of each construction team is at most $((1 + \varepsilon)\alpha + \beta)M$, where the completion time of each team is the total construction weight divided by its construction speed, n is the number of vertices.

3 An Approximation Algorithm with Constant Approximation Ratio

In this section, we consider the heterogeneous rooted tree cover (HRTC) problem. Without loss of generality, we may assume that the considered graph are all connected and $2 \leq k \leq n - 1$.

By Lemma 1, we design the following strategies to solve the HRTC problem.

(1) Find an $(\alpha, \beta)_{M,\varepsilon}$-assignable collection of subtrees to cover all vertices;

(2) Assign subtrees in the above collection to k construction teams to minimize the completion time of any team.

Given an undirected graph $G = (V, E; w, f)$ with two current values M and ε (the precise value to be chosen later), we first give a partition of V that is $\{V_0, V_1, \cdots\}$, where $V_0 = \{v \in V \mid w(rv) + f(v) \leq M\}$, and $V_i = \{v \in V \mid (1 + \varepsilon)^{i-1}M < w(rv) + f(v) \leq (1 + \varepsilon)^i M\}$ for each $i \geq 1$. For each $i \geq 0$, let $V_{\leq i} = \bigcup_{j \leq i} V_j$ and $V_{\geq i} = \bigcup_{j \geq i} V_j$. Similarly, we give a partition of E that is $\{E_0, E_1, \cdots\}$, where $E_i = \{uv \in E \mid u \in V_{\leq i}, v \in V_i\}$ for each $i \geq 0$. For each $i \geq 0$, let $E_{\leq i} = \bigcup_{j \leq i} E_j$ and $E_{\geq i} = \bigcup_{j \geq i} E_j$.

Now, analyzing the lower bound of the HRTC problem, we obtain the following lemma.

Lemma 2. Given a complete graph $G = (V, E; w, f)$ as an instance of the HRTC problem, for any constant $M \geq OPT$, we have $w(T^{MS}_{G/V_{<l}}) + f(V_{\geq l}) \leq M \cdot \Lambda((1 + \varepsilon)^{l-1})$ for each integer $l \geq 0$, where OPT is the optimal value to the given instance, $T^{MS}_{G/V_{<l}}$ is a minimum edge-weight spanning tree of $G/V_{<l}$, i.e., a new graph obtained by contracting a vertex set $V_{<l}$, and $\Lambda((1 + \varepsilon)^{l-1})$ is the total of speeds that exceed $(1 + \varepsilon)^{l-1}$.

Proof. Consider that in an optimal solution to G for the HRTC problem, if any vertex $v \in V_{\geq l}$ can be constructed by a construction team with speed λ', then we have $\lambda' > (1+\varepsilon)^{l-1}$. This is because $w(rv) + f(v) > (1+\varepsilon)^{l-1}M$ holds for each $v \in V_{\geq l}$, and the construction weight of a construction team with speed λ' is at most $\lambda' \cdot OPT \leq \lambda' \cdot M$, implying $\lambda' \cdot M \geq \lambda' \cdot OPT > (1+\varepsilon)^{l-1}M$. Since $\max\{w(rv) + f(v), w(rv') + f(v')\} > (1+\varepsilon)^{l-1}M$ holds for each edge $e = vv' \in E_{\geq l}$, we deduce that any edge in $E_{\geq l}$ must be constructed by some team with speed exceed $(1+\varepsilon)^{l-1}$.

Let E^* denote the set of edges constructed by teams in the optimal solution. By the above arguments, we have $w(E^* \cap E_{\geq l}) + f(V_{\geq l}) \leq OPT \cdot \Lambda((1+\varepsilon)^{l-1}) \leq M \cdot \Lambda((1+\varepsilon)^{l-1})$, meaning $w(E^* \cap E_{\geq l}) + f(V_{\geq l}) \leq M \cdot \Lambda((1+\varepsilon)^{l-1})$. Clearly, $G[E^*]$ is a spanning tree of G, and $E^* \cap E_{\geq l}$ corresponds to a spanning tree of $G/V_{<l}$. Since $T^{MS}_{G/V_{<l}}$ is a minimum edge-weight spanning tree of $G/V_{<l}$, we obtain $w(T^{MS}_{G/V_{<l}}) + f(V_{\geq l}) \leq w(E^* \cap E_{\geq l}) + f(V_{\geq l})$, implying $w(T^{MS}_{G/V_{<l}}) + f(V_{\geq l}) \leq M \cdot \Lambda((1+\varepsilon)^{l-1})$. □

For each $i \geq 0$, let H_i denote the edge subset of G corresponding to a minimum spanning tree of $G[V_{\leq i}]/V_{<i}$. Obviously, $G[\mathcal{H}]$ with $\mathcal{H} = \bigcup_{i \geq 0} H_i$ is a spanning tree of G. Using the similar arguments in [8] to analyze the relation between $\mathcal{H}$ and a minimum edge-weight spanning tree of G, we obtain a result as follows.

Lemma 3. [8] *Given a complete graph $G = (V, E; w, f)$ and a constant $\varepsilon > 0$, a spanning tree $G[\mathcal{H}]$ with $\mathcal{H} = \bigcup_{i \geq 0} H_i$ of G can be constructed to satisfy:*

- *The vertex levels along every root-leaf path are nondecreasing.*
- *For each $i \geq 0$, we have $\sum_{j \geq i} w(H_j) \leq (6 + \frac{6}{\varepsilon}) \cdot w(T^{MS}_{G/V_{<i}})$, where $T^{MS}_{G/V_{<i}}$ is a minimum edge-weight spanning tree of $G/V_{<i}$.*

In Lemma 3, for each $i \geq 0$, it is clear that $\sum_{j \geq i} w(H_j) \leq (6+\frac{6}{\varepsilon}) \cdot w(T^{MS}_{G/V_{<i}})$ means $\sum_{j \geq i}(w(H_j) + f(V_j)) \leq (6+\frac{6}{\varepsilon}) \cdot w(T^{MS}_{G/V_{<i}}) + f(V_{\geq i}) \leq (6+\frac{6}{\varepsilon}) \cdot (w(T^{MS}_{G/V_{<i}}) + f(V_{\geq i}))$. By Lemma 2, we obtain at once that $\sum_{j \geq i}(w(H_j) + f(V_j)) \leq (6 + \frac{6}{\varepsilon}) \cdot (w(T^{MS}_{G/V_{<i}}) + f(V_{\geq i})) \leq (6+\frac{6}{\varepsilon}) \cdot M \cdot \Lambda((1+\varepsilon)^{i-1})$, which is stated in the following

Lemma 4. *Given a complete graph $G = (V, E; w, f)$ with two constants $\varepsilon > 0$ and $M \geq OPT$, where OPT is the optimal value to the instance G for the HRTC problem, then the spanning tree $G[\mathcal{H}]$ with $\mathcal{H} = \bigcup_{i \geq 0} H_i$ of G mentioned-above satisfies:*

- *The vertex levels along every root-leaf path are nondecreasing.*
- *For each $i \geq 0$, we have $\sum_{j \geq i}(w(H_j) + f(V_j)) \leq (6 + \frac{6}{\varepsilon}) \cdot M \cdot \Lambda((1+\varepsilon)^{i-1})$.*

To shorten notation, given each edge $e = uv$ in $\mathcal{H}$, denote $x_e \in \{u, v\}$ to be a vertex farther away from r in $G[\mathcal{H}]$, and $y_e \in \{u, v\}$ to be a vertex closer to r in $G[\mathcal{H}]$. For each subtree $G' = (V', E')$ of $G[\mathcal{H}]$ mentioned above, we define a new function $f_1(\cdot)$ to be $f_1(G') = \sum_{e \in E'} f_1(x_e)$, implying $f_1(G') = f(G') - f(y_{G'})$, where $y_{G'}$ is a vertex in $G[\mathcal{H}]$ closest to r.

Basing from Lemma 1 to Lemma 4, we design a following algorithm, denoted by the algorithm HRTC$_1$, to solve the HRTC problem.

Algorithm: HRTC$_1$

INPUT: An undirected complete graph $G = (V, E; w, f)$ with a root $r \in V$, an edge-weight function $w : E \to R^+$, a vertex-weight function $f : V\backslash\{r\} \to R_0^+$, k construction teams having speeds $\lambda_1, \ldots, \lambda_k$ respectively, a small fixed constant $\delta > 0$, and two constants $\zeta = 2$ and $\varepsilon = 1.3146$ (to be chosen in Theorem 1);

OUTPUT: A set $\mathcal{T} = \{T_i \mid i = 1, 2, \ldots, k\}$ of k trees.

Begin

Step 1 Set $M = \max_{v \in V}\{\frac{w(r,v)+f(v)}{\lambda_{\max}}\}$, where $\lambda_{\max} = \max\{\lambda_i \mid i = 1, 2, \ldots, k\}$;

Step 2 Using M and ε, we can partition the set V into subsets $V_0, V_1, \ldots$, and the set E into subsets $E_0, E_1, \ldots$ as mentioned-above; For convenience, we may actually assume that the number of subsets partitioned is t, i.e., $(1+\varepsilon)^{t-1}M < \max\{w(rv) + f(v) \mid v \in V\} \le (1+\varepsilon)^t M$;

Step 3 If $(w(T_{G/V_{<l}}^{MS})+f(V_{\ge l}) > M \cdot \Lambda((1+\varepsilon)^{l-1})$ holds for some $l \in \{1, 2, \ldots, t\}$) then

   set $M := (1 + \delta)M$, and go to Step 2;

Step 4 Construct a spanning tree $G[\mathcal{H}]$ with $\mathcal{H} = \bigcup_{i \ge 0} H_i$ of G, where H_i is the edge subset of G corresponding to a minimum edge-weight spanning tree of $G[V_{\le i}]/V_{<i}$; Set $\mathcal{S}_0 := \{H_0\}$;

Step 5 For each $i \in \{1, 2, \ldots, t\}$, partition H_i into a set $\mathcal{S}_i$ of subtrees such that each subtree $\eta \in \mathcal{S}_i$ contains exactly one edge $h(\eta)$ from $V_{<i}$ to V_i; Let $\gamma = \frac{\varepsilon}{(2+\varepsilon)(1+\varepsilon)}$ and $\mathcal{S}_0^m = \mathcal{S}_0$; For each $i \in \{1, 2, \ldots, t\}$, set $\mathcal{S}_i^m = \emptyset$ and $\mathcal{S}_i^u = \emptyset$;

Step 6 For all $i \in \{1, 2, \ldots, t\}, \eta \in \mathcal{S}_i$ do:
   If $(w(\eta) + f_1(\eta) \ge \gamma \cdot (1 + \varepsilon)^i M)$ then
      Set $\mathcal{S}_i^m := \mathcal{S}_i^m \cup \{\eta\}$;
   Else
      Set $\mathcal{S}_i^u := \mathcal{S}_i^u \cup \{\eta\}$;

Step 7 For all $i \in \{1, 2, \ldots, t\}, \sigma \in \mathcal{S}_i^u$ do:
   Determine a subtree $\pi(\sigma)$ in $\bigcup_{j<i} \mathcal{S}_j$, having $\pi(\sigma) \cap \sigma \ne \emptyset$;

Step 8 For all $i \in \{0, 1, 2, \ldots, t\}, \tau \in \mathcal{S}_i^m$ do:
   (8.1) Set $\text{Dangle}(\tau) = \{\sigma \in \mathcal{S}_{i+1}^u : \pi(\sigma) = \tau\}$;
   (8.2) If the total weight of $(\tau\backslash h(\tau)) \cup \text{Dangle}(\tau)$ is at most $\zeta(1+\varepsilon)^{i+1}M$, then set $q = 1$ and $F_1' = (\tau\backslash h(\tau)) \cup \text{Dangle}(\tau)$, and go to Step (8.5);
   (8.3) Find an Euler tour in the multigraph $((\tau\backslash h(\tau)) \cup \text{Dangle}(\tau)) + ((\tau\backslash h(\tau)) \cup \text{Dangle}(\tau))$, and transform the tour to a cycle by "short-cutting" previously visited vertices;
   (8.4) Split the resulting cycle into maximal paths of total weight, including edge weights and vertex weights, at most $\zeta(1+\varepsilon)^{i+1}M$ each, denoted by $F_1', F_2', \ldots, F_q'$;
   (8.5) For each $j \in \{1, \ldots, q\}$, augment F_j' by adding an edge from r to the vertex in F_j' closest to r, to obtain a set of subtrees starting at r, denoted by $\mathcal{F}_i(\tau) = \{F_1, F_2, \ldots, F_q\}$;

Step 9 For each $i \in \{0, 1, \ldots, t\}$, set $\mathcal{F}_i = \bigcup_{\tau \in \mathcal{S}_i^m} \mathcal{F}_i(\tau)$; Using the ASSIGN algorithm, combine the set $\{\mathcal{F}_i\}_{i \ge 0} = \bigcup_{i \ge 0} \mathcal{F}_i$ into k trees $\mathcal{T} = \{T_i \mid i = 1, 2, \ldots, k\}$ corresponding to k construction teams;

Step 10 Output k trees $\mathcal{T} = \{T_i \mid i = 1, 2, \ldots, k\}$ corresponding to k teams.
End

Using Step 4, we obtain that each connected component of the subgraph in G corresponding to H_i ($i \geq 1$) is a subtree. Note that such a subtree contains at least one edge from $V_{<i}$ to V_i, it follows that the partition at Step 5 is indeed executed.

Using Steps 6–7 in the algorithm HRTC$_1$, we can obtain the following

Lemma 5. *For any $i \geq 1$ and $\sigma \in \mathcal{S}_i^u$, there exists $\pi(\sigma) \in \mathcal{S}_{i-1}$. Moreover, $\pi(\sigma) \in \mathcal{S}_{i-1}^m$.*

Proof. For an $\sigma \in \mathcal{S}_i^u$, it is clear that $w(h(\sigma)) + f_1(h(\sigma)) \leq w(\sigma) + f_1(\sigma) < \gamma \cdot (1 + \varepsilon)^i M$. By the definition of $\mathcal{S}_i^u$, we have $v \in V_{\leq i}$ for every $v \in V(\sigma)$. Let $h(\sigma) = yx$, where $y \in \pi(\sigma)$ and $x \in V_i$. Moreover, we deduce that $y \in V_{i-1}$. Otherwise, we assume $x \in V_i$ and $y \in V_{<i-1}$, it follows that $w(\sigma) + f_1(\sigma) \geq w(yx) + f(x) \geq w(rx) + f(x) - (w(ry) + f(y)) > (1 + \varepsilon)^{i-1} M - (1 + \varepsilon)^{i-2} M > \gamma \cdot (1 + \varepsilon)^i M$, which contradicts $\sigma \in \mathcal{S}_i^u$. Thus, we obtain $y \in V_{i-1}$, implying $\pi(\sigma) \in \mathcal{S}_{i-1}$.

For the second part of the lemma, if $i = 1$, then clearly $\pi(\sigma) = \mathcal{S}_0$ and $\pi(\sigma) \in \mathcal{S}_0^m$. When $i \geq 2$, from the above arguments, we have $\pi(\sigma) \in \mathcal{S}_{i-1}$. Similar to the above arguments, since G is connected, we conclude that there is a vertex $z \in \pi(\sigma)$ satisfying $z \in V_{<i-1}$. Using the triangle inequality twice, we have the following

$$w(zy) + f(y) + w(yx) + f(x) \geq w(zx) + f(x) \geq w(rx) + f(x) - (w(rz) + f(z)).$$

Since $x \in V_i$ and $z \in V_{<i-1}$, we obtain $w(rx) + f(x) - (w(rz) + f(z)) > (1 + \varepsilon)^{i-1} M - (1+\varepsilon)^{i-2} M = \varepsilon(1+\varepsilon)^{i-2} M$, meaning $w(rx) + f(x) - (w(rz) + f(z)) > \varepsilon(1+\varepsilon)^{i-2} M$. Since $\sigma \in \mathcal{S}_i^u$, we have $w(yx) + f(x) \leq w(\sigma) + f_1(\sigma) < \gamma \cdot (1+\varepsilon)^i M$. Hence, we have the following

$$w(\pi(\sigma)) + f_1(\pi(\sigma)) \geq w(zy) + f(y) > \varepsilon(1+\varepsilon)^{i-2} M - \gamma \cdot (1+\varepsilon)^i M = \gamma \cdot (1+\varepsilon)^{i-1} M.$$

This shows that the subtree $\pi(\sigma) \in \mathcal{S}_{i-1}^m$.   □

Employing the similar argument as in [8], we obtain the following two lemmas by executing Step 8.

Lemma 6. *For any $F \in \mathcal{F}_i(\tau)$, we have $w(F) + f(F) \leq (\zeta + 1 + (\zeta + 1)\varepsilon)(1 + \varepsilon)^i M$.*

Proof. For each $F \in \mathcal{F}_i(\tau)$, note that F consists of some subtree F_j' ($1 \leq j \leq q$) and an edge rv_j' connecting r to v_j', where v_j' is a vertex in F_j' closest to r. Based on the construction of F_j', we obtain $w(F_j') + f(F_j') \leq \zeta(1 + \varepsilon)^{i+1} M$. Since F_j' only contains vertices in $V_{\leq i+1}$, we have $w(rv_j') \leq w(rv_j') + f(v_j') \leq (1+\varepsilon)^{i+1} M$. Hence, it follows that $w(F) + f(F) = w(F_j') + f(F_j') + w(rv_j') \leq (\zeta + 1 + (\zeta + 1)\varepsilon)(1 + \varepsilon)^i M$.   □

Lemma 7. $\sum_{F \in \mathcal{F}_i(\tau)}(w(F) + f(F)) \leq \max\{2 + \frac{2}{\varepsilon}, \frac{4}{\zeta} + 2\} \cdot (w(\tau \cup \mathrm{Dangle}(\tau)) + f_1(\tau \cup \mathrm{Dangle}(\tau)))$.

Proof. We break the analysis into two cases depending on q.

Case 1: $q = 1$, *i.e.*, $\mathcal{F}_i(\tau)$ only contains a subtree F.

If $i = 0$, then τ includes r and $w(F) + f(F) \leq w(\tau \cup \mathrm{Dangle}(\tau)) + f_1(\tau \cup \mathrm{Dangle}(\tau))$. If $i > 0$, it is clear that there exists $u \in \tau$ having $u \in V_{<i}$. Based on the construction of subtree, we obtain $w(F) + f(F) \leq w(rx_{h(\tau)}) + w(F') + f_1(\tau \cup \mathrm{Dangle}(\tau)) \leq w(ry_{h(\tau)}) + w(y_{h(\tau)}x_{h(\tau)}) + w(F') + f_1(\tau \cup \mathrm{Dangle}(\tau)) \leq (1 + \varepsilon)^{i-1}M + w(\tau \cup \mathrm{Dangle}(\tau)) + f_1(\tau \cup \mathrm{Dangle}(\tau))$. Since $\tau \in \mathcal{S}_i^m$, implying $w(\tau) + f_1(\tau) \geq \gamma \cdot (1 + \varepsilon)^i M$, we have $(1 + \varepsilon)^{i-1}M \leq \frac{2+\varepsilon}{\varepsilon} \cdot (w(\tau) + f_1(\tau))$. Thus, we obtain $w(F) + f(F) \leq (1 + \varepsilon)^{i-1}M + w(\tau \cup \mathrm{Dangle}(\tau)) + f_1(\tau \cup \mathrm{Dangle}(\tau)) \leq (\frac{2+\varepsilon}{\varepsilon} + 1) \cdot (w(\tau \cup \mathrm{Dangle}(\tau)) + f_1(\tau \cup \mathrm{Dangle}(\tau))) = (2 + \frac{2}{\varepsilon}) \cdot (w(\tau \cup \mathrm{Dangle}(\tau)) + f_1(\tau \cup \mathrm{Dangle}(\tau)))$, which implies $\sum_{F \in \mathcal{F}_i(\tau)}(w(F) + f(F)) = w(F) + f(F) \leq (2 + \frac{2}{\varepsilon}) \cdot (w(\tau \cup \mathrm{Dangle}(\tau)) + f_1(\tau \cup \mathrm{Dangle}(\tau)))$.

Case 2: $q \geq 2$.

By Step 8, we obtain that $4(w(\tau \cup \mathrm{Dangle}(\tau)) + f_1(\tau \cup \mathrm{Dangle}(\tau))) > (q - 1) \cdot \zeta(1 + \varepsilon)^{i+1}M$, that is $(1 + \varepsilon)^{i+1}M < \frac{4(w(\tau \cup \mathrm{Dangle}(\tau)) + f_1(\tau \cup \mathrm{Dangle}(\tau)))}{\zeta(q-1)}$. Since $V(\tau \cup \mathrm{Dangle}(\tau)) \subseteq V_{\leq i+1}$, each edge added from r to subtree F'_j ($1 \leq j \leq q$) has weight at most $(1 + \varepsilon)^{i+1}M$. Therefore, we conclude that $\sum_{F \in \mathcal{F}_i(\tau)}(w(F) + f(F)) \leq q \cdot (1 + \varepsilon)^{i+1}M + 2w(\tau \cup \mathrm{Dangle}(\tau)) + f_1(\tau \cup \mathrm{Dangle}(\tau)) \leq (\frac{4q}{\zeta(q-1)} + 2) \cdot (w(\tau \cup \mathrm{Dangle}(\tau)) + f_1(\tau \cup \mathrm{Dangle}(\tau))) \leq (\frac{4}{\zeta} + 2) \cdot (w(\tau \cup \mathrm{Dangle}(\tau)) + f_1(\tau \cup \mathrm{Dangle}(\tau)))$.

Combining the two preceding arguments in Cases 1–2, we obtain $\sum_{F \in \mathcal{F}_i(\tau)}(w(F) + f(F)) \leq \max\{2 + \frac{2}{\varepsilon}, \frac{4}{\zeta} + 2\} \cdot (w(\tau \cup \mathrm{Dangle}(\tau)) + f_1(\tau \cup \mathrm{Dangle}(\tau)))$. □

Applying Lemmas 5–7, we obtain the following

Lemma 8. *If* $w(T_{G/V_{<i}}^{MS}) + f(V_{\geq i}) \leq M \cdot \Lambda((1 + \varepsilon)^{i-1})$ *holds for each integer* $i \geq 0$, *then the collection* $\{\mathcal{F}_i\}_{i \geq 0}$ *obtained at Step 8 is* $(\alpha, \beta)_{M,\varepsilon}$*-assignable, where* $\alpha = \zeta + 1 + (\zeta + 1)\varepsilon$, $\beta = (6 + \frac{6}{\varepsilon}) \max\{2 + \frac{2}{\varepsilon}, \frac{4}{\zeta} + 2\}$ *and* $\mathcal{F}_i = \bigcup_{\tau \in \mathcal{S}_i^m} \mathcal{F}_i(\tau)$.

Proof. We shall prove that the collection $\{\mathcal{F}_i\}_{i \geq 0}$ satisfies the two properties in Definition 1. By Lemma 6, it is clear that the property (1) in Definition 1 holds. Recall that in Lemma 4, $\sum_{j \geq i}(w(H_j) + f(V_j)) \leq (6 + \frac{6}{\varepsilon}) \cdot M \cdot \Lambda((1 + \varepsilon)^{i-1})$ holds for each $i \geq 0$. Now, the proof is completed by showing that $\sum_{j \geq i}(w(\mathcal{F}_j) + f(\mathcal{F}_j)) \leq \max\{2 + \frac{2}{\varepsilon}, \frac{4}{\zeta} + 2\} \cdot \sum_{j \geq i}(w(H_j) + f(V_j))$.

In the algorithm HRTC₁, we observe that

$$\sum_{j \geq i}(w(\mathcal{S}_j^m) + w(\mathcal{S}_{j+1}^u)) \leq \sum_{j \geq i} w(\mathcal{S}_j) = \sum_{j \geq i} w(H_j).$$

By Lemma 5, $\{\mathrm{Dangle}(\tau) \mid \tau \in \mathcal{S}_j^m\}$ is a partition of $\mathcal{S}_{j+1}^u$, which means $\sum_{\tau \in \mathcal{S}_j^m}(w(\tau \cup \mathrm{Dangle}(\tau)) + f_1(\tau \cup \mathrm{Dangle}(\tau))) = w(\mathcal{S}_j^m) + w(\mathcal{S}_{j+1}^u) + f_1(\mathcal{S}_j^m \cup \mathcal{S}_{j+1}^u)$. By Lemma 7, we have $w(\mathcal{F}_j) + f(\mathcal{F}_j) = \sum_{\tau \in \mathcal{S}_j^m}(w(\mathcal{F}_j(\tau)) + f(\mathcal{F}_j(\tau))) = \sum_{\tau \in \mathcal{S}_j^m} \sum_{F \in \mathcal{F}_j(\tau)}(w(F) + f(F)) \leq \max\{2 + \frac{2}{\varepsilon}, \frac{4}{\zeta} + 2\} \cdot \sum_{\tau \in \mathcal{S}_j^m}(w(\tau \cup \mathrm{Dangle}(\tau)) +$

$f_1(\tau \cup \text{Dangle}(\tau)))$, implying $w(\mathcal{F}_j) + f(\mathcal{F}_j) \leq \max\{2 + \frac{2}{\varepsilon}, \frac{4}{\zeta} + 2\} \cdot (w(\mathcal{S}_j^m) + w(\mathcal{S}_{j+1}^u) + f_1(\mathcal{S}_j^m \cup \mathcal{S}_{j+1}^u))$. By Steps 6–8, for each edge $e \in \mathcal{S}_i^m$ $(i \geq 0)$, we see that $x_e \in V_{\geq i}$, and any two subtrees in $\{\mathcal{F}_i\}_{i \geq 0}$ are disjoint except for root r. Thus, we obtain $\sum_{j \geq i}(w(\mathcal{F}_j) + f(\mathcal{F}_j)) \leq \max\{2 + \frac{2}{\varepsilon}, \frac{4}{\zeta} + 2\} \cdot \sum_{j \geq i}(w(\mathcal{S}_j^m) + w(\mathcal{S}_{j+1}^u) + f_1(\mathcal{S}_j^m \cup \mathcal{S}_{j+1}^u) \leq \max\{2 + \frac{2}{\varepsilon}, \frac{4}{\zeta} + 2\} \cdot \sum_{j \geq i}(w(H_j) + f(V_j))$.   □

Using the above lemmas, we obtain the following result.

Theorem 1. *The algorithm $HRTC_1$ is a $58.3286(1 + \delta)$-approximation algorithm to solve the HRTC problem, and it runs in time $O(n^3(1 + \frac{1}{\delta}) + \log(w(E) + f(V\backslash\{r\})))$, where $w(E) = \sum_{e \in E} w(e)$ and $f(V\backslash\{r\}) = \sum_{v \in V\backslash\{r\}} f(v)$, respectively.*

Proof. By Lemma 2, the decision condition in Step 3 does not hold whenever $M \geq OPT$. Based on the update rule for M, we deduce that Steps 4–9 is executed with $M \leq (1 + \delta) \cdot OPT$. When fixing $\zeta = 2$ and $\varepsilon = 1.3146$, using Lemma 8, we obtain a $(6.9438, 42.2565)_{M,1.3146}$-assignable collection $\{\mathcal{F}_i\}_{i \geq 0}$. Using Lemma 1 at Step 9, we can assign $\{\mathcal{F}_i\}_{i \geq 0}$ into k construction teams in time $O(n^3 + \log(w(E) + f(V\backslash\{r\})))$, such that the completion time of any construction team is at most $58.3286 \cdot M$, which implies $OUT \leq 58.3286 \cdot M \leq 58.3286(1 + \delta) \cdot OPT$.

Notice that every step in the algorithm $HRTC_1$ can be executed in polynomial time. We shall bound the number of iterations. As mentioned above, the algorithm $HRTC_1$ halts before $M > (1 + \delta)OPT$, where $(1 + \delta)OPT \leq (1 + \delta) \cdot \frac{\sum_{v \in V}(w(rv) + f(v))}{\lambda_{\max}} \leq (1 + \delta) \cdot |V| \cdot \frac{\max_{v \in V}\{w(rv) + f(v)\}}{\lambda_{\max}}$. Since M is initialized at $\frac{\max_{v \in V}\{w(r,v) + f(v)\}}{\lambda_{\max}}$, and increased by an $(1 + \delta)$-factor for each iteration, we deduce that the number of iterations is at most $O(\frac{1}{\delta} \log |V|)$. This implies that Steps 1–3 run in at most time $O(\frac{n^3}{\delta})$. By Lemma 1, it is easy to check that Steps 4–10 execute in time $O(n^3 + \log(w(E) + f(V\backslash\{r\})))$. Hence, the algorithm $HRTC_1$ can be implemented in time $O(n^3(1 + \frac{1}{\delta}) + \log(w(E) + f(V\backslash\{r\})))$.   □

4   An Approximation Algorithm with Lower Time Complexity

In practice, we observe a fact that $\frac{\lambda_{\max}}{\lambda_{\min}}$ is generally small, where $\lambda_{\max} = \max\{\lambda_i \mid i = 1, 2, \ldots, k\}$ and $\lambda_{\min} = \min\{\lambda_i \mid i = 1, 2, \ldots, k\}$. Thus, we intend to design a better approximation algorithm to resolve the HRTC problem under the above fact.

Different from the method in [10] for solving the RTC problem, we modify a splitting technique in [6] to design an approximation algorithm to resolve the HRTC problem, which is described as follows.

Algorithm: $HRTC_2$

INPUT: An undirected complete graph $G = (V, E; w, f)$ with a root $r \in V$, an edge-weight function $w : E \rightarrow R^+$, a vertex-weight function $f : V\backslash\{r\} \rightarrow R_0^+$ and k construction teams having speeds $\lambda_1, \ldots, \lambda_k$, respectively;

OUTPUT: A set $\mathcal{T} = \{T_i \mid i = 1, 2, \ldots, k\}$ of trees.

Begin

Step 1 Find a minimum edge-weight spanning tree T^{MS} in G; Determine an Euler tour in $(V, E_{T^{MS}} + E_{T^{MS}})$ traversing each edge exactly once, and use "short-cutting" to transform such tour to a cycle $C = rv_{i_1}v_{i_2}\cdots r$ traversing each vertex $v \in V$ exactly once;

Step 2 Set $w_0 = \max\{w(rv) \mid v \in V\}$; For each edge $uv \in E$, set $w'(uv) = w(uv) + f(u) + f(v)$; For each $i \in \{1, 2, \ldots, k\}$, set $\lambda_{1,i} = \sum_{j=1}^{i} \lambda_j$;

Step 3 For $j = 1$ to $k - 1$ do:
Set $L_j = \frac{\lambda_{1,j}}{\lambda_{1,k}}(w'(C) - 2w_0) + w_0$, find the last vertex $v_{i_{p(j)}}$ such that $w'(C[r, v_{i_{p(j)}}]) \leq L_j$, where $C[r, v_{i_{p(j)}}] = rv_{i_1}v_{i_2}\cdots v_{i_{p(j)}}$;

Step 4 Set $T_1' = C[r, v_{i_{p(1)}}] = rv_{i_1}\cdots v_{i_{p(1)}}$, $T_2' = C[v_{i_{p(1)}+1}, v_{i_{p(2)}}]$, $\ldots$, $T_k' = C[v_{i_{p(k-1)}+1}, r]$; For each $j \in \{1, 2, \ldots, k\}$, augment T_j' by connecting r to a vertex in T_j' closest to r with edge, where the resulting tree is denoted by T_j;

Step 5 Output k trees $\mathcal{T} = \{T_i \mid i = 1, 2, \ldots, k\}$ corresponding to k teams.

End

Analyzing the lower bound of the optimal value for the HRTC problem, we obtain the following result.

Lemma 9. *Given a complete graph $G = (V, E; w, f)$ as an instance of the HRTC problem, we have $OPT \geq \max\{\frac{f_0}{\lambda_{max}}, \frac{w_0}{\lambda_{max}}, \frac{w'(C)}{2\lambda_{1,k}}\}$, where OPT is the optimal value to the given instance, $f_0 = \max\{f(v) \mid v \in V\}$, $w_0 = \max\{w(rv) \mid v \in V\}$ and C is produced at Step 1 of the algorithm $HRTC_2$.*

Proof. Note that any feasible solution for the instance G cover all vertices in V, it is clear that $OPT \geq \frac{f_0}{\lambda_{max}}$ and $OPT \geq \frac{w_0}{\lambda_{max}}$. We shall prove $OPT \geq \frac{w'(C)}{2\lambda_{1,k}}$.

Consider any optimal solution $\mathcal{T}^* = \{T_i^* \mid i = 1, 2, \ldots, k\}$ for the instance G. By the construction of C at Step 1, we have $w(T^{MS}) + f(T^{MS}) \geq \frac{w(C)}{2} + f(C)$, where T^{MS} is a minimum edge-weight spanning tree of G. Since all subtrees in $\mathcal{T}^*$ can be merged into a spanning tree, we obtain $\sum_{i=1}^{k}(w(T_i^*) + f(T_i^*)) \geq w(T^{MS}) + f(T^{MS})$, implying $\sum_{i=1}^{k}(w(T_i^*) + f(T_i^*)) \geq \frac{w(C)}{2} + f(C)$. Since $OPT = \max\{\frac{w(T_i^*) + f(T_i^*)}{\lambda_i} \mid i = 1, 2, \ldots, k\}$, it follows that $OPT \cdot \sum_{i=1}^{k} \lambda_i \geq \sum_{i=1}^{k}(w(T_i^*) + f(T_i^*)) \geq \frac{w(C)}{2} + f(C)$, meaning $OPT \geq \frac{w(C) + 2f(C)}{2\sum_{i=1}^{k} \lambda_i} = \frac{w(C) + 2f(C)}{2\lambda_{1,k}}$. This implies $OPT \geq \frac{w(C) + 2f(C)}{2\lambda_{1,k}} = \frac{w'(C)}{2\lambda_{1,k}}$. $\square$

By the algorithm $HRTC_2$, we obtain the following result.

Theorem 2. *The algorithm $HRTC_2$ is a $\max\{\frac{2\lambda_{max}}{\lambda_{min}}, 2 + \frac{\lambda_{max}}{\lambda_{min}} - \frac{2}{k}\}$-approximation algorithm for resolving the HRTC problem, and it runs in time $O(n^2)$, where n is the number of vertices.*

Proof. Given an instance $G = (V, E; w, f)$ of the HRTC problem, we may assume that $\mathcal{T}^* = \{T_i^* \mid i = 1, 2, \ldots, k\}$ is an optimal solution with the optimal value $OPT = \max\{\frac{w(T_i^*) + f(T_i^*)}{\lambda_i} \mid i = 1, \ldots, k\}$, and $\mathcal{T}$ is trees outputted by the algorithm $HRTC_2$ with the output value $OUT = \max\{\frac{w(T_i) + f(T_i)}{\lambda_i} \mid i = 1, \ldots, k\}$.

Now, we consider the jth tree T_j $(1 \le j \le k)$ in $\mathcal{T}$. Using the algorithm HRTC$_2$, we obtain the following

$$\frac{w(T_j) + f(T_j)}{\lambda_j} \le \frac{w(T'_j) + f(T'_j) + w_0}{\lambda_j}$$

$$\le \max\{\frac{f_0}{\lambda_j}, \frac{w'(T'_j)}{\lambda_j}\} + \frac{w_0}{\lambda_j}$$

$$= \max\{\frac{f_0 + w_0}{\lambda_j}, \frac{w'(T'_j) + w_0}{\lambda_j}\}.$$

From Lemma 9, it may be concluded that $\frac{f_0 + w_0}{\lambda_j} = \frac{f_0}{\lambda_j} + \frac{w_0}{\lambda_j} \le \frac{f_0}{\lambda_{min}} + \frac{w_0}{\lambda_{min}} \le \frac{2\lambda_{max}}{\lambda_{min}}OPT$. On account of the construction of $\mathcal{T}$ in algorithm, we have the following

$$\frac{w'(T'_j) + w_0}{\lambda_j} \le \frac{\frac{\lambda_j}{\lambda_{1,k}} \cdot (w'(C) - 2w_0) + w_0}{\lambda_j}$$

$$= \frac{w_0}{\lambda_j} + \frac{w'(C) - 2w_0}{\lambda_{1,k}}$$

$$\le \frac{w_0}{\lambda_{min}} + \frac{w'(C)}{\lambda_{1,k}} - \frac{2w_0}{\lambda_{1,k}}$$

$$= \frac{\lambda_{max} \cdot w_0}{\lambda_{min} \cdot \lambda_{max}} + \frac{w'(C)}{\lambda_{1,k}} - \frac{2w_0}{\lambda_{1,k}}$$

$$\le \frac{\lambda_{max} \cdot w_0}{\lambda_{min} \cdot \lambda_{max}} + \frac{w'(C)}{\lambda_{1,k}} - \frac{2w_0}{k \cdot \lambda_{max}}$$

$$= \frac{w'(C)}{\lambda_{1,k}} + (\frac{\lambda_{max}}{\lambda_{min}} - \frac{2}{k}) \cdot \frac{w_0}{\lambda_{max}}$$

$$\le 2OPT + (\frac{\lambda_{max}}{\lambda_{min}} - \frac{2}{k}) \cdot OPT$$

$$= (2 + \frac{\lambda_{max}}{\lambda_{min}} - \frac{2}{k}) \cdot OPT,$$

implying $\frac{w(T_j) + f(T_j)}{\lambda_j} \le \max\{\frac{2\lambda_{max}}{\lambda_{min}}, 2 + \frac{\lambda_{max}}{\lambda_{min}} - \frac{2}{k}\} \cdot OPT$.

Thus, for each $j \in \{1, \ldots, k\}$, we have $\frac{w(T_j) + f(T_j)}{\lambda_j} \le \max\{\frac{2\lambda_{max}}{\lambda_{min}}, 2 + \frac{\lambda_{max}}{\lambda_{min}} - \frac{2}{k}\} \cdot OPT$ by using the above arguments. This shows that

$$OUT \le \max\{\frac{2\lambda_{max}}{\lambda_{min}}, 2 + \frac{\lambda_{max}}{\lambda_{min}} - \frac{2}{k}\} \cdot OPT.$$

The time complexity of the algorithm HRTC$_2$ can be determined as follows. (1) Using Prim algorithm for solving the minimum spanning tree problem, Step 1 execute in time $O(n^2)$; (2) Step 2 needs $O(m)$ time to compute w_0 and define $w'(\cdot)$, where $m = |E|$; (3) Step 3 needs time $O(n^2)$ to split a cycle; (4) Step 4 needs time $O(m)$ to construct the trees $\mathcal{T} = \{T_i \mid i = 1, 2, \ldots, k\}$. Therefore, the running time of the algorithm HRTC$_2$ is $O(n^2)$. □

5   Conclusion and Further Work

In this paper, we consider the heterogeneous rooted tree cover problem (the HRTC problem), and design two approximation algorithms for solving the HRTC problem.

In further work, it is a challenge for us to design some approximation algorithms with constant approximation ratios to solve the HRTC problem in strongly polynomial time, and we shall study other versions of the cover problems with nonuniform speeds.

References

1. Averbakh, I., Berman, O.: A heuristic with worst-case analysis for minimax routing of two travelling salesmen on a tree. Discret. Appl. Math. **68**(1–2), 17–32 (1996)
2. Campbell, A.M., Vandenbussche, D., Hermann, W.: Routing for relief efforts. Transp. Sci. **42**(2), 127–145 (2008)
3. Christofides, N.: Worst-case analysis of a new heuristic for the travelling salesman problem, Report 388. Carnegie Mellon University, Graduate School of Industrial Administration (1976)
4. Even, G., Garg, N., Koemann, J., Ravi, R., Sinha, A.: Min-max tree covers of graphs. Oper. Res. Lett. **32**(4), 309–315 (2004)
5. Farbstein, B., Levin, A.: Min-max cover of a graph with a small number of parts. Discret. Optim. **16**, 51–61 (2015)
6. Frederickson, G.N., Hecht, M.S., Kim, C.E.: Approximation algorithms for some routing problems. SIAM J. Comput. **7**(2), 178–193 (1978)
7. Golden, B.L., Raghavan, S., Wasil, E.A.: The Vehicle Routing Problem: Latest Advances and New Challenges. Springer, New York (2008). https://doi.org/10.1007/978-0-387-77778-8
8. Gørtz, I.L., Molinaro, M., Nagarajan, V., Ravi, R.: Capacitated vehicle routing with nonuniform speeds. Math. Oper. Res. **41**(1), 318–331 (2016)
9. Nagamochi, H., Okada, K.: Polynomial time 2-approximation algorithms for the minmax subtree cover problem. In: Ibaraki, T., Katoh, N., Ono, H. (eds.) ISAAC 2003. LNCS, vol. 2906, pp. 138–147. Springer, Heidelberg (2003). https://doi.org/10.1007/978-3-540-24587-2_16
10. Nagamochi, H.: Approximating the minmax rooted-subtree cover problem. IEICE Trans. Fundamentals Electron. Commun. Comput. Sci. **E88-A**(5), 1335–1338 (2005)
11. Schwartz, S.: An overview of graph covering and partitioning. Discret. Math. **345**(8), 112884 (2022)
12. Toth, P., Vigo, D.: Vehicle Routing: Problems. Methods and Applications. MOS-SIAM, Philadelphia (2014)
13. Wu, W., Zhang, Z., Lee, W., Du, D.Z.: Optimal Coverage in Wireless Sensor Networks. Springer, Heidelberg (2020). https://doi.org/10.1007/978-3-030-52824-9
14. Xu, Z., Wen, Q.: Approximation hardness of min-max tree covers. Oper. Res. Lett. **38**(3), 169–173 (2010)
15. Xu, W., Liang, W., Lin, X.: Approximation algorithms for min-max cycle cover problems. IEEE Trans. Comput. **64**(3), 600–613 (2015)
16. Yu, W., Liu, Z.: Better approximability results for min-max tree/cycle/path cover problems. J. Comb. Optim. **37**, 563–578 (2019)

The Hardness of Optimization Problems on the Weighted Massively Parallel Computation Model

Hengzhao Ma[✉] and Jianzhong Li[✉]

Shenzhen Institute of Advanced Technology, Chinese Academy of Sciences,
Shenzhen, China
{hz.ma,lijzh}@siat.ac.cn

Abstract. The topology-aware Massively Parallel Computation (MPC) model is proposed and studied recently, which enhances the classical MPC model by the awareness of network topology. The work of Hu et al. on topology-aware MPC model considers only the tree topology. In this paper a more general case is considered, where the underlying network is a weighted complete graph. We then call this model Weighted Massively Parallel Computation (WMPC) model, and study the problem of minimizing communication cost under it. Two communication cost minimization problems are defined based on different pattern of communication, which are the Data Redistribution Problem and Data Allocation Problem. We also define four kinds of objective functions for communication cost, which consider the total cost, bottleneck cost, maximum of send and receive cost, and summation of send and receive cost, respectively. Combining the two problems in different communication pattern with the four kinds of objective cost functions, 8 problems are obtained. The hardness results of the 8 problems make up the content of this paper. With rigorous proof, we prove that some of the 8 problems are in P, some FPT, some NP-complete, and some W[1]-complete.

Keywords: massively parallel computation · Weighted MPC model · communication cost optimization

1 Introduction

The Massively Parallel Computation model [14], MPC for short, has been a well acknowledged model to study parallel algorithms [2–5,9,11,16,19] ever since it was proposed. Compared to other parallel computation models such as PRAM [15], BSP [20], LogP [8] and so on, the advantage of the MPC model lies in its simplicity and the power to capture the essence of computation procedure of modern share-nothing clusters. In the MPC model, computation proceeds in

This work was supported by the National Natural Science Foundation of China under grants 61832003, 62273322, 61972110, and National Key Research and Development Program of China under grants 2021YFF1200100 and 2021YFF1200104.

W. Wu and G. Tong (Eds.): COCOON 2023, LNCS 14423, pp. 106–117, 2024.
https://doi.org/10.1007/978-3-031-49193-1_9

synchronous rounds, where in each round the computation machines first communicate with each other, then conduct local computation. Any pair of machines can communicate in a point-to-point manner, and all the communication messages can be transferred without congestion.

Although the MPC model is simple and powerful, one of its most important shortcomings is revealed by some recent works [6,13], which is the strong assumption of *homogeneity*. All the machines in MPC model are considered as identical, and the communication bandwidth between any pair of machines are identical too [14]. In realistic parallel environment, the assumption of identical computation machines can be satisfied in most cases, but the assumption of identical communication bandwidth can not. Typically, a cluster consists of several racks connected by slower communication channels, and each rack includes several machines connected by faster communication channels. Thus, the communication bandwidth of in-rack and across-rack communication differ significantly, which refutes the assumption of homogeneous communication network in MPC model.

In order to tackle this shortcoming of the MPC model, a new *topology aware massively parallel computation* model was proposed and studied in [6,13]. The computation machines are still identical in this model[1], but the communication bandwidth between different pair of machines are different. This model was first proposed in recent works [6], where the underlying communication network is represented as a graph, and the edges are assigned with a weight which represents the communication bandwidth. However, the paper [6] only declared the new model but did not give any theoretical results. The other work [13] considered three data processing tasks on this model, which are set intersection, Cartesian product and sorting. Algorithms and lower bounds about the communication cost optimization problems for the three tasks were proposed. However, the authors of [13] restricted the underlying communication network to trees, and the algorithm and lower bounds given in that paper can not be generalized to graphs other than trees.

In this paper, we follow the line of research started by [6,13], and consider the topology aware massively parallel computation model in a more general case, where the underlying communication network is a complete weighted graph. In this sense, our work is a complement to the work in [13]. The goal of this paper is also to minimize the communication cost. However, unlike the work in [13] which considers specific computation tasks, in this paper we define general communication cost minimization problems that capture the characteristics of a variety of computation tasks.

1.1   Description of the Research Problems in This Paper

The WMPC Model. We first give a more detailed description of the computational model considered in this paper, which is called Weighted Massively Parallel Computation (WMPC) model.

In WMPC model, there are n computation machines with identical computational power. The communication network is modeled as a weighted complete

---

[1] There may be non-computational machines in this model, though.

graph represented by a $n \times n$ matrix C. C is called the communication cost matrix from now on, and it is considered as a known parameter of the WMPC model. $C[i,j]$ is the communication cost from computation machine i to j for $1 \leq i, j \leq n$, where larger value implies larger communication cost or communication latency. $C[i,i]$ is set to 0 for $1 \leq i \leq n$. It is assumed that all pairs of machines can communicate in a point-to-point way which is in accordance with the original MPC model, and thus $C[i,j] < \infty$ holds for $1 \leq i, j \leq n$. The matrix C is not necessary to be symmetric, i.e., $C[i,j]$ may not be equal to $C[j,i]$.

The computation on WMPC proceeds in synchronous rounds which behaves the same with the original MPC model. In each round, the computation machines first communicate with each other, then conduct local computation.

The initial data distribution is important in the problems studied in this paper. A lot of former research works on MPC model assume that the data are uniformly split across the machines [4,11]. In this paper, it is assumed that the data can be arbitrarily distributed, and the amount of data placed at each machine is known in advance. This is also the same assumption adopted in [5,13].

Objective Functions. The goal of this paper is to minimize the communication cost under WMPC model, which is divided into *send cost* and *receive cost*. If α amount of data is transferred from machine i to machine j, it incurs $\alpha \cdot C[i,j]$ send cost to machine i, and $\alpha \cdot C[i,j]$ receive cost to machine j. Denote $send_i$ and rcv_i to be the send and receive cost of machine i for $1 \leq i \leq n$, then we define the following four objective functions.

Total cost (TOTAL): $\sum_{i=1}^{n} send_i$.

Bottleneck cost (BTNK): $\max_{i=1}^{n} rcv_i$.

Maximum of send and receive cost (MSR): $\max_{i=1}^{n}\{send_i, rcv_i\}$.

Sum of send and receive cost (SSR): $\max_{i=1}^{n}\{send_i + rcv_i\}$.

Note that the send and receive cost is defined based on the amount of data transferred between two machines. For different computation task and different communication pattern, the way of calculating the amount of transferred data will be different. Next we will use parallel sorting as the introducing example, analyze their communication patterns, and define the problems to be studied in this paper. We will introduce two problems, named Data Redistribution Problem and Data Allocation Problem.

The Data Redistribution Problem. Consider the following parallel sorting algorithm on classical MPC model, which is often referred as TeraSort [18]. The algorithm first selects $n-1$ splitters $s_1 \leq s_2 \leq \cdots \leq s_{n-1}$ and broadcast the splitters to all machines. The $n-1$ splitters form n intervals $I_i = (s_{i-1}, s_i]$ where $s_0 = -\infty$ and $s_n = \infty$. After obtaining the splitters, each machine sends the local data falling in the i-th interval to the i-th machine. In such way the data is ordered across the machines. Note that the label of the machines are fixed before the algorithm starts. Then the machines conduct local sorting, and the sorting task can be finished.

Now consider running the parallel sorting algorithm on the WMPC model, and assume that the splitters have been determined. The algorithm described above asks the data in the i-th interval to be sent to the i-th machine. However, this operation may lead to non-optimal communication cost. Consider the following extreme case. The data are initially inversely sorted across the machines, i.e., for machine $i < j$, the data in machine i are always no less than the data in machine j. In such a case, if the i-th interval is assigned to the $(n-i)$-th machine, there would be no need to conduct communication. However, if the algorithm asks to send the data in the i-th interval to the i-th machine, all the data will be totally redistributed, incurring large amount of communication.

Actually, there exist two shortcomings for the above TeraSort algorithm on classical MPC model. First, it neglects the initial data distribution, and neglects the importance of the way to assign the intervals to the machines to minimize the communication cost. Second, it does not consider the difference of communication costs between different pair of machines. By tackling these two points together, the first research problem to be studied in this paper is formed, which is called the Data Redistribution Problem (DRP).

The input of DRP is two $n \times n$ matrices T and C. $T[i,j]$ represents the amount of data in the i-th machine that fall in the j-th interval. The C matrix is the communication cost matrix of the WMPC model. The output is to assign the intervals to the machines, such that the communication cost is minimized. By applying the four communication cost functions introduced in Sect. 1.1, we get four problems denoted as DRP-TOTAL, DRP-BTNK, DRP-MSR and DRP-SSR, respectively. The four problems are studied in Sect. 2.

The Data Allocation Problem. In the above case of parallel sorting, it is assumed that the splitters are known in advance. However, how to select the splitters to minimize the communication cost is also an important research problem [19], and even a new problem under the WMPC model. For a formal description, let N be the total number of data records to be sorted, and n be the number of machines. Under the assumption that the initial data distribution is known in advance, let $S_i = \{s_{i,1}, s_{i,2}, \cdots, s_{i,l_i}\}$, $1 \le i \le n$, which is the data initially residing in machine i. l_i is the number of data records in machine i, and $\sum_{i=1}^{n} l_i = N$. If the splitters are chosen as $s_1, s_2, \cdots, s_{n-1}$, they will form n intervals $(s_{j-1}, s_j]$, where $s_0 = -\infty$ and $s_n = \infty$. Let $T[i,j] = |S_i \cap (s_{j-1}, s_j]|$, which is the number of data records in machine i that falls into the j-th interval $(s_{j-1}, s_j]$. To minimize the communication cost, the problem is to select $n-1$ splitters $s_1 \le s_2 \le \cdots \le s_{n-1}$ which split the data into n intervals, then find an assignment from the intervals to the machines, such that the communication cost is minimized. This problem is called Data Allocation Problem (DAP).

Remark. Although DRP and DAP are introduced based on sorting, they can be defined using the idea of virtual machines and physical machines. For DRP, the input $T[i,j]$ can be considered as the amount of data initially residing in physical machine i to be processed by virtual machine j, and the output is a permutation which assigns virtual machines to physical machines so that the communication cost is minimized. For DAP, choosing the splitters can be regarded as deciding the data distribution across the virtual machines. In such a point of view, DRP and DAP can be applied to a wide range of concrete problems. Also, DRP and DAP reflect only the problems that can be solved using one synchronous round. It will the future work to study multi-round algorithms on WMPC model.

1.2 Summary of Results and Paper Organization

Summarizing the above descriptions, we have two kinds of problems including DRP and DAP. We also have four kinds of communication cost functions including TOTAL, BTNK, MSR and SSR. 8 problems are obtained by combining two kinds of problems with four kinds of cost functions. The hardness for the 8 problems make up the content of this paper. Table 1 summarizes all the proposed results. Note that the parameterized complexities consider the number of machines as the parameter.

Table 1. Summary of results

	TOTAL	BTNK	MSR	SSR
DRP	P	P	NP-complete	NP-complete
DAP	FPT	FPT	W[1]-complete	W[1]-complete

In the rest of this paper, we first introduce some denotations in Sect. 1.3, then present the theoretical results in Sects. 2 and 3. The future work are discussed in Sect. 4. Finally Sect. 5 concludes this paper.

1.3 Denotations

A $m \times n$ matrix A is denoted as $A^{m \times n}$. The element in A at row i and column j is denoted as $A[i, j]$. The set of consecutive integers $\{i, i+1, i+2, \cdots, j\}$ is denoted as $[i, j]$. The set of integers $\{1, 2, \cdots, n\}$ is denoted as $[n]$.

A permutation on $[n]$ is a one-to-one mapping from $[n]$ to $[n]$, and it is usually denoted as π. The set of all permutations on $[n]$ is denoted as $\Pi(n)$. Denote π_i as the image of i under π. If $\pi_i = j$, it is also said that i is assigned to j by permutation π. We also use π^{-1} to denote the inverse permutation of π, i.e., if $\pi_i = j$ then $\pi_j^{-1} = i$.

2 The Data Redistribution Problem Series

Definition 2.1 (DRP). *Input: A $n \times n$ transmission matrix $T^{n \times n}$ and a $n \times n$ communication cost matrix $C^{n \times n}$, where $C[i, i] = 0$ for $i \in [n]$.*
Output: find a permutation $\pi \in \Pi(n)$ such that the communication cost function chosen from TOTAL, BTNK, MSR and SSR is minimized. Formally,

$DRP\text{-}TOTAL:$ $\min\limits_{\pi \in \Pi(n)} \sum\limits_{i=1}^{n} \sum\limits_{j=1}^{n} T[i,j]C[i,\pi_j]$

$DRP\text{-}BTNK:$ $\min\limits_{\pi \in \Pi(n)} \max\limits_{i \in [n]} \sum\limits_{j=1}^{n} T[j,\pi_i^{-1}]C[j,i]$

$DRP\text{-}MSR:$ $\min\limits_{\pi \in \Pi(n)} \max\limits_{i \in [n]} \left\{ \sum\limits_{j=1}^{n} T[i,j]C[i,\pi_j], \sum\limits_{j=1}^{n} T[j,\pi_i^{-1}]C[j,i] \right\}$

$DRP\text{-}SSR:$ $\min\limits_{\pi \in \Pi(n)} \max\limits_{i \in [n]} \left\{ \sum\limits_{j=1}^{n} T[i,j]C[i,\pi_j] + \sum\limits_{j=1}^{n} T[j,\pi_i^{-1}]C[j,i] \right\}$

Theorem 2.1. *DRP-TOTAL can be solved in $O(n^3)$ time.*

Theorem 2.2. *DRP-BTNK can be solved in $O(n^3)$ time.*

Theorem 2.3. *DRP-MSR is NP-complete.*

Theorem 2.4. *DRP-SSR is NP-complete.*

The proofs for the four theorems are omitted due to space limitation. See our full paper online [17].

3 The Problem Series of Data Allocation Problem

In this section we study the parameterized hardness and algorithms for the DAP problem series, parameterized by the number of machines. We will use N to denote the size of the input, and n to denote the number of machines.

Definition 3.1 (DAP). *Input: a set S of N integers divided into n subsets $S_1 = \{s_{1,1}, s_{1,2}, \cdots, s_{1,l_1}\}, \cdots, S_n = \{s_{n,1}, s_{n,2}, \cdots, s_{n,l_n}\}$, where $n > 1$ is the number of machines, and l_i is the size of S_i satisfying $\sum\limits_{i=1}^{n} l_i = N$.*

Output: find $n-1$ integers $s_1^*, \cdots s_{n-1}^* \in S$ and a permutation $\pi \in \Pi(n)$, such that the communication cost function chosen from TOTAL, BTNK, MSR and SSR is minimized. Formally,

$DAP\text{-}TOTAL:$ $\min\limits_{s_1^*, \cdots s_{n-1}^* \in S} \min\limits_{\pi \in \Pi(n)} \sum\limits_{i=1}^{n} \sum\limits_{j=1}^{n} T[i,j]C[i,\pi_j]$

$DAP\text{-}BTNK:$ $\min\limits_{s_1^*, \cdots s_{n-1}^* \in S} \min\limits_{\pi \in \Pi(n)} \max\limits_{i \in [n]} \sum\limits_{j=1}^{n} T[j,\pi_i^{-1}]C[j,i]$

$DAP\text{-}MSR:$ $\min\limits_{s_1^*, \cdots s_{n-1}^* \in S} \min\limits_{\pi \in \Pi(n)} \max\limits_{i \in [n]} \left\{ \sum\limits_{j=1}^{n} T[i,j]C[i,\pi_j], \sum\limits_{j=1}^{n} T[j,\pi_i^{-1}]C[j,i] \right\}$

$DAP\text{-}SSR:$ $\min\limits_{s_1^*, \cdots s_{n-1}^* \in S} \min\limits_{\pi \in \Pi(n)} \max\limits_{i \in [n]} \left\{ \sum\limits_{j=1}^{n} T[i,j]C[i,\pi_j] + \sum\limits_{j=1}^{n} T[j,\pi_i^{-1}]C[j,i] \right\}$

where $T[i,j] = |S_i \cap (s_{j-1}^*, s_j^*]|$ and $s_0^* = -\infty, s_n^* = \infty$.

3.1   The Splitter-Graph

We introduce the splitter-graph, which transforms the problem of choosing splitters to choosing a path in a special graph. Given a set $S = \{s_1, s_2, \cdots, s_N\}$ of integers, assuming $s_1 \leq s_2 \leq \cdots \leq s_N$, and a parameter n, construct a graph $G(V, E)$ as follows. For $i \in [n-1], j \in [N]$, construct a vertex $v_{i,j}$. Let $v_{0,0}$ be the starting vertex $-\infty$, and $v_{n,N+1}$ be the end vertex ∞. Let $(v_{i,j}, v_{i',j'}) \in E$ iff $i+1 = i'$ and $j < j'$. In such way, a vertex $v_{i,j}$ represents a splitter s_j placed in the i-th position, and a path $(-\infty, v_{1,i_1}, v_{2,i_2} \cdots v_{n-1,i_{n-1}}, \infty)$ represents selecting $s_{i_1}, s_{i_2}, \cdots s_{i_{n-1}}$ as splitters. A splitter-graph based on set S with parameter n will be denoted as $G_s(V, E, S, n)$.

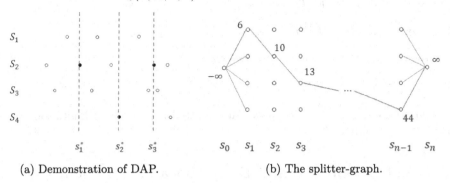

(a) Demonstration of DAP.   (b) The splitter-graph.

3.2   FPT Algorithm of DAP-TOTAL and DAP-BTNK

The FPT algorithm of DAP-TOTAL and DAP-BTNK is based on the following transformation. Given an instance of DAP-TOTAL, denote $S = \{s_1, s_2, \cdots, s_N\}$ and assume $s_1 \leq s_2 \leq \cdots \leq s_N$. Let $s_0 = -\infty$ and $s_{N+1} = \infty$. Let $Acc[i, j] = |S_i \cap (-\infty, s_j]|, i \in [n], j \in [0, N+1]$. Slightly abusing denotation, let $\pi^{n \times n}$ be a matrix defined based on the permutation π, such that $\pi[i, j] = 1$ if $\pi_i = j$, and $\pi[i, j] = 0$ otherwise, $i, j \in [n]$. Under the above denotations, DAP-TOTAL can be transformed into

$$\min_{s_1^*, \cdots s_{n-1}^* \in S} \min_{\pi \in \Pi(n)} \sum_{i=1}^{n} \sum_{j=1}^{n} \sum_{k=1}^{n} (Acc[i, s_j^*] - Acc[i, s_{j-1}^*])C[i, k]\pi[j, k] \quad (1)$$

where $s_0^* = -\infty$ and $s_n^* = \infty$. Let $F[j, k] = \sum_{i=1}^{n} Acc[i, j]C[i, k], j \in [0, N+1], k \in [n]$, then the above equation is transformed into

$$\min_{s_1^*, \cdots s_{n-1}^* \in S} \min_{\pi \in \Pi(n)} \sum_{j=1}^{n} \sum_{k=1}^{n} (F[s_j^*, k] - F[s_{j-1}^*, k])\pi[j, k] \quad (2)$$

Let $Cost[i, j, k] = F[i, k] - F[j, k], 0 \leq j < i \leq N+1, k \in [n]$, and we get

$$\min_{s_1^*, \cdots s_{n-1}^* \in S} \min_{\pi \in \Pi(n)} \sum_{j=1}^{n} \sum_{k=1}^{n} Cost[s_j^*, s_{j-1}^*, k]\pi[j, k] \quad (3)$$

If π is represented by a permutation, we get

$$\min_{s_1^*, \cdots s_{n-1}^* \in S} \min_{\pi \in \Pi(n)} \sum_{j=1}^{n} Cost[s_j^*, s_{j-1}^*, \pi_j] \tag{4}$$

Now we can associate the above $Cost$ function to the spiltter-graph. For each edge $(v_{i,j}, v_{i',j'})$ in the splitter-graph and each $l \in [n]$, let $\omega(v_{i,j}, v_{i',j'}, l) = Cost[j', j, l]$, and we have the following splitter-graph formation of DAP-TOTAL.

Definition 3.2. *Input: a splitter-graph $G_s(V, E, S, n)$, the weight function ω : $V \times V \times [n] \to \mathcal{R}$ of DAP-TOTAL.*
Output: a path $(-\infty, v_{1,i_1}, v_{2,i_2} \cdots v_{n-1,i_{n-1}}, \infty)$, and a permutation π, to minimize $\sum_{j=1}^{n} \omega(v_{j,i_j}, v_{j-1,i_{j-1}}, \pi_j)$.

FPT Algorithm for Decision-DAP-TOTAL. We prove the following decision version of DAP-TOTAL is FPT.

Definition 3.3 (Decision-DAP-TOTAL). *Input: a splitter-graph $G_s(V, E, S, n)$, the weight function $\omega : V \times V \times [n] \to \mathcal{R}$ of DAP-TOTAL, a threshold value α, and parameter n.*
Output: Is the optimum value of DAP-TOTAL less than α?

We need the following definition of partial permutations. A partial permutation π is a function defined on $[i]$ where $i \in [n]$, such that $\pi_j, \pi_k \in [n]$ and $\pi_j \neq \pi_k$ for $1 \leq j \neq k \leq i$. Here π_j is the image of j under π. Given a partial permutation π whose definition domain is $[i]$, and an integer $l \in [n]$, let $l \in \pi$ denote that there exists some $j \in [i]$ such that $\pi_j = l$. Given an integer $l \notin \pi$, let $\pi \cup \{l\}$ be a new partial permutation π' defined on $[i+1]$ such that $\pi'_{i+1} = l$ and $\pi'_j = \pi_j$ for $j \in [i]$. Given an integer $l = \pi_i$, let $\pi \setminus \{l\}$ be a partial permutation π' defined on $[i-1]$, such that $\pi'_j = \pi_j$ for all $j \in [i-1]$. Denote Φ as the empty partial permutation.

Algorithm 1 is the FPT algorithm for Decision-DAP-TOTAL. The algorithm maintains two arrays of length $O(n!)$ for each vertex $v_{i,j}$, namely Perm$(v_{i,j})$ and Cost$(v_{i,j}, \pi)$. Perm $(v_{i,j})$ stores all the feasible partial permutations for the path from $-\infty$ to $v_{i,j}$, and Cost$(v_{i,j}, \pi)$ stores the partial accumulated cost value corresponding to the partial permutation π. The proof for the correctness and complexity of the algorithm is omitted due to space limitation.

FPT Algorithm for DAP-BTNK. Using a transformation similar with that for DAP-TOTAL, we have the following splitter-graph formation for DAP-BTNK.

Definition 3.4. *Input: a splitter-graph $G_s(V, E, S, n)$, the weight function ω : $V \times V \times [n] \to \mathcal{R}$ of DAP-BTNK, and parameter n.*
Output: a path $(-\infty, v_{1,i_1}, v_{2,i_2} \cdots v_{n-1,i_{n-1}}, \infty)$, and a permutation π, to minimize $\max_{j \in [n]} \omega(v_{j,i_j}, v_{j-1,i_{j-1}}, \pi_j)$.

Algorithm 1: Decision version of DAP-TOTAL

1 $Perm(-\infty) \leftarrow \{\Phi\}, Cost(-\infty, \Phi) \leftarrow 0$;
2 **for** $1 \le i \le n, 1 \le j \le N, 1 \le k \le N$ **do**
3   **if** *edge* $(v_{i-1,k}, v_{i,j})$ *exists* **then**
4   **for** $1 \le l \le n$ **do**
5   **foreach** *partial permutation* $\pi \in Perm(v_{i-1,k})$ **do**
6   **if** $l \notin \pi$ **then**
7   **if** $Cost(v_{i-1,k}, \pi) + \omega(v_{i-1,k}, v_{i,j}, l) \le \alpha$ **then**
8   Add $\pi \cup \{l\}$ into $Perm(v_{i,j})$;
9                          $Cost(v_{i,j}, \pi \cup \{l\}) \leftarrow Cost(v_{i-1,k}, \pi) + \omega(v_{i-1,k}, v_{i,j}, l)$;
10   **else**
11   Update $Cost(v_{i,j}, \pi)$ if
                    $Cost(v_{i-1,k}, \pi \setminus \{l\}) + \omega(v_{i-1,k}, v_{i,j}, l) < Cost(v_{i,j}, \pi)$
12 Return *Yes* if $Perm(\infty)$ is non-empty, and *No* otherwise.

The decision version of DAP-BTNK has an extra value α as input, and asks whether the optimum value of DAP-BTNK is less than α. We propose the FPT algorithm for the decision version, which is given as Algorithm 2. It needs one array for each vertex $v_{i,j}$ which is $Perm(v_{i,j})$. The algorithm is similar with that for Decision-DAP-TOTAL, only changing the sum-check (Line 7 in Algorithm 1) to maximum check (Line 6 in Algorithm 2). The correctness proof of this algorithm is omitted.

Algorithm 2: Decision version of DAP-BTNK

1 $Perm(-\infty) \leftarrow \{\Phi\}$;
2 **for** $1 \le i \le n, 1 \le j \le N, 1 \le k \le N$ **do**
3   **if** *edge* $(v_{i-1,k}, v_{i,j})$ *exists* **then**
4   **for** $1 \le l \le n$ **do**
5   **foreach** *partial permutation* $\pi \in Perm(v_{i-1,k})$ **do**
6   **if** $l \notin \pi$ *and* $\omega(v_{i-1,k}, v_{i,j}, l) \le \alpha$ **then**
7   Add $\pi \cup \{l\}$ into $Perm(v_{i,j})$;
8 Return *Yes* if $Perm(\infty)$ is non-empty, and *No* otherwise.

3.3 W[1]-Completeness of DAP-MSR and DAP-SSR

Due to space limitation, the proof for DAP-MSR and DAP-SSR are in W[1] is omitted. See our full paper online [17]. We then prove the W[1]-hardness of the two problems. We first transform DAP-MSR and DAP-SSR into a splitter-graph formation. We only describe the transformation for DAP-MSR, and it is similar for the other.

$$\min_{s_1^*,\cdots s_{n-1}^*\in S}\min_{\pi\in\Pi(n)}\max_{i\in[n]}\left\{\begin{array}{l}\sum_{j=1}^{n}\sum_{k=1}^{n}(Acc[i,s_j^*]-Acc[i,s_{j-1}^*])C[i,k]\pi[j,k]\\\sum_{j=1}^{n}\sum_{k=1}^{n}(Acc[j,s_k^*]-Acc[j,s_{k-1}^*])C[j,i]\pi[k,i]\end{array}\right\}\quad(5)$$

Let V be a $N\times N$ matrix where each element is a vector of length n, and let $V[j,k][i]=Acc[i,j]-Acc[i,k]$, $i\in[n],j,k\in[0,N+1]$, then Eq. 5 is transformed into

$$\min_{s_1^*,\cdots s_{n-1}^*\in S}\min_{\pi\in\Pi(n)}\max_{i\in[n]}\left\{\sum_{j=1}^{n}\sum_{k=1}^{n}V[s_j^*,s_{j-1}^*][i]C[i,k]\pi[j,k],\sum_{j=1}^{n}\sum_{k=1}^{n}V[s_k^*,s_{k-1}^*][j]C[j,i]\pi[k,i]\right\}$$

Next we give the following splitter-graph formation of DAP-MSR.

Definition 3.5. *Input: splitter-graph $G_s(V,E,S,n)$, edge weight function ω : $V\times V\to\mathcal{R}^n$, communication cost matrix $C^{n\times n}$, and parameter n.*
Output: a path $(-\infty,v_{1,k_1},v_{2,k_2}\cdots v_{n-1,k_{n-1}},\infty)$, which corresponds to a matrix $T^{n\times n}$ where $T[i,j]=\omega(v_{j,k_j},v_{j-1,k_{j-1}})[i]$, and a permutation π, such that the following MSR cost function is minimized:

$$\max_{i\in[n]}\left\{\sum_{j=1}^{n}T[i,j]C[i,\pi_j],\sum_{j=1}^{n}T[j,\pi_i^{-1}]C[j,i]\right\}$$

To prove the W[1]-hardness of the problem, we introduce an intermediate problem called Selecting-PARTITION. The idea is to reduce k-clique, which is W[1]-complete, to Selecting-PARTITION, and reduce Selecting-PARTITION to DAP-MSR (and similarly to DAP-SSR).

Definition 3.6 (Selecting-PARTITION). *Input: n integers $S = \{s_1, s_2, \cdots, s_n\}$, target sum value B, and parameter k.*
Output: decide whether there exists a set $A\subset S$ with $|A|=k$, such that A is a Yes-instance of PARTITION, i.e., there exists A_1,A_2 such that $A_1\cap A_2=\emptyset, A_1\cup A_2=A$ and $\sum_{s_i\in A_1}s_i=\sum_{s_i\in A_2}s_i=B/2$.

Theorem 3.1. *There is a parameterized reduction from k-clique to Selecting-PARTITION, and from Selecting-PARTITION to DAP-MSR and DAP-SSR.*

Proof. See the full version of this paper [17]. □

4 Future Works

Recall that the problem series of DRP and DAP are introduced using parallel sorting as the representing example. They reflect the communication pattern of problems that can be solved in one synchronous round. However, there are many problems that need multiple rounds to solve. For example, joining multiple relations can be solved using one round [5] or multiple rounds [1]. Computing the graph coloring [7], maximum matching [12], shortest path [10], etc., must use multiple rounds. The problem to minimize the communication cost on WMPC model with multiple rounds is left as future work.

5   Conclusion

In this paper we proposed the WMPC (Weighted Massively Parallel Computation) model based on the existing works of topology-aware Massively Parallel Computation model [6,13]. The WMPC model considers the underlying computation network as a complete weighted graph, which is a complement to the work in [13] where the network topology is restricted to trees. Based on the WMPC model the DRP and DAP problem series are defined, each representing a set of problems with the same pattern of communication. We also defined four kinds of objective functions for communication cost which are TOTAL, BTNK, MSR and SSR, and obtained 8 problems combining the four objective functions with two communication pattern problems. We studied the hardness of the 8 problems, and provided substantial theoretical results. In conclusion, this paper studied the communication minimization problem on WMPC model with a scope both deep and wide, but we must point out that this paper only investigated a small portion of the research area on the WMPC or topology-aware MPC model. There are a lot of problems to be studied following what was studied in this paper.

References

1. Afrati, F.N., Joglekar, M.R., Re, C.M., Salihoglu, S., Ullman, J.D.: GYM: a multiround distributed join algorithm. In: Leibniz International Proceedings in Informatics, LIPIcs, vol. 68, pp. 4:1–4:18 (2017)
2. Afrati, F.N., Ullman, J.D.: Optimizing joins in a map-reduce environment. In: Advances in Database Technology - EDBT 2010–13th International Conference on Extending Database Technology, Proceedings, pp. 99–110. ACM Press, New York (2010)
3. Andoni, A., Nikolov, A., Onak, K., Yaroslavtsev, G.: Parallel algorithms for geometric graph problems. In: Shmoys, D.B. (ed) Symposium on Theory of Computing, STOC 2014, New York, NY, USA, May 31 - June 03, 2014, pp. 574–583. ACM, (2014)
4. Beame, P., Koutris, P., Suciu, D.: Communication steps for parallel query processing. In: Proceedings of the ACM SIGACT-SIGMOD-SIGART Symposium on Principles of Database Systems, vol. 64, pp. 273–284. ACM Press, New York, USA (2013)
5. Beame, P., Koutris, P., Suciu, D.: Skew in parallel query processing. In: Proceedings of the ACM SIGACT-SIGMOD-SIGART Symposium on Principles of Database Systems, pp. 212–223. Association for Computing Machinery (2014)
6. Blanas, S.: Topology-aware parallel data processing : models, algorithms and systems at scale. In: 10th Annual Conference on Innovative Data Systems Research (CIDR 2020) (2020)
7. Chang, Y.-J., Fischer, M., Ghaffari, M., Uitto, J., Zheng, Y.: The complexity of $(\delta + 1)$ coloring in congested clique, massively parallel computation, and centralized local computation. In: Proceedings of the 2019 ACM Symposium on Principles of Distributed Computing, pp. 471–480 (2019)
8. Culler, D.E., et al.: LogP: a practical model of parallel computation. Commun. ACM **39**(11), 78–85 (1996)

9. Dean, J., Ghemawat, S.: MapReduce: simplified data processing on large clusters. Commun. ACM **51**(1), 107–113 (2008)
10. Dory, M., Fischer, O., Khoury, S., Leitersdorf, D.: Constant-round spanners and shortest paths in congested clique and MPC. In: Proceedings of the 2021 ACM Symposium on Principles of Distributed Computing, pp. 223–233 (2021)
11. Ghaffari, M., Gouleakis, T., Konrad, C., Mitrović, S., Rubinfeld, R.: Improved massively parallel computation algorithms for MIS, matching, and vertex cover. In: Proceedings of the Annual ACM Symposium on Principles of Distributed Computing, pp. 129–138. ACM, New York, NY, USA (2018)
12. Ghaffari, M., Kuhn, F.: Distributed minimum cut approximation. In: Lecture Notes in Computer Science (including subseries Lecture Notes in Artificial Intelligence and Lecture Notes in Bioinformatics), volume 8205 LNCS, pp. 1–15 (2013)
13. Hu, X., Koutris, P., Blanas, S.: Algorithms for a topology-aware massively parallel computation model. In: Proceedings of the ACM SIGACT-SIGMOD-SIGART Symposium on Principles of Database Systems, pp. 199–214. Association for Computing Machinery (2021)
14. Karloff, H., Suri, S., Vassilvitskii, S.: A model of computation for MapReduce. In: Proceedings of the Annual ACM-SIAM Symposium on Discrete Algorithms, pp. 938–948. Society for Industrial and Applied Mathematics, Philadelphia, PA (2010)
15. Karp, R.M.: A Survey of Parallel Algorithms for Shared-memory Machines (1988)
16. Koutris, P., Suciu, D.: Parallel evaluation of conjunctive queries. In: Lenzerini, M., Schwentick, T. (eds) Proceedings of the 30th ACM SIGMOD-SIGACT-SIGART Symposium on Principles of Database Systems, PODS 2011, June 12–16, 2011, Athens, Greece, pp. 223–234. ACM (2011)
17. Ma, H., Li, J., Gao, X.: Optimization problems on the weighted massively parallel computation model: hardness and algorithms. arXiv preprint arXiv:2302.12953 (2023)
18. O'malley, O.: Terabyte sort on apache hadoop, pp. 1–3 (2008). https://sortbenchmark.org/Yahoo-Hadoop.pdf
19. Tao, Y., Lin, W., Xiao, X.: Minimal MapReduce algorithms. In: Proceedings of the ACM SIGMOD International Conference on Management of Data, number June, pp. 529–540. ACM Press, New York, New York, USA (2013)
20. Valiant, L.G.: A bridging model for parallel computation. Commun. ACM **33**(8), 103–111 (1990)

The Regularized Submodular Maximization via the Lyapunov Method

Xin Sun[1], Congying Han[1], Chenchen Wu[2], Dachuan Xu[3], and Yang Zhou[4(✉)]

[1] School of Mathematical Sciences, University of Chinese Academy of Sciences, Beijing 100049, People's Republic of China
{sunxin,hancy}@ucas.ac.cn
[2] College of Science, Tianjin University of Technology, Tianjin 300072, People's Republic of China
[3] Beijing Institute for Scientific and Engineering Computing, Beijing University of Technology, Beijing 100124, People's Republic of China
xudc@bjut.edu.cn
[4] School of Mathematics and Statistics, Shandong Normal University, Jinan 250014, People's Republic of China
zhouyang@sdnu.edu.cn

Abstract. In the paper, we study Regularized Submodular Maximization (RegularizedSM) problem over a down-closed family of sets by applying the Lyapunov method. The Regularized Submodular Maximization can be viewed as a generalization of the Submodular Maximization since it adds an extra linear regular term (possibly negative) to the objective so that it may no longer be non-negative. For Regularized Nonmonotone Submodular Maximization (RegularizedNSM), we systematically design an algorithm framework in two phases. With a proper choice of coefficients in the framework, a $(1/e, \frac{\gamma - \gamma/e - 1}{\gamma - 1})$-approximation algorithm is obtained in the continuous-time phase, where $\gamma \in [0, 1) \cup (1, +\infty)$ is a parameter reflecting the relative dominance of the positive and negative parts of the linear optimal value. In the second phase, we make the algorithm implemented by discretization with almost the same approximation guarantee and $\mathcal{O}\left(\frac{n^3}{\epsilon}\right)$ time complexity. Moreover, the Lyapunov method can also be applied in Regularized Monotone Submodular Maximization (RegularizedMSM) with $(1 - 1/e, 1)$-approximation performance, which coincides with the state-of-the-art result given by Feldman [9]. This observation implies that the algorithm framework designed by the Lyapunov method can unify some of the existing approximation algorithms.

Keywords: Regularized submodular maximization · Lyapunov function · Approximation algorithm design and analysis · Down-closed family of sets

W. Wu and G. Tong (Eds.): COCOON 2023, LNCS 14423, pp. 118–143, 2024.
https://doi.org/10.1007/978-3-031-49193-1_10

1 Introduction

Recently, plenty of works related to RegularizedSM have been published from both theoretical and practical perspectives, since the objective, which consists of submodular and linear terms, owns very general properties. Formally, the RegularizedSM can be formulated as $\max_{S \subseteq E, S \in \mathcal{I}} f(S) + \ell(S)$ with a given groundset E and a certain constraint $\mathcal{I}$. Usually, the first part is a non-negative submodular set function $f : 2^E \to \mathbb{R}_{\geq 0}$. *Submodularity* is an important economic concept that quantifies the degree of fungibility between objects. The second part of RegularizedSM is a linear set function $\ell : 2^E \to \mathbb{R}$ defined by the summation of all single-element value in the field of real numbers, therefore it makes the objective not necessarily non-negative. The linear term enhances the application versatility compared with submodular maximization. One example is in the design of approximation algorithms, where RegularizedSM can be viewed as a subproblem of the curvature optimization problems, where the goal is to improve the constant approximation ratio by introducing a parameter curvature [4] of the objective function. Another popular scenario is the overfitting in machine-learning. When the linear term is non-positive, RegularizedSM generalizes the regularization model. It is worth noting that the two functions in the objective of RegularizedSM tend to be valued differently in applications, where the first term normally is more important since it is the initial target (e.g. the quantity function in machine-learning) and the second is just the implementation of regularization.

1.1 Lyapunov Function Approach

In response to the above situation, a new method called *Lyapunov function* is introduced in submodular maximization not only as a technique for analysis and proof, but also as a systematic instruction for *designing* approximation algorithm. The "designing" here indicates that it can directly start the analysis process and derive the approximation guarantee and complexity without knowing the algorithm. Besides, we could gradually learn the details of the algorithm during this procedure. The Lyapunov function was originally used to study the stability of an ODE equilibrium as a technique for analysis and is therefore widely used in the stability and control theory of dynamic system [15]. This method normally works in the *continuous-time setting*, where a vecter-valued function $\mathbf{x}(t) \in \mathbb{R}^n$ varies itself by following an evolution equation $\dot{\mathbf{x}}(t) = \phi(\mathbf{x}(t))$ with $t \in [0, T]$. Denote $V(\mathbf{x}(t)) \in \mathbb{R}$ as the Lyapunov function of $\mathbf{x}(t)$ such that the stable criteria $\dot{V}(\mathbf{x}(t)) = (\nabla V(\mathbf{x}(t)))^\top \cdot \phi(\mathbf{x}(t))$ is non-negative on $t \in [0, T]$. This implies that $V(\mathbf{x}(t))$ is a non-decreasing function during this time interval. With this property, we could obtain surprisingly simple sufficient conditions for the problem we consider. The Lyapunov function method also has many applications in other areas, such as optimization. Frequently, with a given algorithm, this method is used as a proof technique to show the convergence [14, 17, 20–22]. For example, Bansal and Gupta [1] discussed the convergence rate for gradient methods such as smooth and non-smooth gradient descent, mirror descent and some accelerated variants. Recently, Diaknonikolas and Orecchia [5] dealt with it in a reverse direction, i.e., the continuous-time algorithm emerges itself from the derivation, to present a general analysis framework of first-order methods. Similarly, Du [7,8] firstly gave a two-phase systematical framework for various submodular maximization problems by applying the Lyapunov method as an algorithm-design technique. The first phase still focuses on the continuous-time analysis. It begins with specifying the parametric form of the Lyapunov function. Then, it derives the sufficient conditions of the stable criteria by the help of a proper bound of the optimal value. Next, it produces the algorithm by solving the ordinary differential equation related to update of the solution in the conditions. Finally, it maximizes the approximation guarantee by determining the parameters given before. Since submodular maximization is a kind of combinatorial optimization, the second phase should fulfill the request of an implementable algorithm and output an integral solution. To do so, the discrete counterpart of the Lyapunov function should be derived and the discretization of the continuous-time algorithm is necessary. At the end of the whole procedure, a similar approximation ratio with only a small loss and the time complexity can be obtained automatically. Running this two-phase paradigm, Du reproduced the state-of-the-art results for maximizing monotone DR-submodular over a solvable convex set, non-monotone DR-submodular subject to a down-closed solvable convex set and non-monotone DR-submodular constrained by a solvable convex set. The more important achievement for this work is that it essentially explains the design philosophy of the commonly used methods mentioned above.

1.2   Related Work

In RegularizedSM, an algorithm is called (α, β)-approximation for some coefficients $\alpha, \beta \geq 0$ if its output $S \subseteq E$ satisfying $\mathbb{E}[f(S) + \ell(S)] \geq \max_{O \subseteq E}[\alpha \cdot f(O) + \beta \cdot \ell(O)]$. For the Regularized Monotone Submodular Maximization (RegularizedMSM) problem with matroid constraints, Sviridenko et al. [19] presented two algorithms that are a modified non-oblivious local search and an adapted continuous-greedy, respectively. They both achieved an $(1 - 1/e, 1)$-approximation with $\mathcal{O}(\epsilon)$ error term. Unfortunately, a guessing step is necessary which remarkably damages its query complexity. To avoid the guessing step, Feldman [9,10] described a clean alternative algorithm called distorted continuous greedy for solving the multilinear relaxation of the above problem over a *down-monotone* and solvable polytope. This constraint is more general and includes many common polytopes such as matroid polytope. By applying the adaptive weight $(1 + \delta)^{(t-1)/\delta}$ related to the time interval $\delta \in \mathcal{O}(\epsilon/n^2)$ on the first-order information of the submodular term, this variant of continuous-greedy yielded $(1 - 1/e, 1)$ guarantee for the submodular and linear term, respectively. Also, a hardness result of this problem is showed in [9]. It states that there exists no polynomial time algorithm with bi-factor better than $(1 - e^{-\lambda} + \epsilon, \lambda)$, where $\lambda \in [0, 1]$ is a calibrating parameter. Now, researchers pay attention on a more challenging setting, where the submodular component of the objective is not necessarily monotone. In this case, the continuous-greedy normally fails, since the marginal gains right now could be negative. For non-monotone submodular maximization, a powerful technique named *measured continuous-greedy* is introduced by Feldman et al. [11] as a unified algorithm. It compensates for the difference between the residual value of the current fractional solution and its gradient. This is achieved by multiplying an adaptive ratio in the update process. The ratio slows down the evolution of the fractional solution by diverging the direction obtained by linear programming, so as to mimic the gradient value. Fortunately, the first-order property of the multilinear extension can ensure this intention. Based on this, Lu et al. [16] designed a variant of the measured continuous-greedy with distorted objective for *Regularized Non-Monotone Submodular Maximization* (RegularizedNSM) subject to a matroid constraint. The performance guarantee of the algorithm is $(1/e - \epsilon, 1)$. However, this result is built only when ℓ is non-positive. For the linear term without any limitation, Qi [18] presented a $(\frac{te^{-t}}{t+e^{-t}} - \epsilon, \frac{t}{t+e^{-t}})$-approximation algorithm for RegularizedNSM subject to a matroid constraint, where $t \in [0,1]$. Furthermore, many inapproximability results for RegularizedSM were also provided in this work. Besides, for *Regularized Unconstrained Submodular Maximization* (RegularizedUSM), Bodek and Feldman [3] offered the first non-trivial guarantee and proved that non-oblivious local search yields $(\alpha(\beta) - \epsilon, \beta - \epsilon)$-approximation for all $\beta \in [0,1]$, $\alpha(\beta) := \beta(1 - \beta)/(1 + \beta)$. They also showed some negative conclusions for the special case of RegularizedUSM in which the linear function is non-positive and non-negative, respectively.

For applying the Lyapunov function method as not only an analysis but also a design technique in submodular maximization, Du [7] used three problems of DR-submodular maximization with different constraints as examples to propose a two-phase systematical framework. After determining the form of the Lyapunov function, a continuous-time algorithm is designed in the first stage, and afterwards the theoretical approximation guarantee could be derived. Then, for the purpose of implementation, a *discrete-time algorithm* and the corresponding complexity are obtained in the second stage. For the three problems they consider, the performance guarantees and complexities coincides with the current best results. Moreover, Du et al. [6] improved the approximation ratio of maximizing a DR-submodular function over a general convex set from 0.19 to 0.25 by using the Lyapunov method to design a Frank-Wolfe type algorithm with the same order of time-complexity.

1.3   Our Contribution

The informal conclusions for RegularizedNSM and RegularizedMSM are presented below by denoting $F : [0,1]^n \to \mathbb{R}_{\geq 0}$ and $L : [0,1]^n \to \mathbb{R}$, which are the multi-linear extensions of f and ℓ, respectively.

Theorem 1.1. *For RegularizedNSM over a down-closed family of sets, there exists a continuous-time approximation algorithm designed by a certain Lyapunov function. Given $\gamma \in [0,1) \cup (1,+\infty)$, it outputs a feasible $\mathbf{x} \in [0,1]^E$*

obeying

$$F(\mathbf{x}) + L(\mathbf{x}) \geq \frac{1}{e} f(\mathbf{x}^\star) + \frac{\gamma - \gamma/e - 1}{\gamma - 1} \ell(\mathbf{x}^\star).$$

Theorem 1.2. *For RegularizedNSM over a down-closed family of sets, there exists a discrete-time approximation algorithm designed by a certain Lyapunov function. Given $\epsilon > 0$ and $\gamma \in [0,1) \cup (1, +\infty)$, it outputs a feasible $\mathbf{x} \in [0,1]^E$ obeying*

$$F(\mathbf{x}) + L(\mathbf{x}) \geq \frac{1}{e} f(\mathbf{x}^\star) + \frac{\gamma - \gamma/e - 1}{\gamma - 1} \ell(\mathbf{x}^\star) - \mathcal{O}(\epsilon).$$

Moreover, the time complexity of the algorithm is $\mathcal{O}\left(\frac{nD}{\epsilon}\right)$, where n is the element's number of the ground set and D is the smoothness of F.

Moreover, we can also take the Lyapunov function method to deal with RegularizedMSM with the same constraint. With a similar derivation process, we can obtain the following results.

Theorem 1.3. *For RegularizedMSM over a down-closed family of sets, there exists a continuous-time approximation algorithm designed by a certain Lyapunov function, whose output $\mathbf{x} \in [0,1]^E$ is feasible and satisfies*

$$F(\mathbf{x}) + L(\mathbf{x}) \geq \left(1 - \frac{1}{e}\right) f(\mathbf{x}^\star) + \ell(\mathbf{x}^\star).$$

Theorem 1.4. *For RegularizedMSM over a down-closed family of sets, there exists a discrete-time approximation algorithm designed by a certain Lyapunov function. Given $\epsilon > 0$, it outputs a feasible $\mathbf{x} \in [0,1]^E$ obeying*

$$F(\mathbf{x}) + L(\mathbf{x}) \geq \left(1 - \frac{1}{e}\right) f(\mathbf{x}^\star) + \ell(\mathbf{x}^\star) - \mathcal{O}(\epsilon).$$

Moreover, the time complexity of the algorithm is $\mathcal{O}\left(\frac{nD}{\epsilon}\right)$, where n is the element's number of the ground set and D is the smoothness of F.

1.4   Organization

In Sect. 2, we introduce some necessary conceptions. In Sect. 3, we explain why the Lyapunov method can systematically design an algorithm framework and analyze its approximation guarantee for RegularizedNSM with no limitation on the linear term. In Sect. 4, we finally conclude the paper. Due to page limitation, we present the discrete-time phase of the Lyapunov method for RegularizedNSM in Appendix B, the results for RegularizedMSM in Appendix C and all missing proofs of Sect. 3 in Appendix D.

2 Preliminaries

Given a ground set E with n elements and a set function $f : 2^E \to \mathbb{R}_{\geq 0}$, we use $f_S(\{e\}):=f(S \cup \{e\}) - f(S)$ to denote the marginal gains of adding an element e to a set $S \in E$ w.r.t. f. We say f is *submodular* if $f_S(\{e\}) \geq f_T(\{e\})$ holds for any $S \subseteq T \subseteq E$ and an element $e \in E \backslash T$. Also, the set function f is *non-negative* if $f(S) \geq 0$ for any $S \subseteq E$. Besides, we say a set function $\ell : 2^E \to \mathbb{R}$ is *linear* if $\ell(S) = \sum_{e \in S} \ell(\{e\})$ for every set $S \in E$, where $\ell(\{e\}) \in \mathbb{R}$ for each $e \in E$. For easy of implementation, we omit the brace when viewing an element as a set and define $S \cup \{e\}$ and $S \backslash \{e\}$ by the shorthand $S \cup e$ and $S - e$, respectively.

We study RegularizedNSM with no limitation on the linear term. The solutions are required to be feasible to an instance of a *down-closed family of subsets* $\mathcal{I} \subseteq 2^E$, which is quite general including several regular and natural constraints such as matroid and knapsack. $\mathcal{I}$ is down-closed (or down-monotone) if every subset of $S \in \mathcal{I}$ also belongs to $\mathcal{I}$. Since there are no lossless rounding techniques for this setting, we instead aim at the relaxation of the above problem. For a submodular function f, its multilinear extension is defined as $F : [0,1]^n \to \mathbb{R}_{\geq 0}$, which constructs a mapping between a point $\mathbf{x} = (x_{e_1}, \ldots, x_{e_n}) \in [0,1]^n$ and the expected function value of a random set $R_\mathbf{x} \subseteq S$ w.r.t. $\mathbf{x}$. The random set $R_\mathbf{x}$ includes each element $e \in E$ with probability x_e independently. The formal mathematical expression is $F(\mathbf{x}):=\mathbb{E}[f(R_\mathbf{x})] = \sum_{S \subseteq E} f(S) \prod_{e \in S} x_e \prod_{e \notin S}(1 - x_e)$. We denote the join, meet and product operations for any two vectors $\mathbf{x}, \mathbf{y} \in [0,1]^n$ by $(\mathbf{x} \vee \mathbf{y})_e = \max\{x_e, y_e\}$, $(\mathbf{x} \wedge \mathbf{y})_e = \min\{x_e, y_e\}$ and $(\mathbf{x} \odot \mathbf{y})_e = x_e \cdot y_e$ respectively. Also, we refer to the polytope $\mathcal{P}_\mathcal{I} \subseteq [0,1]^n$ as the feasible field in the relaxation problem. It is the convex hull of characteristic vectors of all the feasible sets of $\mathcal{I}$. Its formal definition is $\mathcal{P}_\mathcal{I} = \text{conv}\{1_S, S \in \mathcal{I}\}$. Similarly, we say $\mathcal{P}_\mathcal{I}$ is down-monotone if $0 \leq \mathbf{y} \leq \mathbf{x}$ ($0 \leq y_i \leq x_i$ for each $i = 1, ..., n$) and $\mathbf{x} \in \mathcal{P}_\mathcal{I}$ imply $\mathbf{y} \in \mathcal{P}_\mathcal{I}$. Moreover, $\mathcal{P}_\mathcal{I}$ is solvable if linear functions can be maximized over it in polynomial time. The formal description of the relaxation (which is the real problem we consider in this paper) can be stated as $\max_{\mathbf{x} \in \mathcal{P}_\mathcal{I}} F(\mathbf{x}) + L(\mathbf{x})$, where $L(\cdot)$ is the extension form of ℓ. It is actually the dot product of the input variable and the regularized (weight) vector $\vec{\ell} = (\ell_{e_1}, \ldots, \ell_{e_n})$. Formally, $L(\mathbf{x}):=\langle \vec{\ell}, \mathbf{x} \rangle = \sum_{e \in E} \ell_e \cdot x_e$ and $\nabla L(\mathbf{x}) = \vec{\ell}$. For the regularized vector $\vec{\ell}$, it is obvious that it can be split into $\vec{\ell}_+$ and $\vec{\ell}_-$, where all the components are non-negative for the former and non-positive for the latter. Therefore, we have $\vec{\ell} = \vec{\ell}_+ + \vec{\ell}_-$. Moreover, we denote $\mathbf{x}^\star \in \{0,1\}^n$ as the optimal integral solution for RegularizedNSM with down-monotone. One advantage of denoting $\mathbf{x}^\star$ is that we can directly build the connection between the relaxation's solution and the optimal value of RegularizedNSM without any anxiety about the rounding errors.

To simplify the proof process, we next present some assumptions about the multilinear extension. Note that these hypotheses are not necessary for the conclusions. We can estimate the relevant variables by sampling, so that the same results can be obtained with a high probability.

Assumption 1. The value oracles of the multilinear extension $F(\cdot)$ and its gradient $\nabla F(\cdot)$ are given, which means that the feedback of any vector consultation could be provided immediately.

Assumption 2. The multilinear extension $F : [0,1]^n \to \mathbb{R}_{\geq 0}$ is D-smooth, i.e., for any $\mathbf{x}, \mathbf{y} \in [0,1]^n$, we have

$$F(\mathbf{x}) + \langle \nabla F(\mathbf{x}), \mathbf{y} - \mathbf{x} \rangle - \frac{D}{2}\|\mathbf{y} - \mathbf{x}\|^2 \leq F(\mathbf{y}) \leq F(\mathbf{x}) + \langle \nabla F(\mathbf{x}), \mathbf{y} - \mathbf{x} \rangle + \frac{D}{2}\|\mathbf{y} - \mathbf{x}\|^2.$$

The D-smoothness above is widely assumed in first-order methods in convex optimization [17] and in DR-submodular maximization [2]. Moreover, with Lemma 3.3 in [9] we have

$$D = \mathcal{O}(n^2)f(OPT),$$

where OPT denotes the optimal solution for the submodular maximization problem with certain constraints.

3   Lyapunov Method – Continuous-Time Phase

In this section, we apply the Lyapunov function method to design an algorithm framework and analyze it for RegularizedNSM with no limitation on the linear term. The output approximation algorithm yields the best guarantee for the submodular term. Inspired by [7,8], we also present the whole process in two-stage, where the theoretical algorithm and the approximation guarantee are derived in the continuous-time phase and the practicable counterpart of the algorithm and the complexity are showed in the discrete-time setting.

The purpose of the continuous-time phase is to give a guideline on algorithm design and an informal expression of the analysis, where continuous-time means that the solution $\mathbf{x}(t)$ varies continuously about time t and an update rule could be described as a dynamical system $\dot{\mathbf{x}}(t) = \phi(\mathbf{x}(t))$. In this stage, the Lyapunov function will be taken as an input and an approximation algorithm is going to be produced with a provable ratio. The first task is to determine the specific parametric form of the Lyapunov function, which may be quiet different related to the problem on hand. In this study, the Lyapunov function can be defined as:

$$V(\mathbf{x}(t)) = a(t)F(\mathbf{x}(t)) - b(t)F(\mathbf{x}^\star) + c(t)L(\mathbf{x}(t)) - d(t)L(\mathbf{x}^\star), \qquad (1)$$

where the coefficient functions $a(t), b(t), c(t), d(t) \in \mathbb{R}_{\geq 0}$ are non-decreasing, non-negative and differentiable for $t \in [0, T]$ and we demand $a(T) = c(T)$. For ease of notation, we scale the time interval by $1/T$, which only affects the complexity order. We could easily recover the general result by multiplying T. As we mentioned above, the most critical property of the Lyapunov function is that in the given time interval it is a non-decreasing function, i.e., the stable criteria

$$\dot{V}(\mathbf{x}(t)) = \dot{a}(t)F(\mathbf{x}(t)) + \dot{c}(t)L(\mathbf{x}(t)) + \langle a(t)\nabla F(\mathbf{x}(t)) + c(t)\nabla L(\mathbf{x}(t)), \dot{\mathbf{x}}(t) \rangle$$
$$- \left(\dot{b}(t)F(\mathbf{x}^\star) + \dot{d}(t)L(\mathbf{x}^\star) \right) \qquad (2)$$
$$\geq 0.$$

Then, we could easily have $V(\mathbf{x}(1)) - V(\mathbf{x}(0)) \geq 0$ and obtain the approximation ratio below by rearranging the inequality

$$F(\mathbf{x}(1)) + L(\mathbf{x}(1)) \geq \frac{b(1) - b(0)}{a(1)} F(\mathbf{x}^\star) + \frac{d(1) - d(0)}{a(1)} L(\mathbf{x}^\star)$$

$$= \frac{b(1) - b(0)}{a(1)} f(\mathbf{x}^\star) + \frac{d(1) - d(0)}{a(1)} \ell(\mathbf{x}^\star),$$

where the equality holds since we assume $\mathbf{x}^\star$ is the optimal integral solution. Finally, we solve a maximization problem with approximation ratio of the first term as the objective, since it is normally the initial target (e.g. the quantity function in machine-learning) and the second term is just the implementation of regularization. We seek feasible coefficient-functions $a(t), b(t), c(t), d(t)$ so that the objective value is as large as possible.

After stating the high-level ideas, we formally start to derive the sufficient conditions for meeting the stable criteria of the Lyapunov function $V(\mathbf{x}(t))$ and design a conceptual algorithm at the same time. To do so, one of the main obstacles is the optimal sum $\left(\dot{b}(t)F(\mathbf{x}^\star) + \dot{d}(t)L(\mathbf{x}^\star)\right)$ in the derivative $\dot{V}(\mathbf{x}(t))$, since we have no information of it. A natural thought for it in dealing with maximization problems in optimization is to find a proper upper bound, which is constructed mainly by the current solution and its gradient information so that the bound and the optimal solution $\mathbf{x}^\star$ are irrelevant or at least not highly correlated. The following lemma gives a such bound.

Lemma 3.1. *Assuming $\mathbf{x}_e(t) \leq \theta(t)$ for every $e \in E$ and $t \in [0, 1]$, there exists an upper bound*

$$U(t) = \dot{b}(t) \cdot \frac{\langle \nabla F(\mathbf{x}(t)), \mathbf{v}(\mathbf{x}(t)) \odot (1 - \mathbf{x}(t)) \rangle + F(\mathbf{x}(t))}{1 - \theta(t)}$$

$$+ \dot{d}(t) \cdot \frac{\gamma - 1}{\gamma(1 - \theta(t)) - 1} \langle \vec{\ell}, \mathbf{v}(\mathbf{x}(t)) \odot (1 - \mathbf{x}(t)) \rangle$$

such that $U(t) \geq \dot{b}(t) F(\mathbf{x}^\star) + \dot{d}(t) L(\mathbf{x}^\star)$, where $\gamma = \frac{\langle \vec{\ell}_+, \mathbf{x}^\star \rangle}{-\langle \vec{\ell}_-, \mathbf{x}^\star \rangle}$ and

$$\mathbf{v}(\mathbf{x}(t)) = \arg \max_{\mathbf{v} \in \mathcal{P}_\mathcal{I}} \left\langle \frac{\dot{b}(t)}{1 - \theta(t)} \nabla F(\mathbf{x}(t)) + \frac{\gamma - 1}{\gamma(1 - \theta(t)) - 1} \dot{d}(t)\vec{\ell}, \mathbf{v} \odot (1 - \mathbf{x}(t)) \right\rangle.$$

Note that the parameter γ is unavoidable according to the proof of the last lemma, which we present in Appendix D. Now we could show the sufficient conditions that guarantee the stable criteria of the Lyapunov function $V(\mathbf{x}(t))$.

Lemma 3.2. *For any $t \in [0, 1]$, the defined Lyapunov function $V(\mathbf{x}(t))$ is non-decreasing if the coefficient functions $(a(t), b(t), c(t), d(t)) \in \mathcal{C}_{con}$ where $\mathcal{C}_{con}*

Algorithm 1. CONTINUOUS-TIME ALGORITHM FOR REGULARIZEDNSM

Input: multilinear extension F, regularized vector $\vec{\ell}$, polytope $\mathcal{P}_{\mathcal{I}}$ and coefficient functions $(a(t), b(t), c(t), d(t)) \in \mathcal{C}_{\text{con}}$
Output: $\mathbf{x}(1)$
1: Set: $\mathbf{x}(0) = \mathbf{0}$
2: **for** $t \in [0, 1]$ **do**
3:    $\mathbf{v}(\mathbf{x}(t)) = \arg\max_{\mathbf{v} \in \mathcal{P}_{\mathcal{I}}} \left\langle \dot{a}(t)\nabla F(\mathbf{x}(t)) + \frac{\gamma-1}{\gamma^{\frac{a(0)}{a(t)}}-1}\dot{d}(t)\vec{\ell}, \mathbf{v} \odot (1 - \mathbf{x}(t)) \right\rangle$
4:    $\dot{\mathbf{x}}(t) = \frac{\dot{a}(t)}{a(t)}\left(\mathbf{v}(\mathbf{x}(t)) \odot (1 - \mathbf{x}(t))\right)$
5: **end for**

---

includes

$$1 > \theta(t) \geq \|\mathbf{x}(t)\|_\infty \geq 0$$

$$\dot{a}(t) - \frac{\dot{b}(t)}{1 - \theta(t)} \geq 0$$

$$\dot{c}(t) = 0$$

$$a(t)\dot{\mathbf{x}}(t) - \frac{\dot{b}(t)}{1 - \theta(t)}\left(\mathbf{v}(\mathbf{x}(t)) \odot (1 - \mathbf{x}(t))\right) = 0$$

$$c(t)\dot{\mathbf{x}}(t) - \frac{\dot{d}(t)(\gamma - 1)}{\gamma(1 - \theta(t)) - 1}\left(\mathbf{v}(\mathbf{x}(t)) \odot (1 - \mathbf{x}(t))\right) = 0,$$

where $\theta(t)$ is the upper bound of $\mathbf{x}_e(t)$ for every $e \in E$.

According to the constraints $\mathcal{C}_{\text{con}}$, we could derive the update rule of the solution in continuous-time setting and guarantee its feasibility.

Lemma 3.3. *Given the Lyapunov function $V(\mathbf{x}(t))$, the update rule in continuous-time is*

$$\dot{\mathbf{x}}(t) = \frac{\dot{a}(t)}{a(t)}\left(\mathbf{v}(\mathbf{x}(t)) \odot (1 - \mathbf{x}(t))\right)$$

for $t \in [0, 1]$. Moreover, assuming that $\ln a(t)$ is a cumulative distribution function on $[0, 1]$, the output solution $\mathbf{x}(1) \in \mathcal{P}_{\mathcal{I}}$.

Due to the update rule given in Lemma 3.3, we can quantify the coordinate-wise upper bound of the solution by the following lemma.

Lemma 3.4. *For the coordinate-wise upper bound $\theta(t)$ of the solution $\mathbf{x}(t)$ with $t \in [0, 1]$, we have $\theta(t) \leq 1 - \frac{a(0)}{a(t)}$.*

Under the guidance of the Lyapunov function $V(\mathbf{x}(t))$ and combining all lemmas above, our continuous-time algorithm for RegularizedNSM with no limitation on the linear term could be automatically designed and shown as Algorithm 1. And its utility guarantee is shown by the following theorem.

Theorem 3.1. *For RegularizedNSM with no limitation on the linear term, the algorithm* CONTINUOUS-TIME ALGORITHM FOR REGULARIZEDNSM *achieves an* $(1/e, \frac{\gamma - \gamma/e - 1}{\gamma - 1})$-*approximation with feasible range* $\gamma \in [0, 1) \cup (\frac{1}{1-1/e}, +\infty)$ *and* $\gamma \in (1, \frac{1}{1-1/e})$ *for* $\ell(\mathbf{x}^\star) \geq 0$ *and* $\ell(\mathbf{x}^\star) < 0$ *respectively, where* $\gamma = \frac{\langle \vec{\ell}_+, \mathbf{x}^\star \rangle}{-\langle \vec{\ell}_-, \mathbf{x}^\star \rangle}$.

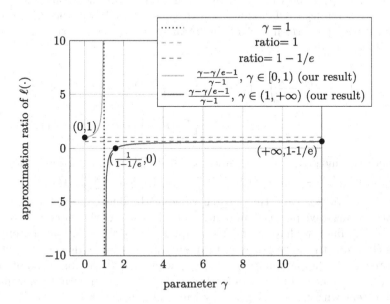

Fig. 1. The approximation ratio of the linear term. The light blue ($\gamma \in [0, 1)$) and dark blue ($\gamma \in (1, +\infty)$) solid lines are the curves of $\frac{\gamma - \gamma/e - 1}{\gamma - 1}$ (our result), which illustrates the approximation ratio of the linear term with various regularized weight vector. The green dash line is the result given by Lu et al. [16] for the non-positive setting of $\ell(\cdot)$. The red dash line is the result presented by Qi [18] for the non-negative setting of $\ell(\cdot)$.(Color figure online)

The Discussion of the Approximation Ratio of the Linear Term. Since the parameter $\gamma \in [0, +\infty)$ according to its definition, the curve of $\frac{\gamma - \gamma/e - 1}{\gamma - 1}$ is represented by light ($\gamma \in [0, 1)$) and dark blue ($\gamma \in (1, +\infty)$) solid lines in Fig. 1. There are several cases to discuss.

(a) $\gamma = 1$ (the black dotted vertical line in Fig. 1.) In this case, the optimal value of the linear term is $\ell(\mathbf{x}^\star) = 0$ since there is no dominator between the positive and negative parts.

(b) $\gamma \in (1, +\infty)$. In this case, the optimal value of the linear term is non-negative since the positive part is dominant. From the curve of this interval (the dark blue solid line), we have a non-negative guarantee when $\gamma \in [\frac{1}{1-1/e}, +\infty]$.

Moreover, the limitation of the guarantee is $(1 - 1/e)$ when $\gamma \to +\infty$ (the black point $(+\infty, 1-1/e)$ in Fig. 1) implying that the linear function is nearly non-negative. Hence, the bi-ratio is $(1/e, 1 - 1/e)$ for this special situation, which coincides with the result given by Qi [18]. Then, we focus on the interval $\gamma \in (1, \frac{1}{1-1/e})$, where the guarantee of $\ell(\mathbf{x}^\star)$ is negative. However, it is not unacceptable since there is no algorithm whose output can make a positive $\ell(\cdot)$ value while keeping $1/e$ guarantee for the submodular term, when the non-negativity assumption of $\ell(\cdot)$ is violated. Now the only problem is how bad the performance of the linear term will be. Notice that we have the following result when $\gamma \in (1, \frac{1}{1-1/e})$

$$\left(\frac{\gamma - \gamma/e - 1}{\gamma - 1} \right) \cdot \ell(\mathbf{x}^\star) = \left(1 - \frac{1}{e(1-\frac{1}{\gamma})} \right) (1 - \gamma) \cdot \langle \vec{\ell}_-, \mathbf{x}^\star \rangle$$

$$\geq \left| \sum_{k=1}^n (\vec{\ell}_-)_k \right| \cdot ((1 - 1/e)\gamma - 1),$$

where the inequality is true since $\mathbf{x}^\star \leq \mathbf{1}$. Therefore, we have the lower bound of $\left(\frac{\gamma-\gamma/e-1}{\gamma-1} \cdot \ell(\mathbf{x}^\star) \right)$ if there is a limitation of $\sum_{k=1}^n (\vec{\ell}_-)_k$.

(c) $\gamma \in [0, 1)$. In this case, the optimal value of the linear term is non-positive since the negative part is dominant. From the curve of this interval (the light blue solid line), we have a positive guarantee for $\ell(\mathbf{x}^\star) < 0$. Fortunately, the conclusion is the same since the inequality above also holds when $\gamma \in [0, 1)$. Moreover, the approximation ratio is 1 when $\gamma = 0$ (the black point $(0, 1)$ in Fig. 1). Hence, the bi-ratio is $(1/e, 1)$ for this special situation, which is coincident with the result given by Lu et al. [16] but worse than $(0.385, 1)$ presented by Qi [18].

(d) Comparing with $(0.29, 0.59)$-approximation given by Qi [18], our result given in Theorem 3.1 is strictly better from the perspective of the performance of the submodular function, which is more important in a real scenario of RegularizedNSM since the linear function is normally just a regular term.

4   Conclusion

In the paper, we present algorithm frameworks for RegularizedNSM subject to down-monotone family of sets with the help of the Lyapunov function. By properly choosing the coefficients, our algorithms yield $(1/e, \frac{\gamma-\gamma/e-1}{\gamma-1})$ approximation guarantees with polynomial-time complexity for the above problems respectively, where $\gamma \in [0, 1) \cup (1, +\infty)$ is a parameter reflecting the relative dominance of the positive and negative parts of the linear optimal value. Notably, our results are quiet general since it could go back to many existed conclusions when the linear function own special property like non-negativity and non-positivity. Moreover, our results are strictly better from the perspective of the performance of the submodular function comparing with $(0.29, 0.59)$-approximation given by Qi [18] when there is no limitation on the linear term. At the end, we also show a $(1 - 1/e, 1)$-approximation algorithm for RegularizedMSM by the Lyapunov method with a similar analysis process, which coincides with that the state-of-the-art result. In the future, we focus on eliminating the parameter γ in the bicriteria approximation without losing the approximation guarantee of the submodular term as much as possible.

Acknowledgments. The first author is supported by National Natural Science Foundation of China (No. 12301419), the Fundamental Research Funds for the Central Universities and the National key research and development program of China (2021YFA1000403). The second author is supported by National Natural Science Foundation of China (No. 11991022) and the National key research and development program of China (2021YFA1000403). The third author is supported by National Natural Science Foundation of China (No. 11971149). The fourth author is supported by National Natural Science Foundation of China (No. 12131003). The fifth author is supported by the National Science Foundation of China (No. 12001335).

Appendices

A   Technical Lemmata

Finally, we give two rephrased technical lemmata about the multilinear extension, which are frequently used in later sections. The first is a well-known bound, which build the connection of any two feasible vectors with coordinate-wise operations and the dot product between the gradient and their difference.

Lemma A.1 ([13]).   *For any differentiable DR-submodular function F : $[0,1]^n \to \mathbb{R}_{\geq 0}$, we have the following result for any $\mathbf{x}, \mathbf{y} \in [0,1]^n$,*

$$\langle \nabla F(\mathbf{x}), \mathbf{y} - \mathbf{x} \rangle \geq F(\mathbf{x} \vee \mathbf{y}) + F(\mathbf{x} \wedge \mathbf{y}) - 2F(\mathbf{x}).$$

The second lemma can be viewed as a typical technique, which is normally applied in non-monotone setting for obtaining a binary-valued vector.

Lemma A.2 ([11]).   *For any DR-submodular function $F : [0,1]^n \to \mathbb{R}_{\geq 0}$ and any $\mathbf{y} \in \{0,1\}^n$, we have $F(\mathbf{x} \vee \mathbf{y}) \geq (1 - \|\mathbf{x}\|_\infty) \cdot F(\mathbf{y})$.*

B   Discrete-Time Phase for RegularizedNSM

In the last subsection, we introduce Algorithm 1, which is automatically designed by the given Lyapunov function for solving RegularizedNSM in continuous-time. Although the algorithm can yield state of the art approximation ratio, it is only theoretically illustrative and hard to be implemented on a discrete computer. Therefore, the goal of this phase is to discretize the algorithm with acceptable guarantee losses so that it is executable with provable time complexity. The whole process is not as intuitive as we think. One major obstacle is that the Lyapunov function may not be strictly monotone due to the discretization errors. The discrete-time counter-part of the Lyapunov function remains the following form

$$V(\mathbf{x}(t_i)) = a(t_i)F(\mathbf{x}(t_i)) - b(t_i)F(\mathbf{x}^\star) + c(t_i)L(\mathbf{x}(t_i)) - d(t_i)L(\mathbf{x}^\star), \qquad (3)$$

where for every $i = 1, \ldots, K$, $a(t_i), b(t_i), c(t_i), d(t_i)$ are the point-mass sequences sampled by the coefficient functions in the last phase. Since it is impossible to verify the stable criteria through the derivative of V, we focus on the value increment per unit of time of two adjacent iteration and expect that it has a controllable lower bound even if it is negative, i.e.,

$$\frac{V(\mathbf{x}(t_{i+1})) - V(\mathbf{x}(t_i))}{t_{i+1} - t_i} \geq -B_i \in -\mathcal{O}\left(\frac{\epsilon}{K^2}\right), \qquad (4)$$

where the error term $B_i \geq 0$ for every $i = 1, \ldots, K$ and the total number of rounds K decides the time complexity. Then, by telescoping over all time stamps, we have

$$\begin{aligned} V(\mathbf{x}(t_K)) - V(\mathbf{x}(t_0)) &= a(t_K)F(\mathbf{x}(t_K)) - (b(t_K) - b(t_0))F(\mathbf{x}^\star) \\ &\quad + c(t_K)L(\mathbf{x}(t_K)) - (d(t_K) - d(t_0))L(\mathbf{x}^\star) \\ &\geq -\sum_{i=0}^{K-1} B_i(t_{i+1} - t_i) \in -\mathcal{O}(\epsilon). \end{aligned}$$

Finally, the approximation guarantee could be obtained with requiring $a(t_K) = c(t_K)$

$$F(\mathbf{x}(t_K)) + L(\mathbf{x}(t_K)) \geq \frac{b(t_K) - b(t_0)}{a(t_K)}F(\mathbf{x}^\star) + \frac{d(t_K) - d(t_0)}{a(t_K)}L(\mathbf{x}^\star) - \mathcal{O}(\epsilon).$$

After the high-level thinking, we formally give the analysis process of the discrete-time phase. Similar with the continuous-time phase, we begin with the derivation of all sufficient constraints that the sampled sequences should satisfy.

Lemma B.1. *For every $i = 1, \ldots, K$, the increment of two successive solutions has the following error term*

$$B_i = \frac{a(t_{i+1})D}{2(t_{i+1} - t_i)}\|\mathbf{x}(t_{i+1}) - \mathbf{x}(t_i)\|^2,$$

when the point-mass sequences $(a(t_i), b(t_i), c(t_i), d(t_i)) \in \mathcal{C}_{dis}$, where $\mathcal{C}_{dis}$ includes

$$a(t_{i+1}) - a(t_i) \geq \frac{b(t_{i+1}) - b(t_i)}{1 - \theta(t_i)}$$

$$c(t_{i+1}) - c(t_i) = 0$$

$$\mathbf{x}(t_{i+1}) - \mathbf{x}(t_i) = \frac{b(t_{i+1}) - b(t_i)}{a(t_{i+1})(1 - \theta(t_i))}\left(\mathbf{v}(\mathbf{x}(t_i)) \odot (1 - \mathbf{x}(t_i))\right)$$

$$\mathbf{x}(t_{i+1}) - \mathbf{x}(t_i) = \frac{(\gamma - 1)(d(t_{i+1}) - d(t_i))}{c(t_i)(\gamma(1 - \theta(t_i)) - 1)}\left(\mathbf{v}(\mathbf{x}(t_i)) \odot (1 - \mathbf{x}(t_i))\right),$$

where $\mathbf{v}(\mathbf{x}(t_i)) = \arg\max_{\mathbf{v}\in\mathcal{P}_\mathcal{I}}\langle w_f(i)\nabla F(\mathbf{x}(t_i)) + \vec{\ell}, \mathbf{v}\odot(1-\mathbf{x}(t_i))\rangle$, and $w_f(i) = \frac{b(t_{i+1})-b(t_i)}{(t_{i+1}-t_i)(1-\theta(t_i))}$, $w_\ell(i) = \frac{(d(t_{i+1})-d(t_i))(\gamma-1)}{((t_{i+1}-t_i))(\gamma(1-\theta(t_i))-1)}$.

Proof. The value increment per unit of time is,

$$\frac{V(\mathbf{x}(t_{i+1})) - V(\mathbf{x}(t_i))}{t_{i+1}-t_i} = \frac{a(t_{i+1})F(\mathbf{x}(t_{i+1})) - a(t_i)F(\mathbf{x}(t_i)) - (b(t_{i+1})-b(t_i))F(\mathbf{x}^\star)}{t_{i+1}-t_i}$$
$$+ \frac{c(t_{i+1})L(\mathbf{x}(t_{i+1})) - c(t_i)L(\mathbf{x}(t_i)) - (d(t_{i+1})-d(t_i))L(\mathbf{x}^\star)}{t_{i+1}-t_i}.$$

Due to Lemma 3.1, the upper bound of the optimal sum can be denoted as

$$U(t_i) = \frac{b(t_{i+1})-b(t_i)}{(t_{i+1}-t_i)(1-\theta(t_i))}\left(\langle\nabla F(\mathbf{x}(t_i)), \mathbf{v}(\mathbf{x}(t_i))\odot(1-\mathbf{x}(t_i))\rangle + F(\mathbf{x}(t_i))\right)$$
$$+ \frac{(d(t_{i+1})-d(t_i))(\gamma-1)}{(t_{i+1}-t_i)(\gamma(1-\theta(t_i))-1)}\langle\vec{\ell}, \mathbf{v}(\mathbf{x}(t_i))\odot(1-\mathbf{x}(t_i))\rangle$$
$$\geq \frac{(b(t_{i+1})-b(t_i))F(\mathbf{x}^\star) + (d(t_{i+1})-d(t_i))L(\mathbf{x}^\star)}{t_{i+1}-t_i},$$

where $\mathbf{v}(\mathbf{x}(t_i)) = \arg\max_{\mathbf{v}\in\mathcal{P}_\mathcal{I}}\langle w_f(i)\nabla F(\mathbf{x}(t_i)) + w_\ell(i)\vec{\ell}, \mathbf{v}\odot(1-\mathbf{x}(t_i))\rangle$ and $w_f(i) = \frac{b(t_{i+1})-b(t_i)}{(t_{i+1}-t_i)(1-\theta(t_i))}$, $w_\ell(i) = \frac{(d(t_{i+1})-d(t_i))(\gamma-1)}{(t_{i+1}-t_i)(\gamma(1-\theta(t_i))-1)}$.

Then,

$$\frac{V(\mathbf{x}(t_{i+1})) - V(\mathbf{x}(t_i))}{t_{i+1}-t_i}$$
$$\geq \frac{a(t_{i+1})\left(F(\mathbf{x}(t_{i+1})) - F(\mathbf{x}(t_i))\right) + (a(t_{i+1})-a(t_i))F(\mathbf{x}(t_i))}{t_{i+1}-t_i}$$
$$+ \frac{c(t_{i+1})L(\mathbf{x}(t_{i+1})) - c(t_i)L(\mathbf{x}(t_i))}{t_{i+1}-t_i} - U(t_i).$$

Applying the D-smoothness assumption of F, we obtain

$$\frac{a(t_{i+1})}{t_{i+1}-t_i}\left(F(\mathbf{x}(t_{i+1})) - F(\mathbf{x}(t_i))\right) \geq \langle\nabla F(\mathbf{x}(t_i)), \frac{a(t_{i+1})}{t_{i+1}-t_i}(\mathbf{x}(t_{i+1})-\mathbf{x}(t_i))\rangle$$
$$- \frac{a(t_{i+1})D}{2(t_{i+1}-t_i)}\|\mathbf{x}(t_{i+1})-\mathbf{x}(t_i)\|^2.$$

Combining with U_t, the above inequality could be further derived as

$$\frac{V(\mathbf{x}(t_{i+1})) - V(\mathbf{x}(t_i))}{t_{i+1} - t_i}$$

$$\geq \langle \nabla F(\mathbf{x}(t_i)), \frac{a(t_{i+1})}{t_{i+1} - t_i}(\mathbf{x}(t_{i+1}) - \mathbf{x}(t_i)) - \frac{b(t_{i+1}) - b(t_i)}{(1 - \theta(t_i))(t_{i+1} - t_i)}\mathbf{v} \odot (1 - \mathbf{x}(t_i)))\rangle$$

$$+ \left(\frac{a(t_{i+1}) - a(t_i) - \frac{b(t_{i+1}) - b(t_i)}{1 - \theta(t_i)}}{t_{i+1} - t_i} \right) F(\mathbf{x}(t_i))$$

$$+ \langle \vec{\ell}, \frac{c(t_{i+1})\mathbf{x}(t_{i+1}) - c(t_i)\mathbf{x}(t_i)}{t_{i+1} - t_i} - \frac{(d(t_{i+1}) - d(t_i))(\gamma - 1)}{(\gamma(1 - \theta(t_i)) - 1)((t_{i+1} - t_i))}\mathbf{v} \odot (1 - \mathbf{x}(t_i)))\rangle$$

$$- \frac{a(t_{i+1})D}{2(t_{i+1} - t_i)}\|\mathbf{x}(t_{i+1}) - \mathbf{x}(t_i)\|^2.$$

Since F is non-negative and non-monotone and the range of L is read domain, the sufficient conditions $\mathcal{C}_{dis}$ that the point-mass sequences $(a(t_i), b(t_i), c(t_i), d(t_i))$ should satisfy are listed below

$$a(t_{i+1}) - a(t_i) \geq \frac{b(t_{i+1}) - b(t_i)}{1 - \theta(t_i)}$$

$$c(t_{i+1}) - c(t_i) = 0$$

$$\mathbf{x}(t_{i+1}) - \mathbf{x}(t_i) = \frac{b(t_{i+1}) - b(t_i)}{a(t_{i+1})(1 - \theta(t_i))}(\mathbf{v}(\mathbf{x}(t_i)) \odot (1 - \mathbf{x}(t_i)))$$

$$\mathbf{x}(t_{i+1}) - \mathbf{x}(t_i) = \frac{(\gamma - 1)(d(t_{i+1}) - d(t_i))}{c(t_i)(\gamma(1 - \theta(t_i)) - 1)}(\mathbf{v}(\mathbf{x}(t_i)) \odot (1 - \mathbf{x}(t_i))).$$

Therefore, the increment per unit of time has the lower bound

$$\frac{V(\mathbf{x}(t_{i+1})) - V(\mathbf{x}(t_i))}{t_{i+1} - t_i} \geq -\frac{a(t_{i+1})D}{2(t_{i+1} - t_i)}\|\mathbf{x}(t_{i+1}) - \mathbf{x}(t_i)\|^2 = -B_i.$$

$\square$

With a properly choice of the coeffcients in continuous-time, we meet the sufficient constraints in discrete-time and make the error term controllable with the next lemma. Moreover, the time complexity will also be automatically decided by the total number of rounds K.

Lemma B.2. *Setting* $1 - \theta(t_i) = \frac{a(t_0)}{a(t_i)}$, $a(t_i) = e^{t_i - 1}$, $b(t_i) = \frac{t}{e}$, $c(t_i) = 1$, $d(t_i) = -\frac{\gamma e^{-t_i} + t}{\gamma - 1}$, *where* $t_i = \frac{i}{K}$ *for* $i = 1, \ldots, K$ *and* $K = \mathcal{O}\left(\frac{nD}{\epsilon}\right)$ *with* $\epsilon > 0$. *Then, the sufficient conditions* $\mathcal{C}_{dis}$ *can be satisfied and the total error is*

$$\sum_{i=0}^{K-1} B_i(t_{i+1} - t_i) \leq \sum_{i=0}^{K-1} \frac{a(t_{i+1})D}{2}\|\mathbf{x}(t_{i+1}) - \mathbf{x}(t_i)\|^2 \leq \mathcal{O}(\epsilon).$$

Proof. We first testify that the choice of all point-mass sequence meets the sufficient conditions. Since $1 - \theta(t_i) = \frac{a(t_0)}{a(t_i)}$, $a(t_i) = e^{t_i-1}$, $b(t_i) = \frac{t}{e}$, $c(t_i) = 1$, $d(t_i) = -\frac{\gamma e^{-t_i}+t}{\gamma-1}$, we have $c(t_{i+1}) - c(t_i) = 0$ and

$$a(t_{i+1}) - a(t_i) - \frac{b(t_{i+1}) - b(t_i)}{1 - \theta(t_i)} = e^{t_{i+1}-1} - e^{t_i-1} - (t_{i+1} - t_i)e^{t_i-1}$$
$$= e^{t_i-1}\left[e^{t_{i+1}-t_i} - 1 - (t_{i+1} - t_i)\right]$$
$$\geq 0,$$

where the inequality holds since $e^x \geq 1 + x$ for $x \in \mathbb{R}$. Moreover,

$$\frac{b(t_{i+1}) - b(t_i)}{a(t_{i+1})(1 - \theta(t_i))} = \frac{t_{i+1} - t_i}{e^{t_{i+1}-t_i}} \in \mathcal{O}\left(\frac{1}{K}\right) = \mathcal{O}(\epsilon),$$

and

$$\frac{(\gamma - 1)(d(t_{i+1}) - d(t_i))}{c(t_i)(\gamma(1 - \theta(t_i)) - 1)} = \frac{t_{i+1} - t_i - \gamma e^{-t_i}(1 - e^{-t_{i+1}+t_i})}{1 - \gamma e^{-t_i}}$$
$$\geq \frac{t_{i+1} - t_i - \gamma e^{-t_i}(t_{i+1} - t_i)}{1 - \gamma e^{-t_i}}$$
$$= t_{i+1} - t_i \in \mathcal{O}\left(\frac{1}{K}\right) = \mathcal{O}(\epsilon),$$

where the inequality holds since $x \geq 1 - e^{-x}$ for $x \in \mathbb{R}$.

Then, we focus on the total error by telescoping over all iterations

$$\sum_{i=0}^{K-1} B_i(t_{i+1} - t_i) = \sum_{i=0}^{K-1} \frac{a(t_{i+1})D}{2}\|\mathbf{x}(t_{i+1}) - \mathbf{x}(t_i)\|^2$$

By plugging the relevant sufficient constraint in $\mathcal{C}_{\text{dis}}$ and $\mathbf{v}(\mathbf{x}(t_i)) \odot (1 - \mathbf{x}(t_i)) \leq \mathbf{1}$, we get

$$\|\mathbf{x}(t_{i+1}) - \mathbf{x}(t_i)\|^2 \leq \left\|\mathcal{O}\left(\frac{1}{K}\right) \cdot \mathbf{1}\right\|^2 = \mathcal{O}\left(\frac{n}{K}\right).$$

Therefore, the total error can be bounded by

$$\sum_{i=0}^{K-1} B_i(t_{i+1} - t_i) \leq \mathcal{O}\left(\frac{a(t_{i+1})nD}{K}\right) \leq \mathcal{O}(\epsilon),$$

where the inequality holds due to $a(t_{i+1}) \leq 1$. $\qquad\square$

After presenting the discrete-time phase analysis, the DISCRETE-TIME ALGORITHM FOR REGULARIZEDNSM could be naturally introduced as Algorithm 2. The designed algorithm produces a sequential feasible vectors $\mathbf{x}(t_i)$ for all time stamps $i = 1, \ldots, K$ in polynomial-time complexity, and it yields an identical approximation ratio with arbitrarily small loss.

Algorithm 2. DISCRETE-TIME ALGORITHM FOR REGULARIZEDNSM

Input: multilinear extension F, regularized vector $\vec{\ell}$, polytope $\mathcal{P}_\mathcal{I}$ and coefficient functions $(a(t_i), b(t_i), c(t_i), d(t_i)) \in \mathcal{C}_{\mathrm{dis}}$ for every $i = 1, \ldots, K$
Output: $\mathbf{x}(t_K)$
1: Set: $\mathbf{x}(t_0) = \mathbf{0}$, $K = \frac{nD}{\epsilon} = \frac{n^3}{\epsilon}$
2: **for** $i = 0, \ldots, K - 1$ **do**
3:     $t_i = \frac{i}{K}$
4:     $\mathbf{v}(\mathbf{x}(t_i)) = \arg\max_{\mathbf{v} \in \mathcal{P}_\mathcal{I}} \langle w_f(i)\nabla F(\mathbf{x}(t_i)) + w_\ell(i)\vec{\ell}, \mathbf{v} \odot (1 - \mathbf{x}(t_i)) \rangle$
5:     $\mathbf{x}(t_{i+1}) = \mathbf{x}(t_i) + \frac{b(t_{i+1}) - b(t_i)}{a(t_i)(1 - \theta(t_i))}(\mathbf{v}(\mathbf{x}(t_i)) \odot (1 - \mathbf{x}(t_i)))$
6: **end for**

Theorem B.1. *For arbitrary $\epsilon > 0$, the* DISCRETE-TIME ALGORITHM FOR REGULARIZEDNSM *outputs $\mathbf{x}(t_K) \in \mathcal{P}_\mathcal{I}$ and it satisfies*

$$F(\mathbf{x}(t_K)) + L(\mathbf{x}(t_K)) \geq \frac{1}{e}f(\mathbf{x}^\star) + \frac{\gamma - \gamma/e - 1}{\gamma - 1}\ell(\mathbf{x}^\star) - \mathcal{O}(\epsilon),$$

with feasible range $\gamma \in [0, 1) \cup (\frac{1}{1 - 1/e}, +\infty)$ and $\gamma \in (1, \frac{1}{1 - 1/e})$ for $\ell(\mathbf{x}^\star) \geq 0$ and $\ell(\mathbf{x}^\star) < 0$ respectively, where $\gamma = \frac{\langle \vec{\ell}_+, \mathbf{x}^\star \rangle}{-\langle \vec{\ell}_-, \mathbf{x}^\star \rangle}$. Moreover, the time complexity of is $\mathcal{O}\left(\frac{n^3}{\epsilon}\right)$.

Proof. Since the algorithm runs K rounds, we trivially get the $\mathcal{O}\left(\frac{n^3}{\epsilon}\right)$ time complexity. For the increment value of two iterative solution, its telescoping sum over all time stamps t_i for $i = 1, \ldots, K$ is

$$\sum_{i=0}^{K-1} V(\mathbf{x}(t_{i+1})) - V(\mathbf{x}(t_i)) = F(\mathbf{x}(t_K)) - \frac{1}{e}F(\mathbf{x}^\star) + L(\mathbf{x}(t_K)) - \frac{\gamma - \gamma/e - 1}{\gamma - 1}L(\mathbf{x}^\star)$$

$$\geq -\sum_{i=0}^{K-1} B_i(t_{i+1} - t_i) \geq -\mathcal{O}(\epsilon),$$

where the inequality holds due to Lemma B.1 and Lemma B.2.

Since $\mathbf{x}^\star \in \{0, 1\}^n$, we obtain the following result by rearranging the inequality,

$$F(\mathbf{x}(t_K)) + L(\mathbf{x}(t_K)) \geq \frac{1}{e}f(\mathbf{x}^\star) + \frac{\gamma - \gamma/e - 1}{\gamma - 1}\ell(\mathbf{x}^\star) - \mathcal{O}(\epsilon).$$

$\square$

C   The Lyapunov Method for RegularizedMSM

For RegularizedMSM, where the submodular function in the objective is monotone, i.e. $f(T) \geq f(S)$ for any $S \subseteq T \subseteq E$, the Lyapunov function method can also be applied to design continuous-time and discrete-time algorithms. The approximation ratio and the time-complexity coincide with the result given by Feldman [9]. Since the analysis process is similar to Sect. 3 and Appendix B, we will only show the algorithms designed by this systematical framework and the conclusions about guarantee and time complexity without proofs.

Continuous-Time Phase. For RegularizedMSM, we also define the Lyapunov function in this phase as Eq. (1) with non-decreasing, non-negative and differentiable coefficient functions $a(t), b(t), c(t), d(t) \in \mathbb{R}_{\geq 0}$ for $t \in [0, 1]$ and require $a(1) = c(1)$. Hence, its derivative is the same as Eq. (2). The analysis process of the Lyapunov function method in continuous-time setting is consistent. The key point is deriving the sufficient conditions of the stable criteria. To do so, we still begin with find a proper upper bound for the optimal sum $\left(\dot{b}(t)F(\mathbf{x}^\star) + \dot{d}(t)L(\mathbf{x}^\star)\right)$ in the derivative $\dot{V}(\mathbf{x}(t))$. The following lemma gives a such bound.

Lemma C.1. *For any $t \in [0,1]$, there exists an upper bound*

$$U(t) = \dot{b}(t)\left(\langle \nabla F(\mathbf{x}(t)), \mathbf{v}(\mathbf{x}(t))\rangle + F(\mathbf{x}(t))\right) + \dot{d}(t)\langle \vec{\ell}, \mathbf{v}(\mathbf{x}(t))\rangle$$

such that $U(t) \geq \dot{b}(t)F(\mathbf{x}^\star) + \dot{d}(t)L(\mathbf{x}^\star)$, where

$$\mathbf{v}(\mathbf{x}(t)) = \arg\max_{\mathbf{v} \in \mathcal{P}_\mathcal{I}} \left\langle \dot{b}(t)\nabla F(\mathbf{x}(t)) + \dot{d}(t)\vec{\ell}, \mathbf{v}\right\rangle.$$

The following lemma shows the sufficient conditions of the stable criteria.

Lemma C.2. *For RegularizedMSM, the defined Lyapunov function $V(\mathbf{x}(t))$ is non-decreasing if the coefficient functions $(a(t), b(t), c(t), d(t)) \in \mathcal{C}'_{con}$ for any $t \in [0, 1]$, where $\mathcal{C}'_{con}$ includes*

$$\dot{a}(t) - \dot{b}(t) \geq 0$$
$$\dot{c}(t) = 0$$
$$a(t)\dot{\mathbf{x}}(t) - \dot{b}(t)\mathbf{v}(\mathbf{x}(t)) \geq 0$$
$$c(t)\dot{\mathbf{x}}(t) - \dot{d}(t)\mathbf{v}(\mathbf{x}(t)) = 0,$$

where $\mathbf{v}(\mathbf{x}(t)) = \arg\max_{\mathbf{v} \in \mathcal{P}_\mathcal{I}} \left\langle \dot{b}(t)\nabla F(\mathbf{x}(t)) + \dot{d}(t)\vec{\ell}, \mathbf{v}\right\rangle$.

According to the constraints $\mathcal{C}'_{con}$, we could derive the update rule of the solution in continuous-time setting and guarantee its feasibility.

Lemma C.3. *Given the Lyapunov function $V(\mathbf{x}(t))$, the update rule in continuous-time is*

$$\dot{\mathbf{x}}(t) = \frac{\dot{a}(t)}{a(t)}\mathbf{v}(\mathbf{x}(t))$$

for $t \in [0, 1]$. Moreover, assuming that $\ln a(t)$ is a cumulative distribution function on $[0, 1]$, the output solution $\mathbf{x}(1) \in \mathcal{P}_\mathcal{I}$.

Algorithm 3. CONTINUOUS-TIME ALGORITHM FOR REGULARIZEDMSM

Input: multilinear extension F, regularized vector $\vec{\ell}$, polytope $\mathcal{P}_{\mathcal{I}}$ and coefficient functions $(a(t), b(t), c(t), d(t)) \in \mathcal{C}'_{con}$
Output: $\mathbf{x}(1)$
1: Set: $\mathbf{x}(0) = \mathbf{0}$
2: **for** $t \in [0, 1]$ **do**
3:     $\mathbf{v}(\mathbf{x}(t)) = \arg\max_{\mathbf{v} \in \mathcal{P}_{\mathcal{I}}} \langle a(t)\nabla F(\mathbf{x}(t)) + \dot{d}(t)\vec{\ell}, \mathbf{v} \rangle$
4:     $\dot{\mathbf{x}}(t) = \frac{\dot{a}(t)}{a(t)} \mathbf{v}(\mathbf{x}(t))$
5: **end for**

Then, the continuous-time algorithm for RegularizedMSM with no limitation on the linear term could be automatically designed and shown as Algorithm 3. The following theorem presents its utility guarantee, which coincides with the result given in [9]. The approximation ratio is derived with a careful choice of the coefficient functions, i.e. $a(t) = e^{t-1}, b(t) = e^{t-1}, c(t) = 1, d(t) = t$ and therefore the Lyapunov function is finally given by

$$V(\mathbf{x}(t)) = e^{t-1}\left(F(\mathbf{x}(t)) - F(\mathbf{x}^\star)\right) + L(\mathbf{x}(t)) - tL(\mathbf{x}^\star).$$

Theorem C.1. *For RegularizedMSM with no limitation on the linear term, the algorithm* CONTINUOUS-TIME ALGORITHM FOR REGULARIZEDMSM *achieves an* $(1 - 1/e, 1)$-*approximation.*

Discrete-Time Phase. In discrete-time phase, we take advantage of the obtained results in the last phase to develop an implemented algorithm with yielding almost the same approximation guarantee along with polynomial-time complexity. The counter-part of the Lyapunov function in this setting keeps Eq. (3) unchanged, where for every $i = 1, \ldots, K$, $a(t_i), b(t_i), c(t_i), d(t_i)$ are the point-mass sequences sampled by the coefficient functions in the last phase with requiring $a(t_K) = c(t_K)$. Instead of verifying the stable criteria strictly, we consider the difference operation (shown as Eq. (4)) of the Lyapunov function in discrete-time setting, although it may cause discretization errors. Fortunately, we could get an identical approximation ratio with arbitrarily small loss as long as the iterative accumulation error can be bounded by properly choosing the number of rounds.

The following lemma gives the sufficient conditions of the increment per unit of time with certain error.

Lemma C.4. *For every* $i = 1, \ldots, K$, *the increment of two successive solutions has the following error term*

$$B_i = \frac{a(t_{i+1})D}{2(t_{i+1} - t_i)}\|\mathbf{x}(t_{i+1}) - \mathbf{x}(t_i)\|^2,$$

when the point-mass sequences $(a(t_i), b(t_i), c(t_i), d(t_i)) \in \mathcal{C}'_{dis}$, where $\mathcal{C}'_{dis}$ includes

$$a(t_{i+1}) - a(t_i) \geq b(t_{i+1}) - b(t_i)$$
$$c(t_{i+1}) - c(t_i) = 0$$
$$\mathbf{x}(t_{i+1}) - \mathbf{x}(t_i) \geq \frac{b(t_{i+1}) - b(t_i)}{a(t_{i+1})} \mathbf{v}(\mathbf{x}(t_i))$$
$$\mathbf{x}(t_{i+1}) - \mathbf{x}(t_i) = \frac{d(t_{i+1}) - d(t_i)}{c(t_i)} \mathbf{v}(\mathbf{x}(t_i)),$$

where $\mathbf{v}(\mathbf{x}(t_i)) = \arg\max_{\mathbf{v} \in \mathcal{P}_{\mathcal{I}}} \left\langle \frac{a(t_{i+1}) - a(t_i)}{t_{i+1} - t_i} \nabla F(\mathbf{x}(t_i)) + \frac{d(t_{i+1}) - d(t_i)}{t_{i+1} - t_i} \vec{\ell}, \mathbf{v} \right\rangle$.

Our next lemma shows that the sufficient constraints in discrete-times can be satisfied with a properly choices of the coefficient functions in continuous-time. Moreover, the discretization errors are manageable if we set the number of rounds reasonable.

Lemma C.5. *Setting* $a(t_i) = e^{t_i - 1}, b(t_i) = e^{t_i - 1}, c(t_i) = 1, d(t_i) = t$, *where* $t_i = \frac{i}{K}$ *for* $i = 1, \ldots, K$ *and* $K = \mathcal{O}\left(\frac{n^3}{\epsilon}\right)$ *with* $\epsilon > 0$. *Then, the sufficient conditions* $\mathcal{C}'_{dis}$ *can be satisfied and the accumulative error is*

$$\sum_{i=0}^{K-1} B_i(t_{i+1} - t_i) \leq \frac{a(t_{i+1})KD}{2} \|\mathbf{x}(t_{i+1}) - \mathbf{x}(t_i)\|^2 \leq \mathcal{O}(\epsilon).$$

The DISCRETE-TIME ALGORITHM FOR REGULARIZEDMSM designed by this framework could be naturally introduced as Algorithm 4, which produces a sequential feasible vectors $\mathbf{x}(t_i)$ for all time stamps $i = 1, \ldots, K$ in polynomial-time complexity. Given the specific form of the Lyapunov function,

$$V(\mathbf{x}(t_i)) = e^{t_i - 1} (F(\mathbf{x}(t_i)) - F(\mathbf{x}^\star)) + L(\mathbf{x}(t_i)) - tL(\mathbf{x}^\star),$$

for every $i = 1, \ldots, K$ with $K = \mathcal{O}\left(\frac{n^3}{\epsilon}\right)$, it yields an identical approximation ratio with arbitrarily small loss, which coincides with the result presented in [9].

Theorem C.2. *For arbitrary* $\epsilon > 0$, *the* DISCRETE-TIME ALGORITHM FOR REGULARIZEDMSM *outputs* $\mathbf{x}(t_K) \in \mathcal{P}_{\mathcal{I}}$ *and it satisfies*

$$F(\mathbf{x}(t_K)) + L(\mathbf{x}(t_K)) \geq \left(1 - \frac{1}{e}\right) f(\mathbf{x}^\star) + \ell(\mathbf{x}^\star) - \mathcal{O}(\epsilon).$$

Moreover, the time complexity is of $\mathcal{O}\left(\frac{n^3}{\epsilon}\right)$.

Algorithm 4. DISCRETE-TIME ALGORITHM FOR REGULARIZEDMSM

Input: multilinear extension F, regularized vector $\vec{\ell}$, polytope $\mathcal{P}_{\mathcal{I}}$ and coefficient functions $(a(t_i), b(t_i), c(t_i), d(t_i)) \in \mathcal{C}'_{\text{dis}}$ for every $i = 1, \ldots, K$

Output: $\mathbf{x}(t_K)$

1: Set: $\mathbf{x}(t_0) = \mathbf{0}$, $K = \frac{nD}{\epsilon} = \frac{n^3}{\epsilon}$
2: **for** $i = 0, \ldots, K - 1$ **do**
3:     $t_i = \frac{i}{K}$
4:     $\mathbf{v}(\mathbf{x}(t_i)) = \arg\max_{\mathbf{v} \in \mathcal{P}_{\mathcal{I}}} \left\langle \frac{a(t_{i+1}) - a(t_i)}{t_{i+1} - t_i} \nabla F(\mathbf{x}(t_i)) + \frac{d(t_{i+1}) - d(t_i)}{t_{i+1} - t_i} \vec{\ell}, \mathbf{v} \right\rangle$
5:     $\mathbf{x}(t_{i+1}) = \mathbf{x}(t_i) + \frac{d(t_{i+1}) - d(t_i)}{c(t_i)} \mathbf{v}(\mathbf{x}(t_i))$
6: **end for**

D Missing Proofs in Section 3

Lemma 3.1. *Assuming* $\mathbf{x}_e(t) \leq \theta(t)$ *for every* $e \in E$ *and* $t \in [0, 1]$*, there exists an upper bound*

$$U(t) = \dot{b}(t) \cdot \frac{\langle \nabla F(\mathbf{x}(t)), \mathbf{v}(\mathbf{x}(t)) \odot (1 - \mathbf{x}(t)) \rangle + F(\mathbf{x}(t))}{1 - \theta(t)}$$

$$+ \dot{d}(t) \cdot \frac{\gamma - 1}{\gamma(1 - \theta(t)) - 1} \langle \vec{\ell}, \mathbf{v}(\mathbf{x}(t)) \odot (1 - \mathbf{x}(t)) \rangle$$

such that $U(t) \geq \dot{b}(t) F(\mathbf{x}^\star) + \dot{d}(t) L(\mathbf{x}^\star)$*, where* $\gamma = \frac{\langle \vec{\ell}_+, \mathbf{x}^\star \rangle}{-\langle \vec{\ell}_-, \mathbf{x}^\star \rangle}$ *and*

$$\mathbf{v}(\mathbf{x}(t)) = \arg\max_{\mathbf{v} \in \mathcal{P}_{\mathcal{I}}} \left\langle \frac{\dot{b}(t)}{1 - \theta(t)} \nabla F(\mathbf{x}(t)) + \frac{\gamma - 1}{\gamma(1 - \theta(t)) - 1} \dot{d}(t) \vec{\ell}, \mathbf{v} \odot (1 - \mathbf{x}(t)) \right\rangle.$$

Proof. We first consider the $F(\mathbf{x}^\star)$ term in the optimal sum $\dot{b}(t) F(\mathbf{x}^\star) + \dot{d}(t) L(\mathbf{x}^\star)$. Since the non-negative multilinear extension F is a kind of DR-submodular function and $\mathbf{x}^\star \in \{0, 1\}^n$, we build the following connection for $F(\mathbf{x}^\star)$ in the optimal sum by Lemma A.2

$$F(\mathbf{x}^\star) \leq \frac{F(\mathbf{x}(t) \vee \mathbf{x}^\star)}{1 - \|\mathbf{x}(t)\|_\infty} \leq \frac{F(\mathbf{x}(t) \vee \mathbf{x}^\star)}{1 - \theta(t)}.$$

Due to Lemma A.1, we can further derive the inequality by taking $\mathbf{x} = \mathbf{x}(t)$ and $\mathbf{y} = \mathbf{x}(t) \vee \mathbf{x}^\star$,

$$F(\mathbf{x}^\star) \leq \frac{\langle \nabla F(\mathbf{x}(t)), \mathbf{x}(t) \vee \mathbf{x}^\star - \mathbf{x}(t) \rangle + F(\mathbf{x}(t))}{1 - \theta(t)}$$

$$= \frac{\langle \nabla F(\mathbf{x}(t)), \mathbf{x}^\star \odot (1 - \mathbf{x}(t)) \rangle + F(\mathbf{x}(t))}{1 - \theta(t)},$$

where the equality holds since $\mathbf{x} \vee \mathbf{y} - \mathbf{x} = \mathbf{y} \odot (1 - \mathbf{x})$ for any $\mathbf{x} \in [0, 1]^n$ and $\mathbf{y} \in \{0, 1\}^n$.

With oracle assumptions of F, it is obvious that the only uncertainty is $\mathbf{x}^\star$. Since $\mathbf{x}^\star \in \mathcal{P}_\mathcal{I}$ and $\mathcal{P}_\mathcal{I}$ is solvable, we could bound the right-side above by maximizing the linear programming in polynomial time.

For the term $L(\mathbf{x}^\star)$, the situation is more complicated since there is no non-negativity. To do so, we define a parameter $\gamma = \frac{\langle \vec{\ell}_+, \mathbf{x}^\star \rangle}{-\langle \vec{\ell}_-, \mathbf{x}^\star \rangle}$, which characterizes the relative dominance of the positive and negative parts of $L(\mathbf{x}^\star)$. It is trivial that $\langle \vec{\ell}, \mathbf{x}^\star \rangle = (1 - \frac{1}{\gamma}) \cdot \langle \vec{\ell}_+, \mathbf{x}^\star \rangle = (1 - \gamma) \cdot \langle \vec{\ell}_-, \mathbf{x}^\star \rangle$. Now, we present the upper bound for $L(\mathbf{x}^\star)$ in a similar manner. Due to the definition of γ, we have

$$
\begin{aligned}
L(\mathbf{x}^\star) &= \frac{\gamma - 1}{\gamma(1 - \theta(t)) - 1} \left(\frac{1 - \theta(t)}{1 - 1/\gamma} + \frac{1}{1 - \gamma} \right) \langle \vec{\ell}, \mathbf{x}^\star \rangle \\
&= \frac{\gamma - 1}{\gamma(1 - \theta(t)) - 1} \left((1 - \theta(t)) \langle \vec{\ell}_+, \mathbf{x}^\star \rangle + \langle \vec{\ell}_-, \mathbf{x}^\star \rangle \right) \\
&\leq \frac{\gamma - 1}{\gamma(1 - \theta(t)) - 1} \left(\langle \vec{\ell}_+, \mathbf{x}^\star \odot (1 - \mathbf{x}(t)) \rangle + \langle \vec{\ell}_-, \mathbf{x}^\star \rangle \right) \\
&\leq \frac{\gamma - 1}{\gamma(1 - \theta(t)) - 1} \langle \vec{\ell}, \mathbf{x}^\star \odot (1 - \mathbf{x}(t)) \rangle,
\end{aligned}
$$

where the first inequality holds due to the assumption and the second is guaranteed by $\vec{\ell}_- \leq \vec{0}$.

Combining with the coefficients, we obtain an upper bound

$$
\begin{aligned}
U(t) = \dot{b}(t) \cdot &\frac{\langle \nabla F(\mathbf{x}(t)), \mathbf{v}(\mathbf{x}(t)) \odot (1 - \mathbf{x}(t)) \rangle + F(\mathbf{x}(t))}{1 - \theta(t)} \\
&+ \dot{d}(t) \cdot \frac{\gamma - 1}{\gamma(1 - \theta(t)) - 1} \langle \vec{\ell}, \mathbf{v}(\mathbf{x}(t)) \odot (1 - \mathbf{x}(t)) \rangle,
\end{aligned}
$$

where $\mathbf{v}(\mathbf{x}(t)) = \arg\max_{\mathbf{v} \in \mathcal{P}_\mathcal{I}} \left\langle \frac{\dot{b}(t)}{1 - \theta(t)} \nabla F(\mathbf{x}(t)) + \frac{(\gamma - 1)\dot{d}(t)}{\gamma(1 - \theta(t)) - 1} \vec{\ell}, \mathbf{v} \odot (1 - \mathbf{x}(t)) \right\rangle$. $\square$

Lemma 3.2. *For any $t \in [0,1]$, the defined Lyapunov function $V(\mathbf{x}(t))$ is non-decreasing if the coefficient functions $(a(t), b(t), c(t), d(t)) \in \mathcal{C}_{con}$ where $\mathcal{C}_{con}$ includes*

$$
1 > \theta(t) \geq \|\mathbf{x}(t)\|_\infty \geq 0
$$

$$
\dot{a}(t) - \frac{b(t)}{1 - \theta(t)} \geq 0
$$

$$
\dot{c}(t) = 0
$$

$$
a(t)\dot{\mathbf{x}}(t) - \frac{\dot{b}(t)}{1 - \theta(t)} (\mathbf{v}(\mathbf{x}(t)) \odot (1 - \mathbf{x}(t))) = 0
$$

$$
c(t)\dot{\mathbf{x}}(t) - \frac{\dot{d}(t)(\gamma - 1)}{\gamma(1 - \theta(t)) - 1} (\mathbf{v}(\mathbf{x}(t)) \odot (1 - \mathbf{x}(t))) = 0,
$$

where $\theta(t)$ is the upper bound of $\mathbf{x}_e(t)$ for every $e \in E$.

Proof. By replacing the optimal sum in the derivative of the defined Lyapunov function with the upper bound $U(t)$ given by Lemma 3.1, we have

$$\dot{V}(\mathbf{x}(t)) \geq \dot{a}(t)F(\mathbf{x}(t)) + \dot{c}(t)L(\mathbf{x}(t)) + \langle a(t)\nabla F(\mathbf{x}(t)) + c(t)\nabla L(\mathbf{x}(t)), \dot{\mathbf{x}}(t) \rangle - U(t).$$

Rearranging the inequality, we get

$$\dot{V}(\mathbf{x}(t)) \geq \left(\dot{a}(t) - \frac{\dot{b}(t)}{1 - \theta(t)} \right) F(\mathbf{x}(t)) + \dot{c}(t)L(\mathbf{x}(t))$$

$$+ \left\langle \nabla F(\mathbf{x}(t)), a(t)\dot{\mathbf{x}}(t) - \frac{\dot{b}(t)}{1 - \theta(t)} \left(\mathbf{v}(\mathbf{x}(t)) \odot (1 - \mathbf{x}(t)) \right) \right\rangle$$

$$+ \left\langle \nabla L(\mathbf{x}(t)), c(t)\dot{\mathbf{x}}(t) - \frac{\dot{d}(t)(\gamma - 1)}{\gamma(1 - \theta(t)) - 1} \left(\mathbf{v}(\mathbf{x}(t)) \odot (1 - \mathbf{x}(t)) \right) \right\rangle.$$

Since F is non-negative and non-monotone and the range of L is real domain, the sufficient conditions that makes $\dot{V}(\mathbf{x}(t)) \geq 0$ can be naturally listed below

$$1 > \theta(t) \geq \|\mathbf{x}(t)\|_\infty \geq 0$$

$$\dot{a}(t) - \frac{\dot{b}(t)}{1 - \theta(t)} \geq 0$$

$$\dot{c}(t) = 0$$

$$a(t)\dot{\mathbf{x}}(t) - \frac{\dot{b}(t)}{1 - \theta(t)} \left(\mathbf{v}(\mathbf{x}(t)) \odot (1 - \mathbf{x}(t)) \right) = 0$$

$$c(t)\dot{\mathbf{x}}(t) - \frac{\dot{d}(t)(\gamma - 1)}{\gamma(1 - \theta(t)) - 1} \left(\mathbf{v}(\mathbf{x}(t)) \odot (1 - \mathbf{x}(t)) \right) = 0.$$

□

Lemma 3.3. *Given the Lyapunov function $V(\mathbf{x}(t))$, the update rule in continuous-time is*

$$\dot{\mathbf{x}}(t) = \frac{\dot{a}(t)}{a(t)} \left(\mathbf{v}(\mathbf{x}(t)) \odot (1 - \mathbf{x}(t)) \right)$$

for $t \in [0, 1]$. Moreover, assuming that $\ln a(t)$ is a cumulative distribution function on $[0, 1]$, the output solution $\mathbf{x}(1) \in \mathcal{P_I}$.

Proof. Due to Lemma 3.2, we get

$$\dot{\mathbf{x}}(t) = \frac{\dot{a}(t)}{a(t)} \left(\mathbf{v}(\mathbf{x}(t)) \odot (1 - \mathbf{x}(t)) \right).$$

Therefore,

$$\mathbf{x}(1) = \int_0^1 \mathbf{v}(\mathbf{x}(t)) \odot (1 - \mathbf{x}(t)) d(\ln a(t)).$$

Since $\mathbf{v}(\mathbf{x}(t)) \in \mathcal{P_I}$ for any $t \in [0, 1]$ and $\mathcal{P_I}$ is down-monotone, we have $\mathbf{v}(\mathbf{x}(t)) \odot (1 - \mathbf{x}(t)) \in \mathcal{P_I}$.

By the assumption that $\ln a(t)$ is a cumulative distribution function on $[0, 1]$, the output $\mathbf{x}(1)$ is the convex combination of feasible solutions.   □

Lemma 3.4. *For the coordinate-wise upper bound $\theta(t)$ of the solution $\mathbf{x}(t)$ with $t \in [0,1]$, we have $\theta(t) \leq 1 - \frac{a(0)}{a(t)}$.*

Proof. Due to the update rule, we have

$$\dot{\mathbf{x}}(t) = \frac{\dot{a}(t)}{a(t)} \left(\mathbf{v}(\mathbf{x}(t)) \odot (1 - \mathbf{x}(t)) \right) \leq \frac{\dot{a}(t)}{a(t)} (1 - \mathbf{x}(t)),$$

where the inequality is true since $\mathbf{v}(\mathbf{x}(t)) \leq 1$ and $a(t), \dot{a}(t) \geq 0$.

An equivalent expression could be obtained for the above inequality by applying the Grönwall's inequality [12], i.e., for any $i = 1, \ldots, n$,

$$1 - x_i(t) \geq e^{-\int_0^t \frac{\dot{a}(s)}{a(s)} ds} = e^{-\ln \frac{a(t)}{a(0)}} = \frac{a(0)}{a(t)}.$$

Therefore, we have $\theta(t) \leq 1 - \frac{a(0)}{a(t)}$.   □

Theorem 3.1. *For RegularizedNSM with no limitation on the linear term, the algorithm* CONTINUOUS-TIME ALGORITHM FOR REGULARIZEDNSM *achieves an $(1/e, \frac{\gamma - \gamma/e - 1}{\gamma - 1})$-approximation with feasible range $\gamma \in [0,1) \cup (\frac{1}{1 - 1/e}, +\infty)$ and $\gamma \in (1, \frac{1}{1 - 1/e})$ for $\ell(\mathbf{x}^\star) \geq 0$ and $\ell(\mathbf{x}^\star) < 0$ respectively, where $\gamma = \frac{\langle \vec{\ell}_+, \mathbf{x}^\star \rangle}{-\langle \vec{\ell}_-, \mathbf{x}^\star \rangle}$.*

Proof. Since obtaining a better performance for the first term in the objective of RegularizedNSM is the priority, the approximation ratio of Algorithm 1 could be yielded by solving the maximization problem below

$$
\begin{aligned}
&\sup \frac{b(1) - b(0)}{a(1)} \\
&s.t. \ (a(t), b(t), c(t), d(t)) \in \mathcal{C}_{\mathrm{con}} \\
&\quad\quad a(1) = c(1) \\
&\quad\quad a(0), b(0), c(0), d(0) \geq 0 \\
&\quad\quad \dot{a}(t), \dot{b}(t), \dot{c}(t), \dot{d}(t) \geq 0 \\
&\quad\quad t \in [0,1].
\end{aligned}
\tag{5}
$$

where $\mathcal{C}_{\mathrm{con}}$ is given by Lemma 3.2.

A feasible solution for the above ODE is

$$a(t) = e^{t-1}, b(t) = t/e, c(t) = 1, d(t) = -\frac{\gamma/e^t + t}{\gamma - 1}.$$

Thus, the Lyapunov function is

$$V(\mathbf{x}(t)) = e^{t-1} F(\mathbf{x}(t)) - \frac{t}{e} F(\mathbf{x}^\star) + L(\mathbf{x}(t)) + \frac{\gamma/e^t + t}{\gamma - 1} L(\mathbf{x}^\star).$$

Guaranteed by Lemma 3.2, the stable criteria of $V(\mathbf{x}(t))$ yields the following result for the output $\mathbf{x}(1)$ of Algorithm 1

$$F(\mathbf{x}(1)) + L(\mathbf{x}(1)) \geq \frac{1}{e} \cdot f(\mathbf{x}^\star) + \left(\frac{\gamma - \gamma/e - 1}{\gamma - 1} \right) \cdot \ell(\mathbf{x}^\star).$$

□

References

1. Bansal, N., Gupta, A.: Potential-function proofs for gradient methods. Theory Comput. **15**(1), 1–32 (2019)
2. Bian, Y., Buhmann, J.M., Krause, A.: Continuous submodular function maximization. Preprint arXiv:2006.13474 (2020)
3. Bodek, K., Feldman, M.: Maximizing sums of non-monotone submodular and linear functions: Understanding the unconstrained case. Preprint arXiv:2204.03412. (2022)
4. Conforti, M., Cornuejols, G.: Submodular set functions, matroids and the greedy algorithm: tight worst-case bounds and some generalizations of the Rado-Edmonds theorem. Discret. Appl. Math. **7**(3), 251–274 (1984)
5. Diakonikolas, J., Orecchia, L.: The approximate duality gap technique: a unified theory of first-order methods. SIAM J. Optim. **29**(1), 660–689 (2019)
6. Du, D., Liu, Z., Wu, C., Xu, D., Zhou, Y.: An improved approximation algorithm for maximizing a DR-submodular function over a convex set. Preprint arXiv:2203.14740 (2022)
7. Du, D.: Lyapunov function approach for approximation algorithm design and analysis: with applications in submodular maximization. Preprint arXiv:2205.12442 (2022)
8. Du, D.: Submodularity and lattice: theory, algorithms and applications. Unpublished book (in preparation) (2022)
9. Feldman, M.: Guess free maximization of submodular and linear sums. Algorithmica **83**(3), 853–878 (2021)
10. Feldman, M.: Correction to: guess free maximization of submodular and linear sums. Algorithmica **84**(10), 3101–3102 (2022)
11. Feldman, M., Naor, J., Schwartz, R.: A unified continuous greedy algorithm for submodular maximization. In: Proceedings of the 52nd Annual Symposium on Foundations of Computer Science, pp. 570–579 (2011)
12. Grönwall, T.H.: Note on the derivatives with respect to a parameter of the solutions of a system of differential equations. Ann. Math. **20**, 292–296 (1919). https://doi.org/10.2307/1967124
13. Hassani, H., Soltanolkotabi, M., Karbasi, A.: Gradient methods for submodular maximization. In: Proceedings of the 30th International Conference on Advances in Neural Information Processing Systems, pp. 5841–5851 (2017)
14. Krichene K, Bayen A, Bartlett PL. Accelerated mirror descent in continuous and discrete time. In Proceedings of the 28th International Conference on Advances in Neural Information Processing Systems, pp. 2845–2853 (2015)
15. Lyapunov, A.M.: The general problem of the stability of motion. Int. J. Control **55**(3), 531–534 (1992)
16. Lu, C., Yang, W., Gao, S.: Regularized non-monotone submodular maximization. Preprint arXiv:2103.10008
17. Nemirovsky, A., Yudin, D.B.: Problem Complexity and Method Efficiency in Optimization. John Wiley, New York (1983)
18. Qi, B.: On maximizing sums of non-monotone submodular and linear functions. Preprint arXiv:2205.15874 (2022)
19. Sviridenko, M., Vondrák, J., Ward, J.: Optimal approximation for submodular and supermodular optimization with bounded curvature. Math. Oper. Res. **42**(4), 1197–1218 (2017)

20. Su, W., Boyd, S., Candes, E.J.: A differential equation for modeling Nesterov's accelerated gradient method: theory and insights. J. Mach. Learn. Res. **17**, 1–43 (2016)
21. Wibisono, A., Wilson, A.C., Jordan, M.I.: A variational perspective on accelerated methods in optimization. Proc. Natl. Acad. Sci. **113**(47), E7351–E7358 (2016)
22. Wilson, A.C., Recht, B., Jordan, M.I.: A Lyapunov analysis of momentum methods in optimization. J. Mach. Learn. Res. **22**, 1–34 (2021)

Topological Network-Control Games

Zihui Liang$^{(\boxtimes)}$, Bakh Khoussainov, and Haidong Yang

University of Electronic Science and Technology of China, Chengdu, China
`zihuiliang.tcs@gmail.com`, `bmk@uestc.edu.cn`

Abstract. The paper introduces new combinatorial games, called topological network-control games, played on graphs. These games model the influence of competing two parties aiming to control a given network. In a such game given the network, the players move alternatively. At each turn, a player selects an unclaimed vertex and its unclaimed neighbours within distance t. The players obey the topological condition that all claimed vertices stay connected. The goal is to decide which player claims the majority of the vertices at the end of the play. We study greedy, symmetric and optimal strategies. We solve the topological network-control games on various classes of graphs. This progresses our understanding of combinatorial games played on graphs. We prove that finding an optimal winning strategy is a PSPACE-complete problem.

Keywords: combinatorial games · strategies · algorithms · PSPACE

1 Introduction, Preliminary Definitions and Results

We introduce new combinatorial games played on finite graphs. These games are called topological network-control games. These games model the influence of competing two parties aiming to control the network by preserving connectedness property. Below we present basic definitions, preliminary concepts, related work and our contribution.

1.1 Statement of the Problem

Let $G = (V, E)$ be a finite graph. We assume that the graphs are simple, that is, the graphs are undirected, have no loops and multiple edges. For a vertex $x \in V$ in $G = (V, E)$, and an integer $t \geq 0$, the t-neighbourhood of x, denoted by $\mathcal{N}_t(x)$, is the set of all vertices at distance at most t from x. So, $\mathcal{N}_0(x) = x$ and $\mathcal{N}_t(x) \subseteq \mathcal{N}_{t+1}(x)$ for all $t \geq 0$. Vertices in $\mathcal{N}_t(x)$ are called t-neighbours of x. For a set of vertices $X \subseteq V$ in $G = (V, E)$, and an integer $t \geq 0$, the t-neighbourhood of X is denoted by $\mathcal{N}_t(X) = \cup_{x \in X} \mathcal{N}_t(x)$. We fix t that will be our parameter.

Now we define topological network-control game played on $G = (V, E)$. There are two players: Player 1 and Player 2. The opponent of Player ϵ, where $\epsilon \in \{0, 1\}$,

B. Khoussainov—Acknowledges the National Science Foundation of China under Grant No. 62172077.

is denoted by Player $\bar{\epsilon}$. Each play consists of rounds. At odd rounds Player 1 moves. At even rounds Player 2 moves. Player 1 starts the first round. At this round the player selects a vertex x and claims (all vertices in) $\mathcal{N}_t(x)$. Let X_{2i-1} be the set of all vertices claimed by the players by the end of round $2i-1$, $i > 0$. At round $2i$, where $i > 0$, Player 2 selects a vertex from $\mathcal{N}_{t+1}(X_{2i-1})/X_{2i-1}$, and then claims the selected vertex and its unclaimed t-neighbour vertices. Let X_{2i} be the set of all vertices claimed by the players by the end of round $2i$, $i > 0$. At round $2i + 1$, where $i > 0$, Player 1 selects a vertex in $\mathcal{N}_{t+1}(X_{2i})/X_{2i}$, and then claims the selected vertex and its unclaimed t-neighbour vertices.

Let t be the round at which no vertices can be claimed. If $X_t = V$, then the play stops. If $V \neq X_t$, then the player whose turn it is at round t, selects a new vertex (from a component with unclaimed vertices) and claims its t-neighbourhood, and the play continues on just as above.

To define a winner, we need a few notations. By $C_{1,2i+1}$ we denote the set of vertices claimed by Player 1 at the end of round $2i + 1$, $i \geq 0$. Similarly, $C_{2,2i+2}$ denotes the set of vertices claimed by Player 2 at the end of round $2i + 2$. If S_{2i+1} is the set of vertices claimed by Player 1 at round $2i + 1$, then $C_{1,2i+1} = C_{1,2i-1} \cup S_{2i+1}$. Similarly, $C_{2,2i+2} = C_{2,2i} \cup S_{2i+2}$.

Once the play stops, let C_1 and C_2 be all vertices claimed by Player 1 and Player 2, respectively. The sets C_1 and C_2 partition G.

Definition 1. *We say that Player ϵ* **wins** *the play if $|C_\epsilon| > |C_{\bar{\epsilon}}|$. If $|C_\epsilon| = |C_{\bar{\epsilon}}|$, then we say that the play is a* **draw**.

If G is connected, then the set of claimed vertices is connected at any round. Connectedness is a topological property, and hence we call our games *topological network-control game*. We assume the reader is familiar with the notion of strategy. A player is **the winner** of the topological network-control game played on G if the player has a winning strategy. Our goal is twofold. First, we want to **solve** games by designing algorithms that given a game decide the winner. Second, we want to extract winning strategies for the winners. Note that when $t = 0$, Player 1 wins iff $|V|$ is odd. So, we always assume that $t > 0$.

1.2 Preliminaries and Basic Results

Assume that the players play the game on G. By (S, v) we denote a move where Player ϵ selects vertex v and claims the set of vertices S on G. Thus, a **play** is a sequence (S_1, v_1), ..., (S_i, v_i) of selected nodes and claimed vertices at each round. The **configuration** determined by this play is the tuple (G_i, C_i), where

1. The set G_i is the sub-graph of G that consists of all unclaimed vertices at the end of the play, and thus $G_i = V \setminus (S_1 \cup S_2 \cup \ldots S_i)$
2. The set C_i is a set of vertices which players can select in G_i. Thus, $C_i = \mathcal{N}_{t+1}(S_1 \cup S_2 \cup \ldots S_i) \setminus (S_1 \cup S_2 \cup \ldots S_i)$.

Greedy and monotonic strategies. During a game, the players might follow a greedy strategy.

Definition 2. *A* **greedy strategy** *is one that, at any round, selects a vertex with the maximal number of unclaimed t-neighbours.*

Greedy strategies could be losing strategies.

Example 1. Consider the graph in Fig 1. Assume $t = 1$. Player 1's first greedy move is the vertex of degree 5. Player 2 responds by selecting the vertex of degree 4. Then Player 1 greedily selects the vertex of degree 3. Player 2 selects the vertex of degree 4 and wins.

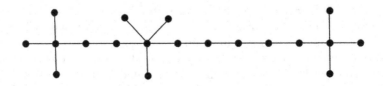

Fig. 1. Greedy strategy does not always help

However, greedy strategies might be useful if they satisfy monotonicity property. Let (G, C) be a configuration of a play. By $max\text{-}deg_t(G, C)$ we denote the maximal cardinality among cardinalities of the sets $\mathcal{N}_t(v)$, where $v \in C$. This corresponds to a greedy move in the configuration (G, C).

Definition 3. *A strategy is* **monotonic** *if for any play* $(S_1, v_1), \ldots, (S_i, v_i), \ldots$ *consistent with the strategy,* $max\text{-}deg_t(G_i, C_i) \geq max\text{-}deg_t(G_{i+1}, C_{i+1})$, *where* (G_i, C_i) *is the configuration determined by the play* $(S_1, v_1), (S_2, v_2), \ldots, (S_i, v_i)$.

Example 2. If $G = C$ then Player 1 has a monotonic greedy strategy on (G, C). Moreover, Player 2 can guarantee to lose at most $max\text{-}deg_t(G, C)$ vertices.

Lemma 1. *Assume that Player 1 has a monotonic greedy strategy on G. Then Player 1 can guarantee to not lose the game.*

Proof. Fix a monotonic greedy strategy for Player 1. Let $(S_1, v_1), \ldots, (S_n, v_n)$ be a play consistent with the strategy and $(G_0, C_0), \ldots, (G_n, C_n)$ be the corresponding configurations. For all $i = 1, \ldots, \lfloor \frac{n}{2} \rfloor$, $|S_{2i-1}| = max\text{-}deg_t(G_{2i-2}, C_{2i-2}) \geq max\text{-}deg_t(G_{2i-1}, C_{2i-1}) \geq |S_{2i}|$. Therefore, $\sum_{i=1}^{\lceil \frac{n}{2} \rceil} |S_{2i-1}| \geq \sum_{i=1}^{\lfloor \frac{n}{2} \rfloor} |S_{2i}|$ and Player 1 doesn't lose.

We apply this lemma to a few examples of graphs. For this, we recall several specific classes of graphs. These graphs will further be studied in this paper.

The **path** graph of size n, $Path_n$, has vertices $\{1, \ldots, n\}$ with the edge relation between the consecutive integers. The **cycle** graph of size $n \geq 3$, $Cycle_n$, is obtained from $Path_n$ by adding the edge between 1 and n. A **caterpillar** is a tree in which all non-leaf vertices are on a path and leaves are at distance 1 from the path. We call the path the *central path*. Let G be a caterpillar with central path $Path_n$. A caterpillar is degree-homogeneous if degrees of all vertices along the central path are the same. By $Caterpillar_{n,k}$ we denote the degree-homogeneous caterpillar whose central path is $Path_n$ in which every vertex has degree k. Not hard to see that Player 1 has a monotonic greedy strategy on $Path_n$, $Cycle_n$, and $Caterpillar_{n,m}$. Therefore, we have the following corollary.

Corollary 1. *Player 1 never loses $Path_n$, $Cycle_n$, and $Caterpillar_{n,m}$.*

Symmetric strategies. Two configurations (G_1, C_1) and (G_2, C_2) are *isomorphic* if there exists an isomorphism $f : G_1 \rightarrow G_2$ from $G_1 = (V_1, E_1)$ to $G_2 = (V_2, E_2)$ such that for all x in G_1 we have $x \in C_1$ iff $f(x) \in C_2$.

Definition 4. *Let $f : G_1 \rightarrow G_2$ be an isomorphism between two disjoint configurations (G_1, C_1) and (G_2, C_2). Let $(G, C) = (G_1, C_1) \cup (G_2, C_2)$ be the union of (G_1, C_1) and (G_2, C_2): $V = V_1 \cup V_2$, $E = E_1 \cup E_2$ and $C = C_1 \cup C_2$. The **symmetric** strategy for Player ϵ is this: if Player $\bar{\epsilon}$ selects x from G_1 then Player ϵ selects $f(x)$; if Player $\bar{\epsilon}$ selects x from G_2 then Player ϵ selects $f^{-1}(x)$.*

Lemma 2. *If (G_1, C_1) and (G_2, C_2) are isomorphic, then Player 2 guarantees a draw in the configuration $(G, C) = (G_1, C_1) \cup (G_2, C_2)$.*

Optimal strategies. Here is a definition of optimal strategies.

Definition 5. *Let $w \geq 0$ be an integer. A strategy for Player ϵ is w-**optimal** if*

1. *The strategy can guarantee that Player ϵ wins any play with at least w more vertices independent on the opponent's strategies.*
2. *The opponent, Player $\bar{\epsilon}$, has a strategy that guarantees that no more than w vertices are lost independent on strategies of Player ϵ.*

Let (G, C) be a configuration. We define the integer $Opt(G, C)$ such that player ϵ faced with the configuration can guarantee a win with at least $Opt(G, C)$ vertices starting at (G, C):

1. If $G = \emptyset$ then $Opt(G, C) = 0$.
2. If $C = \emptyset$, Player ϵ can select any vertex $v \in G$ and make corresponding moves (S, v). Let $S_1, \ldots, S_m$ be all possible moves of player ϵ at (G, C). Let (G_i, C_i) be updated configuration after moving S_i and $n(S_i)$ be the cardinality of S_i, $i = 1, \ldots, m$. Set $Opt(G, C) = \max_{1 \leq i \leq m} \{n(S_i) - Opt(G_i, C_i)\}$.

Lemma 3. *Player ϵ has an $Opt(G, C)$-optimal strategy at (G, C). Moreover, the computation of $Opt(G, C)$-optimal strategy is in PSPACE.*

Proof. The case $G = \emptyset$ is clear. Assume that for all (G', C') with $0 \leq n(G') \leq k$ the lemma is true. Let (G, C) be a configuration with $n(G) = k + 1$. Let $S_1, \ldots, S_m$ be possible moves at (G, C) . By induction, Player $\bar{\epsilon}$ has an $Opt(G_i, C_i)$-optimal strategy on (G_i, C_i). So Player $\bar{\epsilon}$ guarantees to lose no more than $\max_{1 \leq i \leq m} \{n(S_i) - Opt(G_i, C_i)\}$ vertices, and Player ϵ guarantees to win at least $\max_{1 \leq i \leq m} \{n(S_i) - Opt(G_i, C_i)\}$ vertices. This implies that $Opt(G, C) = \max_{1 \leq i \leq m} \{n(S_i) - Opt(G_i, C_i)\}$. Showing that $Opt(G, C)$-optimal strategy can be computed in PSPACE is standard.

1.3   Related Work and Our Contribution

This work belongs to the area of combinatorial games. Typically these games are of finite duration played on spaces of finite configurations. Combinatorial games are usually classified as scoring games and non-scoring games. In scoring games, a player wants to collect a certain amount of points to win the game. Examples of scoring games are graph-grabbing game [13,23], median game [12], orthogonal colouring game [3], Vertex-Capture Game [8] and the largest connected subgraph game [6,7]. In non-scoring games, the players aim to put their opponent into a deadlock (e.g., the opponent makes the last move). Examples of non-scoring games are GRIM [1], Nim on graphs [11,17], Kayles on graphs [9,10,18–20,24], game 0.33 [4] Weighted Arc and generalized Kayles [15]. Our games are obviously scoring games as the winner is the one who claims the most number of vertices of the graph. However, one unexpected side of our games is that they exhibit a behaviour of non-scoring games when the game graph G consists of more than one component. Indeed, once players start playing in a component C of G, the players need to claim all the vertices of C before they move to the other components. Therefore, if a player makes the last move on C, then the opponent will start a new component and might gain an advantage. This implies that the players need to take into account the parity of the last move in C. So, when the players play the games on graphs that consists of more than one component, the parties of the last moves on the components matter. This is a typical feature of non-scoring games. Here now we briefly list the main contributions of this work:

1. We introduce topological network-control games. These games model the influence of competing parties to control a given network. The restraint is that the players need to ensure connectivity of the set of claimed vertices.
2. In combinatorial game theory, understanding games in algebraically simple graphs (such as paths, cycles, trees, etc.) usually constitutes bottleneck problems [2,5,14,16]. That is why we start by studying our games in these classes of graphs. We succeed to characterize some classes of graphs where Player 1 wins. Our proofs are combinatorial and are based on careful analysis of configuration spaces of games. These are presented in Theorems 1, 2, and 3 of Sect. 2.
3. As we mentioned above, one novel side of our games is that they exhibit characteristics of both scoring games and non-scoring games. We study this interplay between scoring condition (the number of vertices claimed) and the parity condition (the player moving the last loses) on graphs through two recursively defined functions F_{even} and S_{odd}. We then fully describe a non-trivial behaviour of these functions on path graphs. This is presented in Theorem 4 of Sect. 3.
4. We fully characterize the graphs $Path_{\ell_1} \cup Path_{\ell_2}$ (disjoint union of two path graphs), where Player 1 wins. Our charcaterization makes use of a computer program that lists all graphs $Path_{\ell_1} \cup Path_{\ell_2}$, where $\ell_1 \leq 100$ and $\ell_2 \leq 100$, won by Player 1. See Theorem 5 in Sect. 4. The proof takes into account the interplay between the parity and the number claimed vertices.

5. We prove that finding optimal strategies is a PSPACE-complete problem. There are two key difficulties in the proof. The first is that the cardinalities of t-neighbourhoods of unclaimed vertices depend on configurations. In comparison, this does not happen in the Competetive Facility Allocation problem, where coding of a quantified Boolean Formula becomes easy [21]. The second is the connectivity condition put on the set of claimed vertices. These two conditions make it challenging to code PSPACE-complete problems into our games.

We finally mention the recent work by Z. Liang, B. Khoussainov, and M. Xiao on network-control games played on graphs [22]. The work of this paper is a natural and independent follow-up of the control-network games. As opposed to topological network-control games, network-control games lack the connectivity condition of the set of claimed vertices. In particular, in network-control games, players can select and claim vertices with no regards (in terms of connectivity) to already claimed vertices. In this sense, network-control games from [22] do not exhibit characteristics of non-scoring games as we described above. Hence, the proof methods and ideas in this work are, in many ways, orthogonal to those in the study of network-control games. For instance, greedy strategies in network-control games suffice to prove that Player 1 never loses any network-control game. In contrast, we already showed that the greedy strategies in topological network-control games can be losing for Player 1. Furthermore, we do not know if there is graph G where Player 1 loses the topological network-control game.

2   Games on Paths, Cycles, and Caterpillars

The strategies defined above can be used to analyze the topological network-control games on paths, cycles, and caterpillars. Note that by Corollary 1 Player 1 never loses these graphs.

Theorem 1. *Player 1 wins the topological network-control games on paths.*

Proof. Player 1 wins $Path_n$ with $n \leq 2t + 1$ by claiming all vertices in the first move. Assume $n > 2t + 1$. If $n = 2k + 1$ then Player 1 has a monotonic greedy strategy on $Path_n$ and by Lemma 1, Player 1 wins $Path_n$. Assume $n = 2k$. Player 1 selects the middle vertex k and claims its t-neighbours $\mathcal{N}_t(k)$. Both players make greedy moves. In the remaining configuration, Player 2 claims one more vertex at most. Therefore, Player 1 wins $2t$ vertices. If Player 2 makes non-greedy moves more times, Player 2 will lose more vertices.

Theorem 2. *Player 1 wins the game on $Cycle_n$ iff $n \neq 0 \mod (4t + 2)$.*

Proof. It is clear that Player 1 wins $Cycle_n$ if $n < 4t + 2$. Assume $n \geq 4t + 2$. If $n = 0 \mod (4t + 2)$, then both of the players must follow greedy strategies. Otherwise, the player making a non-greedy move loses. Therefore, the last move will be made by Player 2. This guarantees a draw for Player 2 due to the condition on n. Assume $n \neq 0 \mod (4t + 2)$. Player 1 selects p, say $p = 1$, and claimed its t-neighbours $\mathcal{N}_t(p)$. Assume both players must follow greedy strategies. Because $n \neq 0 \mod (4t + 2)$, Player 1 can claim more vertices than Player 2 in last two rounds. If Player 2 makes non-greedy moves until Player 1 makes more moves, Player 1 will claim more vertices than before.

The proof of the next theorem is more involved but uses the same ideas as the proofs of the theorems above.

Theorem 3. *Player 1 wins the games on* $Caterpillar_{n,k}$.

3   Parity Vs the Number of Claimed Vertices

Suppose that the game graph G consists of more than one component, say C_1 and C_2. Even if Player 1 wins topological network-control game on each of these components C_1 and C_2, this does not guarantee that Player 1 wins G. The reason is that Player 1 might make the last move playing on each of these components. Therefore, once all vertices in one of the components are claimed, Player 2 continues the game by starting the remaining component. Thus, each player has two, somewhat opposing, aims. On the one hand, the player would like to win as many nodes as possible in a given component. On the other hand, the player would like the opponent to make the last move in the current component to take the advantage of the remaining component.

A natural question is thus the following. Assume that the players play the game on connected graph G. How many nodes should a player give up to ensure that the opponent makes the last move? To answer this question, below we provide a general framework, and apply it to the case when the underlying graphs are paths graphs. Even the case of paths turns out to involve non-trivial combinatorial and inductive arguments.

Let (G, C) be a configuration. Let us define two functions $F_{even}(G, S)$ and $S_{odd}(G, C)$. The function $F_{even}(G, S)$ computes the maximal number of vertices won by the first player who starts the game at (G, S) under the assumption that the player can force the opponent to make the last move in the game. Similarly, $S_{even}(G, S)$ computes the maximal number of vertices won by the second player in the game (G, C) under the assumption that the player can force the opponent to make the last move in the game. Note that the value $F_{even}(G, C)$ and $S_{odd}(G, C)$ can be undefined if the player can not force the opponent to make the last move. We defined these functions through mutual recursion as follows:

- Set $F_{even}(\emptyset, \emptyset) = 0$ and $S_{odd}(\emptyset, \emptyset) = -\infty$.
- Let $S_1, \ldots, S_m$ be all possible moves of a player at (G, C). Let (G_i, C_i) be the configuration after move S_i, $i = 1, \ldots, m$. Set

$$F_{even}(G, C) = \max\{n(S_i) + S_{odd}(G_i, S_i) \mid i = 1, \ldots, m\}, \text{ and}$$
$$S_{odd}(G, C) = \min\{-n(S_i) + F_{even}(G_i, S_i) \mid i = 1, \ldots, m\}$$

As an example, on graphs $Path_{13}$ and $Path_{18}$, one can compute that we have $F_{even}(Path_{13}, \emptyset) = -1$ and $F_{18}(Path_{18}, \emptyset) = -2$, respectively. Also, from the definitions of F_{even} and S_{odd}, we have the following corollary.

Corollary 2. $F_{even}(G, C) \neq -\infty$ if and only if $S_{odd}(G, C) = -\infty$.

It turns out guaranteeing that the opponent makes the last move can be costly even in such graphs as $Path_n$. The proof of the next theorem is based on a careful analysis of F_{even} and S_{odd} functions.

Theorem 4. Consider a game played on $Path_n$ with $t = 1$. If $n < 3$ then $F_{even}(Path_n) = -\infty$ and $S_{odd}(Path_n) = -n$. Otherwise:

$$S_{odd}(Path_n) = \begin{cases} -1 - \lfloor \frac{n}{5} \rfloor \text{ if } n = 1 \mod 5 \\ -\lfloor \frac{n}{5} \rfloor \text{ if } n = 2 \mod 5 \text{ and } n \neq 7 \\ -3 \text{ if } n = 7 \\ -\infty \text{ otherwise} \end{cases}$$

$$F_{even}(Path_n) = \begin{cases} -\lfloor \frac{n}{5} \rfloor + 4 \text{ if } n = 0 \mod 5, n \neq 0 \text{ and } n \neq 5 \\ -\lfloor \frac{n}{5} \rfloor + 1 \text{ if } n = 3 \mod 5 \\ -\lfloor \frac{n}{5} \rfloor + 2 \text{ if } n = 4 \mod 5 \\ 0 \text{ if } n = 0 \\ 1 \text{ if } n = 5 \\ -\infty \text{ otherwise} \end{cases}$$

4   Topological Network-Control Games on $Path_{\ell_1} \cup Path_{\ell_2}$

From the previous section, we see that controlling the parity is a challenging task even on path graphs. Our goal is to settle the topological network-control games problem for the class of graphs of the type $Path_{\ell_1} \cup Path_{\ell_2}$. The previous section shows that the players can not rely solely on only greedy strategies or parity control strategies in $Path_{\ell_1} \cup Path_{\ell_2}$. The players need to adapt different strategies, e.g., mixing greedy and parity strategies. Note that Player 1 never loses the game on $Path_{\ell_1} \cup Path_{\ell_2}$. Thus, we aim to characterize those graphs $Path_{\ell_1} \cup Path_{\ell_2}$ where Player 1 guarantees a win. For this section we assume that $t = 1$.

Our next Lemma 4 is proved through a computer-assisted technique. The code for the lemma computes the function $Opt(Path_{\ell_1} \cup Path_{\ell_2}, \emptyset)$[1]

---

[1] The code is at https://github.com/ZihuiLiang/Topological-Network-Control-Game..

Lemma 4. *Player 2 can guarantee a draw in the game $Path_{\ell_1} \cup Path_{\ell_2}$, where $1 \le \ell_1 \le \ell_2 < 100$, if and only if $\{\ell_1, \ell_2\} \in \{\{1,1\}, \{2,2\}, \{7,7\}, \{7, 6x+5\}, \{2, 6x\}, \{6x, 6y\}, \{6x+5, 6y+5\}\}_{1 \le x,y}$.*

This lemma can be used to fully characterize those games on $Path_{\ell_1} \cup Path_{\ell_2}$ where Player 2 guarantees a draw. We prove the theorem below based on the analysis of several conditions on ℓ_1 and ℓ_2. We showcase our proofs on several cases.

Theorem 5. *Player 2 guarantees a draw on $Path_{\ell_1} \cup Path_{\ell_2}$ iff $\{\ell_1, \ell_2\}$ belongs to $\{\{1,1\}, \{2,2\}, \{7,7\}, \{7, 6x+5\}, \{2, 6x\}, \{6x, 6y\}, \{6x+5, 6y+5\}\}_{1 \le x,y}$.*

Proof. Our proof is based on the analysis of 9 cases (and their subcases). By the lemma above we can always assume that either $\ell_1 \ge 100$ or $\ell_2 \ge 100$.

Case 1: The parities of ℓ_1 and ℓ_2 are different. Without loss of generality, assume ℓ_1 is odd, ℓ_2 is even, with $\ell_1 \ge 3$ and $\ell_2 \ge 2$. Then Player 1 wins as follows. The player selects the middle node in the path of odd length (greater than 1) and then uses a greedy strategy. Player 1 wins at least 3 vertices. Even if Player 2 starts the line of even length, the play can win at most 2 vertices. If the length of the odd path graph is 1, then Player starts with the even length path graph and wins two nodes guaranteeing the overall win.

Case 2: $\ell_1 = 4 \pmod 6$. Player 1 uses a greedy strategy on $Path_{\ell_1}$ starting from one end of $Path_{\ell_1}$. If Player 2 uses a greedy strategy, then the last move on $Path_{\ell_1}$ is made by Player 2. Hence, in this case Player 1 wins the game. In order to ensure that Player 1 makes the last move in $Path_{\ell_1}$, Player 2 must give up the greedy strategy at least twice. Thus, either Player 1 starts the second line $Path_{\ell_2}$ or wins at least 3 vertices on $Path_{\ell_1}$. In either case, Player 1 wins the game.

Case 3: $\ell_1 = 3 \pmod 6$. Player 1 starts using the greedy strategy on $Path_{\ell_1}$ from one end of the graph. If Player 2 also plays a greedy strategy, then in the last move on $Path_{\ell_1}$, Player 1 will claim 2 vertices instead of greedy 3. In this way, Player 2 makes the last move in $Path_{\ell_1}$. Hence, Player 1 moves to the next line and wins the game. This implies that Player 2 must abandon its greedy strategy on $Path_{\ell_1}$ at least four times. If this happens, Player 1 will have won at least 4 vertices when the play moves to $Path_{\ell_2}$. So Player 1 wins.

Case 4: $\ell_1 = 2 \pmod 6$ with $\ell_1 > 2$. Player 1 makes a move from one side of $Path_{\ell_1}$. In the first round, Player 1 only claims 2 nodes. After that Player 1 will always use a greedy strategy. If Player 2 uses a greedy strategy, then in the last move on $Path_{\ell_1}$ Player 1 claims two vertices (instead of 3). In this case, Player 2 makes the last move in $Path_{\ell_1}$. Hence, Player 1 starts $Path_{\ell_2}$ (with a draw on $Path_{\ell_1}$) and wins the game. Therefore, Player 2 must abandon its greedy strategy while playing on $Path_{\ell_1}$ at least 4 times. In this game Player 2 moves to $Path_{\ell_2}$, where Player 1 won at least 3 vertices in $Path_{\ell_1}$. So, Player 1 wins.

Case 5: $\ell_1 = 1 \pmod 6$ with $\ell_1 > 7$. Player 1 makes a move from the middle of $Path_{\ell_1}$, dividing the $Path_{\ell_1}$ into two parts. The length of two parts are $6x_1 + 3, 6x_2 + 1$ respectively. In previous rounds, Player 1 gives up once in first part. After that Player 1 will always use a greedy strategy. If Player 2 uses a greedy strategy, Player 2 makes last move in $Path_{\ell_1}$. Hence, Player 1 starts $Path_{\ell_1}$ (loses 1 node on $Path_{\ell_1}$ and wins the game). Therefore Player 2 must abandon its greedy strategy while playing on $Path_{\ell_1}$ at least 3 times. In this game the game moves to $Path_{\ell_2}$, where Player 1 won at least 5 vertices in $Path_{\ell_1}$. So, Player 1 wins.

Case 6. $\ell_1 = 5$ and $\ell_2 = 5$ (mod 5). In this case, we show that Player 1 wins on $Path_{\ell_1} \cup Path_{\ell_2}$. Player 1 starts by selecting a vertex in $Path_{\ell_1}$ so that Player 1 wins 1 vertex and Player 2 makes the last move in $Path_{\ell_1}$. Since Player 1 wins on $Path_{\ell_2}$, Player 1 wins the game.

Case 7. $\ell_1 = 5$ (mod 6) and $\ell_2 = 5$ (mod 6) with $\ell_1 > 7$ and $\ell_2 > 7$. In this case we show that Player 2 doesn't lose. W.L.O.G, assume Player 1 starts by selecting a vertex in $Path_{\ell_1}$. Then there are 4 subcases.

Subcase 1: The unclaimed vertices of $Path_{\ell_1}$ consist of two parts L, R where $|L| = 6x_1$ and $|R| = 6x_2 + 2$. Consider the play in the unclaimed parts of $Path_{\ell_1}$. If Player 1 applies a greedy strategy all the time, then Player 2 also follows a greedy strategy until the number of unclaimed vertices of L is less than 10. By abandoning greedy strategy for one time in L, Player 2 guarantees that the opponent wins three vertices and makes the last move in $Path_{\ell_1}$. Since Player 2 wins at least 3 vertices on $Path_{\ell_2}$, Player 2 doesn't lose the game. Therefore Player 1 needs to abandon its greedy strategy at least 6 times in $Path_{\ell_1}$ so that Player 2 makes the last move in $Path_{\ell_1}$. Correspondingly, Player 2 wins 3 vertices in $Path_{\ell_1}$. Since Player 1 wins at most 3 vertices on $Path_{\ell_2}$, Player 2 doesn't lose the game.

Subcase 2: The unclaimed vertices of $Path_{\ell_1}$ consist of two parts L, R where $|L| = 6x_1 + 3$ and $|R| = 6x_2 + 5$. If Player 1 applies a greedy strategy all the time, then Player 2 also follows a greedy strategy until the number of unclaimed vertices of L is less than 7. By abandoning greedy strategy for one time, Player 2 guarantees that the opponent wins three vertices and makes the last move in $Path_{\ell_1}$. Since Player 2 wins at least 3 vertices in $Path_{\ell_2}$, Player 2 doesn't lose the game. Therefore Player 1 needs to abandon at least 6 times in $Path_{\ell_1}$ so that Player 2 makes the last move in $Path_{\ell_1}$. Correspondingly, Player 2 wins 3 vertices in $Path_{\ell_1}$. Since Player 1 wins at most 3 vertices on $Path_{\ell_2}$, Player 2 doesn't lose the game.

Subcase 3: The unclaimed vertices of $Path_{\ell_1}$ consist of two parts L, R where $|L| = 6x_1 + 1$ and $|R| = 6x_2 + 1$. If both players apply greedy strategy, then Player 1 wins 3 vertices and makes the last move in. Since Player 2 wins at least 3 vertices in $Path_{\ell_2}$, Player 2 doesn't lose the game. Therefore Player 1 needs to abandon at least 6 times in $Path_{\ell_1}$ so that Player 2 makes the last move in $Path_{\ell_1}$. Correspondingly, Player 2 wins 3 vertices in $Path_{\ell_1}$. Since Player 1 wins at most 3 vertices on $Path_{\ell_2}$, Player 2 doesn't lose the game.

Subcase 4: The unclaimed vertices of $Path_{\ell_1}$ consist of two parts L, R where $|L| = 6x_1 + 4$ and $|R| = 6x_2 + 4$. If both players apply greedy strategy, then Player 1 wins 3 vertices and makes the last move in. Since Player 2 wins at least

3 vertices in $Path_{\ell_2}$, Player 2 doesn't lose the game. Therefore Player 1 needs to abandon at least 6 times in $Path_{\ell_1}$ so that Player 2 makes the last move in $Path_{\ell_1}$. Correspondingly, Player 2 wins 3 vertices in $Path_{\ell_1}$. Since Player 1 wins at most 3 vertices on $Path_{\ell_2}$, Player 2 doesn't lose the game.

Case 8: $\ell_1 = 0$ (mod 6) and $\ell_2 = 0$ (mod 6). In this case, we show that Player 2 doesn't lose. W.L.O.G, assume Player 1 starts by selecting a vertex in $Path_{\ell_1}$. Then there are 2 subcases.

Subcase 1: The unclaimed vertices of $Path_{\ell_1}$ consist of two parts L, R where $|L| = 6x_1$ and $|R| = 6x_2 + 3$. Assume both players apply greedy strategy and Player 2 abandons its greedy strategy one time in $Path_{\ell_1}$. Then Player 1 wins 2 vertices and makes the last move in $Path_{\ell_1}$. Since Player 2 wins at least 2 vertices on $Path_{\ell_2}$, Player 2 doesn't lose the game. Therefore Player 1 needs to abandon its greedy strategy at least two times in $Path_{\ell_1}$ so that Player 2 makes the last move in $Path_{\ell_1}$. Correspondingly, Player 2 wins 2 vertices in $Path_{\ell_1}$. Since Player 1 wins at most 2 vertices on $Path_{\ell_2}$, Player 2 doesn't lose the game.

Subcase 2: The unclaimed vertices of $Path_{\ell_1}$ consist of two parts L, R where $|L| = 6x_1 + 1$ and $|R| = 6x_2 + 2$. If both players apply greedy strategy, then Player 1 wins two vertices and makes the last move in $Path_{\ell_1}$. Since Player 2 wins at least 2 vertices in $Path_{\ell_2}$, Player 2 doesn't lose the game. Therefore Player 1 needs to abandon its greedy strategy at least two times in $Path_{\ell_1}$ so that Player 2 makes the last move in $Path_{\ell_1}$. Correspondingly, Player 2 wins 2 vertices in $Path_{\ell_1}$. Since Player 1 wins at most 2 vertices on $Path_{\ell_2}$, Player 2 doesn't lose the game.

Case 9: Player 2 doesn't lose on the following subcases.

Subcases 1: $\ell_1 = 2$ and $\ell_2 = 0$ (mod 6). Note that $\ell_2 \geq 100$. If Player 1 starts by selecting a vertex in $Path_{\ell_1}$, then Player 1 wins 2 vertices in $Path_{\ell_1}$. Since Player 2 wins at least 2 vertices in $Path_{\ell_2}$, Player 2 doesn't lose the game. Therefore, assume Player 1 starts by selecting a vertex in $Path_{\ell_2}$. If both players apply greedy strategy, Player 1 wins 2 vertices and makes the last move on $Path_{\ell_2}$. Since Player 2 wins 2 vertices in $Path_{\ell_1}$, Player 2 doesn't lose the game. Therefore Player 1 needs to abandon its greedy strategy at least two times in $Path_{\ell_2}$ so that Player 2 makes the last move in $Path_{\ell_2}$. Correspondingly, Player 2 wins 2 vertices in $Path_{\ell_2}$. Since Player 1 wins at most 2 vertices on $Path_{\ell_1}$, Player 2 doesn't lose the game.

Subcases 2: $\ell_1 = 7$ and $\ell_2 = 5$ (mod 6). Note that $\ell_2 \geq 100$. If Player 1 starts by selecting a vertex in $Path_{\ell_1}$, then Player 2 can forces Player 1 to select the last move in $Path_{\ell_1}$ and guarantee to lose at most 3 vertices. Since Player 2 wins $Path_{\ell_2}$ with at least 3 vertices, Player 2 doesn't lose the game. Therefore, assume Player 1 starts by selecting a vertex in $Path_{\ell_2}$. Note that $Opt(Path_7) = Opt(Path_{6x+5}) = 2$ where $x > 0$. Therefore, following similar proofs of Case 7, one can prove that Player 2 doesn't lose the game.

Corollary 3. *Player 1 wins topological network-control games on $Path_{\ell_1} \cup Path_{\ell_2}$ if and only if $\{\ell_1, \ell_1\}$ satisfies one of the following conditions: (1) ℓ_1 and ℓ_2 have different parities, (2) $\ell_1 = 4 \ (mod \ 6)$, (3) $\ell_1 = 3 \ (mod \ 6)$, (4)*

$\ell_1 = 2$ *(mod 6) except for* $\{2,2\}$, $\{2,6x\}$, *(5)* $\ell_1 = 1$ *(mod 6) except for* $\{1,1\}$, $\{7,7\}$, $\{7,6x+5\}$. *In all other cases, Player 2 can guarantee a draw and cannot win.*

5 PSPACE-Completeness

The alternating 3-TQBF problem is to determine if a fully quantified Boolean formula $\psi = \exists x_1 \forall x_2 \exists x_3 \dots \exists x_n \phi(x_1, x_2, \dots, x_n)$ is true, where ϕ is a conjunction of clauses with three literals. The problem is PSPACE-complete [25]. The optimal topological network-control game problem with parameter t is this:

$$OTNC(t) = \{\langle G, w \rangle \mid \text{Player 1 has a } w\text{-optimal strategy on game } G\}.$$

Let ψ be an alternating 3-TQBF formula with clauses $C_1, \dots, C_m$ and variables $x_1, \dots, x_n$. W.L.O.G, we assume $n > 1$, n is odd and for each pair of distinct variables x_i and x_j, the clauses $x_i \vee \overline{x_i} \vee x_j$ and $x_i \vee \overline{x_i} \vee \overline{x_j}$ are in $\{C_i\}_{i \leq m}$. We build a connected graph G_ψ with $2n + 6n(n+1)m + 3m$ vertices and $n + 12n(n+1)m + 3m + 2m^2 + m(m-1)/2$ edges such that the formula $\psi = \exists x_1 \forall x_2 \exists x_3 \dots \exists x_n \phi(x_1, x_2, \dots, x_n)$ is true iff $F(G_\psi) \geq 6nm + 5m + 2$. Therefore, we can prove the following theorem.

Theorem 6. *The $OTNC(1)$ problem is PSPACE-complete.*

References

1. Adams, R., Dixon, J., Elder, J., Peabody, J., Vega, O., Willis, K.: Combinatorial analysis of a subtraction game on graphs. Int. J. Comb. **2016**, 1–9 (2016)
2. Ahn, H.K., Cheng, S.W., Cheong, O., Golin, M., Van Oostrum, R.: Competitive facility location: the Voronoi game. Theoret. Comput. Sci. **310**(1–3), 457–467 (2004)
3. Andres, S.D., Huggan, M., Mc Inerney, F., Nowakowski, R.J.: The orthogonal colouring game. Theoret. Comput. Sci. **795**, 312–325 (2019)
4. Beaudou, L., et al.: Octal games on graphs: the game 0.33 on subdivided stars and bistars. Theoret. Comput. Sci. **746**, 19–35 (2018)
5. Bensmail, J., Fioravantes, F., Mc Inerney, F., Nisse, N.: The largest connected subgraph game. In: Kowalik, L., Pilipczuk, M., Rzążewski, P. (eds.) WG 2021. LNCS, vol. 12911, pp. 296–307. Springer, Cham (2021). https://doi.org/10.1007/978-3-030-86838-3_23
6. Bensmail, J., Fioravantes, F., Mc Inerney, F., Nisse, N.: The largest connected subgraph game. Algorithmica **84**(9), 2533–2555 (2022)
7. Bensmail, J., Fioravantes, F., Mc Inerney, F., Nisse, N., Oijid, N.: The maker-breaker largest connected subgraph game. Theoret. Comput. Sci. **943**, 102–120 (2023)
8. Bensmail, J., Mc Inerney, F.: On a vertex-capturing game. Theoret. Comput. Sci. **923**, 27–46 (2022)
9. Bodlaender, H.L., Kratsch, D.: Kayles and nimbers. J. Algorithms **43**(1), 106–119 (2002)

10. Brown, S., et al.: Nimber sequences of node-kayles games. J. Integer Seque. **23**, 1–43 (2020)
11. Calkin, N.J., et al.: Computing strategies for graphical Nim. In: Proceedings of the Forty-First Southeastern International Conference on Combinatorics, Graph Theory and Computing, vol. 202, pp. 171–185. Citeseer (2010)
12. Changat, M., Lekha, D.S., Peterin, I., Subhamathi, A.R., Špacapan, S.: The median game. Discret. Optim. **17**, 80–88 (2015)
13. Cibulka, J., Kynčl, J., Mészáros, V., Stolař, R., Valtr, P.: Graph sharing games: complexity and connectivity. Theoret. Comput. Sci. **494**, 49–62 (2013)
14. Cohen, N., Martins, N.A., Mc Inerney, F., Nisse, N., Pérennes, S., Sampaio, R.: Spy-game on graphs: complexity and simple topologies. Theoret. Comput. Sci. **725**, 1–15 (2018)
15. Dailly, A., Gledel, V., Heinrich, M.: A generalization of ARC-KAYLES. Int. J. Game Theory **48**(2), 491–511 (2019)
16. Duchene, E., Gonzalez, S., Parreau, A., Rémila, E., Solal, P.: Influence: a partizan scoring game on graphs. Theoret. Comput. Sci. **878**, 26–46 (2021)
17. Duchêne, É., Renault, G.: Vertex Nim played on graphs. Theoret. Comput. Sci. **516**, 20–27 (2014)
18. Fleischer, R., Trippen, G.: Kayles on the Way to the Stars. In: van den Herik, H.J., Björnsson, Y., Netanyahu, N.S. (eds.) CG 2004. LNCS, vol. 3846, pp. 232–245. Springer, Heidelberg (2006). https://doi.org/10.1007/11674399_16
19. Guignard, A., Sopena, É.: Compound node-KAYLES on paths. Theoret. Comput. Sci. **410**(21–23), 2033–2044 (2009)
20. Huggan, M.A., Stevens, B.: Polynomial time graph families for ARC KAYLES. Integers **16**, A86 (2016)
21. Kleinberg, J., Tardos, E.: Algorithm Design. Pearson Education, India (2006)
22. Liang, Z., Khoussainov, B., Xiao, M.: Who controls the network? SSRN 4291268
23. Micek, P., Walczak, B.: A graph-grabbing game. Comb. Probab. Comput. **20**(4), 623–629 (2011)
24. Schaefer, T.J.: On the complexity of some two-person perfect-information games. J. Comput. Syst. Sci. **16**(2), 185–225 (1978)
25. Stockmeyer, L.J., Meyer, A.R.: Word problems requiring exponential time (preliminary report). In: Proceedings of the Fifth Annual ACM Symposium on Theory of Computing, pp. 1–9 (1973)

Lower Bounds of Functions on Finite Abelian Groups

Jianting Yang[1,2($\boxtimes$)], Ke Ye[1,2], and Lihong Zhi[1,2]

[1] Key Lab of Mathematics Mechanization, AMSS, Beijing 100190, China
{yangjianting,keyk}@amss.ac.cn, lzhi@mmrc.iss.ac.cn
[2] University of Chinese Academy of Sciences, Beijing 100190, China

Abstract. The problem of computing the optimum of functions on finite abelian groups is an important problem in mathematics and computer science. Many combinatorial problems, such as MAX-SAT, MAX-CUT and the knapsack problem, can be recognized as optimization problems on the group $C_2^n = \{-1, 1\}^n$. This paper proposes an algorithm that efficiently computes verifiable lower bounds of functions on finite abelian groups by the technique of the Fourier sum of squares with error. Moreover, we propose a new rounding method to obtain a feasible solution that minimizes the objective function as much as possible. We also implement the algorithm and test it on MAX-SAT benchmark problems and random functions. These experiments demonstrate the advantage of our algorithm over previously known methods.

1 Introduction

Combinatorial problems can usually be formulated as optimizaiton problems on finite sets, thus many of them are notoriously difficult. Examples include the Knapsack problem [19,20], the set cover problem [4,13], the k-SAT problem [3,11] and their numerous variants. Although each of these problems can be regarded as an integer programming problem, there do not exist polynomial time algorithms for most of them unless $\mathbf{P} = \mathbf{NP}$ [11,13,14,22,23]. Therefore various approximation algorithms are employed to resolve the issue [2,8,17,28,31,32]. Semidefinite programming (SDP) based relaxation methods, particularly the sum of squares (SOS) relaxation is one of the most powerful and extensively studied techniques to design and analyze an approximation algorithm for polynomial optimization problems [6,10,12,16,25,29,31,40]. Among those successful applications of the SDP technique, the most well-known ones are MAX-2SAT [9], MAX-3SAT [10] and MAX-CUT [12]. On the other side, it is noticed in [5,27,38,39] that if a finite set is equipped with an abelian group structure, then one can efficiently certify nonnegative functions on it by Fourier sum of squares (FSOS). Motivated by previous works, this paper is concerned with establishing a framework to solve Problem 1 below by techniques of FSOS with error and semidefinite programming.

L. Zhi—This research is supported by the Nsational Natural Science Foundation of China 12071467 (Zhi).

Problem 1 (lower bound by FSOS). Given a function f on a finite abelian group, find a lower bound of f efficiently.

Let $S \subseteq \mathbb{C}^n$ be an algebraic variety. Algebraically, identifying the ring $\mathbb{C}[S]$ of polynomial functions on S with $\mathbb{C}[z_1, \ldots, z_n]/I(S)$ is a favorable perspective as the latter ring is endowed with rich geometric and algebraic structures. For computational purposes, however, regarding a function as an equivalence class is not convenient, on account of the fact that an equivalence class can be represented by infinitely many different polynomials. If $S = G$ is a finite group, then there is an alternative algebraic structure on $\mathbb{C}[G]$ which is extremely useful for computations [5,27,38,39]. Namely, one can identify $\mathbb{C}[G]$ with the group ring of G via the Fourier transform [7]. The advantage of such a point of view is that a function f on G can be expanded as $f = \sum_{\chi \in \widehat{G}} \widehat{f}(\chi)\chi$, where $\widehat{G}$ is the dual group of G and $\widehat{f}(\chi)$ is the Fourier coefficient of f at $\chi \in \widehat{G}$. This well-known viewpoint enables us to introduce analytic tools to solve Problem 1.

Related Works and Our Contributions

A method for general-purpose sparse polynomial optimization called the TSSOS hierarchy is proposed in [35]. The new method follows the well-known methodology established in [15], but it exploits the sparsity of polynomials to reduce the size of SDP. Combing the TSSOS hierarchy and the method in [33] for correlative sparsity, [34] introduces the CS-TSSOS hierarchy for large scale sparse polynomial optimization. In particular, both TSSOS and CS-TSSOS hierarchies are applicable to optimization problems on finite abelian groups.

Due to its great importance in computer science, there are various solvers for MAX-SAT. For instance, in [31], the quotient structure of $\mathbb{C}[C_2^n]$ is explored for support selection strategies, from which one can solve MAX-SAT by SDP; by combing several optimization techniques specifically designed for MAX-SAT, [36] provides an efficient SDP-based MIXSAT algorithm; based on the resolution refutation, a solver called MS-builder is proposed in [24].

On the one hand, TSSOS and CS-TSSOS hierarchies can handle general polynomial optimization problems, while specially designed solvers such as MIXSAT [36] and MS-builder [24] can only deal with MAX-SAT problems. On the other hand, however, it is natural to expect that these specially designed solvers would outperform general-purpose methods on MAX-SAT problems.

Our framework balances the universality and efficiency. Indeed, our method is applicable to optimization problems on any finite abelian groups, including the hypercube C_2^n, cyclic group $\mathbb{Z}_N$ and their product. We briefly summarize our main contributions below.

- We present an efficient approximation algorithm to minimize a function on a finite abelian group by computing its lower bounds. (Algorithm 1), which is validated by Theorems 1 and 2.
- We propose a new rounding method to obtain high-quality feasible solutions of binary optimization problems (Sect. 3.4).

– We test our algorithm on MAX-2SAT, MAX-3SAT benchmark problems and randomly generated functions. These numerical experiments demonstrate the advantage of our algorithm over aforementioned polynomial optimization methods.

2   Preliminaries

In this section, we review the Fourier analysis on abelian groups and provide a brief introduction to Fourier sum of squares (FSOS) on finite abelian groups.

2.1   Fourier Analysis on Groups

We briefly summarize fundamentals of group theory and representation theory in this subsection. For more details, we refer interested readers to [7,21,26].

Let G be a finite abelian group. A *character* of G is a group homomorphism $\chi : G \to \mathbb{C}^\times$. Here $\mathbb{C}^\times$ is $\mathbb{C} \setminus \{0\}$ endowed with the multiplication of complex numbers as the group operation. The set of all characters of G is denoted by $\widehat{G}$, called the *dual group*[1] of G. According to [7, Chapter 1], any function f on G admits the *Fourier expansion*:

$$f = \sum_{\chi \in \widehat{G}} \widehat{f}(\chi)\chi,$$

where $\widehat{f}(\chi) := \frac{1}{|G|} \sum_{g \in G} f(g)\overline{\chi(g)}$ is called the *Fourier coefficient* of f at $\chi \in \widehat{G}$. The *support* of f is $\mathrm{supp}(f) := \left\{ \chi \in \widehat{G} : \widehat{f}(\chi) \neq 0 \right\}$. As an example, the dual group of the hypercube $C_2^n = \{-1, 1\}^n$ is

$$\widehat{C_2^n} = \{z^\alpha : \alpha = (\alpha_1, \dots, \alpha_n) \in \mathbb{Z}_2^n\} \simeq \mathbb{Z}_2^n.$$

Here $\mathbb{Z}_2 = \mathbb{Z}/2\mathbb{Z} = \{0, 1\}$ is the additive group and $z^\beta := z_1^{\beta_1} \dots z_n^{\beta_n}$ for each $\beta \in \mathbb{N}^n$. Thus a function $f : C_2^n \to \mathbb{C}$ can be expressed as a linear combination of multilinear monomials: $f = \sum_{\alpha \in \mathbb{Z}_2^n} f_\alpha z^\alpha$.

2.2   Sparse Fourier Sum of Squares on Finite Abelian Groups

This subsection briefly reviews the theory of FSOS developed in [5,27,38]. Let f be a nonnegative function on a finite abelian group G. A *Fourier sum of squares (FSOS)* of f is a finite family $\{h_i\}_{i \in I}$ of complex valued functions on G such that $f = \sum_{i \in I} |h_i|^2$. According to [5], a function $f : G \to \mathbb{R}$ admits an FSOS if and only if f is nonnegative. The *sparsity* of $\{h_i\}_{i \in I}$ is defined to be $|\bigcup_{i \in I} \mathrm{supp}(h_i)|$. We say that $\{h_i\}_{i \in I}$ is a *sparse FSOS of f* if its sparsity is small.

---

[1] Since G is an abelian group, $\widehat{G}$ is indeed a group and $\widehat{G} \simeq G$.

A function f on G is nonnegative if and only if there exists a Hermitian positive semidefinite matrix $Q = (Q_{\chi,\chi'})_{\chi,\chi' \in \widehat{G}} \in \mathbb{C}^{\widehat{G} \times \widehat{G}}$ such that

$$\sum_{\chi' \in \widehat{G}} Q_{\chi',\chi'\chi} = \widehat{f}(\chi), \quad \forall \chi \in \widehat{G}. \tag{1}$$

Here we index columns and rows of Q by elements of $\widehat{G}$. We call Q a *Gram matrix* of f. In fact, if Q is a Gram matrix of f and $Q = M^* M$ for some matrix $M = (M_{j,\chi})_{1 \le j \le r, \chi \in \widehat{G}} \in \mathbb{C}^{r \times \widehat{G}}$, then we must have $f = \sum_{j=1}^{r} |\sum_{\chi \in \widehat{G}} M_{j,\chi} \chi|^2$. Clearly this construction provides us a correspondence between (sparse) Gram matrices and (sparse) FSOS of f. Because of the above correspondence, a sparse FSOS of f is always preferable to reduce the cost of computations involving Gram matrices. We remark that since G is finite, any nonnegative function f on G can be written as $f = |\sqrt{f}|^2$, which gives an FSOS of f. Here $\sqrt{f}$ denotes the pointwise square root of f. Moreover, $\sqrt{f}$ gives an optimal solution to the convex relaxation of the problem of finding an FSOS with the minimal support [38]. However, $\sqrt{f}$ is usually not sparse and thus the computation of $\sqrt{f}$ becomes challenging. In [39], we propose a polynomial approximation method for computing efficiently $\sqrt{f}$ approximately: suppose $0 \le l \le f \le m$ is a function on G, and a univariate polynomial $p(t)$ approximates $\sqrt{t}$ at points between l and m with pointwise error at most ε, then $p \circ f$ is an estimate of $\sqrt{f}$ with coefficient error bounded by ε.

3 Main Results

In this section, we present our solution to Problem 1 for integer-valued functions.

3.1 Lower Bounds by FSOS

Let $f : G \mapsto \mathbb{Z}$ be a function on a finite abelian group G and let α be a real number. For each $S \subseteq \widehat{G}$, we define

$$F_S := \Big\{ \{h_i\}_{i \in I} : f - \alpha = \sum_{i \in I} |h_i|^2 \text{ for some } \alpha \in \mathbb{R}, \bigcup_{i \in I} \operatorname{supp}(h_i) \subseteq S \Big\},$$

and consider the following optimization problem:

$$\max_{\{h_i\}_{i \in I} \in F_S} \alpha,$$
$$\text{s.t. } f - \alpha = \sum_{i \in I} |h_i|^2. \tag{2}$$

According to the discussion in Subsect. 2.2, α is a lower bound of f if and only if $f - \alpha$ admits an FSOS $\{h_i\}_{i \in I}$. As a consequence, an optimal solution to (2) for any $S \subseteq \widehat{G}$ provides a lower bound of f. Moreover, the larger S we choose for (2), the better lower bound we obtain, since $S_1 \subseteq S_2 \subseteq \widehat{G}$ implies $F_{S_1} \subseteq F_{S_2}$.

Since the quality of the lower bound obtained by solving (2) depends on the choice of $S \subseteq \widehat{G}$, in order to efficiently compute a high-quality lower bound, we need to choose a small subset S which contains as many elements in $\bigcup_{i \in I} \text{supp}(h_i)$ as possible, where $\{h_i\}_{i \in I}$ is an FSOS of $f - \min_{g \in G} f(g)$. Clearly such $\{h_i\}_{i \in I}$ is an optimal solution to (2) for $S = \widehat{G}$.

The correspondence between FSOS and Gram matrices (cf. Subsect. 2.2) enables us to reformulate (2) as the following SDP problem:

$$\max_{Q \in \mathbb{C}^{S \times S}} \widehat{f}(\chi_0) - \text{trace}(Q), \tag{3}$$

$$\text{s.t.} \quad \sum_{\chi' \in \widehat{G}} Q_{\chi', \chi' \chi} = \widehat{f}(\chi), \ \chi \neq \chi_0 \in \widehat{G} \tag{4}$$

$$Q \succeq 0. \tag{5}$$

Here χ_0 denotes the identity element in $\widehat{G}$.

The next theorem provides us a simple but effective method to pick $S \subseteq \widehat{G}$ such that an optimal solution to (3)–(5) gives a high-quality lower bound of f.

Theorem 1. *Let f be a function on G and let $a > \widetilde{a} \geq 0$ be lower bounds of f. Assume that c_χ is the coefficient of χ in $\sqrt{f - a}$. If $|c_\chi| > \sqrt{a - \widetilde{a}}$, then $\chi \in \text{supp}(\sqrt{f - \widetilde{a}})$. Moreover, if $G = C_2^n$ and $f \leq m$ for some $m \in \mathbb{R}$, then $|c_\chi| > \frac{a - \widetilde{a}}{2} \left(\frac{1}{\sqrt{a - \widetilde{a}}} - \frac{1}{\sqrt{m - a} + \sqrt{m - \widetilde{a}}} \right)$ implies $\chi \in \text{supp}(\sqrt{f - \widetilde{a}})$.*

Proof. We recall that $c_\chi := \frac{1}{|G|} \sum_{g \in G} \overline{\chi(g)} \sqrt{f(g) - a}$. Similarly, we may write $\sqrt{f - \widetilde{a}} = \sum_{\chi \in \widehat{G}} \widetilde{c}_\chi \chi$. For each $\chi \in \widehat{G}$, we notice that

$$|c_\chi - \widetilde{c}_\chi| = \frac{1}{|G|} \left| \sum_{g \in G} \overline{\chi(g)} \left(\sqrt{f(g) - a} - \sqrt{f(g) - \widetilde{a}} \right) \right|$$

$$= \frac{a - \widetilde{a}}{|G|} \left| \sum_{g \in G} \frac{\overline{\chi(g)}}{\sqrt{f(g) - a} + \sqrt{f(g) - \widetilde{a}}} \right|$$

$$\leq \sqrt{a - \widetilde{a}}.$$

Therefore, if $|c_\chi| > \sqrt{a - \widetilde{a}}$ then $\widetilde{c}_\chi \neq 0$. If $G = C_2^n$, then $\chi \in \widehat{G} \simeq \mathbb{Z}_2^n$ is a multilinear monomial. Thus $\chi(g) = \pm 1$ for any $g \in C_2^n = \{-1, 1\}^n$. Moreover, it is straightforward to verify that we can construct a bijective map $\psi : C_2^n \to C_2^n$ such that $\chi(g) = -\chi(\psi(g))$ and $\psi^2 = \text{Id}$. Thus we have

$$|c_\chi - \widetilde{c}_\chi| = \frac{a - \widetilde{a}}{|G|} \left| \sum_{g \in G} \frac{\overline{\chi(g)}}{\sqrt{f(g) - a} + \sqrt{f(g) - \widetilde{a}}} \right|$$

$$\leq \frac{a - \widetilde{a}}{|G|} \sum_{g \in G, \chi(g) = 1} \left(\frac{1}{\sqrt{a - \widetilde{a}}} - \frac{1}{\sqrt{m - a} + \sqrt{m - \widetilde{a}}} \right)$$

$$= \frac{a - \widetilde{a}}{2} \left(\frac{1}{\sqrt{a - \widetilde{a}}} - \frac{1}{\sqrt{m - a} + \sqrt{m - \widetilde{a}}} \right).$$

An important implication of Theorem 1 is that for each $\alpha \leq f_{\min} :=$ $\min_{g \in G} f(g)$, $\operatorname{supp}(\sqrt{f - \alpha})$ contains elements in $\operatorname{supp}(\sqrt{f - f_{\min}})$ whose Fourier coefficients in $\sqrt{f - \alpha}$ are large. In particular, if f is nonnegative, then we can even take $\operatorname{supp}(\sqrt{f})$ as an estimate of $\operatorname{supp}(\sqrt{f - f_{\min}})$. Therefore, we may construct S of cardinality k by taking the first k terms of an approximation of $\sqrt{f}$ and Theorem 1 ensures that S is a good estimate of $\operatorname{supp}(\sqrt{f - f_{\min}})$.

3.2   FSOS with Error

We discuss in this subsection a remarkable feature of the SDP problem (3)–(5): a solution that violates conditions (4) and (5) may still provide us a tight lower bound of f. This is the content of the next theorem.

Theorem 2. *Let G be a finite abelian group and let S be a subset of $\widehat{G}$. Given a function $f : G \to \mathbb{R}$ and a Hermitian matrix $Q \in \mathbb{C}^{S \times S}$, we have*

$$\min_{g \in G} f(g) \geq -\|\widehat{e}\|_1 + \lambda_{\min}(Q)|S|,$$

where $\lambda_{\min}(Q)$ is the minimal eigenvalue of Q, $e = f - v_S^* Q v_S$, $v_S = (\chi)_{\chi \in S}$ is the column vector consisting of all characters in S and $\|\widehat{e}\|_1 = \sum_{\chi \in \widehat{G}} |\widehat{e}(\chi)|$. Furthermore, $Q - \lambda_{\min}(Q)\operatorname{Id} \succeq 0$ is a Gram matrix of $f - e - \lambda_{\min}(Q)|S|$.

Proof. For any $g \in G$, we have $f(g) - e(g) = v_S(g)^* Q v(g) \geq \lambda_{\min}(Q) v_S(g)^* v(g)$. We observe that $|\chi(g)| = 1$ for each $g \in G$ and $\chi \in \widehat{G}$. Thus $v_S(g)^* v_S(g) = |S|$ and $|e(g)| \leq \|\widehat{e}\|_1$ for any $g \in G$. This implies

$$f(g) \geq -\|\widehat{e}\|_1 + \lambda_{\min}(Q)|S|, \quad \forall g \in G.$$

Theorem 2 implies that $-\|\widehat{e}\|_1 + \lambda_{\min}(Q)|S|$ is a lower bound of f even if $Q \not\succeq 0$ or $e \neq 0$.

Given a Hermitian matrix $M \in \mathbb{C}^{n \times n}$, we may define its associated real-valued polynomial $F_M(z) := z^* M z$. Then we have $F_M = F_{M - \lambda \operatorname{Id}} + F_{\lambda \operatorname{Id}}$ for any $\lambda \in \mathbb{R}$. Therefore, if $M - \lambda \operatorname{Id} \succeq 0$ then $\min_{z \in \mathbb{C}^n} F_{\lambda \operatorname{Id}}(z)$ provides a lower bound of F_M. Unfortunately, $\min_{z \in \mathbb{C}^n} F_{\lambda \operatorname{Id}}(z) = -\infty$ is a trivial lower bound of F_M for $\lambda < 0$, as $F_{\lambda \operatorname{Id}}(z) = \lambda \|z\|^2$. This phenomenon distinguishes optimization problems on finite abelian groups from the usual polynomial optimizations.

For instance, we consider $f(z_1, z_2) = \begin{bmatrix} z_1 & z_2 \end{bmatrix} \begin{bmatrix} 0 & 1 \\ 1 & 0 \end{bmatrix} \begin{bmatrix} z_1 & z_2 \end{bmatrix}^{\mathsf{T}} = 2 z_1 z_2$ on C_2^2.

Since $S = \{z_1, z_2\}$ and $\lambda_{\min}(Q) = -1$ where $Q = \begin{bmatrix} 0 & 1 \\ 1 & 0 \end{bmatrix}$, Theorem 2 implies that $f \geq -2$ on C_2^2. As a comparison, we notice that f is the restriction of F_Q to C_2^2 thus a lower bound of F_Q is also a lower bound of f. However, the above discussion indicates that the lower bound of F_Q obtained by the SOS technique is trivial, since $F_Q = F_{Q + \operatorname{Id}} + F_{-\operatorname{Id}} = |z_1 + z_2|^2 - (|z_1|^2 + |z_2|^2) \geq \min_{(z_1, z_2) \in \mathbb{C}^2} -(|z_1|^2 + |z_2|^2) = -\infty$.

3.3   Computation of Lower Bounds

In this subsection, we present an algorithm for computing a lower bound of a function f on a finite abelian group G. For simplicity, for any function h, we denote $h_0 := \widehat{h}(\chi_0)$, where χ_0 is the identity element in $\widehat{G}$.

Let S be a subset of $\widehat{G}$ and let $Q \in \mathbb{C}^{S \times S}$ be a Hermitian matrix. We define $e := f - f_0 - v_S^* Q v_S$, where v_S is the $|S|$-dimensional column vector consisting of characters in S. Applying Theorem 2 to $f - f_0 - e_0$, we have

$$\min_{g \in G} f(g) \geq f_0 + e_0 - \|\widehat{e - e_0}\|_1 + \lambda_{\min}(Q)|S|. \tag{6}$$

Since $e_0 = -\sum_{\chi = \chi'} Q(\chi, \chi') = -\operatorname{trace}(Q)$, we can rewrite the right side of (6) as $f_0 - F(Q)$ where $F(Q) := \operatorname{trace}(Q) + \|E(Q)\|_1 - \lambda_{\min}(Q)|S|$ and $E : \mathbb{C}^{S \times S} \mapsto \mathbb{C}^{|G|}$ is the affine map sending Q to the (sparse) vector consisting of Fourier coefficients of $e - e_0$. Therefore we may obtain a lower bound of f by solving the following unconstrained convex optimization problem:

$$\min_{Q \in \mathbb{C}^{S \times S}, \, Q = Q^*} F(Q), \tag{7}$$

Combining our discussions on the SDP problem (3)–(5), the support selection method obtained from Theorem 1 and the convex optimization problem (7), we obtain Algorithm 1.

Algorithm 1 Lower Bounds of Functions on Finite Abelian Groups.

Input a nonnegative function f on G, integers d and k, l, m such that $0 \leq l \leq f \leq m$.
Output a lower bound of f
1: approximate $\sqrt{t}$ by a polynomial $p(t)$ of degree at most d at integer points in $[l, m]$.

2: compute the composition $p \circ f = \sum_{i=1}^{|G|} a_i \chi_i$, where $|a_1| \geq \cdots \geq |a_{|G|}|$ and $\deg(\chi_i) \geq$ $\deg(\chi_{i+1})$ if $|a_i| = |a_{i+1}|, 1 \leq i \leq |G| - 1$.
3: Select $S := \{\chi_1, \ldots, \chi_k\}$.
4: Solve (3)-(5) for Q_0.   ▷ by SDPNAL+
5: Solve (7) for Q.   ▷ by gradient descent with initial point Q_0
6: **return** $f_0 - F(Q)$.

We remark that a lower and upper bound l and m of f are required as a part of input for Algorithm 1. In most applications, they are in fact readily available. In rare cases where no non-trivial bounds can be obtained beforehand, we may simply use the trivial bounds: $l = 0$ and $m = \|\widehat{f}\|_1$ for nonnegative function f. The goal of Algorithm 1 is to find a lower bound that is much better than l. Steps 1–3 of Algorithm 1 are validated by Theorem 1. In practice, it suffices to use $d = 1$ or 2, as it is shown by experiments in Sect. 4. Due to Theorem 2, we do not need to solve the SDP problem (3)–(5) exactly in step 4. Instead, we are allowed to compute an approximate solution, which may violate conditions

(4) and (5). Moreover, the ADMM method used by SDPNAL+ augments an objective function by an ℓ^2-norm, while our algorithm actually aims to minimize the ℓ^1-norm. Thus Step 5 in Algorithm 1 is necessary for computing a better lower bound. We solve (7) by the gradient descent method. Although F is not smooth, we are able to compute its subgradient by

$$\partial F = \mathrm{Id} + (\partial E)^* \, \mathrm{sign}(E(Q)) - |S| uu^*,$$

where ∂E is the gradient of E, $\mathrm{sign}(x)$ is the sign function and u is the unit eigenvector of Q corresponding to $\lambda_{\min}(Q)$.

We conclude this subsection by briefly summarizing the main advantages of Algorithm 1. Interested readers are referred to Sect. 4 for numerical examples which demonstrate these advantages.

1. *Early termination*: existing methods [31,35] need to wait the algorithm converges to a solution satisfying conditions (4) and (5). However, Theorem 2 ensures that our method can find a lower bound even if these conditions are not satisfied. This feature enables us to set time limits on solving the SDP problem (3)–(5) instead of waiting for converging to a feasible solution.
2. *Adaptivity to more SDP solvers*: since the conditions (4) and (5) are not required to be satisfied exactly, we can use more efficient SDP solvers, such as SDPNAL+ [30]. As a consequence, MAX-SAT problems of much larger sizes from the benchmark set [2] can be solved efficiently.
3. *Size reduction*: given a subset $S \subseteq \widehat{G}$, the SOS based algorithms may not get a lower bound if f has no FSOS supported on S. However, Example 1 indicates that an FSOS supported on S with a small error may provide us a tight lower bound. This feature leads to a reduction on the size of the SDP problem (3)–(5).

Example 1. Let $f : C_2^3 \to \mathbb{R}$ be the function defined by

$$f(z_1, z_2, z_3) = 4 + z_1 + z_2 + z_3 + z_1 z_2 z_3.$$

We can check that problem (3)–(5) has no feasible solution for $S = \{1, z_1, z_2, z_3\}$. However, we have $f - e = 1 + \frac{1}{2} \sum_{i=1}^{3} (1 + z_i)^2$ where $e(z_1, z_2, z_3) := z_1 z_2 z_3$, from which we obtain $f \geq 1 - \|\widehat{e}\|_1 = 0$. We remark that in this case, the matrix size in (3)–(5) reduces from $8 = |\widehat{G}|$ to $4 = |S|$.

3.4   Rounding

Rounding is an important step in SDP-based algorithms for combinatorial optimization problems, especially when both the optimum value and optimum point are concerned. The purpose of rounding is to find a high-quality feasible solution. There exist several rounding techniques in the literature. Examples include rounding by random hyperplanes [8] together with its improved version [6], the

---

[2]   http://www.maxsat.udl.cat/16/index.html.

skewed rounding procedure [18] and the randomized rounding technique [31]. Among all these rounding strategies, the one proposed in [31] can be easily adapted to our situation.

Our rounding method is based on the null space of the Gram matrix. Let $Q \in \mathbb{R}^{S \times S}$ be a solution to (3)–(5). For a given $f : C_2^n \to \mathbb{R}$ and $S \subseteq \widehat{C_2^n}$ containing all characters of degree at most one, we assume $f - \alpha = v_S^* Q v_S$, where α is the minimum value of f, v_S is the $|S|$-dimensional column vector consisting of characters in S and $Q \succeq 0$.

In practice, the matrix Q obtained in Algorithm 1 may not be positive semidefinite. We need to update it by $Q - \lambda_{\min}(Q) \operatorname{Id}$. It is a little surprise for us to notice that the numerical corank of $Q - \lambda_{\min}(Q) \operatorname{Id}$ is 1 very often, which makes it possible to recover the optimal solution efficiently from its null vector. By normalizing the element indexed by $\chi_0 \in \widehat{G}$ of the null vector v of $Q - \lambda_{\min}(Q) \operatorname{Id}$ to be one, we obtain the desired solution $\tilde{g} \in G$ by rounding elements of the normalized null vector.

To conclude this subsection, we illustrate our rounding procedure by example.

Example 2 (rounding). The polynomial $f(z_1, z_2, z_3) = 4 + z_1 + z_2 + z_3 + z_1 z_2 z_3$ given in Example 1 is a nonnegative function on C_2^3. For $S = \{1, z_1, z_2, z_3, z_1 z_2 z_3\}$, we obtain a Hermitian matrix by SDPNAL+:

$$
Q = \begin{array}{c} \\ 1 \\ z_1 \\ z_2 \\ z_3 \\ z_1 z_2 z_3 \end{array}
\begin{array}{ccccc} 1 & z_1 & z_2 & z_3 & z_1 z_2 z_3 \\ \left[\begin{array}{ccccc} 1.9546 & 0.5000 & 0.5000 & 0.5000 & 0.5000 \\ 0.5000 & 0.4536 & 0.0000 & 0.0000 & 0.0000 \\ 0.5000 & 0.0000 & 0.4536 & 0.0000 & 0.0000 \\ 0.5000 & 0.0000 & 0.0000 & 0.4536 & 0.0000 \\ 0.5000 & 0.0000 & 0.0000 & 0.0000 & 0.4536 \end{array}\right] \end{array},
$$

whose eigenvalues are $-0.0462, 0.4536, 0.4536, 0.4536, 2.4544$. The normalized null vector of $Q + 0.0462 \operatorname{Id}$ is

$$
\begin{array}{ccccc} 1 & z_1 & z_2 & z_3 & z_1 z_2 z_3 \end{array}
$$
$$
v = \begin{bmatrix} 1 & -1.000447 & -1.000447 & -1.000447 & -1.000447 \end{bmatrix}.
$$

We recover the solution $z_1 = z_2 = z_3 = -1$ by rounding the elements of v.

4   Numerical Experiments

In this section, we conduct numerical experiments to test Algorithm 1 and the rounding technique discussed in Subsect. 3.4. We implement our algorithm in Matlab (2016b) and invoke the ADMM algorithm in SDPNAL+ [30] to solve SDP problems. The code is available on github.com/jty-AMSS/Fast-Lower-Bound-On-FAG. All experiments are performed on a desktop computer with Intel Core i9-10900X@3.70 GHz CPU and 128 GB RAM memory. Due to the page limit, we only present three numerical experiments here. Interested readers are referred to the full-length version [37] on arXiv for more numerical experiments.

4.1   Upper Bounds of MAX-2SAT Problems

Recall that any CNF-formula ϕ in n variables with m clauses can be transformed into the characteristic function f_ϕ on C_2^n such that the minimum number of simultaneously falsified clauses in ϕ is equal to $\min_{g \in C_2^n} f_\phi(g)$. Hence solving the MAX-SAT problem for ϕ is equivalent to computing the minimum value of the function f_ϕ on C_2^n. For comparison purposes, we compute lower bounds of the number of falsified clauses of benchmark MAX-SAT problems respectively by Algorithm 1, TSSOS [35] and CS-TSSOS [34][3]. The MAX-SAT problems are drawn from the randomly generated unweighted MAX-2SAT benchmark problem set in 2016 MAX-SAT competition[4]. Such a problem has 120 variables, in which the number of clauses ranges from 1200 to 2600. We apply Algorithm 1 to compute lower bounds of the corresponding characteristic functions, with parameters $d = 1$, $k = |\mathrm{supp}(f)|$, $l = 0$, and m = number of clauses. We also apply TSSOS and CS-TSSOS of the first order relaxation to these functions[5].

Numerical results are reported in Table 1, in which "clause" denotes the number of clauses in each CNF formula, "min" is the minimum of the characteristic function and "bound" means the lower bound of the characteristic function obtained by each method. From Table 1, we see that lower bounds obtained by Algorithm 1 are very close to minimum values of characteristic functions, which are better than those obtained by TSSOS and CS-TSSOS.

Table 1. Unweighted MAX-2SAT problems

No	clause	min	Algorithm 1		TSSOS		CS-TSSOS	
			bound	time	bound	time	bound	time
1	1200	161	**159.5**	370	146.7	45	146.7	52
2	1200	159	**156.7**	327	143.1	49	143.1	55
3	1200	160	**159.0**	362	146.8	46	146.8	64
4	1300	180	**177.5**	450	162.4	52	162.4	73
5	1300	172	**170.6**	417	156.2	47	156.2	65
6	1300	173	**171.6**	432	158.8	44	158.8	58
7	1400	197	**194.8**	506	179.8	46	179.8	75
8	1400	191	**189.3**	499	174.3	51	174.3	87
9	1400	189	**187.2**	504	172.1	58	172.1	78

---

[3] To solve SDP problems, we use SDPNAL+ [30] in Algorithm 1 and MOSEK [1] in TSSOS and CS-TSSOS.

[4] http://www.maxsat.udl.cat/16/index.html.

[5] Both of them fail to compute the second and higher order relaxations due to the insufficient memory. In fact, there are $\sum_{j=0}^{k} \binom{120}{k}$ monomials involved in the k-th relaxation. In particular, the size of the Gram matrix in the first relaxation is 121×121 while it is 7261×7261 in the second relaxation.

4.2   Rounding Techniques

We compare the rounding technique presented in Subsect. 3.4 with rounding techniques with scaling factors ρ_i^N and $2^{-(i-1)}$ in [31] on the same benchmark problems used in the previous experiment. For the rounding method in [31], we solve the SDP problem with MOSEK and basis M_p, where M_p is the monomial basis containing M_1 and monomials $z_i z_j$ whenever x_i and x_j appear in the same clause (cf. [31, Definition 1]) and M_1 is the set of monomials with degree at most 1. For the rounding method proposed in Subsect. 3.4, we select basis with input parameters $l = 0$, $d = 2$, $m =$ number of clauses and k is chosen so that the cardinality of the union of M_1 and the basis obtained by steps 1–3 in Algorithm 1, is equal to the cardinality of M_p. The maximum number of iterations in SDPNAL+ is set to be $|M_p|$.

We record results in Table 2, which indicate that our rounding method outperforms the method presented in [31]. Furthermore, rounding techniques in [31] usually take at least 6000 seconds by the interior point method based solvers as conditions (4) and (5) are required to be satisfied. However, our method takes less than 1000 seconds, as (4) and (5) can be violated.

Table 2. rounding on MAX-2SAT benchmarks

No	clause	min	Gram	ρ_i^N	$2^{-(i-1)}$
1	1200	161	**162**	225	227
2	1200	159	**159**	215	194
3	1200	160	**160**	162	**160**
4	1300	180	**180**	226	243
5	1300	172	**173**	225	230
6	1300	173	**173**	245	253
7	1400	197	**198**	234	270
8	1400	191	**192**	255	246
9	1400	189	**189**	227	231

4.3   Lower Bounds of Random Functions

The goal of this experiment is to exhibit the correctness and efficiency of Algorithm 1. We randomly generate nonnegative integer-valued functions on C_2^{25}, C_3^{15} and C_5^{10} and compute their lower bounds by Algorithm 1, TSSOS [35] and CS-TSSOS [34] respectively. Without loss of generality, we only consider functions whose minimum values are 1.

We generate random functions by the following steps. For group C_2^{25}, we generate a polynomial $f = f_0 + \sum_i c_i z^{\alpha_i}$ of degree three and sparsity around 450 on C_2^{25} by randomly picking multilinear monomials z^{α_i} with degree at most

3 and coefficients $c_i \in \{c \in \mathbb{Z} : -5 \leq c \leq 5\}$. The constant term f_0 is chosen so that the minimum values is 1. Clearly, $f \leq m := \sum_i |c_i| + f_0$. For group C_3^{15} and C_5^{10}, we generate functions f with sparsity around 200 on group C_3^{15} or C_5^{10} by the following procedure:

1. Set $f = 0$;
2. Randomly generate an integer-valued function h on C_3^2 (resp. C_5^2), such that $0 \leq h \leq 10$;
3. Randomly pick a projection map $\tau : C_3^{15} \to C_3^2$ (resp. $\tau : C_5^{10} \to C_5^2$);
4. Update $f \leftarrow f + h \circ \tau$;
5. Repeat steps (2.)–(4.) until the sparsity of f is greater than 200.
6. Update $f \leftarrow f - \min_{g \in C_3^{15}} f(g) + 1$ (resp. $f \leftarrow f - \min_{g \in C_5^{10}} f(g) + 1$).

Then clearly $f \leq m :=$ sum of maximum value of h in step (2.).

For each of these functions, we perform Algorithm 1 with parameters $d = 2$, $k = 3 |\operatorname{supp}(f)|$, $l = 0$, and m as discussed earlier. We also apply TSSOS and CS-TSSOS of the second (resp. fourth, eighth) order relaxation on C_2^{25} (resp. C_3^{15}, C_5^{10}) with the term sparsity parameter TS = "MD" to these functions[6].

Results are shown in Table 3, where "sp" means the sparsity of the function and "bound" means the lower bound obtained by the corresponding algorithm. "MO" in the table indicates that the program was terminated due to insufficient memory. We do not test TSSOS on functions on C_3^{15} and C_5^{10} as it is only applicable to functions with real variables and real coefficients[7]. Thus we place "–" in corresponding positions of the table. It is clear from Table 3 that Algorithm 1 outperforms both TSSOS and CS-TSSOS on random examples.

Table 3. Random examples on C_2^{25}, C_3^{15} and C_5^{10}

No	group	sp	Algorithm 1		TSSOS		CS-TSSOS	
			bound	time	bound	time	bound	time
1	C_2^{25}	451	1.00	1058.00	1.00	1027.03	1.00	1451.53
2	C_2^{25}	451	0.67	867.38	−8.72	853.47	−7.23	1483.55
3	C_2^{25}	451	0.75	773.06	1.00	1442.03	1.00	1846.23
4	C_2^{25}	451	0.99	906.15	1.00	1519.08	1.00	1831.33
5	C_2^{25}	451	0.02	718.21	1.00	1364.66	1.00	1710.47
6	C_3^{15}	203	0.97	327.58	–	–	1.00	7336.23
7	C_3^{15}	203	1.00	223.73	–	–	1.00	2876.34
8	C_3^{15}	203	1.00	212.43	–	–	1.00	1353.14
9	C_5^{10}	201	0.97	191.54	–	–	MO	MO
10	C_5^{10}	201	0.94	236.33	–	–	1.00	559.68
11	C_5^{10}	213	0.90	233.82	–	–	MO	MO

---

[6] CS-TSSOS fails to complete the computation if we use lower order relaxations.
[7] $C_n = \{\exp(2k\pi i/n)\}_{k=0}^{n-1} \not\subseteq \mathbb{R}$ if $n \geq 3$.

References

1. ApS, M.: The MOSEK optimization toolbox for MATLAB manual. Version 9.3.11 (2019). https://docs.mosek.com/9.3/toolbox/index.html
2. Christofides, N.: Worst-case analysis of a new heuristic for the travelling salesman problem. Technical report, Carnegie-Mellon Univ Pittsburgh Pa Management Sciences Research Group (1976)
3. Cook, S.A.: The complexity of theorem-proving procedures. In: Proceedings of the Third Annual ACM Symposium on Theory of Computing, pp. 151–158 (1971)
4. Dinur, I., Steurer, D.: Analytical approach to parallel repetition. In: STOC'14–Proceedings of the 2014 ACM Symposium on Theory of Computing, pp. 624–633. ACM, New York (2014)
5. Fawzi, H., Saunderson, J., Parrilo, P.A.: Sparse sums of squares on finite abelian groups and improved semidefinite lifts. Math. Program. **160**(1–2), 149–191 (2016)
6. Feige, U., Goemans, M.: Approximating the value of two power proof systems, with applications to MAX 2SAT and MAX DICUT. In: Proceedings Third Israel Symposium on the Theory of Computing and Systems, pp. 182–189 (1995)
7. Fulton, W., Harris, J.: Representation Theory: A First Course, vol. 129. Springer Science & Business Media, New York (2013). https://doi.org/10.1007/978-1-4612-0979-9
8. Goemans, M.X., Williamson, D.P.: Improved approximation algorithms for maximum cut and satisfiability problems using semidefinite programming. J. ACM (JACM) **42**(6), 1115–1145 (1995)
9. Goemans, M.X., Williamson, D.P.: Improved approximation algorithms for maximum cut and satisfiability problems using semidefinite programming. J. Assoc. Comput. Mach. **42**(6), 1115–1145 (1995)
10. Karloff, H., Zwick, U.: A 7/8-approximation algorithm for MAX 3SAT? In: Proceedings 38th Annual Symposium on Foundations of Computer Science, pp. 406–415 (1997)
11. Karp, R.M.: Reducibility among combinatorial problems. In: Miller, R.E., Thatcher, J.W., Bohlinger, J.D. (eds.) Complexity of Computer Computations, pp. 85–103. The IBM Research Symposia Series. Springer, Boston, MA (1972). https://doi.org/10.1007/978-1-4684-2001-2_9
12. Khot, S., Kindler, G., Mossel, E., O'Donnell, R.: Optimal inapproximability results for MAX-CUT and other 2-variable CSPs? SIAM J. Comput. **37**(1), 319–357 (2007)
13. Korte, B.H., Vygen, J., Korte, B., Vygen, J.: Combinatorial Optimization, vol. 1. Springer, Berlin, Heidelberg (2011). https://doi.org/10.1007/978-3-642-24488-9
14. Krentel, M.W.: The complexity of optimization problems. J. Comput. Syst. Sci. **36**(3), 490–509 (1988)
15. Lasserre, J.B.: Global optimization with polynomials and the problem of moments. SIAM J. Optim. **11**(3), 796–817 (2001)
16. Lasserre, J.B.: A max-cut formulation of 0/1 programs. Oper. Res. Lett. **44**(2), 158–164 (2016)
17. Laurent, M.: A comparison of the sherali-adams, lovász-schrijver, and lasserre relaxations for 0–1 programming. Math. Oper. Res. **28**(3), 470–496 (2003)
18. Lewin, M., Livnat, D., Zwick, U.: Improved rounding techniques for the MAX 2-SAT and MAX DI-CUT problems. In: Cook, W.J., Schulz, A.S. (eds.) IPCO 2002. LNCS, vol. 2337, pp. 67–82. Springer, Heidelberg (2002). https://doi.org/10.1007/3-540-47867-1_6

19. Martello, S., Toth, P.: Knapsack Problems: Algorithms and Computer Implementations. John Wiley & Sons, Inc., Hoboken (1990)
20. Mathews, G.B.: On the partition of numbers. Proc. Lond. Math. Soc. $1(1)$, 486–490 (1896)
21. O'Donnell, R.: Analysis of Boolean functions. Cambridge University Press, New York (2014)
22. Papadimitriou, C.H.: The Euclidean traveling salesman problem is NP-complete. Theor. Comput. Sci. $4(3)$, 237–244 (1977)
23. Papadimitriou, C.H.: Computational Complexity. Addison-Wesley Publishing Company, Reading, Boston, MA (1994)
24. Py, M., Cherif, M.S., Habet, D.: A proof builder for Max-SAT. In: Li, C.-M., Manyà, F. (eds.) SAT 2021. LNCS, vol. 12831, pp. 488–498. Springer, Cham (2021). https://doi.org/10.1007/978-3-030-80223-3_33
25. Raghavendra, P.: Optimal algorithms and inapproximability results for every CSP? [extended abstract]. In: STOC'08, pp. 245–254. ACM, New York (2008)
26. Rudin, W.: Fourier Analysis on Groups, vol. 121967. Wiley Online Library, Hoboken (1962)
27. Sakaue, S., Takeda, A., Kim, S., Ito, N.: Exact semidefinite programming relaxations with truncated moment matrix for binary polynomial optimization problems. SIAM J. Optim. $27(1)$, 565–582 (2017)
28. Slavík, P.: A tight analysis of the greedy algorithm for set cover. In: Proceedings of the Twenty-Eighth Annual ACM Symposium on the Theory of Computing (Philadelphia, PA, 1996), pp. 435–441. ACM, New York (1996)
29. Slot, L., Laurent, M.: Sum-of-squares hierarchies for binary polynomial optimization. Math. Program. $197(2)$, 621–660 (2023)
30. Sun, D., Toh, K.C., Yuan, Y., Zhao, X.Y.: SDPNAL+: a Matlab software for semidefinite programming with bound constraints (version 1.0). Optim. Methods Softw. $35(1)$, 87–115 (2020)
31. van Maaren, H., van Norden, L., Heule, M.: Sums of squares based approximation algorithms for max-sat. Discret. Appl. Math. $156(10)$, 1754–1779 (2008)
32. Vazirani, V.V.: Approximation Algorithms, vol. 1. Springer, Berlin, Heidelberg (2001). https://doi.org/10.1007/978-3-662-04565-7
33. Waki, H., Kim, S., Kojima, M., Muramatsu, M.: Sums of squares and semidefinite program relaxations for polynomial optimization problems with structured sparsity. SIAM J. Optim. $17(1)$, 218–242 (2006)
34. Wang, J., Magron, V., Lasserre, J.B., Mai, N.H.A.: CS-TSSOS: correlative and term sparsity for large-scale polynomial optimization. ACM Trans. Math. Softw. $48(4)$, 1–26 (2022)
35. Wang, J., Magron, V., Lasserre, J.B.: TSSOS: a moment-SOS hierarchy that exploits term sparsity. SIAM J. Optim. $31(1)$, 30–58 (2021)
36. Wang, P.W., Kolter, J.Z.: Low-rank semidefinite programming for the MAX2SAT problem. In: Proceedings of the AAAI Conference on Artificial Intelligence, vol. 33, pp. 1641–1649 (2019)
37. Yang, J., Ye, K., Zhi, L.: Lower bounds of functions on finite abelian groups (2023)
38. Yang, J., Ye, K., Zhi, L.: Computing sparse Fourier sum of squares on finite abelian groups in quasi-linear time. arXiv preprint arXiv:2201.03912 (2022)
39. Yang, J., Ye, K., Zhi, L.: Short certificates for MAX-SAT via Fourier sum of squares. arXiv preprint arXiv:2207.08076 (2022)
40. Zhang, R.Y., Lavaei, J.: Sparse semidefinite programs with guaranteed near-linear time complexity via dualized clique tree conversion. Math. Program. $188(1)$, 351–393 (2021)

A Discharging Method: Improved Kernels for Edge Triangle Packing and Covering

Zimo Sheng and Mingyu Xiao$^{(\boxtimes)}$ (iD)

School of Computer Science and Engineering, University of Electronic Science and Technology of China, Chengdu, China
`myxiao@uestc.edu.cn`

Abstract. EDGE TRIANGLE PACKING and EDGE TRIANGLE COVERING are dual problems extensively studied in the field of parameterized complexity. Given a graph G and an integer k, EDGE TRIANGLE PACKING seeks to determine whether there exists a set of at least k edge-disjoint triangles in G, while EDGE TRIANGLE COVERING aims to find out whether there exists a set of at most k edges that intersects all triangles in G. Previous research has shown that EDGE TRIANGLE PACKING has a kernel of $(3 + \epsilon)k$ vertices, while EDGE TRIANGLE COVERING has a kernel of $6k$ vertices. In this paper, we show that the two problems allow kernels of $3k$ vertices, improving all previous results. A significant contribution of our work is the utilization of a novel discharging method for analyzing kernel size, which exhibits potential for analyzing other kernel algorithms.

1 Introduction

Preprocessing is a fundamental and commonly used step in various algorithms. However, most preprocessing has no theoretical guarantee on the quality. Kernelization, originating from the field of parameterized algorithms [1], now has been found to be an interesting way to analyze the quality of preprocessing. Consequently, kernelization has received extensive attention in both theoretical and practical studies.

Given an instance (I, k) of a problem, a kernelization (or a kernel algorithm) runs in polynomial time and returns an equivalent instance (I', k') of the same problem such that (I, k) is a yes-instance if and only if (I', k') is a yes-instance, where $k' \leq k$ and $|I'| \leq g(k)$ for some computable function g only of k. The new instance (I', k') is called a *kernel* and $g(k)$ is the size of the kernel. If g($\cdot$) is a polynomial or linear function, we classify the problem as having a polynomial or linear kernel, respectively.

EDGE TRIANGLE PACKING (ETP), to check the existence of k edge-disjoint triangles in a given graph G is NP-hard even on planar graphs with maximum degree 5 [2]. The optimization version of this problem is APX-hard on general graphs [3]. A general result of [4] leads to a polynomial-time $(3/2+\epsilon)$ approximation algorithm for any constant $\epsilon > 0$. When the graphs are restricted to planar graphs, the result can be improved. A polynomial-time approximation scheme for the vertex-disjoint triangle packing problem on planar graphs was given by [5], which can be extended to ETP on planar graphs. In terms of parameterized complexity, a $4k$-vertex kernel and an $O^*(2^{\frac{9k}{2}\log k+\frac{9k}{2}})$-time parameterized algorithm for ETP were developed in [6]. Later, the size of the kernel was improved to $3.5k$ [7]. The current best-known result is $(3+\epsilon)k$ [8], where $\epsilon > 0$ can be any positive constant. On tournaments, there is also a kernel of $3.5k$ vertices [9].

Another problem considered in this paper is EDGE TRIANGLE COVERING (ETC). ETC is the dual problem of ETP, which is to check whether we can delete at most k edges from a given graph such that the remaining graph has no triangle. ETC is also NP-hard even on planar graphs with maximum degree 7 [10]. In terms of kernelization, a $6k$-vertex kernel for ETC was developed [10]. On planar graphs, the result was further improved to $\frac{11k}{3}$ [10].

In this paper, we will deeply study the structural properties of EDGE TRIANGLE PACKING and EDGE TRIANGLE COVERING and give some new reduction rules by using a variant of crown decomposition. After that, we will introduce a new technology called the discharging method to analyze the size of problem kernels. Utilizing the new discharging method, we obtain improved kernel sizes of $3k$ vertices for both ETP and ETC. Notably, our results even surpass the previously best-known kernel size for ETC on planar graphs [10]. Due to the page limitation, proofs of lemmas and theorems marked with '*' are omitted, which can be found in the full version of this paper.

2   Preliminaries

Let $G = (V, E)$ denote a simple and undirected graph with $n = |V|$ vertices and $m = |E|$ edges. A vertex is a *neighbor* of another vertex if there is an edge between them. The set of neighbors of a vertex v is denoted by $N(v)$, and the degree of v is defined as $d(v) = |N(v)|$. For a vertex subset $V' \subseteq V$, we let $N(V') = \cup_{v \in V'} N(v) \setminus V'$ and $N[V'] = N(V') \cup V'$. The subgraph induced by a vertex subset $V' \subseteq V$ is denoted by $G[V']$ and the subgraph spanned by an edge set $E' \subseteq E$ is denoted by $G[E']$. The vertex set and edge set of a graph H are denoted by $V(H)$ and $E(H)$, respectively.

A complete graph on 3 vertices is called a *triangle*. We will use vuw to denote the triangle formed by vertices v, u, and w. If there is a triangle vuw in G, we say that vertex v *spans* edge uw. An *edge triangle packing* in a graph is a set of triangles such that every two triangles in it have no common edge. The EDGE TRIANGLE PACKING problem (ETP) is defined below.

EDGE TRIANGLE PACKING (ETP)   **Parameter:** k
Input: An undirected graph $G = (V, E)$, and an integer k.
Question: Does there exist an edge triangle packing of size at least k in G?

An edge *covers* a triangle if it is contained in the triangle. An *edge triangle covering* in a graph is a set of edges S such that there is no triangle after deleting S from G. The EDGE TRIANGLE COVERING problem (ETC) is defined below.

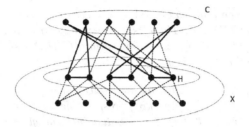

Fig. 1. An illustration for the fat-head crown decomposition

EDGE TRIANGLE COVERING (ETC)   **Parameter:** k
Input: An undirected graph $G = (V, E)$, and an integer k.
Question: Does there exist an edge triangle covering of size at most k in G?

---

3   Fat-Head Crown Decomposition

One important technique in this paper is based on a variant of the crown decomposition. Crown decomposition is a powerful technique for the famous VERTEX COVER problem and it has been extended to solve several related problems [11–14]. Specifically, we employ a specific variant called the fat-head crown decomposition to tackle (ETP) [8]. This variant of the crown decomposition will also be applied in our algorithms for both ETP and ETC. To provide a comprehensive understanding, let us begin by introducing the definition of the fat-head crown decomposition.

A *fat-head crown decomposition* of a graph $G = (V, E)$ is a triple (C, H, X) such that C and X form a partition of V and $H \subseteq E$ is a subset of edges satisfying the following properties:

1. C is an independent set.
2. H is the set of edges spanned by at least one vertex in C.
3. No vertex in C is adjacent to a vertex in $X \setminus V(H)$.
4. There is an edge-disjoint triangle packing P of size $|P| = |H|$ such that each triangle in P contains exactly one vertex in C and exactly one edge in H. The packing P is also called the *witness packing* of the fat-head crown decomposition.

An illustration of the fat-head crown decomposition is shown in Fig. 1. To determine the existence of fat-head crown decompositions in a given graph structure, we present three lemmas.

Lemma 1 (Lemma 2 in [8]). *Let $G = (V, E)$ be a graph such that each edge and each vertex is contained in at least one triangle. Given a non-empty independent set $I \subseteq V$ such that $|I| > |S(I)|$, where $S(I)$ is the set of edges spanned by at least one vertex in I. A fat-head crown decomposition (C, H, X) of G with $C \subseteq I$ and $H \subseteq S(I)$ together with a witness packing P of size $|P| = |H| > 0$ can be found in polynomial time.*

Lemma 2 (*). *Given a graph $G = (V, E)$, a vertex set $A \subseteq V$, and an edge set $B \subseteq E$, where $A \cap V(B) = \emptyset$. There is a polynomial-time algorithm that checks whether there is a fat-head crown decomposition (C, H, X) such that $\emptyset \neq C \subseteq A$ and $H \subseteq B$ and outputs one if yes.*

Lemma 3 (*). *If there is a fat-head crown decomposition (C, H, X) in G, then G has an edge-disjoint triangle packing (resp., edge triangle covering) of size k if and only if the graph G' has an edge-disjoint triangle packing (resp., edge triangle covering) of size $k - |H|$, where G' is the graph obtained from G by deleting vertex set C and deleting edge set H.*

4   The Algorithms

In this section, we present our kernelization algorithms for the EDGE TRIANGLE PACKING (ETP) and EDGE TRIANGLE COVERING (ETC) problems. Our algorithms involve a set of reduction rules that are applied iteratively until no further reduction is possible. Each reduction rule is applied under the assumption that all previous reduction rules have already been applied and cannot be further applied to the current instance. A reduction rule is *correct* if the original instance (G, k) is a yes-instance if and only if the resulting instance (G', k') after applying the reduction rule is a yes-instance.

We have one algorithm for ETP and ETC, respectively. The two algorithms are similar. We will mainly describe the algorithm for ETP and introduce the difference for ETC. In total, we have nine reduction rules. The first four rules are simple rules to handle some special structures, while the remaining five rules are based on a triangle packing. Especially, the last rule will use the fat-head crown decomposition. We will show that the algorithms run in polynomial time.

4.1   Simple Rules

Reduction Rule 1. *For ETP, if $k \leq 0$, then return 'yes' to indicate that the instance is a yes-instance; if $k > 0$ and the graph is empty, then return 'no' to indicate that the instance is a no-instance.*
For ETC, if $k \geq 0$ and the graph is empty, then return 'yes' to indicate that the instance is a yes-instance; if $k < 0$, then return 'no' to indicate that the instance is a no-instance.

Reduction Rule 2. *If there is a vertex or an edge not appearing in any triangle, then delete it from the graph.*

Reduction Rule 3. *If there are 4 vertices $u, v, w, x \in V$ inducing a complete graph (i.e., there are 6 edges $uv, uw, ux, vw, vx, wx \in E$) such that none of the 6 edges is in a triangle except $uwv, uvx, uwx,$ and vwx, then*

- *For ETP, delete the 6 edges uv, uw, ux, vw, vx and wx and let $k = k - 1$;*
- *For ETC, delete the 6 edges uv, uw, ux, vw, vx and wx and let $k = k - 2$.*

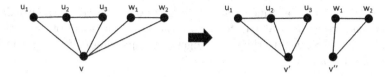

Fig. 2. An Illustration for Reduction Rule 4

The correctness of Reduction Rule 3 is based on the following observation. For ETP, any edge triangle packing can have at most one triangle containing some edge from these 6 edges and we can simply take one triangle from this local structure. For ETC, any edge triangle covering must contain at least two edges from these 6 edges and after deleting vu and wx, none of uw, ux, vw, and vx is contained in a triangle anymore.

Reduction Rule 4. *If there is a vertex $v \in V$ such that all edges incident to v can be partitioned into two parts E_1 and E_2 such no triangle in G contains an edge in E_1 and an edge in E_2, then split v into two vertices v' and v'' such that all edges in E_1 are incident on v' and all edges in E_2 are incident on v''.*

An illustration of Reduction Rule 4 is shown in Fig. 2. This reduction rule will increase the number of vertices in the graph. However, this operation will simplify the graph structure and our analysis.

Lemma 4. *Reduction Rule 4 is correct and can be executed in polynomial time.*

Proof. First, we consider the correctness. Let $G' = (V', E')$ be the graph after applying Reduction Rule 4 on a vertex v. We can establish a one-to-one mapping between the edges in E and the edges in E' by considering the vertices v' and $v'' \in V'$ as $v \in V$. Three edges in E form a triangle in G if and only if the three corresponding edges in E' form a triangle in G' since there is no triangle in G contains an edge in E_1 and an edge in E_2. Thus, an edge triangle packing of size k (resp., an edge triangle covering of size k) in G is also an edge triangle packing of size k (resp., an edge triangle covering of size k) in G'. This implies that Reduction Rule 4 is correct for both ETP and ETC.

We give a simple greedy algorithm to find the edge sets E_1 and E_2 for a given vertex v. Initially, let E_1 contain an arbitrary edge e incident on v. We iteratively perform the following steps until no further updates occur: if there is a triangle containing an edge in E_1 and an edge e' incident on v but not in E_1, then add edge e' to E_1. It is easy to see that all edges in E_1 must be in the same part to satisfy the requirement. If $E_1 \neq E$, then we can split E to two parts E_1 and $E_2 = E \setminus E_1$. Otherwise, the edges incident on v cannot be split. □

4.2   Adjustments Based on a Triangle Packing

After applying the first four rules, our algorithms will find a maximal edge-disjoint triangle packing S by using an arbitrary greedy method. This can be done easily in polynomial time. The following rules are based on the packing S. From now on, we let $F = V \setminus V(S)$ denote the set of vertices not appearing in S and $R = E \setminus E(S)$ denote the set of edges not appearing in S. We begin with the following trivial rule.

Reduction Rule 5. *If $|S| > k$, for ETP, return 'yes' to indicate that the instance is a yes-instance; and for ETC, return 'no' to indicate that the instance is a no-instance.*

The following three rules just update the packing S by replacing some triangles in it and do not change the graph. Illustrations of the three rules are shown in Fig. 3.

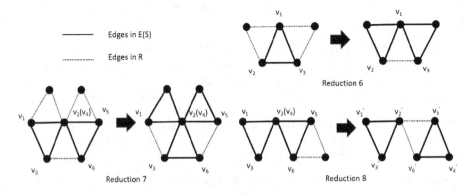

Fig. 3. Illustrations of Reduction Rules 6–8

Reduction Rule 6. *If there is a triangle $v_1v_2v_3 \in S$ such that there are at least two edge-disjoint triangles in the spanned graph $G[R \cup \{v_1v_2, v_1v_3, v_2v_3\}]$, then replace $v_1v_2v_3$ with these triangles in S to increase the size of S by at least 1.*

Reduction Rule 7. *If there are two edge-disjoint triangles $v_1v_2v_3$ and $v_4v_5v_6 \in S$ such that there are at least three edge-disjoint triangles in the spanned graph $G[R \cup \{v_1v_2, v_1v_3, v_2v_3, v_4v_5, v_4v_6, v_5v_6\}]$, then replace $v_1v_2v_3$ and $v_4v_5v_6$ with these triangles in S to increase the size of S by at least 1.*

Reduction Rule 8. *If there are two edge-disjoint triangles $v_1v_2v_3$ and $v_4v_5v_6 \in S$ such that there are two edge-disjoint triangles $v_1'v_2'v_3'$ and $v_4'v_5'v_6'$ in the induced graph $G[F \cup \{v_1, v_2, v_3, v_4, v_5, v_6\}]$ such that $|\{v_1', v_2', v_3'\} \cup \{v_4', v_5', v_6'\}| > |\{v_1, v_2, v_3\} \cup \{v_4, v_5, v_6\}|$, then replace triangles $v_1v_2v_3$ and $v_4v_5v_6$ with triangles $v_1'v_2'v_3'$ and $v_4'v_5'v_6'$ in S to increase the number of vertices appearing in S by at least 1.*

Note that an application of Reduction Rules 6–8 will not change the structure of the graph. Thus, the first four reduction rules will not be applied after executing Reduction Rules 6–8.

4.3   A Reduction Based on Fat-Head Crown Decomposition

After Reduction Rule 8, we obtain the current triangle packing S. An edge in $E(S)$ is called a *labeled edge* if it is spanned by at least one vertex in F. We let L denote the set of labeled edges.

We can find a fat-head crown decomposition (C, H, X) with $C \subseteq V \setminus V(L)$ and $H \subseteq L$ in polynomial time if it exists by Lemma 2. Moreover, we will apply the following reduction rule to reduce the graph, the correctness of which is based on Lemma 3.

Reduction Rule 9. *Use the algorithm in Lemma 2 to check whether there is a fat-head crown decomposition (C, H, X) such that $\emptyset \neq C \subseteq V \setminus V(L)$ and $H \subseteq L$. If yes, then delete vertex set C and edge set H, and let $k = k - |H|$.*

An instance is called *reduced* if none of the nine reduction rules can be applied to it. The corresponding graph is also called a reduced graph.

Lemma 5 (*).   *For any input instance, the kernelization algorithms run in polynomial time to output a reduced instance.*

5   Analysis Based on Discharging

Next, we use a discharging method to analyze the size of a reduced instance. Note that there is no significant difference between ETC and ETP in the analysis. We partition the graph into two parts: one part is the edge-disjoint triangle packing S after applying all the reductions; the other part is the set F of vertices not appearing in S. Before proceeding with the analysis, we will establish some properties that will be utilized.

Lemma 6. *Consider a reduced graph $G = (V, E)$ with triangle packing S. For any triangle $uvw \in S$, at most one of $\{uv, vw, uw\}$ is a labeled edge.*

Proof. Assume to the contrary that there are two edges, say uv and vw are spanned by vertices in F. We show some contradiction.

If edges uv and vw are spanned by two different vertices $x, x' \in F$ respectively, then Reduction Rule 6 could be applied (Case 1 in Fig. 4). Therefore, edges uv and vw are spanned by the same vertex $x \in F$. Since Reduction Rule 3 is not applied on the four vertices $\{u, v, w, x\}$, we know that at least one edge in $\{uw, uw, ux, vw, vx, wx\}$ is contained in a triangle other than uwv, uvx, uwx, and vwx. Due to symmetry, we only need to consider two edges vw and xw.

Assume that edge vw is contained in a triangle vwy, where $y \notin \{u, x\}$. If none of $\{yv, yw\}$ appears in $E(S)$, then Reduction Rule 6 could be applied to replace uvw with two triangles xvu and yvw in S. If at least one edge in $\{yv, yw\}$ is contained in $E(S)$, without loss of generality, assume $yw \in E(S)$ and there is a triangle $ywz \in S$. For this case, Reduction Rule 8 could be applied to replace vuw and ywz with xvu and ywz (Case 2 in Fig. 4).

Assume that edge xw is contained in a triangle xwy, where $y \notin \{u, v\}$. By the maximality of S, we know that at least one of $\{xy, yw\}$ must appear in $E(S)$.

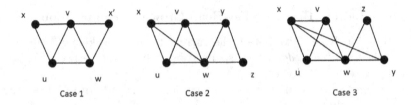

Fig. 4. Three cases in Lemma 6

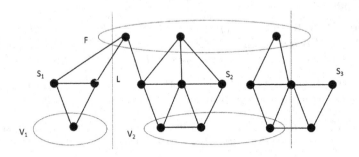

Fig. 5. An illustration for triangles and vertices in G

However, edge xy can not appear in $E(S)$ since $x \in F$. We know that $wy \in E(S)$ and there is a triangle $wyz \in S$. For this case, Reduction Rule 8 could be applied to replace vuw and ywz with xvu and wyz (Case 3 in Fig. 4).

In any of these cases, we can find a contradiction to the fact that the graph is reduced.   □

A triangle $uvw \in S$ is *good* if it contains a labeled edge and *bad* otherwise. By Lemma 6, we know that there is exactly one labeled edge in each good triangle. We let G' be the graph obtained by deleting the set L of labeled edges from G. Consider a good triangle uvw with labeled edge uv. If the two edges vw and wu are not in any triangle in G', we call the triangle *excellent*. Otherwise, we call the triangle *pretty-good*. We let S_1 denote the set of excellent triangles, S_2 denote the set of pretty-good triangles, and S_3 denotes the set of bad triangles in S. The number of triangles in S_1, S_2 and S_3 are denoted by k_1, k_2, and k_3, respectively. Let $V_1 = V(S_1) \setminus (V(L) \cup V(S_2) \cup V(S_3))$ and $V_2 = V(S_2) \setminus V(L)$. See Fig. 5 for an illustration of these concepts.

5.1   The Analysis

The discharging method stands as a renowned technique in graph theory, finding its most notable application in the proof of the famous Four Color Theorem. In this section, we will use the discharging method to analyze the number of vertices present in S_1, S_2, S_3, and F. The idea of the method is as follows.

First, we initially assign some integer values to vertices, edges, and triangles in S. The total value assigned is at most $3k$. Subsequently, we perform steps

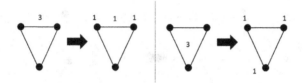

Fig. 6. An illustration for Step 1

to update the values, where certain values on vertices, edges, and triangles are transformed into other vertices, edges, and triangles. In these steps, we never change the structure of the graph and the total value in the graph. After performing these transformations, we demonstrate that each vertex in the graph has a value of at least 1. Consequently, we conclude that the number of vertices in the graph is at most $3k$.

Initialization: Assign value 3 to each edge in L and each triangle in S_3. Edges not in L, vertices, and triangles in $S_1 \cup S_2$ are assigned a value of 0.

By Lemma 6, we know that each of excellent and pretty-good triangles contains exactly one labeled edge in L and each bad triangle in S_3 contains no labeled edge. Thus, the total value in the graph is $3k_1 + 3k_2 + 3k_3 \leq 3k$.

Step 1: For each labeled edge in L, transform a value of 1 to each of its two endpoints; for each triangle in S_3, transform a value of 1 to each of its three vertices.

Figure 6 illustrates the transformation process of Step 1. After Step 1, each labeled edge has a value of 1, and all triangles have values of 0. Note that some vertices may have a value of 2 or more, as they may serve as endpoints of multiple labeled edges and can also be vertices in $V(S_3)$. However, vertices in $F \cup V_1 \cup V_2$ still retain a value of 0.

A *triangle component* is a connected component in the graph $H = (V(S), E(S))$. For a vertex $v \in V(S)$, we let $C(v)$ denote the set of vertices in the triangle component which contain v.

Step 2: For each triangle component in G, we iteratively transform a value of 1 from a vertex with a value of at least 2 to a vertex with a value of 0 in the same triangle component, where vertices in V_1 have a higher priority to get the value.

Lemma 7. *After Step 2, each triangle component has at most one vertex with a value of 0. Moreover,*

(i) *For any triangle component containing a triangle in S_3, each vertex in the triangle component has a value of at least 1;*
(ii) *For any triangle component containing at least one triangle in S_2, if there is a vertex with a value of 0 in the triangle component, then the vertex must be a vertex in V_2.*

Proof. Let Q be a triangle component with x triangles. Since Q is connected, it contains at most $2x + 1$ vertices. Assume that among the x triangles, there

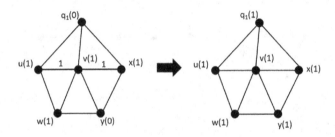

Fig. 7. An illustration for Step 3, where the number in parentheses next to each vertex represents the value of that vertex

are x_1 triangles in $S_1 \cup S_2$ and x_2 triangles in S_3, where $x_1 + x_2 = x$. By the definition, we know that each triangle in $S_1 \cup S_2$ contains a distinct labeled edge. According to the initialization of the assignment, we know that the total value is $2x_1 + 3x_2 = 2x + x_2$. It always holds that $2x + 1 \leq (2x + x_2) + 1$, and $2x + 1 \leq 2x + x_2$ when $x_2 \geq 1$. Thus, Q has at most one vertex with a value of 0. When Q contains some triangles from S_3, i.e., $x_2 \geq 1$, all vertices in Q will get a value of at least 1. The statement (ii) holds because vertices in V_1 have a higher priority to receive the value in Step 2.   □

After Step 2, only vertices in F, some vertices in V_1, and some vertices in V_2 have values of 0. We use the following lemma to transform some values to vertices in V_2 with a value of 0.

Lemma 8 (*). *Consider two triangles vuw and $vxy \in S_2$ sharing a common vertex v, where uv and $vx \in L$. If there is an edge $wy \in E$, then uv and vx are spanned by exactly one vertex in $F \cup V_1 \setminus C(v)$.*

Step 3: If there are two triangles vuw and $vxy \in S_2$ sharing a common vertex v such that uv and vx are two labeled edges in L and there is an edge $wy \in E$, we transform a value of 1 from edge vx to the unique vertex $q_1 \in F$ spanning vx and transform a value of 1 from edge uv to the vertex with a value of 0 in $C(v)$ if this vertex exists.

See Fig. 7 for an illustration of Step 3. We have the following property.

Lemma 9. *Every vertex in V_2 has a value of at least 1 after Step 3.*

Proof. Assume to the contrary there is a vertex $w \in V_2$ with a value of 0. We know that all vertices in $C(w) \setminus \{w\}$ have a value of at least 1 by Lemma 7. Let $wuv \in S_2$ be the triangle containing w, where uv is the labeled edge spanning by a vertex $q \in F$. As shown in Fig. 8. At least one of uw and vw is in a triangle in graph $G - L$ by the definition of S_2. Without loss of generality, we assume that uw is contained in a triangle uwx, where $x \neq v$. At least one of ux and wx is contained in a triangle in S otherwise Reduction Rule 6 could be applied.
Case 1: Edge ux is contained in a triangle $uxy \in S$. See Case 1 in Fig. 8. The triangle uxy is not in S_3 otherwise there is a contradiction that w would have

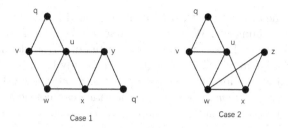

Fig. 8. An illustration for Lemma 9

a value of at least 1 by Lemma 7. Thus, triangle uxy must be in $S_1 \cup S_2$ and there is exactly one of ux, uy, and xy is a labeled edge. Edge ux would not be a labeled edge since triangle uwx is contained in $G - L$. If xy is the labeled edge that is spanned by a vertex $q' \in F$, then Reduction Rule 8 could be applied to replace triangles uvw and uxy with triangles uvw and xyq', a contradiction to the factor that the graph is reduced. If uy is the labeled edge, then triangle $uxy \in S_2$ since ux is contained in a triangle $uxw \in G - L$. For this case, the vertex w would have a value of at least 1 by Lemma 7. We can always find a contradiction.

Case 2: Edge wx is contained in a triangle $wxz \in S$. See Case 2 in Fig. 8. If $z \neq q$, then Reduction Rule 8 could be applied to replace triangles uvw and wxz with triangles wxz and qvu, a contradiction to the factor that the graph is reduced. If $z = q$, then at least two edges in triangle vuw are spanned by vertices in F, a contradiction to Lemma 6.

In either case, a contradiction is reached, which implies that the assumption of a vertex $w \in V_2$ having a value of 0 is incorrect. Therefore, every vertex in V_2 has a value of at least 1 after Step 3.   □

After Step 3, all vertices with a value of 0 are in either F or V_1. We let V_1' denote the set of vertices with a value of 0 in V_1, F' denote the set of vertices with a value of 0 in F, and L' denote the set of edges with a value of 1 in L after Step 3. We give more properties.

Lemma 10. *Set $F' \cup V_1'$ is an independent set.*

Proof. We prove that $F \cup V_1$ is an independent set, which implies $F' \cup V_1'$ is an independent set since $F' \cup V_1' \subseteq F \cup V_1$. Assume to the contrary that there is an edge uv between two vertices in $F \cup V_1$. There is at least one triangle $uvw \in G$ containing uv since Reduction Rule 2 has been applied. At least one of uv, vw, and uw must be in $E(S)$ by the maximality of S.

If $uv \in E(S)$, we let uvx be the triangle in S containing uv. First, we know that u and $v \in V_1$ since $F \cap V(S) = \emptyset$. By the definition of V_1, we get that none of u and v is contained in a triangle in S_3 and none of u and v is an endpoint of a labeled edge. Thus, triangle uvx is not a triangle in S_3 and it does not contain any labeled edge and then it is not a triangle in $S_1 \cup S_2$, which implies triangle uvx is not in S, a contradiction.

Otherwise, one of uw and vw, say uw, is contained in $E(S)$. Let uwy be the triangle in S containing uw. We also have that $u \in V_1$ since $F \cap V(S) = \emptyset$. By the definition of V_1, we know that u is not a vertex in a triangle in S_3, and then uwy is not a triangle in S_3. Thus, triangle uwy can only be in $S_1 \cup S_2$. Note that none of uv, vw and uw can be a labeled edge since u and $v \in F \cup V_1$. Thus, edge uw is still in a triangle uvw in $G - L$, and then uwy can not be a triangle in S_1. However, triangle uvw can not be a triangle in S_2 too since u is a vertex in V_1. We also get a contradiction that triangle uvw is not in S.

Hence, we have shown that no edge exists between any two vertices in $F' \cup V_1'$, which proves that $F' \cup V_1'$ forms an independent set. □

Lemma 11 (*). *Vertices in F' only span edges in L'.*

Lemma 12 (*). *Vertices in V_1' only span edges in L'.*

Lemma 13. *After Step 3, it holds that $|F' \cup V_1'| \leq |L'|$.*

Proof. By Lemma 10, 11, and 12, we know that $F' \cup V_1'$ is an independent set and any vertex $v' \in F' \cup V_1'$ only span edges in L'. If $|F' \cup V_1'| > |L'|$, then by Lemma 1 there is a fat-head crown decomposition (C, H, X) of G with $C \subseteq F' \cup V_1'$ and $H \subseteq L'$. Moreover, the fat-head crown decomposition can be detected by Lemma 2 and will be handled by Reduction Rule 9 since $F' \cup V_1' \subseteq F \cup V_1 \subseteq V \setminus V(L)$ and $L' \subseteq L$. Thus, we know the lemma holds. □

Step 4: We transform value from edges in L' to vertices in $F' \cup V_1'$ such that each vertex in $F' \cup V_1'$ gets value at least 1 by Lemma 13.

After Step 4, each vertex in G has a value of at least 1. Since the total value in G is at most $3k$, we can conclude that the graph has at most $3k$ vertices.

Theorem 1. EDGE TRIANGLE PACKING *and* EDGE TRIANGLE COVERING *admit a kernel of at most $3k$ vertices.*

6   Conclusion

In this paper, we present simultaneous improvements in the kernel results for both EDGE TRIANGLE PACKING and EDGE TRIANGLE COVERING. Our approach incorporates two key techniques to achieve these enhancements. The first technique involves utilizing fat-head crown decomposition, which enables us to effectively reduce various graph structures. By applying this technique, we can simplify the problem instances. The second technique we introduce is the discharging method, which plays a crucial role in analyzing kernel size. This method is simple and intuitive, and we believe it has the potential to be applied to the analysis of other kernel algorithms.

Acknowledgments. The work is supported by the National Natural Science Foundation of China, under the grants 62372095 and 61972070.

References

1. Cygan, M., et al.: Parameterized Algorithms. Springer, Cham (2015). https://doi.org/10.1007/978-3-319-21275-3
2. Holyer, I.: The NP-completeness of some edge-partition problems. SIAM J. Comput. **10**(4), 713–717 (1981)
3. Kann, V.: Maximum bounded h-matching is MAX SNP-complete. Inf. Process. Lett. **49**(6), 309–318 (1994)
4. Hurkens, C.A.J., Schrijver, A.: On the size of systems of sets every t of which have an SDR, with an application to the worst-case ratio of heuristics for packing problems. SIAM J. Discret. Math. **2**(1), 68–72 (1989)
5. Baker, B.S.: Approximation algorithms for np-complete problems on planar graphs. J. ACM **41**(1), 153–180 (1994)
6. Mathieson, L., Prieto, E., Shaw, P.: Packing edge disjoint triangles: a parameterized view. In: Downey, R., Fellows, M., Dehne, F. (eds.) IWPEC 2004. LNCS, vol. 3162, pp. 127–137. Springer, Heidelberg (2004). https://doi.org/10.1007/978-3-540-28639-4_12
7. Yang, Y.: Towards optimal kernel for edge-disjoint triangle packing. Inf. Process. Lett. **114**(7), 344–348 (2014)
8. Lin, W., Xiao, M.: A $(3+\epsilon)k$-vertex kernel for edge-disjoint triangle packing. Inf. Process. Lett. **142**, 20–26 (2019)
9. Yuan, H., Feng, Q., Wang, J.: Improved kernels for triangle packing in tournaments. Sci. China Inf. Sci. **66**(5), 152104 (2023). https://doi.org/10.1007/s11432-021-3551-2
10. Brügmann, D., Komusiewicz, C., Moser, H.: On generating triangle-free graphs. Electron. Notes Discret. Math. **32**, 51–58 (2009)
11. Dehne, F., Fellows, M., Rosamond, F., Shaw, P.: Greedy localization, iterative compression, and modeled crown reductions: new FPT techniques, an improved algorithm for SET SPLITTING, and a novel $2k$ kernelization for VERTEX COVER. In: Downey, R., Fellows, M., Dehne, F. (eds.) IWPEC 2004. LNCS, vol. 3162, pp. 271–280. Springer, Heidelberg (2004). https://doi.org/10.1007/978-3-540-28639-4_24
12. Xiao, M., Kou, S.: Parameterized algorithms and kernels for almost induced matching. Theor. Comput. Sci. **846**, 103–113 (2020)
13. Xiao, M., Kou, S.: Kernelization and parameterized algorithms for 3-path vertex cover. In: Gopal, T.V., Jäger, G., Steila, S. (eds.) TAMC 2017. LNCS, vol. 10185, pp. 654–668. Springer, Cham (2017). https://doi.org/10.1007/978-3-319-55911-7_47
14. Cervený, R., Choudhary, P., Suchý, O.: On kernels for d-path vertex cover. In: 47th International Symposium on Mathematical Foundations of Computer Science, MFCS 2022, 22–26 August 2022, Vienna, Austria, vol. 241 of LIPIcs, Schloss Dagstuhl - Leibniz-Zentrum für Informatik, pp. 29:1–29:14 (2022)

Random Shortening of Linear Codes and Applications

Xue Chen[1], Kuan Cheng[2], Xin Li[3], and Songtao Mao[3]([✉])

[1] University of Science and Technology of China, Anhui, China
[2] Peking University, Beijing, China
ckkcdh@pku.edu.cn
[3] Johns Hopkins University, Baltimore, MD 21218, USA
{lixints,smao13}@jhu.edu

Abstract. Random linear codes (RLCs) are well known to have nice combinatorial properties and near-optimal parameters in many different settings. However, getting explicit constructions matching the parameters of RLCs is challenging, and RLCs are hard to decode efficiently. This motivated several previous works to study the problem of partially derandomizing RLCs, by applying certain operations to an explicit mother code. Among them, one of the most well studied operations is *random puncturing*, where a series of works culminated in the work of Guruswami and Mosheiff (FOCS' 22), which showed that a random puncturing of a low-biased code is likely to possess almost all interesting local properties of RLCs.

In this work, we provide an in-depth study of another, dual operation of random puncturing, known as *random shortening*, which can be viewed equivalently as random puncturing on the dual code. Our main results show that for any small ε, by starting from a mother code with certain weaker conditions (e.g., having a large distance) and performing a random (or even pseudorandom) shortening, the new code is ε-biased with high probability. Our results hold for any field size and yield a shortened code with constant rate. This can be viewed as a complement to random puncturing, and together, we can obtain codes with properties like RLCs from weaker initial conditions.

Our proofs involve several non-trivial methods of estimating the weight distribution of codewords, which may be of independent interest.

1 Introduction

Error correcting codes are fundamental objects in combinatorics and computer science. The study of these objects together with the bounds and parameters that can be achieved, has also helped shape the field of information theory starting from the pioneering work of Shannon and Hamming. In the theory of error-correcting codes, linear codes form a fundamental class of codes that are building blocks of many important constructions and applications. Such codes have simple algebraic structures that are often key ingredients in their performance and analysis. For example, any linear code with message length k and codeword length n over the field $\mathbb{F}_q$ can be described by both a generator matrix in $\mathbb{F}_q^{k \times n}$ and a parity check matrix in $\mathbb{F}_q^{n \times (n-k)}$.

It is well known that random linear codes (RLCs, where one samples each entry of the generator matrix uniformly independently from $\mathbb{F}_q$) enjoy nice combinatorial properties and have near-optimal parameters in many different settings. Specifically, with high probability they achieve Shannon capacity, the Gilbert-Varshamov (GV) bound of rate-distance tradeoff, and are list-decodable up to capacity. However, getting explicit constructions remains a challenging problem in many situations. In addition, random linear codes have little structure, which makes it difficult to design efficient decoding algorithms. Indeed, decoding random linear codes is closely related to the problems of learning parity with noise and learning with errors, whose hardness is the basis of many cryptographic applications (see e.g., [Reg09]). As such, many previous works studied the problem of slightly derandomizing, or equivalently reducing the randomness used in RLCs, while still maintaining their nice properties.

Among these works, random puncturing is one of the most well-studied operations. Here, one takes an explicit mother code, and then randomly punctures some coordinates from the code (or equivalently, punctures some columns from the generator matrix) to get a new, shorter code. Specifically, a $\mathcal{P}$-puncturing of a mother code $\mathcal{C} \subseteq \mathbb{F}_q^n$ randomly chooses a subset $\mathcal{P} \subseteq [n]$ of size p, and for every codeword of $\mathcal{C}$, deletes all symbols with positions in $\mathcal{P}$. Compared to standard RLCs, the number of random bits used is thus reduced from $O(nk \log q)$ to $O(n)$. Furthermore, certain nice structures of the mother code are often inherited by the punctured code, which makes decoding easier.

With sophisticated techniques, previous works have shown that if the mother code satisfies some natural conditions, then after a random puncturing, with high probability the new code has certain properties similar to those of RLCs. For example, motivated by the problem of achieving list-decoding capacity, recent works [Woo13, RW14, FKS22, GST21, BGM22, GZ23, AGL23] studied random puncturing of Reed-Muller (RM) codes and Reed-Solomon (RS) codes. Subsequent works [GM22, PP23] generalized the list-decoding property to all monotone-decreasing local properties. In all these works, the mother code needs to have some special properties, such as being an RS code, an RM code, having a large distance over a large alphabet, or having a low bias over a small alphabet. These properties are not immediately implied by general linear codes, and thus, one of the natural goals is to gradually weaken the requirements of the mother code so that the approach works for a broader class of codes. Indeed, as we shall see later, this is one of the main motivations and themes in previous works.

In this paper we continue this line of work and study the following two natural questions:

1. *If the mother code is not that strong, can we still use some operations to get a new code that has properties similar to random linear codes?*
2. *What other operations, besides random puncturing, are useful in this context?*

Towards answering these questions, we consider a different operation to reduce the randomness of RLCs, called random shortening, previously studied in [BGL17,LDT21,YP17,NvZ15]. Specifically, for an integer s, a random s-shortening of a code $\mathcal{C} \subseteq \mathbb{F}_q^n$ randomly chooses a subset $\mathcal{S} \subseteq [n]$ of size s, and forms a new code by picking all codewords of $\mathcal{C}$ which are zeros at the positions in $\mathcal{S}$, and deleting these zero symbols.

We note that just like random puncturing, the operation of random shortening can in fact be carried out on any code, not just on linear codes. However, for linear codes there is an important, alternative view of random shortening: it is actually the dual version of random puncturing. In particular, one can check that it is equivalent to a random puncturing of size s on the parity check matrix of a linear code C, or the generator matrix of the dual code $\mathcal{C}^\perp$. Thus in this paper, for a linear code, we also call shortening *dual puncturing*.

This view brings some convenience from the viewpoint of the parity check matrix. For example, any puncturing of the parity check matrix (hence also shortening) of a low-density parity check (LDPC) code [Gal62] still results in an LDPC code. Another example is expander codes [SS96]. A binary expander code $\mathcal{C}$ is based on a bipartite expander graph $\Gamma : [N] \times [D] \to [M]$ with N nodes on the left, M nodes on the right, and left degree D. The parity check matrix of $\mathcal{C}$ is defined as follows. Each left node corresponds to a codeword bit and each right node corresponds to a parity check which checks if the parity of its neighboring codeword bits is 0. Such a code has linear time decoding algorithms, and the distance of $\mathcal{C}$ can be lower bounded by using the vertex expansion property of Γ. Specifically, assume that for every left set $A \subseteq [N]$, with $|A| \leq \alpha N$, the neighbors of A, denoted as $\Gamma(A)$ has size at least $(1 - \varepsilon)D|A|$, then [CCLO23] showed that the distance of $\mathcal{C}$ is at least roughly $\frac{\alpha N}{2\varepsilon}$. Notice that an $\mathcal{S}$-shortening of $\mathcal{C}$ actually corresponds to deleting nodes in $\mathcal{S}$ from the left set $[N]$ together with their adjacent edges, thus this does not change the vertex expansion property of the remaining graph. Hence the new code still has a distance of at least roughly $\frac{\alpha N}{2\varepsilon}$, which in fact corresponds to a larger relative distance (since the new code has a shorter length). As we will see shortly, this is actually a general property of any shortening of a code. In summary, just like puncturing, the shortening operation also preserves certain nice properties of the mother code, e.g., being an LDPC code or an expander code. In turn, this makes decoding easier.

Before stating our results, we first review some previous works on random puncturing and random shortening in more detail.

1.1 Previous Work

Recently, random puncturing has drawn a lot of attention in the context of list decoding. In [Woo13], Wootters showed that by applying a random puncturing to a Reed-Muller code and setting the desired rate to $O(\varepsilon^2)$, with high probability one can list-decode the punctured code up to a relative radius of $1/2 - \varepsilon$, with an exponential but non-trivial list size. In [RW14], Rudra and Wootters showed that if the mother code is an RS code, and has a large enough relative distance of $1 - 1/q - \varepsilon^2$, then after puncturing one can get a list-decoding radius of

$1 - 1/q - \varepsilon$ and a rate close to capacity up to a $\mathsf{poly}\log(1/\varepsilon)$ factor, while the list size is $O(1/\varepsilon)$. We remark that a rate upper bound for list-decodable linear codes is given by Shangguan and Tamo [ST20], which is a generalized singleton bound. Specifically, they proved that if $\mathcal{C}$ is a linear code of rate R that is (ρ, L) list decodable, i.e., the code has a relative list decoding radius of ρ and list size L, then $\rho \leq (1 - R)\frac{L}{L+1}$. They conjectured the existence of such codes and proved the case for $L = 2, 3$. Later, towards proving this conjecture, Guo et. al. [GLS+22] showed that there are RS codes that are $(1 - \varepsilon, O(1/\varepsilon))$ list decodable and the rate can be $\Omega(\varepsilon/\log(1/\varepsilon))$, though they mainly use intersection matrices instead of random puncturing. Ferber, Kwan, and Sauermann [FKS22] further showed that through random puncturing one can achieve a rate of $\varepsilon/15$ with list decoding radius $1 - \varepsilon$ and list size $O(1/\varepsilon)$. This was further improved by Goldberg et. al. [GST21] to achieve a rate of $\frac{\varepsilon}{2-\varepsilon}$. Most recently, [BGM22] showed that random puncturing of RS codes can go all the way up to the generalized singleton bound if the field size is $2^{O(n)}$, resolving a main conjecture of [ST20]. This was subsequently improved by [GZ23], which reduced the field size to $O(n^2)$; and again by [AGL23], which further reduced the field size to $O(n)$, although [GZ23, AGL23] can only get close to the generalized singleton bound. We note that all the above works mainly studied RS codes or RM codes, which have strong algebraic structures, and some of them also require a large relative distance (e.g., close to $1 - 1/q$).

On the other hand, Guruswami and Mosheiff [GM22] considered random puncturing of more general codes with weaker properties. Specifically, they considered two cases, where the mother code either has a low bias or has a large distance over a large alphabet (note that the property of a low bias implies a large distance, hence is stronger). For both cases, they showed that the punctured code can achieve list decoding close to capacity. In fact, they showed a stronger result, that all monotone-decreasing local properties of the punctured code are similar to those of random linear codes. Subsequent to [GM22], Putterman and Pyne [PP23] showed that the same results in [GM22] can be achieved by using a pseudorandom puncturing instead, which reduces the number of random bits used in the puncturing to be linear in the block length of the punctured code, even if the mother code has a much larger length.

Unlike puncturing, there are only a handful of previous works on shortening. In [NvZ15], Nelson and Van Zwam proved that all linear codes can be obtained by a sequence of puncturing and/or shortening of a collection of asymptotically good codes. In [YP17], Yardi and Pellikaan showed that all linear codes can be obtained by a sequence of puncturing and/or shortening on some specific cyclic code. In [BGL17], Bioglio et. al. presented a low-complexity construction of polar codes with arbitrary length and rate using shortening and puncturing. In [LDT21], Liu et. al. provided some general properties of shortened linear codes.

1.2 Notation and Definitions

Definition 1. *A linear code $\mathcal{C}$ of length n and dimension k over a finite field $\mathbb{F}_q$ is a k-dimensional subspace of the n-dimensional vector space $\mathbb{F}_q^n$. The dual* code $C^{\perp}$ of a linear code is the dual linear subspace of C. Hence the sum of the rates of C and $C^{\perp}$ is 1. We call $d^{\perp}(C)$ the dual distance of C as the minimum distance of its dual code $C^{\perp}$. The relative dual distance of C is the ratio of its dual distance to its length: $\delta^{\perp}(C) = \frac{d^{\perp}(C)}{n}$. We denote a linear code with these properties as an $[n, k, d]_q$ code or an $[n, k, d, d^{\perp}]_q$ code.

Definition 2. *Let $\mathcal{P}$ be a subset of $[n]$ of size p. A $\mathcal{P}$-puncturing on a code C of length n involves removing p positions indexed by $\mathcal{P}$. The resulted punctured code $C^{(\mathcal{P})}$ has length $n - p$. If $\mathcal{P}$ is a uniformly random subset of size p, we say that $C^{(\mathcal{P})}$ is obtained from C by a random p-puncturing.*

Definition 3. *Let $\mathcal{S}$ be a subset of $[n]$ of size s. An $\mathcal{S}$-shortening on a code C of length n involves selecting all codewords with zero values on positions indexed by $\mathcal{S}$ and removing these positions. The resulted shortened code $C^{[\mathcal{S}]}$ has length $n - s$. If $\mathcal{S}$ is a uniformly random subset of size s, we say that $C^{[\mathcal{S}]}$ is obtained from C by a random s-shortening.*

Definition 4. *The q-ary entropy function is defined as $\mathrm{H}_q(x) = x \log_q(q - 1) - x \log_q x - (1 - x) \log_q(1 - x)$.*

Throughout the paper, we use "with high probability" to mean that when the rate R, relative distance δ, relative dual distance $\delta^{\perp}$ of the code, and other given parameters are fixed, the probability of the event is $1 - O(\exp(-tn))$ for some constant t. Essentially, this means that the probability of the event occurring approaches 1 as the block length n increases, making it increasingly likely that the desired properties hold.

As in [GM22], in this paper we also consider *monotone-decreasing, local* properties. Informally, we call a code property $\mathcal{P}$ monotone-decreasing and local if, the fact that a code C does not satisfy $\mathcal{P}$ can be witnessed by a small "bad set" of codewords in C. For example, some typical properties, such as being list-decodable to capacity and having a small bias, are monotone-decreasing and local properties. More formally, a monotone-decreasing and local property is the opposite of a monotone-increasing and local property, defined below.

Definition 5. *A property $\mathcal{P}$ is said to be*

- *monotone-increasing if, for any code C, whenever one of its subcodes (i.e., a subspace of C) satisfies $\mathcal{P}$, the code C itself also satisfies $\mathcal{P}$ (monotone-decreasing if the complement of $\mathcal{P}$ is monotone-increasing);*
- *b-local for some $b \in \mathbb{N}$ if there exists a family $\mathcal{B}_{\mathcal{P}}$ of sets of words in $\mathbb{F}_q^n$, with the size of the sets at most b, and such that C satisfies $\mathcal{P}$ if and only if there exists an set $B \in \mathcal{B}_{\mathcal{P}}$ satisfying $B \subseteq C$,*
- *row-symmetric if, for any code $C \subseteq \mathbb{F}_q^n$ that satisfies $\mathcal{P}$, the resulting code obtained by performing a permutation on the n positions of C also satisfies $\mathcal{P}$.*

1.3 Main Results

Random Puncturing vs. Random Shortening. Before formally stating our results, we first informally compare the two operations of random puncturing and random shortening. A random p-puncturing of a code of length n involves uniformly selecting p positions randomly from $[n]$, and discarding these positions in the code. One can see that under appropriate conditions, this operation preserves the distinctness of all codewords, and thus can increase the rate of the code. However it may decrease the distance or relative distance of the code. In contrast, a random s-shortening of a code involves picking s positions uniformly randomly from $[n]$, forming a subcode that consists of codewords which contain only zeros at these positions, and then deleting these positions in the subcode. It can be seen that this operation perserves the distance of the code, and thus increases the relative distance of the code, but on the other hand the rate of the code can potentially decrease. Hence, these two operations are indeed "dual" in some sense, and therefore one can apply both operations to adjust both the rate and the relative distance of the code.

A linear code $\mathcal{C} \subseteq \mathbb{F}_q^n$, where $q = p^r$ for some prime p, is called ε-biased, if for every codewords $c \in \mathcal{C}$, $\left|\sum_{i=1}^n \omega^{\mathrm{tr}(a \cdot c_i)}\right| \leq \varepsilon n$ for all $a \in \mathbb{F}_q^*$. where $\omega = e^{\frac{2\pi i}{p}}$ and $\mathrm{tr} : \mathbb{F}_q \to \mathbb{F}_p$ is the field trace map.

Our main results show that random shortening is an effective way to reduce the *bias* of a code. Note that this is stronger than increasing the relative distance, since the former implies the latter. If the mother code satisfies certain conditions, then we show after random shortening the new code can achieve an arbitrarily small bias with high probability. We note that a random linear code has a small bias, and thus in this sense the code after random shortening behaves like random linear codes. Moreover, the condition that the mother code has a low bias is required in several previous works (e.g., [GM22,PP23]), while these works essentially do not care about the rate of the mother code. Thus we can apply a random puncturing to the new code after a random shortening, to get another code where all monotone-decreasing local properties are similar to those of random linear codes. This further weakens the requirements of mother codes in previous works to some extent.

Low-Biased Codes from Codes with Large Distance. A low-biased code must have a large distance. However, the reverse may not hold. The following theorem shows that it is also possible to derive a low-biased code from a code with a large distance by random shortening.

Theorem 1. *For any $0 < \varepsilon < 1$, any $[n, Rn, \delta n]_q$ code $\mathcal{C}$ with $\frac{q-1}{q} - \frac{q}{q-1}\left(\frac{\varepsilon}{2(q-1)}\right)^2 < \delta < \frac{q-1}{q}$ and any constant $0 < \gamma < R$, there exists a number $0 < s < R$ such that the following holds. If we perform a random sn-shortening $\mathcal{S}$ to $\mathcal{C}$, then with high probability, the shortened code $\mathcal{C}^{[\mathcal{S}]}$ is ε-biased and has rate at least $R - \gamma$.*

We note that the theorem only requires a lower bound on the relative distance, but there are no restrictions on the rate of the original code, R. Hence, this requirement is generally easy to satisfy, for example, from simple constructions using code concatenation. Furthermore, we can select an appropriate shortening size to ensure that the rate of the shortened code is arbitrarily close to R.

The distance condition of $\mathcal{C}$ in Theorem 1 can also be relaxed, resulting in the following theorem.

Theorem 2. *Given any $0 < \varepsilon < 1$, if an $[n, Rn, \delta n]_q$ code $\mathcal{C}$ satisfies the condition that there exists some $0 < \beta < 1$, such that $\frac{\delta}{1-(1-\beta)R} > \frac{q-1}{q} - \frac{q}{q-1}\left(\frac{\varepsilon}{2(q-1)}\right)^2$, then there exists a number $0 < s < R$ such that the following holds. If we perform a random sn-shortening $\mathcal{S}$ to $\mathcal{C}$, then with high probability, the shortened code $\mathcal{C}^{[\mathcal{S}]}$ is ε-biased with rate at least βR.*

Indeed, the asymptotic form of the Plotkin bound is given by

$$R \leq 1 - (\frac{q}{q-1}) \cdot \delta + o(1). \tag{1}$$

Thus Theorem 2 implies that as long as the rate-distance trade-off of the original code is close enough to the Plotkin bound, we can obtain a code with an arbitrarily small bias by random shortening. On the other hand, unlike in Theorem 1, the rate of the shortened code may not be arbitrarily close to R, but we can still get a new rate that is only a constant factor smaller.

Low-Biased Codes from Codes with Small Rate and not too Small Dual Distance. In the next theorem, there is no requirement for δ to be very large. Instead, we impose constraints on its dual distance, $\delta^{\perp}$, and the rate, R. If the dual distance is not too small and the rate can be upper bounded, then we can also apply the shortening technique to obtain a low-biased code.

Theorem 3. *Given any $0 < \varepsilon < 1$, if an $[n, Rn, \delta n, \delta^{\perp} n]_q$ code $\mathcal{C}$ satisfies the condition that there exist $0 < \gamma < \frac{1}{4}$, $0 < \delta_0^{\perp} < \min\{\varepsilon^{\frac{1}{\gamma}}, \left(\frac{1+\log_q(1-\delta)}{36}\right)^2, (\frac{1}{q})^{\frac{1}{\gamma}}\}$, such that $\delta^{\perp} > \delta_0^{\perp}$ and $0 < R < \frac{0.5-2\gamma}{1+0.9\cdot\log_q(1-\delta)}H_q(\delta_0^{\perp})$, then there exists a number $0 < s < R$ such that the following holds. If we perform a random sn-shortening $\mathcal{S}$ to $\mathcal{C}$, then with high probability the shortened code $\mathcal{C}^{[\mathcal{S}]}$ is ε-biased with rate at least $0.1R$.*

In Theorem 3, the rate of the dual code must be sufficiently large. Additionally, if the term $\frac{\frac{1}{2}-2\gamma}{1+0.9\cdot\log_q(1-\delta)}$ is less than 1, the rate-distance trade-off of the dual code surpasses the Gilbert-Varshamov (GV) bound. Consequently, when examining the problem within the context of the GV bound, we need to impose specific constraints on δ. This leads to the following corollary.

Corollary 1. *Given any $0 < \varepsilon < 1$, $\delta > 1 - q^{-0.6}$, there exists a number $\gamma > 0$, such that for any $\delta^{\perp} > \delta_0^{\perp}$, $0 < R < (1+\gamma)H_q(\delta_0^{\perp})$ for a certain*

$0 < \delta_0^{\perp} < \min\{\varepsilon^{\frac{1}{7}}, \frac{1}{8100}, (\frac{1}{q})^{\frac{1}{7}}\}$, there exists a number $0 < s < R$ such that the following holds. Let $\mathcal{C}$ be any $[n, Rn, \delta n, \delta^{\perp} n]_q$ code. If we perform a random sn-shortening $\mathcal{S}$ to $\mathcal{C}$, then with high probability the shortened code $\mathcal{C}^{[\mathcal{S}]}$ is ε-biased with rate at least $0.1R$.

Theorem 3 and Corollary 1 show that as long as the mother code and its dual both have a reasonable relative distance, one can use random shortening to get a new code with an arbitrary small bias, while only losing a constant factor in the rate. We note that linear codes such that both the code and its dual have good relative distance are also easily constructible, for example, see [Shp09].

Random-Like Codes by Random Shortening and Puncturing. In [GM22], the authors showed that a random puncturing of a low-biased code results in a new code that behaves like random linear codes. Using our theorems, we present a weaker condition that still achieves similar results. This follows from a combination of random shortening and random puncturing, as briefly discussed before.

Theorem 4. *For any* $0 < \varepsilon < 1$, $b \in \mathbb{N}$, *and prime power* q, *there exists some* $\eta > 0$, *such that the following holds. Let* $\mathscr{P}$ *be a monotone-decreasing, b-local, and row-symmetric property over* $\mathbb{F}_q^n$ *satisfied by a random linear code of length* n *and rate* R'. *There exists some* $\eta > 0$ *such that the following holds. If any one of the following properties is satisfied for* $R, \delta, \delta^{\perp}, q, \eta$:

1. $\delta > (\frac{q-1}{q} - \eta)(1 - R)$, or

2. $\delta^{\perp} > \delta_0^{\perp}$ and $0 < R < \frac{\frac{1}{2} - 2\gamma}{1 + 0.9 \cdot \log_q(1-\delta)} H_q(\delta_0^{\perp})$ for a certain $0 < \delta_0^{\perp} <$
   $\min\{\varepsilon^{\frac{1}{7}}, \left(\frac{1 + \log_q(1-\delta)}{36}\right)^2, (\frac{1}{q})^{\frac{1}{7}}\}$,

then there exists $m, p, s > 0$ *such that for any* $[m, Rm, \delta m]_q$ *code, if we perform a random sm-shortening and then a random pm-puncturing on* $\mathcal{C}$, *the resulted code* $\mathcal{D}$ *has length* n, *rate at least* $R' - \varepsilon$ *and with high probability, satisfies* $\mathscr{P}$.

1.4 Technique Overview

We investigate the effect of shortening as follows. An $\mathcal{S}$-shortening applied to a code $\mathcal{C}$ of length n involves selecting all codewords with zeros at positions indexed by $\mathcal{S}$ and removing these positions. Specifically, if the support of a codeword $c \in \mathcal{C}$ intersects $\mathcal{S}$ (in which case we say $\mathcal{S}$ hits c), then c will not be included in the shortened code; if the support of c does not intersect $\mathcal{S}$, then there is a codeword $c' \in \mathcal{C}^{[\mathcal{S}]}$, which is obtained from c by removing all positions in $\mathcal{S}$. In this way, under a random shortening, each non-zero codeword has a certain probability of being dropped and a certain probability of being retained in $\mathcal{C}^{[\mathcal{S}]}$. If the distance of $\mathcal{C}$ is δn, then the probability of each codeword being hit and dropped is at least $1 - (1 - \delta)^s$, where s is the size of $\mathcal{S}$.

We use $\mathcal{C}_{\varepsilon}$ to denote all codewords in $\mathcal{C}$ which are not ε-biased. If the size of $\mathcal{C}_{\varepsilon}$ is small, then by a union bound, the probability that not all codewords in $\mathcal{C}_{\varepsilon}$

are hit by $\mathcal{S}$ is exponentially small. Thus, with high probability, all codewords in $\mathcal{C}$ that are not hit by $\mathcal{S}$ and inherited to $\mathcal{C}^{[\mathcal{S}]}$ are ε-biased. Hence, a critical part of all our proofs is to upper bound the size of $\mathcal{C}_{\varepsilon}$.

Furthermore, as long as $\mathcal{C}$ is a linear code and s is less than the dimension k of $\mathcal{C}$, we know that the shortened code $\mathcal{C}^{[\mathcal{S}]}$ has dimension at least $k - s$. Consequently, $\mathcal{C}^{[\mathcal{S}]}$ retains a nonzero constant rate as well.

Change of parameters. The shortening results in changes to the parameters of the code. Here, we mainly apply shortening for two purposes: adjusting the bias and amplifying the relative distance.

1. **Adjusting the bias:** Let $\mathcal{C}$ be of length n. When $\mathcal{C}_{\varepsilon'}$ is hit by $\mathcal{S}$, it implies that the codewords in $\mathcal{C}$ not hit by $\mathcal{S}$ are all ε'-biased. However, it doesn't directly imply that the shortened code $\mathcal{C}^{[\mathcal{S}]}$ is also ε'-biased, since the shortening operation changes the length of the code. Nevertheless, the new bias ε of $\mathcal{C}^{[\mathcal{S}]}$ is given by $\varepsilon \leq \frac{\varepsilon'n+s}{n-s}$, where s is the size of the shortening $\mathcal{S}$. If s is small compared to n, ε is close to ε'. In the proof of Theorem 1, we can choose s to be a sufficiently small fraction of n. In the proof of Theorem 3, we provide an upper bound for R, which also enables us to choose a small shortening size. In both cases, we set the shortening size to be less than $0.05\varepsilon'n$, allowing us to choose $\varepsilon' = 0.9\varepsilon$.

2. **Amplifying the relative distance:** We use another technique in the proof of Theorem 2 to first transform a code with a rate-distance trade-off near the Plotkin bound into a code with near-optimal distance. The distance of the shortened code $\mathcal{C}^{[\mathcal{S}]}$ is no less than that of the original code $\mathcal{C}$. However, since $\mathcal{C}^{[\mathcal{S}]}$ has length $n - s$ instead of n, its relative distance becomes $\frac{\delta}{1-\frac{s}{n}}$. This allows us to increase the relative distance of the code. In turn, Theorem 2 follows from Theorem 1.

Estimation of the Size of $\mathcal{C}_{\varepsilon}$. This is the most critical part of all our proofs. For Theorem 1 and Theorem 3, we have two different ways of estimating the upper bound of $|\mathcal{C}_{\varepsilon}|$:

1. **Estimating $|\mathcal{C}_{\varepsilon}|$ with relative distance δ:** We use $J_q(\delta)$ to denote the list decoding radius corresponding to the classical Johnson bound for a code over $\mathbb{F}_q$ with relative distance δ. It is easy to see that when δ is close to the optimal $\frac{q-1}{q}$, so is $J_q(\delta)$. To give an upper bound of $|\mathcal{C}_{\varepsilon}|$, we construct q balls in $\mathbb{F}_q^n$ with radius $J_q(\delta)$ and centered at $t\cdot\mathbf{1}$, where $\mathbf{1}$ is the all-one vector and $t \in \mathbb{F}_q$. By the Johnson bound, the number of codewords covered by these balls is at most $\text{poly}(n)$. We show that, if a codeword c is not covered by these balls, its empirical distribution over $\mathbb{F}_q$ is close to the uniform distribution, which implies c is small biased. This upper bounds $|\mathcal{C}_{\varepsilon}|$ by $\text{poly}(n)$.

2. **Estimating $|\mathcal{C}_{\varepsilon}|$ with relative dual distance $\delta^{\perp}$ and rate R:** If $\mathcal{C}$ has dual distance $d^{\perp}$, then any $d^{\perp} - 1$ columns of the generator matrix of $\mathcal{C}$ are linearly independent, which means that if we uniformly randomly choose a codeword from $\mathcal{C}$, then any $d^{\perp} - 1$ symbols of the codeword are independently

uniform, i.e., the symbols of a random codeword are $d^\perp - 1$-wise independent. We can now use this property to estimate the probability that a codeword randomly chosen from $\mathcal{C}$ is not ε-biased. This is a typical application of the concentration phenomenon from the higher moment method, where we use Hoeffding inequality, Chernoff bound, and Sub-Gaussian property to bound the $(d^\perp - 1)$th moment of the summation of some random variables. Then by Markov's inequality, the probability that a random codeword is not ε-biased can be bounded, which also gives an upper bound on $|\mathcal{C}_\varepsilon|$.

Obtaining Random-Like Codes. To obtain random-like codes, we combine our results with those in [GM22], which state that a randomly punctured low-biased code is likely to possess any monotone-decreasing local property typically satisfied by a random linear code of a similar rate. By our results, we can start from a code with less stringent conditions and achieve the same results as in [GM22], through the operations of a random shortening followed by a random puncturing.

2   Estimation on Low-Biased Codewords

For a random vector $x \in \mathbb{F}_q^n$, it is known from the law of large numbers that its empirical distribution Emp_x is, with high probability, ε-close to the uniform distribution over $\mathbb{F}_q$ for any ε as n goes to infinity. Therefore, for each ε, let $\mathcal{C}$ be a random code; $\mathcal{C}_\varepsilon$ will, with high probability, constitute only a small fraction of $\mathcal{C}$. In the following, we present several estimation methods for the size of $|\mathcal{C}_\varepsilon|$ under general conditions.

Lemma 1. *Let $\mathcal{C}$ be an $[n, Rn, \delta n]_q$ code. For any $\varepsilon \geq 2(q-1)\sqrt{\frac{q-1}{q}(\frac{q-1}{q} - \delta)}$, $|\mathcal{C}_\varepsilon| \leq q^2 \delta n^2$.*

Another approach to approximate $|\mathcal{C}_\varepsilon|$ is the probability method. It is essential to observe that when the dual code of $\mathcal{C}$ has distance $d+1$, every set of d columns within the generator matrix of $\mathcal{C}$ are linearly independent. This observation implies that when examining the distribution of a randomly selected codeword from $\mathcal{C}$, the bits exhibit d-wise independence. Consequently, $\mathcal{C}$ is bound by the constraints of the d-th moment inequality.

Lemma 2. *$x_1, \cdots, x_n$ are independent random variables with $\mu = 0$, and $x_i \in [-1, 1]$. Denote $X_n = \sum_{i=1}^n x_i$. Then for any even d,*

$$\mathbb{E}((X_n)^d) \leq 2 \cdot (2n)^{d/2} \cdot (\frac{d}{2})!. \tag{2}$$

Corollary 2. *Let $x_1, x_2 \cdots, x_n$ be random variables taking values in $[-1, 1]$ which are d-wise independent, $\mathbb{E}(x_i) = 0$. Let $X_n = \sum_{i=1}^n x_i$, $\delta = d/n$, then for any $\varepsilon > 0$,*

$$\Pr(|\sum_{i=1}^n x_i| \geq \varepsilon n) \leq 4\sqrt{\pi d}(\frac{\delta}{\varepsilon^2 e})^{\delta n/2}. \tag{3}$$

Lemma 3. *Let x be a random vector, whose components uniformly take values in $\mathbb{F}_q$ and are d-wise independent. Let $\delta = d/n$. Then*

$$\Pr(x \text{ is not } \varepsilon\text{-biased}) \leq 2\sqrt{2}(q-1)\left(\frac{2\delta}{\varepsilon^2 e}\right)^{\delta n/2}. \tag{4}$$

Corollary 3. *Let $\mathcal{C}$ be a code of length n, rate R and dual distance $d^\perp = \delta^\perp n$ over the field $\mathbb{F}_q$. Then for each $\varepsilon > 0$, the number of codewords which are not ε-biased is not more than*

$$8q\sqrt{\pi\delta^\perp n} \cdot \left(\frac{2\delta^\perp}{\varepsilon^2 e}\right)^{\delta^\perp n/2} \cdot q^{Rn}$$

for sufficiently large n.

3   Proof of the Main Theorems

Before proving Theorem 1, we first give the following theorem.

Theorem 5. *Let $\mathcal{C}$ be an $[n, Rn, \delta n]_q$ code. If we perform a random sn-shortening $\mathcal{S}$ to $\mathcal{C}$, where $s < R$, then with high probability, the shortened code $\mathcal{C}^{[\mathcal{S}]}$ is ε-biased, where $\varepsilon = \frac{2(q-1)\sqrt{\frac{q-1}{q}\left(\frac{q-1}{q} - \delta\right)} + s}{1 - s}$.*

Proof. (of Theorem 1). Let $s = \min\{\frac{\gamma}{1+\gamma}, \frac{R}{2}, \frac{\varepsilon}{2} - (q-1)\sqrt{\frac{q-1}{q}(\frac{q-1}{q} - \delta)}\}$. Since $s \leq \min\{\frac{\gamma}{1+\gamma}, \frac{R}{2}\}$, the rate of $\mathcal{C}^{[\mathcal{S}]}$ is

$$\frac{R - s}{1 - s} > R - \gamma. \tag{5}$$

Rearranging the inequality $\frac{q-1}{q} - \frac{q}{q-1}\left(\frac{\varepsilon}{2(q-1)}\right)^2 < \delta$, we get

$$(q-1)\sqrt{\frac{q-1}{q}(\frac{q-1}{q} - \delta)} < \frac{\varepsilon}{2}, \tag{6}$$

And since $s < \frac{\varepsilon}{2} - (q-1)\sqrt{\frac{q-1}{q}(\frac{q-1}{q} - \delta)}$.

$$\frac{2(q-1)\sqrt{\frac{q-1}{q}(\frac{q-1}{q} - \delta)} + s}{1 - s}$$

$$< \frac{(q-1)\sqrt{\frac{q-1}{q}(\frac{q-1}{q} - \delta)} + \frac{\varepsilon}{2}}{1 - \frac{\varepsilon}{2} + (q-1)\sqrt{\frac{q-1}{q}(\frac{q-1}{q} - \delta)}} \tag{7}$$

$$< \varepsilon.$$

By Lemma 5, $\mathcal{C}^{[\mathcal{S}]}$ is ε-biased.

Now we present an inequality concerning the q-ary entropy function $H_q(x)$ here.

Lemma 4. *For any* $q = p^r$, $0 < \gamma < \frac{1}{4}$, *when* $0 < x < (\frac{1}{q})^{\frac{1}{\gamma}}$, $H_q(x) < -(1 + 2\gamma)x \log_q x$.

Proof. (of Theorem 3). We set $\varepsilon' = 0.9\varepsilon$ and get $\delta_0^\perp < (\frac{\sqrt{e}\varepsilon'}{\sqrt{2}})^{\frac{1}{\gamma}}$. Let $s = \frac{R - (\frac{1}{2} - 2\gamma)H_q(\delta_0^\perp)}{-\log_q(1-\delta)}$. We get

$\Pr(\mathcal{S} \text{ doesn't hit all codewords in } \mathcal{C}_{\varepsilon'})$

$\leq \sum\limits_{c \in \mathcal{C}_{\varepsilon'}} \Pr(c \text{ is not hit by } \mathcal{S})$

$\leq |\mathcal{C}_{\varepsilon'}| \cdot (1 - \delta)^{sn}$

$\leq 8q\sqrt{\pi \delta_0^\perp n} \cdot \left(\frac{2\delta_0^\perp}{(\varepsilon')^2 e} \right)^{\delta_0^\perp n/2} \cdot q^{Rn} \cdot (1-\delta)^{sn}$ $\qquad$ (using Corollary 3)

$= 8q\sqrt{\pi \delta_0^\perp n} \cdot \left(\frac{2\delta_0^\perp}{(\varepsilon')^2 e} \right)^{\delta_0^\perp n/2} \cdot q^{Rn} \cdot q^{\log_q(1-\delta)sn}$

$= 8q\sqrt{\pi \delta_0^\perp n} \cdot \left((\delta_0^\perp)^{1-2\gamma} \right)^{\delta_0^\perp n/2} \cdot q^{(\frac{1}{2}-2\gamma)H_q(\delta_0^\perp) \cdot n}$ $\qquad$ (using $s = \frac{R - (\frac{1}{2} - 2\gamma)H_q(\delta_0^\perp)}{-\log_q(1-\delta)}$)

$\leq 8q\sqrt{\pi \delta_0^\perp n} \cdot \left((\delta_0^\perp)^{1-2\gamma} \right)^{\delta_0^\perp n/2} \cdot q^{-(\frac{1}{2}-2\gamma)(1+2\gamma)\delta_0^\perp \log_q \delta_0^\perp \cdot n}$ $\qquad$ (using Lemma 4)

$= 8q\sqrt{\pi \delta_0^\perp n} \cdot \left((\delta_0^\perp)^{\frac{1}{2}-\gamma} \cdot (\delta_0^\perp)^{-(\frac{1}{2}-2\gamma)(1+2\gamma)} \right)^{\delta_0^\perp n}$

$= 8q\sqrt{\pi \delta_0^\perp n} \cdot (\delta_0^\perp)^{4\gamma^2 \delta_0^\perp n}.$

$$(8)$$

Therefore, this probability in Eq. 8 tends to 0 as n approaches infinity. Moreover, one can verify that $s < 0.9R$ given $s = \frac{R - (\frac{1}{2} - 2\gamma)H_q(\delta_0^\perp)}{-\log_q(1-\delta)}$ and $R \leq \frac{\frac{1}{2} - 2\gamma}{1 + 0.9 \cdot \log_q(1-\delta)}H_q(\delta_0^\perp)$. Hence, the shortened code $\mathcal{C}^{[\mathcal{S}]}$ has rate at least $\frac{R-s}{1-s} > 0.1R$, and

$s < R$

$\quad < \dfrac{0.5 - 2\gamma}{1 + \log_q(1-\delta)} H_q(\delta_0^\perp)$

$\quad < -\dfrac{0.75}{1 + \log_q(1-\delta)} \cdot \delta_0^\perp \cdot \log_q(\delta_0^\perp)$ $\qquad$ (using Lemma 4)

$\quad < \dfrac{0.75}{1 + \log_q(1-\delta)} \cdot (\delta_0^\perp)^{\frac{1}{2}} \cdot (\delta_0^\perp)^{\gamma} \cdot (\delta_0^\perp)^{\frac{1}{4}} \cdot (-\log_q(\delta_0^\perp))$ $\qquad$ (using $\gamma < \frac{1}{4}$)

$\quad < \dfrac{1}{48} \cdot \dfrac{10}{9}\varepsilon' \cdot \dfrac{4}{e \ln(2)}$

$$(9)$$

$\mathcal{C}^{[\mathcal{S}]}$ is $\frac{\varepsilon'+s}{1-s}$-biased and since $\frac{\varepsilon'+s}{1-s} < \frac{10}{9}\varepsilon' < \varepsilon$. So, with high probability, $\mathcal{C}^{[\mathcal{S}]}$ is ε-biased.

Acknowledgments. Xin Li is supported by NSF CAREER Award CCF-1845349 and NSF Award CCF-2127575. Sontao Mao is supported by NSF Award CCF-2127575.

References

AGL23. Alrabiah, O., Guruswami, V., Li, R.: Randomly punctured reed-solomon codes achieve list-decoding capacity over linear-sized fields. arXiv preprint arXiv:2304.09445 (2023)

BGL17. Bioglio, V., Gabry, F., Land, I.: Low-complexity puncturing and shortening of polar codes. In: 2017 IEEE Wireless Communications and Networking Conference Workshops (WCNCW), pp. 1–6. IEEE (2017)

BGM22. Brakensiek, J., Gopi, S., Makam, V.: Generic reed-solomon codes achieve list-decoding capacity. arXiv preprint arXiv:2206.05256, (2022)

CCLO23. Chen, X., Cheng, K., Li, X., Ouyang, M.: Improved decoding of expander codes. IEEE Trans. Inform. Theory (2023)

FKS22. Ferber, A., Kwan, M., Sauermann, L.: List-decodability with large radius for reed-solomon codes. IEEE Trans. Inf. Theory **68**(6), 3823–3828 (2022)

Gal62. Gallager, R.: Low-density parity-check codes. IRE Trans. Inf. Theory **8**(1), 21–28 (1962)

GLS+22. Guo, X., Li, R., Shangguan, C., Tamo, I., Wootters, M.: Improved list-decodability and list-recoverability of reed-solomon codes via tree packings. In: 2021 IEEE 62nd Annual Symposium on Foundations of Computer Science (FOCS), pp. 708–719. IEEE (2022)

GM22. Guruswami, V., Mosheiff, J.: Punctured low-bias codes behave like random linear codes. In: 2022 IEEE 63rd Annual Symposium on Foundations of Computer Science (FOCS), pp. 36–45. IEEE (2022)

GST21. Goldberg, E., Shangguan, C., Tamo, I.: List-decoding and list-recovery of reed-solomon codes beyond the johnson radius for any rate. arXiv preprint arXiv:2105.14754 (2021)

GZ23. Guo, X., Zhang, Z.: Randomly punctured reed-solomon codes achieve the list decoding capacity over polynomial-size alphabets. arXiv preprint arXiv:2304.01403 (2023)

LDT21. Liu, Y., Ding, C., Tang, C.: Shortened linear codes over finite fields. IEEE Trans. Inf. Theory **67**(8), 5119–5132 (2021)

NvZ15. Nelson, P., van Zwam, S.H.M.: On the existence of asymptotically good linear codes in minor-closed classes. IEEE Trans. Inf. Theory **61**(3), 1153–1158 (2015)

PP23. Putterman, A.L., Pyne, E.: Pseudorandom linear codes are list decodable to capacity. arXiv preprint arXiv:2303.17554 (2023)

Reg09. Regev, O.: On lattices, learning with errors, random linear codes, and cryptography. J. ACM **56**(6), 34:1–34:40, 2009

RW14. Rudra, A., Wootters, M.: Every list-decodable code for high noise has abundant near-optimal rate puncturings. In: Proceedings of the Forty-sixth Annual ACM Symposium on Theory of Computing (2014)

Shp09. Shpilka, A.: Constructions of low-degree and error-correcting ε-biased generators. Comput. Complex. **18**(4), 495–525 (2009)

SS96. Sipser, M., Spielman, D.A.: Expander codes. IEEE Trans. Inf. Theory **42**(6), 1710–1722 (1996)

ST20. Shangguan, C., Tamo, I.: Combinatorial list-decoding of reed-solomon codes beyond the johnson radius. In: Proceedings of the 52nd Annual ACM SIGACT Symposium on Theory of Computing, pp. 538–551 (2020)

Woo13. Wootters, M.: On the list decodability of random linear codes with large error rates. In: Proceedings of the Forty-fifth Annual ACM Symposium on Theory of Computing, pp. 853–860 (2013)

YP17. Yardi, A., Pellikaan, R.: On shortened and punctured cyclic codes. arXiv preprint arXiv:1705.09859 (2017)

Algorithms for Full-View Coverage of Targets with Group Set Cover

Jingfang Su and Hongwei Du[(✉)]

School of Computer Science and Technology, Harbin Institute of Technology
(Shenzhen), Shenzhen 518055, China
`sujingfang@stu.hit.edu.cn`, `hongwei.du@ieee.org`

Abstract. Group Set Cover is an optimization problem which its solution that can be used to solve the problem of full-view target coverage. In the context of time slices, two methods (TSC-FTC, FTC-TW) based on Group Set Cover have been proposed to optimize full-view coverage of targets problem. In TSC-FTC, the set of sensors that can cover the most targets are chosen using Group Set Cover in each time slice, and the total number of targets covered throughout time is calculated. This method effectively utilizes the resources in each time slice and enables the evaluation of coverage effectiveness. FTC-TW, an improvement of TSC-FTC, involves using Group Set Cover in each time slice to select the set of sensors that can cover the maximum number of targets, and then calculating the cumulative number of targets covered after a certain period of time. This method enables a quick selection of the set of sensors covering the maximum number of targets and also enables the evaluation of the effectiveness of full-view coverage. Both methods can effectively improve the full-view coverage of targets in different scenarios. We perform approximate solutions for both algorithms separately and provide a global approximation to the local optimality of non-submodular optimization.

Keywords: Camera sensor network · Full-view coverage · Group set cover

1 Introduction

Camera sensor networks (CSNs) have revolutionized the way we monitor and collect data in various applications, ranging from environmental monitoring to security surveillance. A fundamental challenge for this network is to achieve target coverage and ensure sufficient monitoring of the required areas of interest. In more detail, the goal is to obtain full-view coverage, in which case the camera sensors catch the entire target region. This study focuses on developing approximation methods to solve the target full-view coverage issue in camera sensor networks.

Camera sensor networks offer a promising solution for target coverage due to their ability to capture visual information from multiple viewpoints. However, achieving full-view coverage can be a complex optimization problem, especially when considering factors such as limited resources, variable camera ranges, overlapping coverage, and dynamic target movements. Full-view coverage ensures the accuracy of target coverage, provides coverage services for moving targets. It provides more information for subsequent image processing, another unavoidable tendency in the growth of the Internet of Things in the monitoring industry.

The camera sensor is a typical directional antenna equipped sensor. This directed sensing model (Fig. 1) is modeled as a triple $r, \alpha, \vec{f}$. When r is the sensing radius ($0 \leq r \leq \mathcal{R}$), α is the sensing angle, and $\vec{f}$ is the sensing direction. $\vec{f}$ is located on the extension line of the angle bisector of the sensing angle. The camera sensor is a directional sensor that can be rotated and zoomed. At a certain moment, one perceived direction of the camera sensor can simultaneously cover multiple facing directions of multiple targets. The camera sensor may rotate and zoom over time to capture more of the target's facing orientation. Therefore, the direction and position selection of the sensor has always been the focus of research. For the first time, Du et al. [3] studied the problem of maximizing the target full-view coverage by planning the direction and position of the sensor. In this paper, group set cover was created from the full-view coverage maximization problem. And it has been established that this problem is NP-Hard.

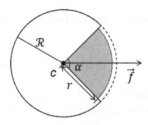

Fig. 1. A Directed Sensing Model for Camera Sensors.

Targets are typically reprented of as being static in latest coverage studies. The target's directionality is not necessary. As long as it is in the camera sensor's field of view, it is assumed to be covered. However, the ever-increasing demands on surveillance quality give new definitions to target coverage. When all views of an object are covered, this object is called full-view coverage. Here, targets can be static or even moving. Under this definition, the camera sensor can cover different viewing angles of the target at different times, so as to realize the monitoring of the moving target. As shown in Fig. 2(a), there is the directional target with only one facing direction. A directed target is a triple $R, \theta, \vec{\mathcal{F}}$ where θ represents facing angle and $\vec{\mathcal{F}}$ is the facing direction ($\vec{\mathcal{F}}$ bisects θ). Figure 3 demonstrates how the sensor covers one direction of the target. The conditions that need to be met for one perspective $\vec{\mathcal{F}}$ of target t to be covered are as follows,

- Target t is inside the sensor's sensing range.

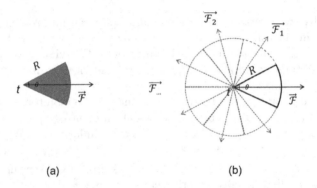

Fig. 2. Target model. (a) The directionality of the goal; (b) Schematic diagram of target full view coverage.

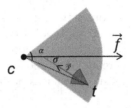

Fig. 3. The direction of the target is covered.

– ct lies in the fan-shaped area of c and the facing area of t at the same time.

Figure 2(b) shows multiple directions of the target. When all views $\vec{\mathcal{F}}_1, \vec{\mathcal{F}}_2, \ldots, \vec{\mathcal{F}}$ are covered, it means that the target is covered by the full view.

Related Works. An upsurge in research has been aroused within the domain of coverage [1,2]. Especially in the area of CSNs, with a focus on resolving issues related to networks with limited resources [4–6]. The research on full-view coverage is mainly aimed at achieving the maximum coverage of moving targets with limited camera sensor resources. For the first time, Wang and Cao [7] proposed the definition of full-view coverage of target. The domain of the coverage model they presented is $[0, 2\pi]$, thence, many efforts [8–10] were concentrated on the basis of this model. They conducted related research focusing on the deployment strategy of camera sensors [11–15]. He et al. [11] proposed that point coverage is the basis of area coverage, so the area coverage problem was reduced to point coverage, and two methods were proposed, one was a greedy algorithm, the other was a set-cover-based algorithm, and analyzed the approximate ratio of the two methods. Finding that achieving full view coverage depends more on sensor coverage area than sensor shape, Wu et al. [12] explored the required and sufficient criteria to accomplish full-view coverage under two random deployment strategies, uniform deployment and Poisson deployment. Shi et al. [13] proposed iterative screening algorithm (ISA) and improved ISA (IISA) to solve the optimal deployment of research full view point coverage (OFP) problem. Wang et al. [14] explored regular-shaped deployment strategies and developed sufficient requirements for the sensor density necessary for full-view coverage in stochastic uniform deployments. To forecast the sensor specifications needed to achieve a specific heterogeneous CSN full-view coverage probability, Liu et al. [15] created a model for estimating sensor parameters, including sensor scale and perception radius. Du et al. [3] divided the target into h ($h = 4\pi/\theta$) parts evenly. They proposed Maximum Group Set Coverage with Size Constraint (MGSC-SC). This algorithm achieved the purpose of maximizing full-view coverage by solving the selection problem of camera sensor position and orientation. The above researches generally converts the covering problem into a group set covering problem to find a local optimal solution. But this is only a solution at a certain moment, and full-view coverage is a process that needs to be completed within a period of time.

Contribution of this Paper. We introduce the concept of time slice into the Group Set Cover problem, thereby transforming the full-view coverage of mobile targets into a complete process for the first time. The time element is added to the group set coverage problem, and two approximation algorithms, TSC-FTCand FTC-TW to obtain the global optimum from the local optimum are proposed Where FTC-TW is an improvement of TSC-FTC.

- Group Set Cover with Time Segmented Cumulative Full-View Target Coverage (TSC-FTC). Select the sensor's position and orientation for each time slice to cover the most targets with a full view and, cumulatively, to cover the most targets overall. TSC-FTC maximizes the use of the target facing directions covered by each time slice. This approximation algorithm has an approximate ratio of $(1 - e^{-1})^{\beta}(1 - \varepsilon)$.
- Group Set Cover with Cumulative Maximum Target Coverage Time Window (FTC-TW). FTC-TW changes the steps of accumulative calculation of full-view coverage. The position and orientation of the sensor is chosen in each time slice to accumulatively cover the maximum number of targets throughout the time window. This approximation algorithm has an approximate ratio of $(1 - e^{-\gamma})^{\beta}(1 - \varepsilon)$.

This paper is organized from the following sections. Section 1 mainly introduces the research problem. Some preliminaries are detailed in Sect. 2. In Sect. 3, two improved group set covering algorithms are studied, and the corresponding approximate solutions are given. The final conclusion is presented in Sect. 4.

2 Preliminaries

The network of camera sensors is defined as heterogeneous. The heterogeneous camera sensor network has different sensing radii and sensing angles, and each sensing direction can cover a set of target facing directions. Through rotation and scaling, each camera covers multiple sets of target orientations. Therefore, the target full-view coverage problem is transformed into group set coverage. But at a time, each sensor has only one direction of perception. This study investigates the greatest target full-view coverage with time slices under the constraint of a finite number of sensors and the coverage problem of moving targets.

At time t, X represents the set of all directions for all moving targets, $|X| = n$. There are k combined targets in X, i.e., $X_1, X_2, \ldots, X_k$ and $X = X_1 \cup X_2 \cup \cdots \cup X_k$. There is no intersection between any two subsets, $X_i \cap X_j = \emptyset, i \neq j, \forall i, j \in \{1, 2, \ldots, k\}$. $G_1, G_2, \ldots, G_m$ are subset groups of X. m groups of finite set X correspond to m camera sensors.

The full-view coverage maximization problem has been shown to be NP-Hard [3], and under the consideration of temporal slices, it remains NP-Hard.   □

Three stages are required to solve the full-view coverage maximization problem: first, choose l groups from the m groups (pick cameras at l places), and then choose S_p subsets from each group G_p from the l groups (choosing the sensor's detecting direction). Finally, we maximize the number of composed targets covered during γ time slices.

The target side is investigated to optimize the desired full-view coverage. The method of judging whether each facing direction is covered or not is as follows, t is located in the fan-shaped sensing area of c, ct lies in the fan-shaped area of c and the facing direction of t at the same time.

Lemma 1. *The angle σ between the extension line facing direction $\vec{\mathcal{F}}$ and the opposite direction of the sensing direction $\vec{f}$ needs to meet the following conditions,*

$$\frac{\theta - \alpha}{2} \leq \sigma \leq \frac{\alpha + \theta}{2}$$

Proof. When t is within the coverage of c, it is guaranteed that ct is in the perception area of c. Therefore we only need to prove whether ct is within the target's facing direction. When the line segment ct is on the left boundary of the target, as shown in Fig. 4(a), we have

$$\alpha' + \theta' = \sigma$$
$$0 \leq \alpha' \leq \frac{\alpha}{2}$$

Hence,

$$\sigma \leq \frac{\alpha + \theta}{2}$$

When the line segment ct is on the right boundary of the target, as shown in Fig. 4(b), we have

$$\alpha' + \sigma = \pi - \theta'$$
$$0 \leq \alpha' \leq \frac{\alpha}{2}$$

Hence,

$$\sigma = \pi - \theta' - \alpha'$$
$$\geq \frac{\theta - \alpha}{2}$$

$$\frac{\theta - \alpha}{2} \leq \sigma \leq \frac{\alpha + \theta}{2}$$

The lemma is proved.   □

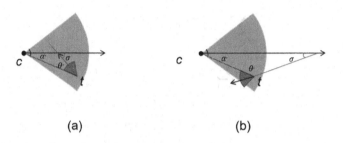

(a)   (b)

Fig. 4. The direction of the target is covered.

3   Problem Formulation

To effectively cover a moving target, in addition to dividing the target into a
full view, time t as another moving feature of the moving target also needs to
be taken into consideration. Therefore, two improved algorithms are proposed
on the basis of MGSC-SC [3].

Assume that the time required for each rotation and scaling of the camera
sensor is Δt. A time slice is defined as Δt, and τth time slice $t_\tau = t_0 + \tau * \Delta t$.

At time t, given m subset groups of a finite set X $G_1, G_2, \ldots, G_m$. The k
composed targets are pairwise disjoint subsets of X, $X_1, X_2, \ldots, X_k$ and $X =
X_1 \cup X_2 \cup \cdots \cup X_k$, $X_i \cap X_j = \emptyset, i \neq j, \forall i, j \in \{1, 2, \ldots, k\}$. First, we select
the position of the sensor, that is, select l groups from m groups. Where l is an
integer, $l > 0$. Then we choose the direction of the camera, that is, select a subset
S_p from each group G_p, $S_1, S_2, \ldots, S_l$. w_p, u_{pS} and v_q are indicator functions. w_p
means whether the group set G_p will be selected, $w_p \in \{0, 1\}$ $\forall p = 1, 2, \ldots, m$.
u_{pS} indicates whether S is selected in group G_p. $v_q = 1$ means that element q is
in the selected subset S. Next we will analyze the proposed two time-slice-based
full-view coverage algorithms.

3.1 Group Set Cover with Time Segmented Cumulative Full-View Target Coverage (TSC-FTC)

To maximize the number of targets covered by the full-view during this time slice, the camera sensor's position and orientation are chosen at time t. However, only a small number of targets can be full-view covered in a time slice. Therefore, we show the cumulative full-view coverage target over a time segment. When t from t_0 to T, $T = t_0 + \gamma * \Delta t$ represents the time spent by the most time-consuming target in a set of composed targets. As depicted in Formula (1),

$$\max \sum_{t=t_0}^{T} \sum_{i=1}^{k} \prod_{q \in X_i} v_q \quad \text{s.t.} \quad v_q \leq \sum_{p=1}^{m} \sum_{S:q \in S \in \mathcal{G}_p} u_{pS} \quad \forall q = 1, ..., n,$$

$$\sum_{S:S \in \mathcal{G}_i} u_{pS} \leq w_p \quad \forall i = 1, ..., m,$$

$$\sum_{p=1}^{m} w_p \leq l \quad \forall p = 1, ..., m \tag{1}$$

$$v_q \in \{0,1\} \quad \forall q = 1, 2, ..., n,$$
$$w_p \in \{0,1\} \quad \forall p = 1, 2, ..., m,$$
$$u_{pS} \in \{0,1\} \quad \forall S \in \mathcal{G} \text{ and } p = 1, 2, ..., m.$$

Its relaxation is as shown below,

$$\max \sum_{t=t_0}^{T} \sum_{i=1}^{k} \prod_{q \in X_i} v_q \quad \text{s.t.} \quad v_q \leq \sum_{p=1}^{m} \sum_{S:q \in S \in \mathcal{G}_p} u_{pS} \quad \forall q = 1, ..., n,$$

$$\sum_{S:S \in \mathcal{G}_i} u_{pS} \leq w_p \quad \forall i = 1, ..., m,$$

$$\sum_{p=1}^{m} w_p \leq l \quad \forall p = 1, ..., m \tag{2}$$

$$0 \leq v_q \leq 1 \quad \forall q = 1, 2, ..., n,$$
$$0 \leq w_p \leq 1 \quad \forall p = 1, 2, ..., m,$$
$$0 \leq u_{pS} \leq 1 \quad \forall S \in \mathcal{G} \text{ and } p = 1, 2, ..., m.$$

$$\mathrm{E}[v_q] = Prob[v_q = 1] \geq (1 - e^{-1})v_q^* \tag{3}$$

$$E\left[\sum_{i=1}^{k} \prod_{q \in X_t} v_q\right] \geq (1 - e^{-1})^{\beta}(1 - \varepsilon)lopt \tag{4}$$

Literature [3] provided the approximate solution as shown in Eq. (3)(4).

Theorem 1.

$$E\left[\sum_{t=t_0}^{T}\sum_{i=1}^{k}\prod_{q\in X_i} v_q\right] \geq \left(1-e^{-1}\right)^{\beta}(1-\varepsilon)lopt_1 \tag{5}$$

Proof. From Formula (3), we have,

$$E[v_q] = Prob[v_q = 1] \geq (1-e^{-1})v_q^*$$

Then,

$$E\left[\sum_{t=t_0}^{T}\sum_{i=1}^{k}\prod_{q\in X_i} v_q\right] = \sum_{t=t_0}^{T}\sum_{i=1}^{k} E\left[\prod_{q\in X_i} v_q\right] = \sum_{t=t_0}^{T}\sum_{i=1}^{k}\prod_{q\in X_i} E[v_q]$$

$$\geq \left(1-e^{-1}\right)^{\beta}\sum_{t=t_0}^{T}\sum_{i=1}^{k}\prod_{q\in X_i} v_q^* = (1-e^{-1})^{\beta}(1-\varepsilon)\cdot lopt_1.$$

The objective value of a locally optimal solution for the Formula (2) is $lopt_1$. The total number of full-view coverage targets that can be reached is β in a time segment T.

The theorem is proved.   □

Algorithm 1 serves as a summary of the above approximation algorithm.

Algorithm 1: TSC-FTC

    Input   : a finite set X; m groups $\mathcal{G}_1, \mathcal{G}_2, \cdots, \mathcal{G}_i, \cdots, \mathcal{G}_m$ of subsets of X;
               a partition of X into composed targets $X = X_1 \cup X_2 \cup \cdots \cup X_k$;
               an integer $r > 0$.
    Output: a collection $\mathcal{S}$; a collection of composed targets $\mathcal{X}$

1 $\mathcal{X} \leftarrow \emptyset$;
2 **for** $t = t_0,\ t \leq T,\ t_{i++}$ **do**
3      $\mathcal{S} \leftarrow \emptyset$;
4   Solve formula (2) to obtain a $(1-\varepsilon)$-approximate solution
        (u_{pS}^*, v_q^*, w_p^*);
5   Select one collection $\mathcal{S}_\ell$ with probability u_{pS}^*/v_q^*;
6   **for** *each* $i \in \mathcal{S}_i$ **do**
7   Select one subset S with probability u_{pS}^*/v_q^* from $\mathcal{G}_i$;
8          $S \leftarrow S \cup \{S\}$.
9   **end**
10   return $\mathcal{S}$;
11 **end**
12 The set of objective functions composed of selected subsets $\mathcal{X}$;
13 $\mathcal{X} \leftarrow \mathcal{X} \cup \{X\}$.
14 return $\mathcal{X}$;

3.2   Group Set Cover with Cumulative Maximum Full-View Target Coverage Time Window (FTC-TW)

TSC-FTC effectively utilizes all coverage resources in a time slice, and realizes the largest coverage in each time slice, but cannot achieve the best coverage effect. FTC-TW enables fast maximum full-view coverage throughout the entire time segment. First calculate all the target facing directions that can be covered in a time slice. Since one time slice cannot cover all target orientations, we maximize the number of objects covered by the full view within a time window T. Multiple time slices form a time window. Choose a camera position and orientation that maximizes full-view target coverage over the entire time window. As illustrated by the formulation in (6),

$$\max \sum_{i=1}^{k} \prod_{q \in X_i} \sum_{t=t_0}^{T} v_q \quad \text{s.t.} \quad v_q \leq \sum_{p=1}^{m} \sum_{S:q \in S \in \mathcal{G}_p} u_{pS} \quad \forall q = 1, ..., n,$$

$$\sum_{S:S \in \mathcal{G}_i} u_{pS} \leq w_p \quad \forall i = 1, ..., m,$$

$$\sum_{p=1}^{m} w_p \leq l \quad \forall i = p, ..., m \tag{6}$$

$$v_q \in \{0, 1\} \quad \forall q = 1, 2, ..., n,$$
$$w_p \in \{0, 1\} \quad \forall p = 1, 2, ..., m,$$
$$u_{pS} \in \{0, 1\} \quad \forall S \in \mathcal{G} \text{ and } p = 1, 2, ..., m.$$

Relax as follows,

$$\max \sum_{i=1}^{k} \prod_{q \in X_i} \sum_{t=t_0}^{T} v_q \quad \text{s.t.} \quad v_q \leq \sum_{p=1}^{m} \sum_{S:q \in S \in \mathcal{G}_p} u_{pS} \quad \forall q = 1, ..., n,$$

$$\sum_{S:S \in \mathcal{G}_i} u_{pS} \leq w_p \quad \forall i = 1, ..., m,$$

$$\sum_{p=1}^{m} w_p \leq l \quad \forall i = p, ..., m \tag{7}$$

$$0 \leq v_q \leq 1 \quad \forall q = 1, 2, ..., n,$$
$$0 \leq w_p \leq 1 \quad \forall p = 1, 2, ..., m,$$
$$0 \leq u_{pS} \leq 1 \quad \forall S \in \mathcal{G} \text{ and } p = 1, 2, ..., m.$$

Theorem 2.

$$E\left[\sum_{i=1}^{k} \prod_{q \in X_i} \sum_{t=t_0}^{T} v_q \right] \geq \left(1 - e^{-\gamma}\right)^{\beta} (1 - \varepsilon) lopt_2 \tag{8}$$

Proof.

$$\text{Prob}\left[v_q = 0\right] = \left[\prod_{i=1}^{m} \prod_{S:q\in S\in\mathcal{G}_i} \left(1 - u_{pS}^*\right)\right]^{\gamma}$$

$$\leq \left[\left(\frac{\sum_{i=1}^{m} \sum_{S:q\in S\in\mathcal{G}_i} \left(1 - u_{pS}^*\right)}{K_q}\right)^{K_q}\right]^{\gamma}$$

$$= \left(1 - \frac{\sum_{i=1}^{m} \sum_{S:q\in S\in\mathcal{G}_i} u_{pS}^*}{K_q}\right)^{\gamma \cdot K_q}$$

$$\leq \left(1 - \frac{v_q^*}{K_q}\right)^{\gamma \cdot K_q}$$

$$K_q = |\{(q,S) \mid q \in S \in \mathcal{G}_i\}|$$

Hence,

$$\text{Prob}[v_q = 1] = 1 - Prob\left[v_q = 0\right]$$

$$\geq 1 - \left(1 - \frac{v_q^*}{K_q}\right)^{\gamma * K_q} .$$

$$\geq \left(1 - e^{-\gamma}\right) v_q^*$$

$$\text{E}\left[\sum_{i=1}^{k} \prod_{q\in X_i} \sum_{t=t_0}^{T} v_q\right] = \sum_{i=1}^{k} \prod_{q\in X_i} \sum_{t=t_0}^{T} E\left[v_q\right]$$

$$\geq \left(1 - e^{-\gamma}\right)^{\beta} \sum_{i=1}^{k} \prod_{q\in X_i} \sum_{t=t_0}^{T} v_q^*$$

$$\geq \left(1 - e^{-\gamma}\right)^{\beta} \left(1 - \varepsilon\right)lopt_2$$

The objective value of a locally optimal solution for the formula (7) is $lopt_2$. The total number of full-view coverage targets that can be reached is β in a time segment T. □

This approximate algorithm is summarized as Algorithm 2.

Algorithm 2: FTC-TW

   Input   : a finite set X; m groups $\mathcal{G}_1, \mathcal{G}_2, \cdots, \mathcal{G}_i, \cdots, \mathcal{G}_m$ of subsets of X;
                a partition of X into composed targets $X = X_1 \cup X_2 \cup \cdots \cup X_k$;
                an integer $r > 0$.
   Output: a collection $\mathcal{S}$; a collection of composed targets $\mathcal{X}$

1 $\mathcal{X} \leftarrow \emptyset$;
2 **for** $t = t_0,\ t \leq T,\ t_{i++}$ **do**
3     $\mathcal{S} \leftarrow \emptyset$;
4   Solve formula (2) to obtain a $(1 - \varepsilon)$-approximate solution (u^*_{pS}, v^*_q, w^*_i);
5   Select one collection $\mathcal{S}_\ell$ with probability u^*_{pS}/v^*_q;
6   **for** *each $i \in \mathcal{S}_i$* **do**
7   Select one subset S with probability u^*_{pS}/v^*_q from $\mathcal{G}_i$;
8         $\mathcal{S} \leftarrow \mathcal{S} \cup \{S\}$.
9   **end**
10   **return** $\mathcal{S}$;
11 **end**
12 The set of objective functions composed of selected subsets $\mathcal{X}$;
13 $\mathcal{X} \leftarrow \mathcal{X} \cup \{X\}$.
14 **return** $\mathcal{X}$;

---

4 Conclusion

In conclusion, this paper proposes two approximation methods based on Group Set Cover to optimize full-view coverage in the context of time slices. Both approaches choose a collection of sensors that can cover the most targets in each time slice, and then calculating the cumulative number of targets covered over time or after a certain period of time. TSC-FTC effectively utilizes the resources in each time slice and enables the evaluation of coverage effectiveness. Based on TSC-FTC, FTC-TW enables a quick selection of the set of sensors covering the maximum number of targets. Overall, the proposed methods offer practical and efficient solutions to the problem of full-view target coverage in the context of time slices. The results demonstrate that these methods can significantly improve the effectiveness of full-view target coverage and provide a useful tool for surveillance systems.

Acknowledgments. This work is supported by National Natural Science Foundation of China (No. 62172124), the Shenzhen Basic Research Program (Project No. JCYJ20190806143011274).

References

1. Liu, C., Du, H.: t, K-Sweep coverage with mobile sensor nodes in wireless sensor networks. IEEE Internet Things J. **8**(18), 13888–13899 (2021)
2. Luo, H., Du, Kim, D., Ye, Q., Zhu, R., Jia, J.: Imperfection better than perfection: beyond optimal lifetime barrier coverage in wireless sensor networks. In: 2014 10th International Conference on Mobile Ad-hoc and Sensor Networks, Maui, HI, USA, pp. 24–29 (2014)
3. Fleischer, R., Trippen, G.: Kayles on the way to the stars. In: van den Herik, H.J., Björnsson, Y., Netanyahu, N.S. (eds.) CG 2004. LNCS, vol. 3846, pp. 232–245. Springer, Heidelberg (2006). https://doi.org/10.1007/11674399_16
4. Wu, P.F., Xiao, F., Sha, C., et al.: Node scheduling strategies for achieving full-view area coverage in camera sensor networks. Sensors **17**(6), 1303 (2017)
5. Sun, Z., Zhao, G., Xing, X.: ENCP: a new energy-efficient nonlinear coverage control protocol in mobile sensor networks. EURASIP J. Wirel. Commun. Netw. **2018**, 1–15 (2018)
6. Zhu, X., Zhou, M.C.: Multiobjective optimized deployment of edge-enabled wireless visual sensor networks for target coverage. IEEE Internet Things J. (2023)
7. Wang, Y., Cao, G.: On full-view coverage in camera sensor networks. In: 2011 Proceedings IEEE INFOCOM, pp. 1781–1789. IEEE (2011)
8. Zhu, X., Zhou, M., Abusorrah, A.: Optimizing node deployment in rechargeable camera sensor networks for full-view coverage. IEEE Internet Things J. **9**(13), 11396–11407 (2021)
9. Liu, Z., Jiang, G.: Sensor density for full-view problem in heterogeneous deployed camera sensor networks. KSII Trans. Internet Inf. Syst. (TIIS) **15**(12), 4492–4507 (2021)
10. Gan, X., Zhang, Z., Fu, L., et al.: Unraveling impact of critical sensing range on mobile camera sensor networks. IEEE Trans. Mob. Comput. **19**(4), 982–996 (2019)
11. He, S., Shin, D.H., Zhang, J., et al.: Full-view area coverage in camera sensor networks: dimension reduction and near-optimal solutions. IEEE Trans. Veh. Technol. **65**(9), 7448–7461 (2015)
12. Wu, Y., Wang, X.: Achieving full view coverage with randomly-deployed heterogeneous camera sensors. In: 2012 IEEE 32nd International Conference on Distributed Computing Systems, pp. 556–565. IEEE (2012)
13. Shi, K., Liu, S., Li, C., et al.: Toward optimal deployment for full-view point coverage in camera sensor networks. IEEE Internet Things J. **9**(21), 22008–22021 (2022)
14. Wang, Y., Cao, G.: Achieving full-view coverage in camera sensor networks. ACM Trans. Sens. Netw. (ToSN) **10**(1), 1–31 (2013)
15. Liu, Z., Jiang, G.: Sensor parameter estimation for full-view coverage of camera sensor networks based on bounded convex region deployment. IEEE Access **9**, 97129–97137 (2021)

Improved Bounds for the Binary Paint Shop Problem

J. Hančl[1,2,3], A. Kabela[1], M. Opler[2], J. Sosnovec[1,2,3], R. Šámal[3], and P. Valtr[3(✉)]

[1] Faculty of Applied Sciences, University of West Bohemia, Pilsen, Czech Republic
kabela@kma.zcu.cz, j.sosnovec@email.cz
[2] Faculty of Information Technology, Czech Technical University, Prague, Czech Republic
michal.opler@fit.cvut.cz
[3] Faculty of Mathematics and Physics, Charles University, Prague, Czech Republic
samal@iuuk.mff.cuni.cz, valtr@kam.mff.cuni.cz

Abstract. We improve bounds for the binary paint shop problem posed by Meunier and Neveu [Computing solutions of the paintshop-necklace problem. Comput. Oper. Res. 39, 11 (2012), 2666-2678]. In particular, we disprove their conjectured upper bound for the number of color changes by giving a linear lower bound. We show that the recursive greedy heuristics is not optimal by providing a tiny improvement. We also introduce a new heuristics, recursive star greedy, that a preliminary analysis shows to be 10% better.

1 Introduction

A *double occurrence word* w is a word (sequence of letters) in which each of its letters occurs exactly twice. A *legal 2-coloring* of w is a coloring of individual letters such that each letter occurs once red and once blue. We let W_n denote the set of all double occurrence words with letters $A_1, \ldots, A_n$, each occurring twice. We denote the first, resp. the second, occurrence of a letter A by $\underline{A}$, resp. $\bar{A}$.

Our goal is to find, for a double occurrence word w, a legal 2-coloring of w with minimal number of color changes – neighboring letters of different colors. We use $\gamma(w)$ to denote this quantity.

To state this more formally, a legal coloring of $w = w_1 w_2 \ldots w_{2n} \in W_n$ is a mapping $c : \{1, \ldots, 2n\} \to \{0, 1\}$ such that $i \neq j$ and $w_i = w_j$ implies $c(i) \neq c(j)$. We define the number of color changes

$$\gamma(w) := \min_{\text{legal } c} |\{i < 2n : c(i) \neq c(i+1)\}|.$$

A. Kabela—Supported by project 20-09525S of the Czech Science Foundation.

R. Šámal—Partially supported by grant 22-17398S of the Czech Science Foundation. This project has received funding from the European Research Council (ERC) under the European Unions Horizon 2020 research and innovation programme (grant agreement No. 810115).

P. Valtr—Supported by the grant no. 23-04949X of the Czech Science Foundation.

W. Wu and G. Tong (Eds.): COCOON 2023, LNCS 14423, pp. 210–221, 2024.
https://doi.org/10.1007/978-3-031-49193-1_16

The motivation for the definition (and for the "paint shop" in the title of the paper) is the following: Imagine a line of cars in a paint shop factory, suppose that there are just two cars of each type, and that we need to paint one of them blue and the other red. To decrease cost, we want to minimize the number of color changes. This problem was introduced by Epping et al. [6] under the name binary paint shop problem (non-binary case considers more than two colors).

Another guise of the same problem is necklace splitting: two thieves stole a necklace with $2n$ gems stones, two of each of n types. As the price of each of the gems is unknown, they want to cut the necklace, so that each of the thieves can get one of each type of the gems. Alon's necklace-splitting theorem [1] gives an upper bound for this (and for much more general) version. Translated to our setting, this result says that $\gamma(w) \leq n$ for every $w \in W_n$. Easily, this is tight for $w = A_1A_1A_2A_2 \ldots A_nA_n$.

It is known [6] that deciding whether $\gamma(w) \leq k$ for given w and k is NP-complete. Moreover, γ is APX-hard [5,8]. Thus, the study of various heuristics is in order. A natural way to evaluate them is to look at the behaviour on random instances of W_n.

This motivates the following notion, which will be of our main interest: We let γ_n be the expectation of $\gamma(w)$ when w is a random element of W_n,

$$\gamma_n := \mathbb{E}_n \, \gamma$$

where $\mathbb{E}_n \gamma$ is a shortcut for $\mathbb{E}_{w \in W_n} \gamma(w)$. To exemplify, W_2 consists of the following words: $AABB$, $ABAB$, $ABBA$ (up to renaming the letters). As $\gamma(AABB) = 2$, $\gamma(ABAB) = \gamma(ABBA) = 1$, we have $\gamma_2 = 4/3$.

Andres and Hochstättler [4] describe two heuristics, greedy and recursive greedy. Use $g(w)$ and $rg(w)$ for the resulting number of color changes. They prove that

$$\mathbb{E}_n \, g = \tfrac{1}{2}n + O(1) \qquad \text{and} \qquad \mathbb{E}_n \, rg = \tfrac{2}{5}n + \tfrac{7}{10}.$$

Consequently, $\gamma_n \leq \tfrac{2}{5}n + \tfrac{7}{10}$. Meunier and Neveu [7] say: "We were not able to propose an interesting lower bound on $\mathbb{E}_n \gamma$, but we conjecture that $\mathbb{E}_n \gamma = o(n)$."

In the second section, we disprove this conjecture, proving $\gamma_n \geq 0.214n - o(n)$. In the third section we slightly improve the recursive greedy heuristics, getting an upper bound of $(0.4 - \varepsilon)n$ for any $\varepsilon < 10^{-6}$. This shows that recursive greedy heuristics is not optimal. In the fourth section we describe a new simple heuristics "recursive star greedy" that probably satisfies $\mathbb{E}_n \, rsg \leq 0.361n$, although we are not able to prove that. (An ad-hoc algorithm gets even better results, but we cannot analyze its performance at all.) Finally, in the fifth section we indicate how to use Azuma inequality to show that the value of $\gamma(w)$ is concentrated on a short interval.

Recently, Andres [3] cites preliminary version of our work [10] and continues the research of the paintshop problem by comparing optimality of various heuristics.

When we finalized our work (after some pause) for publication we became aware of a closely related work. Alon et al. [2] revisit the problem of necklace splitting for randomly constructed necklaces. This has led to parallel study of equivalent problems and as a result their and our manuscripts have a partial overlap. We now briefly describe their results in our terminology. They study the problem in greater generality – each letter occurs in km copies, the goal is to find coloring with k colors such that each letter has m copies of each color and the number of color changes is minimized. In the context of our paper ($k = 2$, $m = 1$) their Theorem 1.4 gives a lower bound (in a similar way as we do in Theorem 1). They rediscover upper bound $0.4n + o(n)$ of Andres and Hochstättler [4]. They also sketch an argument that both upper and lower bound can be improved and that the optimal number of colors is sharply concentrated.

2 Lower Bounds

We could not find in the literature any mention of lower bounds for γ_n. As a warm-up, we present a very simple lower bound using random interval graphs. Scheinerman [9] introduced a model of random interval graphs that is closely related to our problem. He starts by choosing at random, uniformly and independently, $x_i, y_i \in [0, 1]$ for $i = 1, \dots, n$. Then he creates an interval graph G of intervals $[x_i, y_i]$. Explicitly, the vertices of G are $[n] = \{1, \dots, n\}$ and ij is an edge of G whenever intervals $[x_i, y_i]$ and $[x_j, y_j]$ intersect.

With probability 1, no two of the selected $2n$ real numbers coincide. Thus, we may to our random selection of x_i's and y_i's also assign a double-occurrence word: we sort $X = \{x_i, y_i : i \in [n]\}$ and replace both x_i and y_i by A_i (the i-th letter). If w is the resulting word, we call G the interval graph associated with w and use $IG(w)$ to denote it.

It is easy to see that $\gamma(w)$ is equal to the smallest k such that there is a set $S \subset [0, 1]$ of size k, such that for every i the size of $[x_i, y_i) \cap S$ is odd. This leads to an easy lower bound for $\gamma(w)$ in terms of properties of random interval graphs.

Claim. $\gamma(w) \geq \alpha(IG(w))$

As Scheinerman [9] proves that expectation of $\alpha(G)$ is at least $C\sqrt{n}$, we have the following corollary.

Corollary 1. *There is a $C > 0$ such that $\gamma_n \geq C\sqrt{n}$.*

Next, we provide a linear lower bound on γ_n, disproving the conjecture of Meunier and Neveu [7].

Theorem 1. $\gamma_n \geq 0.214n - o(n)$

Proof. We let w be a uniformly random element of W_n. We will show, for an appropriate choice of k and p, that

$$\Pr[\gamma(w) \leq k] \leq p.$$

This will prove that

$$\gamma_n = \mathbb{E}\,\gamma(w) \geq (1 - p)k. \tag{1}$$

Let $C_n^{\leq k}$ be the set of all possible binary colorings of $1, \ldots, 2n$ using n zeros and n ones, starting with a zero, that have at most k color changes. We use union bound and straightforward estimates:

$$\Pr[\gamma(w) \leq k] = \Pr[w \text{ has a legal coloring in } C_n^{\leq k}]$$

$$\leq \sum_{C \in C_n^{\leq k}} \Pr[C \text{ is legal for } w]$$

$$= \sum_{C \in C_n^{\leq k}} \frac{n!^2}{(2n)!/2^n} \leq \sum_{l=0}^{k} \binom{2n-1}{l} \frac{2^n}{\binom{2n}{n}}$$

$$\leq \frac{\sqrt{4n}}{2^n} \sum_{l=0}^{k} \binom{2n}{l} \leq \frac{\sqrt{4n}}{2^n} \left(\frac{e \cdot 2n}{k} \right)^k$$

So we may let p be equal to the last line and we have proved (1).

Put $a = k/n$. Then $p = \sqrt{4n} \left(\left(\frac{2e}{a} \right)^a \frac{1}{2} \right)^n$. Thus if $\left(\frac{2e}{a} \right)^a \frac{1}{2} < 1$, then we have $p = o(1)$ and, thus,

$$\mathbb{E}\,\gamma(w) \geq (1 - o(1))an.$$

Numerical computation shows, that this works whenever $a < 0.214\ldots$. This finishes the proof.

3   Small Improvement of the Upper Bound

In [4] the *recursive greedy* (RG) heuristics was used to prove the upper bound $\gamma_n \leq 0.4n + O(1)$. We slightly improve that result by decreasing the linear term. Main interest of this is showing that $0.4n$ is not the final answer.

In the original proof the following recursive greedy heuristics was used. To obtain a coloring

$$c : [2n] = \{1, 2, \ldots, 2n\} \rightarrow \{0, 1\}$$

of the double occurrence word $w \in W_n$, we first omit both occurrences of the first letter of w, say A. Then we color the shorter word recursively, add both letters A back and color them in the best possible way. Before we give more formal description, recall that we denote the first, resp. second, occurrence of a letter A by $\underline{A}$, resp. $\bar{A}$.

Recursive greedy algorithm: Let 1 and j be the position of $\underline{A}$ and $\bar{A}$ in a word $w \in W_n$ and $c' : [2n - 2] \rightarrow \{0, 1\}$ be the coloring of $w' \in W_{n-1}$. We define the coloring c of w such that it coincides with c' on the letters of w', hence only colors $c(\underline{A}) = c(1)$ and $c(\bar{A}) = c(j)$ are to be determined. Let $N_{c'}(\bar{A}) = \{c'(j-2), c'(j-1)\}$ be the multiset of colors that c' uses in the neighborhood of $\bar{A}$.
(a) If $j = 2$ then $c(\underline{A}) = 1 - c'(1)$.
(b) If $j = 2n$ and $n > 1$ then $c(\underline{A}) = c'(1)$.

(c) If $j \in [3, 2n - 1]$ then

$$c(\underline{A}) = \begin{cases} c'(1) & \text{if } N_{c'}(\bar{A}) \text{ contains } 1 - c'(1), \\ 1 - c'(1) & \text{if } N_{c'}(\bar{A}) = \{c'(1), c'(1)\}. \end{cases}$$

and color $\bar{A}$ accordingly.

Note that in our description of RG we reversed the input of the original algorithm and colored first letter first (in [4] they colored last letter first). We use the following observation.

Lemma 1. *Let $n \geq 1$ and $w \in W_n$ be a fixed word colored by RG. Then the number of neighboring pairs in w that are colored with color c is at least $\lfloor (n - 1)/2 \rfloor$.*

Proof. We prove by induction that $rg(w)$, i.e. the number of color changes in the coloring of w obtained from RG, is at most n. This clearly holds for $n = 1$ and in every recursive step RG adds two letters and at most one color change. Since every color colors the same number of letters we conclude that half of the remaining $n - 1$ neighboring pairs have the color c.

As proved in [4], this algorithm results in $\mathbb{E}_n \, rg \leq 0.4n + O(1)$. However, we can improve the final coloring of RG. We find a set V of pairs of letters X, Y such that $w = \ldots A\underline{X}B \ldots C\underline{Y}D \ldots E\bar{X}\bar{Y}F \ldots$, there are two color changes around both $\underline{X}$ and $\underline{Y}$ and none around the pair $\bar{X}\bar{Y}$. Recoloring both copies of X and Y decreases the number of color changes by two. Moreover, we will prove that expected size of V is linear in n, getting the following theorem.

Theorem 2. *For $\varepsilon \approx 2 \times 10^{-6}$ and sufficiently large n we have*

$$\gamma_n < \left(\frac{2}{5} - \varepsilon \right) n.$$

Proof. Let (for simplicity) n be even and $w \in W_n$. Denote the letters of w as $A_1, A_2, \ldots, A_n$ such that the first occurrence of A_j lies before the first occurrence of A_i whenever $i < j$, i.e. the letters are ordered in descending order by their first occurrences. We denote by $\tau_k(w) \in W_k$ the word obtained from w by removing every letter A_i for $i > k$. Note that $\tau_k(w)$ is the word that the RG algorithm considers at the k-th deepest level of recursion. We split RG into two stages. In the first stage we color the word $\tau_{n/2}(w)$, in the second stage we extend that coloring onto w; in both stages we color $n/2$ pairs of letters.

Let us fix $w' \in W_{n/2}$ and denote by $c' : [n] \to \{0, 1\}$ the coloring in the output of RG for w'. We set

$$T = \{t \in [n - 1] : c'(t) = c'(t + 1) = 0\}.$$

to be the set of positions of two consecutive letters colored by 0. We know that T contains at least $\lfloor (n - 2)/4 \rfloor$ elements due to Lemma 1. We refer to the elements of T as monochromatic pairs. We consider the probability space of all words $w \in W_n$ that were built from w'; that is $\tau_{n/2}(w) = w'$.

Let $t \in T$. We estimate the probability $p = p(t)$ such that:

S1. In the second stage of RG there are exactly two letters, say C and D, inserted in between the monochromatic pair t. Moreover, no letter is ever inserted around C and D (other than the one in the next step).

S2. First occurrences C and D are colored so that there are two color changes around.

Let $U_{i,j}$ be the set of words in W_n such that S1, S2 holds and letter C and D is the i-th and j-th added letter in stage two for indices $i < j$. We denote by $p_{i,j}$ the probability that $w \in U_{i,j}$ which allows us to express p as

$$p = \sum_{1<i<j<n/2} \Pr[w \in U_{i,j}] = \sum_{1<i<j<n/2} p_{i,j}.$$

We remark that if $j = i+1$ then the first occurrences of C and D cannot have two changes around them and $p_{i,j} = 0$. Let us therefore assume that $j - i \geq 2$. We let $U_{i,j}^k$ be the set of words of length $n/2 + k$ that can be extended to some word $w \in U_{i,j}$, that is

$$U_{i,j}^k = \{\tau_{n/2+k}(w) \mid w \in U_{i,j}\}.$$

Observe that $U_{i,j}^{n/2}$ is precisely the set $U_{i,j}$ while $U_{i,j}^0$ contains only the word w'. We define the probability that after adding k letters in the second stage we did not violate any of the desired properties conditioned by the fact that the same holds after adding $k - 1$ letters.

$$p_{i,j}^k = \Pr[\tau_{n/2+k}(w) \in U_{i,j}^k \mid \tau_{n/2+k-1}(w) \in U_{i,j}^{k-1}].$$

We can chain these probabilities and get

$$p_{i,j} = \prod_{k=1}^{n/2} p_{i,j}^k.$$

We proceed by providing lower bounds on $p_{i,j}^k$ for every k between 1 and $n/2$. Let us denote the k-th added letter by A_k.

(1) For $k = i$, we have to insert $\bar{C}$ inside the pair t and thus $p_{i,j}^i = \frac{1}{n+2i-1}$.

(2) For $k = j$, we have to insert $\bar{D}$ next to $\bar{C}$ and thus $p_{i,j}^j = \frac{2}{n+2j-1}$.

(3) For k such that $|k-i| = 1$ or $|k-j| = 1$, we have to guarantee that A_k receives the color opposite to the color of C and D. Due to Lemma 1, we have at least $\lfloor ((n/2)+k-2)/2 \rfloor \geq (n+2k-6)/4$ options and $p_{i,j}^k \geq \frac{n+2k-6}{4(n+2k-1)} > \frac{n+2k-8}{4(n+2k-1)}$.

(4) Otherwise, we can insert $\bar{A}_k$ everywhere except in t or around C and D. There are at most 7 forbidden positions and thus $p_{i,j}^k \geq \frac{n+2k-8}{n+2k-1}$.

There can be either 3 or 4 positions for the case (3) depending on whether $j - i = 2$. However we can assume that there are 4 of them as we are aiming to obtain a lower bound. Putting it all together, we get

$$p_{i,j} \geq \frac{1}{(n+2i-1)} \frac{2}{(n+2j-1)} \frac{1}{4^4} \prod_{\substack{k=1 \\ k \neq i,j}}^{n/2} \frac{n+2k-8}{n+2k-1}$$

$$= \frac{1}{2^7(n+2i-8)(n+2j-8)} \prod_{k=1}^{n/2} \frac{n+2k-8}{n+2k-1}$$

$$> \frac{1}{2^7(2n-3)^2} \left(\frac{n-7}{n}\right)^{n/2}.$$

As we remarked, the lower bound does not hold whenever $j-i=1$. Furthermore, we cannot guarantee the condition S2 if $i=1$. Summing over all other choices of i and j we obtain

$$p > \left(\binom{n/2}{2} - n\right) \frac{1}{2^7(2n-3)^2} \left(\frac{n-7}{n}\right)^{n/2} \to \frac{e^{-7/2}}{2^{12}} > 7 \cdot 10^{-6}.$$

Thus there exists n_0 such that for $n \geq n_0$ and for fixed $w' \in W_{n/2}$ the expected size of $V \subset T$ for which $S1$ and $S2$ holds is

$$\mathbb{E}_n |V| = p \cdot |T| > \left\lfloor \frac{n-2}{4} \right\rfloor p > n \cdot 10^{-6}.$$

By recoloring both letters inserted between a color pair of V we decrease the number of color changes by two. Notice that the condition S2 guarantees that letters inserted into different color pairs of V do not neighbor since their first occurrences are colored with 1. Therefore recoloring all inserted letters for all color pairs of V decreases the number of color changes by $2|V|$ which implies

$$\gamma_n < \frac{2}{5}n + O(1) - 2\mathbb{E}_n|V| = \left(\frac{2}{5} - 2 \cdot 10^{-6}\right) n.$$

Setting $\varepsilon = 2 \cdot 10^{-6}$ we get the bound.

4   Recursive Star Greedy Heuristics

In this section we describe new heuristic for binary paint shop problem called *recursive star greedy* (RSG) and discuss its mean output that appears to be approximately $0.361n$. That bound is better than the previously described RG heuristics; however, we are not able to prove that rigorously.

Let us start with a simple but crucial observation. In a legal coloring of a word w there might be a letter X such that when we flip the colors of X and $\bar{X}$ the total number of color changes remains the same. Then during the recursive coloring process, we might use this color flip to avoid introducing a new color change.

The *recursive star greedy* (RSG) heuristics is the following modification of RG. It introduces an additional color $*$ which marks that the occurrences of a given letter can be colored in any of the two legal ways without changing the total number of color changes. RSG maintains the invariant that for any letter X, either both $\underline{X}$ and $\bar{X}$ are colored by stars, or they are colored by different binary colors, and that there are no two neighboring letters both colored by star. At the point when another letter is inserted next to a letter X colored by star, RSG recolors $\underline{X}$ and $\bar{X}$ with binary colors if it prevents increasing the number of color changes. The final output of RSG is then a binary coloring of w obtained by recoloring all stars with binary colors in an arbitrary (legal) way.

More precisely, let $w \in W_n$. RSG outputs two colorings of w, a coloring c^* using the color set $C = \{0, 1, *\}$ and a binary coloring c; that is,

$$c^* : [2n] \to C \quad \text{and} \quad c : [2n] \to \{0, 1\}.$$

We proceed recursively with adding (both copies $\underline{X}$, $\bar{X}$ of) the first letter A but keep in mind the positions of the last added letter B, resp. record the number of color changes around B to possibly recolor it by a star.

Recursive star greedy: Let 1 and j be the positions of a letter A in a word $w \in W_n$. Let $w' \in W_{n-1}$ be the word that one obtains from w by deleting $\underline{A}$ and $\bar{A}$, B be the first letter of w', and c' be the C-coloring of w' from the recursion. Recall that $N_{c'}(\bar{A}) = \{c'(j-2), c'(j-1)\}$ denotes the multiset of colors that c' uses in the neighborhood of $\bar{A}$. We define the coloring c^* of w by the case analysis below. Note that we use the notation $c^*(\underline{A}) = c^*(1)$ and $c^*(\bar{A}) = c^*(j)$.
  (A) If $j = 2$ and $n > 1$ we set $c^*(\underline{A}) = 1 - c'(B)$.
  (B) If $j = 2n$ we set $c^*(\underline{A}) = 1$.
  (C) If $N_{c'}(\bar{A}) = \{t, t\}$ for some $t \in \{0, 1\}$ we set $c^*(\underline{A}) = 1 - t$.
  (D) If $N_{c'}(\bar{A}) = \{0, 1\}$ we set $c^*(\underline{A}) = c'(B)$.
  (E) If $N_{c'}(\bar{A}) = \{1 - c'(B), *\}$ we set $c^*(\underline{A}) = c'(B)$.
  (F) If $N_{c'}(\bar{A}) = \{c'(B), *\}$ we set $c^*(\underline{A}) = c'(B)$. Moreover, we set $c^*(X_1) = 1 - c'(B)$ and $c^*(X_2) = c'(B)$, where X_1 is the neighbor of $\bar{A}$ with star color and X_2 is the other copy of the same letter.
We color $c^*(\bar{A})$ accordingly. Finally, if $\underline{B}$ and $\bar{B}$ are not neighboring, all their neighbors are colored with binary colors and swapping the colors of $\underline{B}$ and $\bar{B}$ preserves the total number of color changes, we set $c^*(\underline{B}) = c^*(\bar{B}) = *$. The output of RSG for w is the coloring c^* and a binary coloring c that is obtained by coloring every $*$-colored letter X arbitrarily, say $\underline{X}$ with 0 and $\bar{X}$ with 1.

An example of better performance of RSG over RG is for the word $w = ABCBDCAD$, for which one obtains $rg(w) = 3$, but $rsg(w) = 2$. We follow both algorithms in the following table.

On the other hand, there are words for which RG still performs better than RSG. Surprisingly, we can create a word v (e.g. the one in next table) by adding two new letters to w in a way that they both introduce one new color change

algorithm	$ABCBDCAD$	$BCBDCD$	$CDCD$	DD
RG	1 0 1 1 1 0 0 0	0 1 1 1 0 0	1 1 0 0	1 0
RSG	0 0 0 1 1 1 1 0	0 * 1 1 * 0	1 1 0 0	1 0

algorithm	$WXABCBDWCADX$	$XABCBDCADX$
RG	1 1 1 0 1 1 1 0 0 0 0 0	1 1 0 1 1 1 0 0 0 0
RSG	0 1 * 0 0 1 1 1 1 * 0 0	1 * 0 0 1 1 1 * 0 0

to RSG, but not to RG. Hence $rg(v) = 3$ but $rsg(v) = 4$. Observe that the star color is introduced only to color the letter B at the end of the recursive step. Every star is then recolored back to a binary color either at the end with the transformation of C-coloring into a binary coloring, or in case (F) where it allows us to color $\underline{A}$ and $\bar{A}$ without increasing the total number of changes. That is in contrast to RG, for which the number of color changes in such step might have increased.

Computer experiments suggest, that this modification leads to a significant saving in terms of color changes:

Conjecture 1. $\mathbb{E}_n \, rsg = \alpha n + o(n)$, where $\alpha = (\sqrt{37} - 5)/3 \doteq 0.361$ is the positive solution of $3\alpha^2 + 10\alpha = 4$.

We were unable to prove this conjecture. However, we provide below some of the arguments we have tried. They explain where the constant α comes from.

Let c^*, resp. c, be the C-coloring, resp. binary coloring, of a random word $w \in W_n$ produced by RSG. Let s_n be the probability that a random letter A of w is assigned the star color in the coloring c^*. Then ns_n is the expected number of pairs of stars (coupled by the letter they color) in c^*. Furthermore, let $a_n/2$ be the probability that a random pair of neighbouring letters in w constitute a color change in the coloring c. Observe that $\mathbb{E}_n \, rsg = (2n - 1)a_n/2$ is the value we are interested in.

One could express the probabilities of the individual cases (A) – (F) with the variables a_n and s_n, which leads to recurrence relations that allow one to compute $\mathbb{E}_n \, rsg$ for fixed n in time that is polynomial in n. However, we were not able to solve these recurrences in general.

Let us thoroughly analyze the recursive step and support Conjecture 1. For a word $w \in W_n$ and $j \in [1, 2n - 1]$, let $N_c(w, j)$ be the pair of colors $c(j)c(j+1)$. Based on s_n and a_n we can count the probabilities of all variations of pairs of neighboring colors in c^*-coloring. They are displayed in the Pr-column of the Table 1. For example, in case (D2), the probability of a color with condition $c(2) = 1$ is

$$\Pr_{w,j}[(D2)] = \frac{1}{2}\left(\Pr_{w,j}[N_c(w,j) \in \{01, 10\}] - \frac{1}{2}\Pr_{w,j}[* \in N_{c^*}(w,j)]\right) + O(1/n)$$

$$= \frac{1}{2}\left(\frac{a_n}{2} - s_n\right) = \frac{a_n}{4} - \frac{s_n}{2} + O(1/n),$$

where the first fraction $1/2$ refers to the condition $c(2) = 1$ and the second fraction $1/2$ is there because only half of the color changes with stars produce $01/10$ color changes (others produce $00/11$). Finally, the $O(1/n)$ term includes the case when $|N_c(w, j)| = 1$, i.e. when $\bar{A}$ gets inserted either at the beginning or at the end of the word. Other probabilities of Table 1 up to the error term $O(1/n)$ can be calculated similarly. Set $q_n = 1/4 - a_n/8 - s_n/4$ to be the probability of cases (C1)-(C4).

Table 1. Case analysis of recursive star greedy heuristics: $N_{c'}(\bar{A})$ is the multiset of colors around $\bar{A}$, $c(\underline{A}) = c(1)$ the resulting color of A, $N_c(w, 1) = c(1)c(2)$, Δa_n, resp. Δs_n are the increases of number of color changes of c, resp. pairs of stars, Pr is the probability for that case up to $O(1/n)$ error term and $q_n = 1/4 - a_n/8 - s_n/4$.

		Recursive star greedy					
Case	w	$N_{c'}(\bar{A})$	$c(\underline{A})$	$N_c(w, 1)$	Δa_n	Δs_n	Pr[Case]
(A1)	$\underline{A}\bar{A}0\ldots\ldots$	$\{0\}$	1	10	$+1$		0
(A2)	$\underline{A}\bar{A}1\ldots\ldots$	$\{1\}$	0	01	$+1$		0
(B1)	$\underline{A}0\ldots\ldots0\bar{A}$	$\{0\}$	1	10	$+1$		0
(B2)	$\underline{A}1\ldots\ldots0\bar{A}$	$\{0\}$	1	11	$=$		0
(C1)	$\underline{A}0\ldots0\bar{A}0\ldots$	$\{0,0\}$	1	10	$+1$	$+$	q_n
(C2)	$\underline{A}1\ldots1\bar{A}1\ldots$	$\{1,1\}$	0	01	$+1$	$+$	q_n
(C3)	$\underline{A}1\ldots0\bar{A}0\ldots$	$\{0,0\}$	1	11	$=$		q_n
(C4)	$\underline{A}0\ldots1\bar{A}1\ldots$	$\{1,1\}$	0	00	$=$		q_n
(D1)	$\underline{A}0\ldots0\bar{A}1\ldots$	$\{0,1\}$	0	00	$=$		$a_n/4 - s_n/2$
(D2)	$\underline{A}1\ldots0\bar{A}1\ldots$	$\{0,1\}$	1	11	$=$		$a_n/4 - s_n/2$
(E1)	$\underline{A}1\ldots0\bar{A}*\ldots$	$\{0,*\}$	1	11	$=$		$s_n/2$
(E2)	$\underline{A}0\ldots1\bar{A}*\ldots$	$\{1,*\}$	0	00	$=$		$s_n/2$
(F1)	$\underline{A}0\ldots0\bar{A}*\ldots$	$\{0,*\}$	0	00	$=$	-1	$s_n/2$
(F2)	$\underline{A}1\ldots1\bar{A}*\ldots$	$\{1,*\}$	1	11	$=$	-1	$s_n/2$

Observe that the average increase of color changes in c is

$$(2n - 1)\frac{a_n}{2} - (2n - 3)\frac{a_{n-1}}{2} = a_n + (a_n - a_{n-1})\left(n - \frac{3}{2}\right).$$

That increase happens only in cases (C1) and (C2); indeed, they are the only cases with $+1$ in Δa_n-column and nonzero (limit) probability, we have

$$a_n + (a_n - a_{n-1})\left(n - \frac{3}{2}\right) = \Pr\left[(C1) \cup (C2)\right] = 2q_n + O(1/n). \quad (2)$$

By definition, the average increase of the number of stars is $ns_n - (n-1)s_{n-1}$. We may color A by a star in the next step in cases (C1), (C2), (D1), (D2), (F1) and (F2), but only under the condition that the color of the next added letter X will be opposite to the color of $\underline{A}$, which happens only in the new cases (C1) and (C2). Note that here we assume that these events are independent which we were unfortunately not able to prove. On the other hand, we can also decrease the number of stars by recoloring them to binary colors which happens in case (F). Hence

$$s_{n+1}(n+1) - s_n n$$
$$= \Pr\left[(C1) \cup (C2)\right] \Pr\left[(C1) \cup (C2) \cup (D) \cup (F)\right] - \Pr\left[(F)\right]$$
$$= 2q_n\left(2q_{n-1} + \frac{a_{n-1}}{2}\right) - s_n + O(1/n). \tag{3}$$

Our goal is to solve the system of two recurrences above. However, we are not able to do that analytically and we need additional assumptions. We assume that

$$a_n - a_{n-1} = o(1/n) \qquad \text{and} \qquad s_n - s_{n-1} = o(1/n). \tag{4}$$

That allows us to substitute a_n by $a_{n\pm1}$ and s_n by $s_{n\pm1}$. Hence we modify equations (2) and (3) and using (4) obtain system

$$a_n = \frac{1}{2} - \frac{a_n}{4} - \frac{s_n}{2} + o(1)$$
$$s_n = \frac{3}{2}a_n^2 - s_n + o(1)$$

with the only relevant solution

$$a_n = \alpha + o(1) \qquad \text{and} \qquad s_n = .098\ldots + o(1).$$

We were able to compute the values a_n and s_n for n up to 120. The obtained data suggest that our assumptions (4) are reasonable, and that both sequences $(a_n)_{n\geq1}$ and $(s_n)_{n\geq1}$ seem to be monotone and bounded.

Finally, we remark that the RSG heuristics actually uses two somewhat different types of stars since the pair of letters $\underline{X}, \bar{X}$ colored by star can be either surrounded by four neighbors of the same color or both $\underline{X}$ and $\bar{X}$ have one neighbor colored with 0 and the other colored using 1. It might be more feasible to analyze a modified heuristic that only uses the second type of stars. The numerical data suggest that this heuristic still performs better than RG albeit worse than RSG. However, we were not able to rigorously show this either.

5   Concentration Result for the Optimal Number of Color Changes

Theorem 3. *Let w be a random element of W_n. Then*

$$\Pr\left[|\gamma(w) - \gamma_n| \geq \sqrt{n\log n}\right] \leq 2n^{-1/8}.$$

The proof using Azuma–Hoeffding inequality can be found in the full version of this paper. We remark that the function $\sqrt{n \log n}$ in Theorem 3 could be replaced by $f(n)\sqrt{n}$ for an arbitrary function f such that $f(n) \to \infty$ for $n \to \infty$. We would still have the key corollary that for a random word $w \in W_n$,

$$\Pr\left[|\gamma(w) - \gamma_n| \geq f(n)\sqrt{n} \right] \to 0.$$

Acknowledgments. Our attention to the paint shop problem was brought by a nice talk given by Winfried Hochstättler at Midsummer combinatorial workshop in Prague (MCW2017). We thank him and the organizers of the workshop. This research was started during workshop KAMAK 2017, we are grateful to its organizers.

References

1. Alon, N.: Splitting necklaces. Adv. in Math. **63**(3), 247–253 (1987)
2. Alon, N., Elboim, D., Pach, J., Tardos, G.: Random necklaces require fewer cuts. arXiv:2112.14488
3. Andres, S.D.: Greedy versus recursive greedy: uncorrelated heuristics for the binary paint shop problem. Discrete. Appl. Math. **303**, 4–7 (2021). Technical report FernUniversitat in Hagen 2019: https://www.fernuni-hagen.de/MATHEMATIK/DMO/pubs/feu-dmo064-19.pdf
4. Andres, S.D., Hochstättler, W.: Some heuristics for the binary paint shop problem and their expected number of colour changes. J. Discret. Algorithms **9**(2), 203–211 (2011)
5. Bonsma, P., Epping, T., Hochstättler, W.: Complexity results on restricted instances of a paint shop problem for words. Discret. Appl. Math. **154**(9), 1335–1343 (2006)
6. Epping, T., Hochstättler, W., Oertel, P.: Complexity results on a paint shop problem. Discrete Appl. Math. **136**, 2–3 (2004), 217–226. The 1st Cologne-Twente Workshop on Graphs and Combinatorial Optimization (CTW 2001)
7. Meunier, F., Neveu, B.: Computing solutions of the paintshop-necklace problem. Comput. Oper. Res. **39**(11), 2666–2678 (2012)
8. Meunier, F., Sebő, A.: Paintshop, odd cycles and necklace splitting. Discret. Appl. Math. **157**(4), 780–793 (2009)
9. Scheinerman, E.R.: Random interval graphs. Combinatorica **8**(4), 357–371 (1988)
10. Šámal, R., Hančl, J., Kabela, A., Opler, M., Sosnovec, J., Valtr, P.: The binary paint shop problem. In: Slides from workshop MCW in Prague, 30 July 2019. https://kam.mff.cuni.cz/workshops/mcw/slides/samal.pdf

Algorithmic Solution in Applications

Fitch Graph Completion

Marc Hellmuth[1] [ID], Peter F. Stadler[2] [ID],
and Sandhya Thekkumpadan Puthiyaveedu[1(✉)] [ID]

[1] Department Mathematics, Faculty of Science,
Stockholm University, 10691 Stockholm, Sweden
{marc.hellmuth,thekkumpadan}@math.su.se

[2] Bioinformatics Group, Department Computer Science and Interdisciplinary
Center for Bioinformatics, Universität Leipzig, 04107 Leipzig, Germany
studla@bioinf.uni-leipzig.de

Abstract. Horizontal gene transfer is an important contributor to evolution. According to Walter M. Fitch, two genes are xenologs if they are separated by at least one HGT. More formally, the directed Fitch graph has a set of genes as its vertices, and directed edges (x, y) for all pairs of genes x and y for which y has been horizontally transferred at least once since it diverged from the last common ancestor of x and y. Subgraphs of Fitch graphs can be inferred by comparative sequence analysis. In many cases, however, only partial knowledge about the "full" Fitch graph can be obtained. Here, we characterize Fitch-satisfiable graphs that can be extended to a biologically feasible "full" Fitch graph and derive a simple polynomial-time recognition algorithm. We then proceed to showing that finding the Fitch graphs with total maximum (confidence) edge-weights is an NP-hard problem.

Keywords: Directed Cograph · Fitch Graph · Horizontal Gene Transfer · NP-complete · Recognition Algorithm

1 Introduction

Horizontal gene transfer (HGT) is a biological process by which genes from sources other than the parents are transferred into an organism's genome. In particular in microorganism it is an important contributor to evolutionary innovation. The identification of HGT events from genomic data, however, is still a difficult problem in computational biology, both mathematically and in terms of practical applications. Recent formal results are phrased in terms of a binary relation between genes. In most situations it can be assumed that the evolution of genes is tree-like and thus can described by a gene tree T whose leaves correspond to the present-day, observable genes; the interior vertices and edges then model evolutionary events such as gene duplications, speciations, and also HGT.

This work was supported in part by the German Research Foundation (DFG, STA 850/49-1) and the Data-driven Life Science (DDLS) program funded by the Knut and Alice Wallenberg Foundation.

Since HGT distinguishes between the gene copy that continues to be transmitted vertically, and the transferred copy, one associated the transfer with the edge in T connecting the HGT event with its transferred offspring. Focusing on a pair of present-day genes, it is of interest to determine whether or not they have been separated by HGT events in their history. This information is captured by the Fitch (xenology) graph. It contains an edge $x \to y$ whenever a HGT event occurred between y and the least common ancestor of x and y [6]. Fitch graphs form a hereditary sub-class of the directed cographs [4], and admit a simple characterization in terms of eight forbidden induced subgraphs on three vertices (see Fig. 1 below). Moreover, every Fitch graph uniquely determines a least resolved edge-labeled tree by which it is explained. This tree is related to the gene tree by a loss of resolution [6].

Information on HGT events can be extracted from sequence information using a broad array of methods [15], none of which is likely to yield a complete picture. Reliable information to decide whether or not two genes are xenologs, thus, may be available only for some pairs of genes (x, y), but not for others. In this situation it is natural to ask whether this partial knowledge can be used to infer missing information. In [14] the analogous question was investigated for di-cographs. The main result of the present contribution is a characterization of partial Fitch graphs, Theorem 2, and an accompanying polynomial-time algorithm. In addition, we show that the "weighted" version of Fitch graph completion is NP-hard.

2   Preliminaries

Relations. Throughout, we consider only *irreflexive, binary* relations R on V, i.e., $(x, y) \in R$ implies $x \neq y$ for all $x, y \in V$. We write $\overleftarrow{R} := \{(x, y) \mid (y, x) \in R\}$ and $R^{\mathrm{sym}} := R \cup \overleftarrow{R}$ for the *transpose* and the *symmetric extension* of R, respectively. The relation $R_V^{\times} := \{(x, y) \mid x, y \in V, x \neq y\}$ is called the *full* relation. For a subset $W \subseteq V$ and a relation R, we define the *induced* sub-relation as $R[W] := \{(x, y) \mid (x, y) \in R, x, y \in W\}$. Moreover, we consider ordered tuples of relations $\mathcal{R} = (R_1, \ldots, R_n)$. Let $R_1, \ldots, R_n \subseteq R_V^{\times}$, then $\mathcal{R} = (R_1, \ldots, R_n)$ is *full* if $\cup_{i=1}^{n} R_i = R_V^{\times}$ and *partial* if $\cup_{i=1}^{n} R_i \subseteq R_V^{\times}$. Note that a full tuple of relations is also considered to be a partial one. Moreover, we consider component-wise sub-relation and write $\mathcal{R}[W] := (R_1[W], \ldots, R_n[W])$ and $\mathcal{R} = (R_1, \ldots, R_n) \subseteq \mathcal{R}' = (R_1', \ldots, R_n')$ if $R_i \subseteq R_i'$ holds for all $i \in \{1, \ldots, n\}$. In the latter case, we say that $\mathcal{R}'$ *extends* $\mathcal{R}$.

Digraphs and DAGs. A directed graph (digraph) $G = (V, E)$ comprises a vertex set $V(G) = V$ and an irreflexive binary relation $E(G) = E$ on V called the edge set of G. Given two disjoint digraphs $G = (V, E)$ and $H = (W, F)$, the digraphs $G \cup H = (V \cup W, E \cup F)$, $G \bowtie H = (V \cup W, E \cup F \cup \{(x, y), (y, x) \mid x \in V, y \in W\})$ and $G \triangleright H = (V \cup W, E \cup F \cup \{(x, y) \mid x \in V, y \in W\})$ denote the *union, join* and *directed join* of G and H, respectively. For a given subset $W \subseteq V$, the *induced subgraph* $G[W] = (W, F)$ of $G = (V, E)$ is the subgraph for which $x, y \in W$ and $(x, y) \in E$ implies that $(x, y) \in F$. We call $W \subseteq V$ a *(strongly) connected component* of $G = (V, E)$ if $G[W]$ is an *inclusion-maximal* (strongly) connected subgraph of G.

Given a digraph $G = (V, E)$ and a partition $\{V_1, V_2, \ldots, V_k\}$, $k \geq 1$ of its vertex set V, the *quotient digraph* $G/\{V_1, V_2, \ldots, V_k\}$ *has as vertex set* $\{V_1, V_2, \ldots, V_k\}$ and two distinct vertices V_i and V_j form an edge (V_i, V_j) in $G/\{V_1, \ldots, V_k\}$ if there are vertices $x \in V_i$ and $y \in V_j$ with $(x, y) \in E$. Note, that edges (V_i, V_j) in $G/\{V_1, \ldots, V_k\}$ do not necessarily imply that (x, y) form an edge in G for $x \in V_i$ and $y \in V_j$. Nevertheless, at least one such edge (x, y) with $x \in V_i$ and $y \in V_j$ must exist in G given that (V_i, V_j) is an edge in $G/\{V_1, \ldots, V_k\}$

A cycle C in a digraph $G = (V, E)$ of length n is an ordered sequence of $n > 1$ (not necessarily distinct) vertices $(v_1, \ldots, v_n)$ such that $(v_n, v_1) \in E$ and $(v_i, v_{i+1}) \in E$, $1 \leq i < n$. A digraph that does contain cycles is a *DAG* (directed acyclic graph). Define the relation $\preceq_G$ of V such that $v \preceq_G w$ if there is directed path from w to v. A vertex x is a *parent* of y if $(x, y) \in E$. In this case y is *child* of x. Then G is DAG if and only if $\preceq_G$ is a partial order. We write $y \prec_G x$ if $y \preceq_G x$ and $x \neq y$. A *topological order* of G is a total order $\ll$ on V such that $(v, w) \in E$ implies that $v \ll w$. It is well known that a digraph G admits a topological order if and only if G is a DAG. In this case, $x \prec_G y$ implies $y \ll x$, i.e., $\ll$ is a linear extension of $\prec$. Note that $\preceq$ and $\ll$ are arranged in opposite order. The effort to check whether G admits a topological order $\ll$ and, if so, to compute $\ll$ is linear, i.e., in $O(|V| + |E|)$ [11]. If $C_1, \ldots, C_k$, $k \geq 1$ are the strongly connected components of a digraph, then $G/\{C_1, C_2, \ldots, C_k\}$ is a DAG.

A DAG G is *rooted* if it contains a unique $\preceq_G$-maximal element ρ_G called the *root*. Note that ρ_G is $\ll$-minimal. A rooted tree T with vertex set $V(T)$ is a DAG such that every vertex $x \in V(T) \setminus \{\rho_T\}$ has a unique parent. The rooted trees considered here do not contain vertices v with $\mathrm{indeg}(v) = \mathrm{outdeg}(v) = 1$. A vertex x is an *ancestor* of y if $y \preceq_T x$, i.e., if x is located on the unique path from ρ_T to y. A vertex in T without a child is a *leaf*. The set of leaves of T will be denoted by $L(T)$. The elements in $V^0(T) := V(T) \setminus L(T)$ are called the *inner vertices*. We write $T(u)$ for the subtree of T induced by $\{v \in V(T) \mid v \preceq_T u\}$. Note that u is the root of $T(u)$.

For a subset $W \subseteq L(T)$, the *least common ancestor* $\mathrm{lca}_T(W)$ of W is the unique $\preceq_T$-minimal vertex that is an ancestor of each $w \in W$. If $W = \{x, y\}$, we write $\mathrm{lca}_T(x, y) := \mathrm{lca}_T(\{x, y\})$. A rooted tree T is *ordered*, if the children of every vertex in T are ordered. Rooted trees $T_1, \ldots, T_k$, $k \geq 2$ *are joined under a new root in the tree* T if T is obtained by the following procedure: add a new root ρ_T and all trees $T_1, \ldots, T_k$ to T and connect the root ρ_{T_i} of each tree T_i to ρ_T with an edge (ρ_T, ρ_{T_i}).

Directed Cographs. Di-cographs generalize the notion of undirected cographs [2–5] and are defined recursively as follows: (i) the single vertex graph K_1 is a di-cograph, and (ii) if G and H are di-cographs, then $G \cup H$, $G \bowtie H$, and $G \triangleright H$ are di-cographs [7,14]. Every di-cograph $G = (V, E)$ *is explained by* an ordered rooted tree $T = (W, F)$, called a *cotree* of G, with leaf set $L(T) = V$ and a labeling function $t : W^0 \to \{0, 1, \overrightarrow{1}\}$ that uniquely determines the set of edges

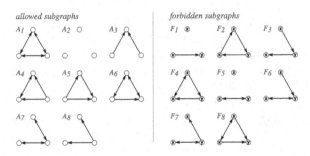

Fig. 1. Of the 16 possible irreflexive binary relations on three vertices, eight (A_1 through A_8) may appear in Fitch graphs, while the remaining eight (F_1 through F_8) form forbidden induced subgraphs.

$E(G) = E_1(T, t) \cup E_{\vec{1}}(T, t)$ and the set of non-adjacent pairs of vertices $E_0(T, t)$ of G as follows:

$$E_1(T, t) = \{(x, y) \mid t(\mathrm{lca}(x, y)) = 1\},$$
$$E_0(T, t) = \{(x, y) \mid t(\mathrm{lca}(x, y)) = 0\}, \text{ and}$$
$$E_{\vec{1}}(T, t) = \{(x, y) \mid t(\mathrm{lca}(x, y)) = \vec{1} \text{ and } x \text{ is left of } y \text{ in } T\}.$$

Note that $E_i(T, t) = E_i(T, t)^{\mathrm{sym}}$ for $i \in \{0, 1\}$ since $\mathrm{lca}(x, y) = \mathrm{lca}(y, x)$. Every di-cograph $G = (V, E)$ is explained by a unique *discriminating* cotree (T, t) satisfying $t(x) \neq t(y)$ for all $(x, y) \in E(T)$. Every cotree (T', t') that explains G is "refinement" of its discriminating cotree (T, t), i.e., (T, t) is obtained by contracting all edges $(x, y) \in E(T')$ with $t'(x) = t'(y)$ [1]. Determining whether a digraph is a di-cograph, and if so, computing its discriminating cotree requires $O(|V| + |E|)$ time [2, 7, 13].

3   Fitch Graphs and Fitch-Satisfiability

Basic Properties of Fitch Graphs. Fitch Graphs are defined in terms of edge-labeled rooted trees T with an *edge-labeling* $\lambda \colon E \to \{0, 1\}$ and leaf set $L(T) = V$. The graph $\mathbb{G}(T, \lambda) = (V, E)$ contains an edge (x, y) for $x, y \in V$ if and only if the (unique) path from $\mathrm{lca}_T(x, y)$ to y contains at least one edge $e \in E(T)$ with label $\lambda(e) = 1$. The edge set of $\mathbb{G}(T, \lambda)$ by construction is a binary irreflexive relation on V. A directed graph G is a *Fitch graph* if there is a tree (T, λ) such that $G \simeq \mathbb{G}(T, \lambda)$. Fitch graphs form a proper subset of directed cographs [6]. Therefore, they can be explained by a cotree.

Definition 1. *A cotree (T, t) is a Fitch-cotree if there are no two vertices $v, w \in V^0(T)$ with $w \prec_T v$ such that either (i) $t(v) = 0 \neq t(w)$ or (ii) $t(v) = \vec{1}$, $t(w) = 1$, and $w \in V(T(u))$ where u is a child of v distinct from the right-most child of v.*

In other words, a Fitch-cotree satisfies:

(a) No vertex with label 0 has a descendant with label 1 or $\overrightarrow{1}$.

(b) If a vertex v has label $\overrightarrow{1}$, then the subtree $T(u)$ rooted at a u child of v — except the right-most one — does not contain vertices with label 1. In particular, if T is discriminating, then $T(u)$ is either a star-tree whose root u has label $t(u) = 0$ or u is a leaf. In either case, the di-cograph $G[L(T(u))]$ defined by the subtree $T(u_1)$ of left-most child u_1 of v, is edge-less.

Fitch graphs have several characterizations that will be relevant throughout this contribution. We summarize [8, L. 2.1] and [6, Thm. 2] in the following

Theorem 1. *For every digraph $G = (V, E)$ the following statements are equivalent.*

1. *G is a Fitch graph.*
2. *G does not contain an induced $F_1, F_2, \ldots, F_8$ (cf. Fig. 1).*
3. *G is a di-cograph that does not contain an induced F_1, F_5 and F_8 (cf. Fig. 1).*
4. *G is a di-cograph that is explained by a Fitch-cotree.*
5. *Every induced subgraph of G is a Fitch graph, i.e., the property of being a Fitch graph is hereditary.*

Fitch graphs can be recognized in $O(|V| + |E|)$ time. In the affirmative case, the (unique least-resolved) edge-labeled tree (T, λ) can be computed in $O(|V|)$ time.

Alternative characterizations can be found in [9]. The procedure cotree2 fitchtree described in [6] can be used to transform a Fitch cotree (T, t) that explains a Fitch graph G into an edge-labeled tree (T', λ) that explains G in $O(|V(T)|)$ time, avoiding the construction of the di-cograph altogether. For later reference, we provide the following simple results. The proofs can be found in [10].

Lemma 1. *The graph obtained from a Fitch graph by removing all bi-directional edges is a DAG.*

Corollary 1. *Every Fitch graph without bi-directional edges is a DAG.*

Removal of the bi-directional edges from the Fitch graph A_6 yields the graph F_1, i.e., although removal of all bi-directional edges from Fitch graphs yields a DAG it does not necessarily result in a Fitch graph.

Corollary 2. *Let G be a directed graph without non-adjacent pairs of vertices and no bi-directional edges. Then, G is a Fitch graph if and only if it is a DAG.*

Characterizing Fitch-satisfiability. Throughout we consider 3-tuples of (partial) relations $\mathcal{E} = (E_0, E_1, E_{\overrightarrow{1}})$ on V such that E_0 and E_1 are symmetric and $E_{\overrightarrow{1}}$ is antisymmetric.

Definition 2 (*Fitch-sat*). *$\mathcal{E}$ is Fitch-satisfiable (in short Fitch-sat), if there is a full tuple $\mathcal{E}^* = \{E_0^*, E_1^*, E_{\overrightarrow{1}}^*\}$ (with E_0^* and E_1^* being symmetric and $E_{\overrightarrow{1}}^*$ being antisymmetric) that extends $\mathcal{E}$ and that is explained by a Fitch-cotree (T, t).*

By slight abuse of notation, we also say that, in the latter case, $\mathcal{E}$ is explained by (T, t). The problem of finding a tuple $\mathcal{E}^*$ that extends $\mathcal{E}$ and that is explained by an arbitrary cotree was investigated in [14]. Theorem 1 together with the definition of cotrees and Def. 2 implies

Corollary 3. $\mathcal{E} = (E_0, E_1, E_{\vec{1}})$ on V is Fitch-sat *precisely if it can be extended to a full tuple* $\mathcal{E}^* = (E_0^*, E_1^*, E_{\vec{1}}^*)$ *for which* $H = (V, E_1^* \cup E_{\vec{1}}^*)$ *is a Fitch graph. In particular, there is a discriminating Fitch-cotree that explains* $\mathcal{E}$, $\mathcal{E}^*$, *and* H.

In [10] we prove that Fitch-satisfiability is a hereditary graph property:

Lemma 2. *A partial tuple* $\mathcal{E}$ *on* V *is Fitch-sat if and only if* $\mathcal{E}[W]$ *is Fitch-sat for all* $W \subseteq V$.

For the proof of Theorem 2, we will need the following technical result, which is proven in [10].

Lemma 3. *Let* $\mathcal{E} = (E_0, E_1, E_{\vec{1}})$ *be a Fitch-sat partial tuple on* V *that is explained by the discriminating Fitch-cotree* (T, t) *and put* $G_0[W] := (W, E_1[W] \cup E_{\vec{1}}[W])$ *for* $W \subseteq V$. *If there is a vertex* $u \in V^0(T)$ *such that* $t(u) = 0$, *then* $G_0[C]$ *is edge-less for all* $C \subseteq L(T(u))$.

We are now in the position to provide a characterization of *Fitch-sat* partial tuples. A detailed version of this proof can be found in [10].

Theorem 2. *The partial tuple* $\mathcal{E} = (E_0, E_1, E_{\vec{1}})$ *on* V *is Fitch-sat if and only if at least one of the following statements hold.*

(S1) $G_0 := (V, E_1 \cup E_{\vec{1}})$ *is edge-less.*
(S2) (a) $G_1 := (V, E_0 \cup E_{\vec{1}})$ *is disconnected and*
     (b) $\mathcal{E}[C]$ *is Fitch-sat for all connected components* C *of* G_1
(S3) (a)(I) $G_{\vec{1}} := (V, E_0 \cup E_1 \cup E_{\vec{1}})$ *contains* $k > 1$ *strongly connected*
         components $C_1, \ldots, C_k$ *collected in* $\mathcal{C}$ *and*
     (II) *there is a* $C \in \mathcal{C}$ *for which the following conditions are satisfied:*
         (i) $G_0[C]$ *is edge-less.*
         (ii) C *is* $\ll$-*minimal for some topological order* $\ll$
            on $G_{\vec{1}} / \{C_1, C_2, \ldots, C_k\}$.
     (b) $\mathcal{E}[V \setminus C]$ *is Fitch-sat.*

Proof. (Sketch) If $\mathcal{E}$ satisfies (S1) then $\mathcal{E}^* = (R_V^\times, \emptyset, \emptyset)$ and the star tree with leaf set V and root label "0" explains $\mathcal{E}^*$. If $\mathcal{E}$ satisfies (S2) then each connected component $\mathcal{E}[C_i]$ can be expanded to a *Fitch-sat* tuple $\mathcal{E}^*[C_i]$ explained by some Fitch-cotree (T_i, t_i). In this way, we obtain the graph $G_1^* = (V, E_0^* \cup E_{\vec{1}}^*)$ whose connected components are the same as those of G_1. The full tuple $\mathcal{E}^*$ is obtained from the union of the $\mathcal{E}^*[C_i]$ and by adding all missing pairs $(x, y), (y, x)$ for all $x \in C_i$ and $y \in C_j$, $i \neq j$. The Fitch cotree is (T, t) is obtained by joining the (T_i, t_i) under a new root with label "1". If $\mathcal{E}$ satisfies (S3), then $G_0[C]$ is edge-less for a $\ll$-minimal component C and thus can be extended to the full tuple $\mathcal{E}^*[C] = (E_0^*[C], \emptyset, \emptyset)$ by adding all pairs $(x, y) \in R_C^\times \setminus C$ to $E_0[C]$. This leaves

$G_{\bar{0}}^*[C] = (C, E_1^*[C] \cup E_{\vec{1}}^*[C])$ edge-less and thus it is explained by the Fitch-cotree (T', t') where T' is a star tree whose root is labeled "0". On the other hand, $\mathcal{E}[V \setminus C]$ is *Fitch-sat* and thus can be extended to a full tuple $\mathcal{E}^*[V \setminus C]$ that is explained by a Fitch-cotree $(\widehat{T}, \widehat{t})$. Then $\mathcal{E}$ is obtained from the union of the $\mathcal{E}^*[C_i]$ and by adding all pairs (x, y) for all $x \in C$ and $y \in V \setminus C$, The Fitch cotree is (T, t) is obtained by placing (T', t') to the left of $(\widehat{T}, \widehat{t})$ under a common root with label $\vec{1}$.

For the *only-if* direction, we start from the discriminating Fitch-cotree (T, t) endowed with the sibling order $<$ that explains $\mathcal{E}$ and $\mathcal{E}^*$ whose existence if guaranteed by Corollary 3. Moreover, the root ρ of T has at $r \geq 2$ children $v_1, \ldots, v_r$ ordered from left to right according to $<$ and has one of the three labels $0, 1, \vec{1}$. Denote by $L_i = \{x \in L(T) \mid x \preceq v_i\}$ the set of all leaves x of T with $x \preceq v_i$ and note that $x \in L_i$ and $y \in L_j$ with $i \neq j$ implies $\operatorname{lca}(x, y) = \rho$. If $t(\rho) = 0$, then Lemma 3 implies that $G_0 = (V, E_1 \cup E_{\vec{1}})$ is edge-less and thus $\mathcal{E}$ satisfies (S1). If $t(\rho) = 1$, then $G_1^* = (V, E_0^* \cup E_{\vec{1}}^*)$ must be disconnected, and Lemma 2 implies that $\mathcal{E}[C]$ is *Fitch-sat* for each connected component. Therefore, $\mathcal{E}$ satisfies (S2). If $t(\rho) = \vec{1}$ then $(x, y) \in E_{\vec{1}}^*$ and $(y, x) \notin E_{\vec{1}}^*$ for all $x \in L_i$ and $y \in L_j$ with $1 \leq i < j \leq r$, and thus $G_{\vec{1}}^* := (V, E_0^* \cup E_1^* \cup E_{\vec{1}}^*) = G_{\vec{1}}^*[L_1] \triangleright \ldots \triangleright G_{\vec{1}}^*[L_r]$ consists of two or more strongly connected components, each of which is contained within some L_i. Since $G_{\vec{1}}$ is a subgraph of $G_{\vec{1}}^*$, it contains more than one strongly connected component and thus $\mathcal{E}$ satisfies (S3.a.I). Since (T, t) is a discriminating Fitch-cotree, the left-most child v_1 of ρ is either a leaf of T or $t(v_1) = 0$, and thus $G_0[L_1]$ is edge-less by Lemma 3, and thus there is a strongly connected component C of $G_{\vec{1}}^*$ for which $G_0[C] \subseteq G_0[L_1]$ is edgeless. Thus (S3.a.II.i) holds. Let C' be any other strongly connected component of $G_{\vec{1}}^*$. If $C' \subseteq L_1$, then there are are no edges between C and C' in $G_{\vec{1}}$. Otherwise, any two adjacent vertices $x \in C$ and $y \in C'$ in $G_{\vec{1}}$ must satisfy $(x, y) \in E_{\vec{1}}$ and $(y, x) \notin E_{\vec{1}}$. It is not difficult to show that that (S3.a.II.ii) is satisfied. Finally, Lemma 2 implies that $\mathcal{E}[V \setminus C]$ is *Fitch-sat* and thus (S3.b) is also satisfied. □

4   Recognition Algorithm and Computational Complexity

The proof of Theorem 2 provides a recipe to construct a Fitch-cotree (T, t) explaining a tuple $\mathcal{E}$. We observe, furthermore, that two or even all three alternatives (S1), (S2.a), and (S3.a) may be satisfied simultaneously, see Fig. 2 for an illustrative example. In this case, it becomes necessary to check stepwisely whether conditions (S2.b) and/or (S3.b) holds. Potentially, this result in exponential effort to determine recursively whether $\mathcal{E}[C]$ or $\mathcal{E}[V \setminus C]$ is *Fitch-sat*. The following simple lemma shows, however, that the alternatives always yield consistent results:

Lemma 4. *Let $\mathcal{E} = (E_0, E_1, E_{\vec{1}})$ be a partial tuple on V. Then*

(S1) *and* (S2.a) *implies* (S2.b);

(S1) and (S3.a) implies (S3.b);
(S2a) and (S3a) implies that (S2.b) and (S3.b) are equivalent.

Proof. If (S1) holds, then Theorem 2 implies that $\mathcal{E}$ is *Fitch-sat*. If (S2.a) holds, then heredity (Lemma 2) implies that $\mathcal{E}[C]$ is *Fitch-sat* and thus (S2.b) is satisfied. Analogously, if (S3.a) holds, then $\mathcal{E}[V \setminus C]$ is *Fitch-sat* and thus (S3.b) holds. Now suppose (S2a) and (S3a) are satisfied but (S1) does not hold. Then $\mathcal{E}$ is *Fitch-sat* if and only if one of (S2.b) or (S3.b) holds; in the affirmative case, heredity again implies that both (S2.b) and (S3.b) are satisfied. □

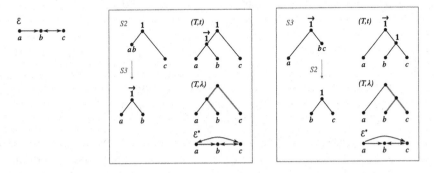

Fig. 2. A partial tuple $\mathcal{E} = (E_0, E_1, E_{\vec{1}})$ on $V = \{a, b, c\}$ with $E_0 = \emptyset$, $E_1 = \{(b, c), (c, b)\}$ (bi-directional arc), and $E_{\vec{1}} = \{(a, b)\}$ (single arc) is shown on the left. Observe that (S1) is not satisfied while (S2.a) and (S3.a) hold for $\mathcal{E}$. Application of the different rules and subsequent construction of the Fitch-cotrees that explain the subgraphs induced by the respective (strongly) connected components results in two Fitch-cotrees that both explain $\mathcal{E}$. Hence, we obtain two different edge-labeled Fitch-trees (T, λ) (with HGT-edges drawn in bold-red) that both explain $\mathcal{E}$.

It follows that testing whether $\mathcal{E}$ can be achieved by checking if any one of the three conditions (S1), (S2), or (S3) holds and, if necessary, recursing down on $\mathcal{E}[C]$ or $\mathcal{E}[V \setminus C]$. This give rise to Algorithm 1.

Lemma 5. *Let $\mathcal{E} = (E_0, E_1, E_{\vec{1}})$ be a partial tuple. Then Alg. 1 either outputs a Fitch-cotree (T, t) that explains $\mathcal{E}$ or recognizes that $\mathcal{E}$ is not Fitch-sat.*

Proof. (Sketch) By Lemma 4, the order in which (S1), (S2), and (S3) are tested is arbitrary. If none of the conditions (S1), (S2), or (S3) is satisfied, Theorem 2 implies that $\mathcal{E}$ is not *Fitch-sat*. If Rule (S1) or (S2.a), resp., is satisfied (Line 4), then Alg. 1 recurses on the connected components defined by G_0 or G_1, respectively. For (S1), correctness follows from the *if* direction of Theorem 2. Otherwise, the Fitch-cotree of connected components, which are *Fitch-sat*, are joined a single cotree (T, t) with root label "1" explaining an extension $\mathcal{E}^*$ of $\mathcal{E}$, which is thus correctly identified as *Fitch-sat*. Finally, if (S3.a) is satisfied, we consider the strongly connected components C_i of $G_{\vec{1}}$. By (S3.a.II) there is component, say C_1 for which $G[C_i]$ is edge-less, and thus I in Line 12 is non-empty and the topological order $\ll$ is well-defined. If $\mathcal{E}[C^*]$ and $\mathcal{E}[\widehat{C}]$ are *Fitch-sat*, the two Fitch-cotrees (T^*, t^*) and $(\widehat{T}, \widehat{t})$ explaining $\mathcal{E}[C^*]$ and $\mathcal{E}[\widehat{C}]$ are returned and joined under a new root resulting in a-cotree (T, t) that explains $\mathcal{E}$. It is not difficult to verify that (T, t) is a Fitch-cotree. For full details [10]. □

Theorem 3. *Let $\mathcal{E} = (E_0, E_1, E_{\overrightarrow{1}})$ be a partial tuple, $n = |V|$ and $m = |E_0 \cup E_1 \cup E_{\overrightarrow{1}}|$. Then, Alg. 1 computes a Fitch-cotree (T, t) that explains $\mathcal{E}$ or identifies that $\mathcal{E}$ is not Fitch-sat in $O(n^2 + nm)$ time.*

Algorithm 1 Recognition of *Fitch-sat* partial tuple $\mathcal{E}$ on V and reconstruction of a cotree (T, t) that explains $\mathcal{E}$.

Input: Partial tuples $\mathcal{E} = (E_0, E_1, E_{\overrightarrow{1}})$
Output: A cotree (T, t) that explains $\mathcal{E}$, if one exists or the statement "$\mathcal{E}$ is not Fitch-satisfiable"

1: Call BUILDFITCHCOTREE$(V, \mathcal{E})$

2: **function** BUILDFITCHCOTREE$(V, \mathcal{E} = (E_0, E_1, E_{\overrightarrow{1}}))$
         ▷ *G_0, G_1 and $G_{\overrightarrow{1}}$ are defined as in Thm. 2 for given $\mathcal{E}$*
3:      **if** $|V| = 1$ **then return** the cotree $((V, \emptyset), \emptyset)$
4:      **else if** G_0 is edge-less (**otherwise if** G_1 is disconnected) **then**
         ▷ *check (S1) (resp., (S2))*
5:          $\mathcal{C} :=$ the set of connected components $\{C_1, \ldots, C_k\}$ of G_0 (resp. G_1)
6:          $\mathcal{T} :=$ set $\{$BUILDFITCHCOTREE$(C_i, \mathcal{E}[C_i]) \mid C_i \in \mathcal{C}\}$
7:          **return** the cotree from joining the cotrees in $\mathcal{T}$ under a new root labeled 0 (resp. 1)
8:      **else if** $G_{\overrightarrow{1}}$ has more than one strongly connected component **then**
         ▷ *check (S3)*
9:          $\mathcal{C} :=$ the set of strongly connected components $\{C_1, \ldots, C_k\}$ of $G_{\overrightarrow{1}}$
10:          $I \leftarrow \emptyset$
11:          **for all** $i \in \{1, \ldots, k\}$ **do**
12:             **if** $G_0[C_i]$ is edge-less **then** $I \leftarrow I \cup \{i\}$
13:          **if** $I = \emptyset$ **then**
14:             Halt and output: "$\mathcal{E}$ is not Fitch-satisfiable"   ▷ *(S3.a.II.i) not satisfied*
15:          **else if** there is no topological order $\ll$ of $G_{\overrightarrow{1}}/\{C_1, \ldots, C_k\}$ with $\ll$-minimal element C_i with $i \in I$ **then**
16:             Halt and output: "$\mathcal{E}$ is not Fitch-satisfiable"   ▷ *(S3.a.II.ii) not satisfied*
17:          **else**
18:             $\ll :=$ a topol. order on the quotient $G_{\overrightarrow{1}}/\{C_1, \ldots, C_k\}$ with $\ll$-minimal element $C^* := C_i$ for some $i \in I$
19:             $\mathcal{T} := \{$BUILDFITCHCOTREE$(C, \mathcal{E}[C]) \mid C \in \{C^*, V \setminus C^*\}\}$
20:             **return** the cotree (T, t) obtained by joining the cotrees in $\mathcal{T}$ under a new root with label $\overrightarrow{1}$, where the tree (T^*, t^*) that explains C^* is placed left of the tree $(\widehat{T}, \widehat{t})$ that explains $V \setminus C$
21:      **else**
22:          Halt and output: "$\mathcal{E}$ is not Fitch-satisfiable"

Proof. Correctness is established in Lemma 5. We first note that in each single call of BUILDFITCHCOTREE, all necessary di-graphs defined in (S1), (S2) and

(S3) can be computed in $O(n+m)$ time. Furthermore, each of the following tasks can be performed in $O(n+m)$ time: finding the (strongly) connected components of each digraph, construction of the quotient graphs, and finding the topological order on the quotient graph using Kahn's algorithm [11]. Moreover, the vertex end edge sets $V[C]$ and $\mathcal{E}[C]$ for the (strongly) connected components C (or their unions) can be constructed in $O(n+m)$ time by going through every element in V and $\mathcal{E}$ and assigning each pair to their respective induced subset. Thus, every pass of BUILDFITCHCOTREE takes $O(n+m)$ time. Since every call of BUILDFITCHCOTREE either halts or adds a vertex to the final constructed Fitch-cotree, and the number of vertices in this tree is bounded by the number n of leaves, it follows that BUILDFITCHCOTREE is called at most $O(n)$ times resulting in an overall running time $O(n(n+m))$.   □

Instead of asking only for the existence of a Fitch-completion of a tuple $\mathcal{E} = (E_0, E_1, E_{\rightarrow})$, it is of interest to ask for the completion that maximizes a total score for the pairs of distinct vertices x, y that are not already classified by $\mathcal{E}$, i.e., $\{x,y\} \in \overline{\mathcal{E}} := \{\{x,y\} \notin (E_0 \cup E_1 \cup E_{\rightarrow})^{sym}\}$. For every pair of vertices there are four possibilities $x :: y \in \{x \rightleftharpoons y, x \rightarrow y, x \leftarrow y, x \smile y\}$. The score $w(x :: y)$ may be a log-odds ratio for observing one of the four possible xenology relationship as determined from experimental data. Let us write $F = F_{\mathcal{E}^*}$ for the Fitch graph defined by the extension $\mathcal{E}^*$ of $\mathcal{E}$ and associate with it the total weight of relations added, i.e.,

$$f(F) = \sum_{\{x,y\} \in \overline{\mathcal{E}}} w(F[\{x,y\}]) \tag{1}$$

The weighted Fitch-graph completion problem can also be seen as special case of the problem with empty tuple $\mathcal{E}^{\emptyset} := (\emptyset, \emptyset, \emptyset)$. To see this, suppose first that an arbitrary partial input tuple $\mathcal{E}$ is given. For each two vertices x, y for which $\{x,y\} \notin \overline{\mathcal{E}}$ the induced graphs $F[\{x,y\}]$ is well-defined and we extend the weight function to all pairs of vertices by setting, for all input pairs, $w(F[\{x,y\}]) = m_0$ and $w(x :: y) = -m_0$ for $(x :: y) \neq F[\{x,y\}]$, where $m_0 \gg |V|^2 \max_{::,\{x,y\} \in \overline{\mathcal{E}}} |w(x :: y)|$. Now consider the weighted Fitch graph completion problem with this weight function and an empty tuple $\mathcal{E}^{\emptyset}$. The choice of weights ensures that any Fitch graph F' maximizing $f(F')$ induces $F'[\{x,y\}] = F[\{x,y\}]$ for all pairs $\{x,y\}$ in the input tuple, because not choosing $F[\{x,y\}]$ reduces the score by $2m_0$ while the total score of all pairs not specified in the input is smaller than m_0. In order to study the complexity of this task, it therefore suffices to consider the following decision problem:

*Problem 1. (*FITCH COMPLETION PROBLEM (FC)*).*

Input: A set V, an assignment of four weights $w_{xy}(x :: y)$ to all distinct $x, y \in V$ where $:: \in \{\rightleftharpoons, \rightarrow, \leftarrow, \smile\}$, and an integer k.
Question: Is there a Fitch graph $F = (V, E)$ such that $f(F) = \sum_{\substack{x,y \in V \\ x \neq y}} w_{xy}(F[\{x,y\}]) \geq k$?

For the NP-hardness reduction, we use the following NP-complete problem [12]

Problem 2. (MAXIMUM ACYCLIC SUBGRAPH PROBLEM (MAS)).

Input: A digraph $G = (V, E)$ and an integer k.
Question: Is there a subset $E' \subseteq E$ such that $|E'| \geq k$ and (V, E') is a DAG?

Theorem 4. FC *is NP-complete.*

Proof. We claim that FC is in NP. To see this, let $F = (V, E)$ be a given digraph. We can check whether $f(F) \geq k$ in polynomial time by iterating over all edges in F. In addition, by Theorem 1, we can check whether F is a Fitch graph in polynomial time by iterating over all 3-subsets of V and verifying that none of the induced a forbidden subgraph of Fitch graphs.

To prove NP-hardness, let $(G = (V, E), k)$ be an instance of MAS. We take as input for FC the set V, the integer k and the following weights for all distinct $x, y \in V$:

  (i) If $(x, y) \in E$, then put $w_{xy}(x \to y) = 1$
  (ii) If $(x, y) \notin E$, then put $w_{xy}(x \to y) = 0$
  (iii) Put $w_{xy}(x \llcorner y) = 0$ and $w_{xy}(x \rightleftharpoons y) = -|V|^2$.

Note that Condition (i) ensures that, for all $x, y \in V$, we have $w_{xy}(y \to x) = 1$ if $(y, x) \in E$ and $w_{xy}(x \to y) = w_{xy}(y \to x) = 1$ whenever both (x, y) and (y, x) are edges in G.

Suppose first that there is a subset $E' \subseteq E$ such that $|E'| \geq k$ and $G' = (V, E')$ is a DAG. Hence, for any $x, y \in V$ not both (x, y) and (y, x) can be contained in E'. This, together with the construction of the weights implies that $f(G') = |E'| \geq k$. We now extend G' to a Fitch graph F. To this end, observe that G' admits a topological order $\ll$. We now add for all pairs x, y with $x \ll y$ and $(x, y) \notin E'$ the edge (x, y) to obtain the di-graph F. Clearly, $\ll$ remains a topological order of F and, therefore, F is a DAG. This with the fact that F does not contain bi-directional edges or non-adjacent vertices together with Corollary 2 implies that F is a Fitch graph. In particular, $f(F) \geq f(G') \geq k$.

Assume now that there is a Fitch graph F such that $f(F) \geq k$. Since $w_{xy}(x \rightleftharpoons y) = -|V|^2$ and the weight for any uni-directed edge and every pair of non-adjacent vertices is 0 or 1 and the maximum number of edges in F is $2 \cdot \binom{|V|}{2} = |V|^2 - |V| < |V|^2$, $f(F) \geq k$ implies that F cannot contain bidirectional edges. By Corollary 1, F is acyclic. Now, take the subgraph G' of F that consists of all edges with weight 1. Clearly, G' remains acyclic and $f(F) = f(G') \geq k$. By construction of the weights, all edges of G' must have been contained in G and thus, $G' \subseteq G$ is an acyclic subgraph of G containing at least k edges. $\square$

5   Concluding Remarks

Since FC is NP-complete, practical approaches will be based on heuristics. The structure of the problem suggests a canonical greedy heuristic, in which the weights for the possible 2-vertex graphs are sorted in descending order. For each proposed insertion of $x :: y$, it suffices to test whether the additional edges produced a forbidden subgraph on $\{x, y, z\}$ for some $z \in V$. The greedy heuristic thus runs in cubic time. We also note that seemingly simpler variants of the problem such as FC without bi-directional edges, and FC without non-adjacent vertices remain NP-complete.

References

1. Böcker, S., Dress, A.W.M.: Recovering symbolically dated, rooted trees from symbolic ultrametrics. Adv. Math. **138**, 105–125 (1998). https://doi.org/10.1006/aima.1998.1743
2. Corneil, D.G., Lerchs, H., Steward Burlingham, L.: Complement Reducible Graphs. Discr. Appl. Math. **3**, 163–174 (1981)
3. Corneil, D.G., Perl, Y., Stewart, L.K.: A linear recognition algorithm for cographs. SIAM J. Comput. **14**, 926–934 (1985)
4. Crespelle, C., Paul, C.: Fully dynamic recognition algorithm and certificate for directed cographs. Discr. Appl. Math. **154**, 1722–1741 (2006)
5. Engelfriet, J., Harju, T., Proskurowski, A., Rozenberg, G.: Characterization and complexity of uniformly nonprimitive labeled 2-structures. Theor. Comp. Sci. **154**, 247–282 (1996)
6. Geiß, M., Anders, J., Stadler, P.F., Wieseke, N., Hellmuth, M.: Reconstructing gene trees from fitch's xenology relation. J. Math. Biol. **77**, 1459–1491 (2018). https://doi.org/10.1007/s00285-018-1260-8
7. Gurski, F.: Dynamic programming algorithms on directed cographs. Stat. Optim. Inf. Comput. **5**(1), 35–44 (2017). https://doi.org/10.19139/soic.v5i1.260
8. Hellmuth, M., Scholz, G.E.: Resolving prime modules: the structure of pseudocographs and galled-tree explainable graphs (2023). http://arxiv.org/abs/10.48550/arXiv.2211.16854
9. Hellmuth, M., Seemann, C.R.: Alternative characterizations of Fitch's xenology relation. J. Math. Biol. **79**(3), 969–986 (2019). https://doi.org/10.1007/s00285-019-01384-x
10. Hellmuth, M., Stadler, P.F., Puthiyaveedu, S.T.: Fitch graph completion (2023). https://doi.org/10.48550/arXiv.2306.06878
11. Kahn, A.B.: Topological sorting of large networks. Commun. ACM **5**, 558–562 (1962). https://doi.org/10.1145/368996.369025
12. Karp, R.M.: Reducibility among combinatorial problems. In: Miller, R.E., Thatcher, J.W., Bohlinger, J.D. (eds.) Complexity of Computer Computations: Proceedings of a symposium on the Complexity of Computer Computations, pp. 85–103. Springer, US, Boston, MA (1972). https://doi.org/10.1007/978-1-4684-2001-2_9
13. McConnell, R.M., De Montgolfier, F.: Linear-time modular decomposition of directed graphs. Discr. Appl. Math. **145**(2), 198–209 (2005). https://doi.org/10.1016/j.dam.2004.02.017

14. Nøjgaard, N., El-Mabrouk, N., Merkle, D., Wieseke, N., Hellmuth, M.: Partial homology relations - satisfiability in terms of di-cographs. In: Wang, L., Zhu, D. (eds.) Computing and Combinatorics. Lecture Notes Comp. Sci., vol. 10976, pp. 403–415. Springer, Cham (2018). https://doi.org/10.1007/978-3-319-94776-1_34
15. Ravenhall, M., Škunca, N., Lassalle, F., Dessimoz, C.: Inferring horizontal gene transfer. PLoS Comp. Biol. **11**, e1004095 (2015). https://doi.org/10.1371/journal.pcbi.1004095

Deterministic Primal-Dual Algorithms for Online k-Way Matching with Delays

Naonori Kakimura[1] and Tomohiro Nakayoshi[2]($\boxtimes$)

[1] Keio University, 3-14-1 Hiyoshi, Kohoku-ku, Yokohama, Kanagawa 223-8522, Japan
kakimura@math.keio.ac.jp
[2] The University of Tokyo, 7-3-1 Hongo, Bunkyo-ku, Tokyo 113-8654, Japan
nakayoshi-tomohiro@g.ecc.u-tokyo.ac.jp

Abstract. In this paper, we study the Min-cost Perfect k-way Matching with Delays (k-MPMD), recently introduced by Melnyk et al. In the problem, m requests arrive one-by-one over time in a metric space. At any time, we can irrevocably make a group of k requests who arrived so far, that incurs the distance cost among the k requests in addition to the sum of the waiting cost for the k requests. The goal is to partition all the requests into groups of k requests, minimizing the total cost. The problem is a generalization of the min-cost perfect matching with delays (corresponding to 2-MPMD). It is known that no online algorithm for k-MPMD can achieve a bounded competitive ratio in general, where the competitive ratio is the worst-case ratio between its performance and the offline optimal value. On the other hand, k-MPMD is known to admit a randomized online algorithm with competitive ratio $O(k^5 \log n)$ for a certain class of k-point metrics called the H-metric, where n is the size of the metric space. In this paper, we propose a deterministic online algorithm with a competitive ratio of $O(mk^2)$ for the k-MPMD in H-metric space. Furthermore, we show that the competitive ratio can be improved to $O(m + k^2)$ if the metric is given as a diameter on a line.

Keywords: Online Matching · Online Algorithm · Competitive Analysis

1 Introduction

Consider an online gaming platform supporting two-player games such as Chess. In such a platform, players arrive one-by-one over time, and stay in a queue to participate in a match. The platform then tries to suggest a suitable opponent for each player from the queue. In order to satisfy the players, the platform aims to maximize the quality of the matched games. Specifically, we aim to minimize the distance of the matched players (e.g., the difference of their ratings) as well as the sum of the players' waiting time.

This work was supported by JSPS KAKENHI Grant Numbers JP20H05795, JP22H05001, and JP21H03397.

The above situation can be modeled as the problem called Online Matching with Delays, introduced by Emek et al. [11]. In the setting, arriving requests (or players) are embedded in a metric space so that the distance of each pair is determined. For the Online Matching with Delays, Emek et al. [11] proposed a randomized algorithm with a competitive ratio of $O(\log^2 n + \log \Delta)$, where n is the number of points in a metric space, and Δ is the ratio of the maximum to minimum distance between two points. The competitive ratio was later improved to $O(\log n)$ by Azar et al. [5]. We remark that both algorithms require that a metric space is finite and all the points in the metric space are known in advance (we note that arriving requests may be embedded into the same point more than once). Bienkowski et al. [9] presented a primal-dual algorithm with a competitive ratio of $O(m)$, where m is the number of requests. Another algorithm with a better competitive ratio of $O(m^{0.59})$ was proposed by Azar et al. [6].

In this paper, we consider a generalization of Online Matching with Delays, called the Min-cost Perfect k-way Matching with Delays (k-MPMD) [22]. In the problem, requests arrive one-by-one over time. At any time, instead of choosing a pair of requests, we make a group of k requests. This corresponds to an online gaming platform that allows more than two players to participate, such as mahjong ($k = 4$), Splatoon ($k = 8$), Apex Legends ($k = 60$), and Fortnite ($k = 100$). Then we aim to partition all the requests into groups of size-k subsets, minimizing the sum of the distance of the requests in the same group and the total waiting time.

To generalize to k-MPMD, it is necessary to measure the distance of a group of $k > 2$ requests. That is, we need to introduce a metric space that defines distances for any subset of k points. Although there are many ways of generalizing a standard distance between two points to $k > 2$ points in the literature [4,16], Melnyk et al. [22] showed that most known generalized metrics on k points cannot achieve a bounded competitive ratio for the k-MPMD. Melnyk et al. [22] then introduced a new interesting class of generalized metric, called H-metric, and proposed a randomized algorithm for the k-MPMD on H-metric with a competitive ratio of $O(k^5 \log n)$, extending Azar et al. [5].

The main contribution of this paper is to propose a deterministic algorithm for the k-MPMD on H-metric with a competitive ratio of $O(mk^2)$, where m is the number of requests. The proposed algorithm adopts a primal-dual algorithm based on a linear programming relaxation of the k-MPMD.

To design a primal-dual algorithm, we first formulate a linear programming relaxation of the offline problem, that is, when a sequence of requests is given in advance. We remark that even the offline setting is NP-hard when $k \geq 3$, as it includes the triangle packing problem. We first show that H-metric can be approximated by a standard metric (Theorem 1). This allows us to construct a linear programming problem with variables for each pair of requests such that the optimal value gives a lower bound on the offline version of the k-MPMD. Using the linear programming problem, we can design a primal-dual algorithm by extending the one by Bienkowski et al. [9] for Online Matching with Delays. We show that, by the observation on H-metric (Theorem 1) again, the cost of the output can be upper-bounded by the dual objective value of our linear programming problem.

An interesting special case of the H-metric is the diameter on a line. That is, points are given on a 1-dimensional line, and the distance of k points is defined to be the maximum difference in the k points. In the context of an online gaming platform, the diameter on a line can be interpreted as the difference of players' ratings. In this case, we show that the competitive ratio of our algorithm can be improved to $O(m + k^2)$. Moreover, we construct an instance such that our algorithm achieves the competitive ratio of $\Omega(m/k)$.

Related Work. An online algorithm for the matching problem was first introduced by Karp et al. [15]. They considered the online bipartite matching problem where arriving requests are required to match upon their arrival. Since then, the problem has been studied extensively in theory and practice. For example, motivated by internet advertising, Mehta et al. [20] generalized the problem to the AdWords problem. See also Mehta [21] and Goel and Mehta [13]. The weighted variant of the online bipartite matching problem is considered in the literature. It includes the vertex-weighted online bipartite matching [1], the problem with metric costs [8,23,25], and the problem with line metric cost [2,12,14,17]. We remark that the edge-weighted online bipartite matching in general has no online algorithm with bounded competitive ratio [1].

This paper deals with a variant of the online matching problem with delays, in which arriving requests are allowed to make decision later with waiting costs. Besides the related work [5,6,9,11] mentioned before, Liu et al. [18] extended the problem to the one with non-linear waiting costs. Other delay costs are studied in [7,10,19]. Ashlagi et al. [3] studied the online matching problem with deadlines, where each arriving request has to make a decision by her deadline. Pavone et al. [24] considered online hypergraph matching with deadlines.

Paper Organization. This paper is organized as follows. In Sect. 2, we formally define the minimum-cost perfect k-way matching problem and H-metric. We also discuss useful properties of H-metrics which will be used in our analysis. In Sect. 3, we present our main algorithm for the k-MPMD on H-metric. In Sect. 4, we show that there exists an instance such that our algorithm admits an almost tight competitive ratio. Due to the space limitation, the proofs of lemmas and theorems are omitted, which may be found in the full version of this paper.

2   Preliminaries

2.1   Minimum-Cost Perfect k-Way Matching with Delays

In this section, we formally define the problem k-MPMD. Let (χ, d) be a generalized metric space where χ is a set and $d : \chi^k \to [0, \infty)$ represents a distance among k elements.

In the problem, m requests $u_1, u_2, \ldots, u_m$ arrive one-by-one in this order. The arrival time of u_i is denoted by $\text{atime}(u_i)$. When u_i arrives, the location $\text{pos}(u_i)$ of u_i in the metric space χ is revealed. Thus, an instance of the problem is given as a tuple $\sigma = (V, \text{atime}, \text{pos})$, where $V = \{u_1, \ldots, u_m\}$, $\text{atime} : V \to \mathbb{R}_+$, and $\text{pos} : V \to \chi$ such that $\text{atime}(u_1) \leq \cdots \leq \text{atime}(u_m)$. We note that m may be unknown in advance, but we assume that m is a multiple of k.

At any time τ, with the only information for requests arrived so far, an online algorithm can make a set of k requests $v_1, \ldots, v_k$ in V, where we say that $v_1, \ldots, v_k$ are *matched*, if they satisfy the following two conditions: (a) The requests $v_1, \ldots, v_k$ have already arrived, that is, $\text{atime}(v_i) \leq \tau$ for any $i = 1, \ldots, k$; (b) None of $v_1, \ldots, v_k$ has been matched to other requests yet. The cost to match $v_1, \ldots, v_k$ at time τ is defined to be

$$d(\text{pos}(v_1), \text{pos}(v_2), \ldots, \text{pos}(v_k)) + \sum_{i=1}^{k} (\tau - \text{atime}(v_i)).$$

The first term means the distance cost among the k requests and the second term is the total waiting cost of the k requests.

The objective of the problem is to design an online algorithm that matches all the requests, minimizing the total cost. In other words, an online algorithm finds a family of disjoint subsets of size k that covers all the requests. We call a family of disjoint subsets of size k a k-*way matching*, and a k-way matching is called *perfect* if it covers all the requests.

To measure the performance of an online algorithm, we define the competitive ratio. For an instance σ, let $\mathcal{ALG}(\sigma)$ be the cost incurred by the online algorithm, and let $\mathcal{OPT}(\sigma)$ be the optimal cost when we know in advance a sequence of requests V as well as $\text{atime}(u_i)$ and $\text{pos}(u_i)$ for each request u_i. The *competitive ratio* of the online algorithm is defined as $\sup_\sigma \frac{\mathcal{ALG}(\sigma)}{\mathcal{OPT}(\sigma)}$.

2.2 H-Metric

In this section, we define H-metric, introduced by Melnyk et al. [22]. Recall that a function $d : \chi^2 \to [0, \infty)$ is called a *distance function* (or a *metric*) if d satisfies the following three axioms:

- **(Symmetry)** $d(p_1, p_2) = d(p_2, p_1)$ for any $p_1, p_2 \in \chi$.
- **(Positive definiteness)** $d(p_1, p_2) \geq 0$ for any $p_1, p_2 \in \chi$, and $d(p_1, p_2) = 0$ if and only if $p_1 = p_2$.
- **(Triangle inequality)** $d(p_1, p_3) \leq d(p_1, p_2) + d(p_2, p_3)$ for any $p_1, p_2, p_3 \in \chi$.

We first define a k-point metric as a k-variable function satisfying generalizations of the symmetry axiom and the positive definiteness axiom.

Definition 1. *We call a function $d : \chi^k \to [0, \infty)$ a k-point metric if it satisfies the following two axioms.*

Π: *For any permutation π of $\{p_1, \ldots, p_k\}$, we have $d(p_1, \ldots, p_k) = d(\pi(p_1), \ldots, \pi(p_k))$.*

O_D: *It holds that* $d(p_1, \ldots, p_k) \geq 0$. *Moreover,* $d(p_1, \ldots, p_k) = 0$ *if and only if* $p_1 = p_2 = \cdots = p_k$.

There are several ways of generalizing the triangle inequality to k-variable functions. One possibility is the following axiom: for any $p_1, \ldots, p_k, a \in \chi$ and any $i \in \{1, \ldots, k\}$, it holds that

$$\Delta_H : d(p_1, \ldots, p_k) \leq d(p_1, \ldots, p_i, \underbrace{a, \ldots, a}_{k-i}) + d(\underbrace{a, \ldots, a}_{i}, p_{i+1}, \ldots, p_k).$$

We note that it is identical to the triangle inequality when $k = 2$.

For a multiset S on χ, we denote by $elem(S)$ the set of all distinct elements contained in S. In addition to the generalized triangle inequality, we consider the relationship between $d(p_1, \ldots, p_k)$ and $d(p'_1, \ldots, p'_k)$ when $elem(\{p_1, \ldots, p_k\}) \subseteq elem(\{p'_1, \ldots, p'_k\})$. The *separation axiom* $\mathcal{S}_H$ says that, for some nonnegative integer $\gamma \leq k - 1$,

$$d(p_1, \ldots, p_k) \leq d(p'_1, \ldots, p'_k) \quad \text{if } elem(\{p_1, \ldots, p_k\}) \subset elem(\{p'_1, \ldots, p'_k\}),$$
$$d(p_1, \ldots, p_k) \leq \gamma \cdot d(p'_1, \ldots, p'_k) \quad \text{if } elem(\{p_1, \ldots, p_k\}) = elem(\{p'_1, \ldots, p'_k\}).$$

The H-metric is a k-point metric that satisfies all the above axioms.

Definition 2 (Melnyk et al. [22]). *A k-point metric $d_H : \chi^k \to [0, \infty)$ is an H-metric with parameter $\gamma \leq k - 1$ if it satisfies Π, O_D, Δ_H and $\mathcal{S}_H$ with parameter γ.*

We remark that there are weaker conditions than Δ_H and $\mathcal{S}_H$, generalizing the triangle inequality, which yields other classes of k-point metrics such as the n-metric [4] and the K-metric [16]. See [22] for the formal definition. Melnyk et al. [22], however, showed that the k-MPMD cannot be solved for such more general metrics. Specifically, they proved that there exists no randomized algorithm for the k-MPMD ($k \geq 5$) problem on n-metric or K-metric agaist an oblivious adversary that has a competitive ratio which is bounded by a function of the number of points n.

2.3   Properties of H-Metric

In this section, we discuss approximating H-metric by a standard metric, and present specific examples of H-metric.

Melnyk et al. proved that H-metric can be approximated by the sum of distances between all pairs [22, Theorem 6]. We refine their results as in the theorem below, which will be used in the next section.

Theorem 1. *Let d_H be an H-metric on χ with parameter γ. Define a metric $d : \chi^2 \to [0, \infty)$ as*

$$d(p_1, p_2) := d_H(p_1, p_2, \ldots, p_2) + d_H(p_2, p_1, \ldots p_1)$$

for any $p_1, p_2 \in \chi$. Then it holds that

$$\frac{1}{\gamma k^2} \cdot \sum_{i=1}^{k-1} \sum_{j=i+1}^{k} d(p_i, p_j) \leq d_H(p_1, \ldots, p_k) \leq \sum_{i=1}^{k} d(v, p_i), \qquad (1)$$

for all $v \in \{p_1, \ldots, p_k\}$.

We conclude this section with providing specific examples of H-metric. We note that the examples below satisfy that $\gamma = 1$, and thus the approximation factor in Theorem 1 becomes small.

Let $d : \chi^2 \to [0, \infty)$ be a distance function. We define a k-point metric $d_{\max}$ by $d_{\max}(p_1, \ldots, p_k) = \max_{i,j \in \{1, \ldots, k\}} d(p_i, p_j)$. Then it turns out to be an H-metric.

Proposition 1. *Let $d : \chi^2 \to [0, \infty)$ be a distance function. Then the k-point metric $d_{\max}$ is an H-metric with $\gamma = 1$.*

For real numbers $p_1, \ldots, p_k \in \mathbb{R}$, we define the *diameter on a line* as $d_D(p_1, \ldots, p_k) = \max_{i,j \in \{1, \ldots, k\}} |p_i - p_j|$. By Proposition 1, d_D is an H-metric.

For a distance function $d : \chi^2 \to [0, \infty)$, we define another H-metric d_{HC} by

$$d_{HC}(p_1, \ldots, p_k) = \min \left\{ \sum_{e \in C} d(e) \mid C \subseteq \binom{\chi}{2}, C \text{ forms a Hamiltonian circuit in } \{p_1, \ldots, p_k\} \right\}$$

where $\binom{\chi}{2} = \{(p, q) \mid p, q \in \chi, p \neq q\}$. This means that $d_{HC}(p_1, \ldots, p_k)$ equals to the minimum cost of a Hamiltonian circuit contained in $\{p_1, \ldots, p_k\}$ with respect to cost d.

Proposition 2. *Let $d : \chi^2 \to [0, \infty)$ be a distance function. Then the k-point metric d_{HC} is an H-metric with parameter $\gamma = 1$.*

3 k-MPMD on H-Metric Space

This section proposes a primal-dual algorithm for k-MPMD on H-metric space. Let (χ, d_H) be an H-metric space with parameter γ.

3.1   Linear Programming Relaxation

This subsection introduces a linear programming relaxation for computing the offline optimal value $\mathcal{OPT}(\sigma)$ for a given instance σ.

We first give some notation. Let $\mathcal{E} = \{F \subseteq V \mid |F| = k\}$. For any subset $S \subseteq V$, we denote $\text{sur}(S) = |S| \mod k$, which is the number of remaining requests when we make a k-way matching of size $\lfloor |S|/k \rfloor$ among S. We denote $\Delta(S) = \{F \in \mathcal{E} \mid F \cap S \neq \emptyset, F \setminus S \neq \emptyset\}$, which is the family of k request sets that intersect both S and $V \setminus S$.

Preparing a variable x_F for any subset $F \in \mathcal{E}$, we define a linear programming problem:

$$(\mathcal{P}) \quad \begin{array}{ll} \text{min.} & \sum_{F \in \mathcal{E}} \text{opt-cost}(F) \cdot x_F \\[2ex] \text{s.t.} & \sum_{F \in \Delta(S)} x_F \geq \left\lceil \dfrac{\text{sur}(S)}{k} \right\rceil, \quad \forall S \subseteq V \qquad (2) \\[2ex] & x_F \geq 0, \qquad\qquad \forall F \in \mathcal{E} \end{array}$$

where, for any $F = (v_1, \ldots, v_k) \in \mathcal{E}$, we define

$$\text{opt-cost}(F) := d_H(\text{pos}(v_1), \ldots, \text{pos}(v_k)) + \sum_{i=1}^{k} \left(\max_j \text{atime}(v_j) - \text{atime}(v_i) \right).$$

Notice that opt-cost(F) is the cost of choosing F at the moment when all the requests in F have arrived.

Let $\mathcal{M}$ be a perfect k-way matching with optimal cost $\mathcal{OPT}(\sigma)$. Define a 0-1 vector $(x_F)_{F \in \mathcal{E}}$ such that $x_F = 1$ if and only if $F \in \mathcal{M}$. Then the vector satisfies the constraint (2). Moreover, the cost incurred by $F \in \mathcal{M}$ is equal to opt-cost(F). This is because the optimal algorithm that returns $\mathcal{M}$ chooses F at the moment when all the requests in F have arrived. Thus the objective value for the vector $(x_F)_{F \in \mathcal{E}}$ is equal to $\mathcal{OPT}(\sigma)$, and hence the optimal value of $(\mathcal{P})$ gives a lower bound of $\mathcal{OPT}(\sigma)$.

We further relax the above LP $(\mathcal{P})$ by replacing x_F's with variables for all pairs of requests. Let $E = \{(u, v) \mid u, v \in V, u \neq v\}$, and we prepare a variable x_e for any $e \in E$. We often call an element in E an *edge*.

We denote by $\delta(S)$ the set of pairs between S and $V \setminus S$. Define the following linear programming problem:

$$(\mathcal{P}') \quad \begin{array}{ll} \text{min.} & \sum_{e \in E} \dfrac{1}{\gamma k^2} \cdot \text{opt-cost}(e) \cdot x_e \\[2ex] \text{s.t.} & \sum_{e \in \delta(S)} x_e \geq \text{sur}(S) \cdot (k - \text{sur}(S)), \quad \forall S \subseteq V \qquad (3) \\[2ex] & x_e \geq 0, \qquad\qquad \forall e \in E \end{array}$$

where, for any $e = (v_1, v_2) \in E$ with $p_1 = \text{pos}(v_1)$ and $p_2 = \text{pos}(v_2)$, we define

$$d(p_1, p_2) := d_H(p_1, p_2, \ldots, p_2) + d_H(p_2, p_1, \ldots, p_1), \text{ and}$$
$$\text{opt-cost}(e) := d(p_1, p_2) + |\text{atime}(v_1) - \text{atime}(v_2)|.$$

The following lemma follows from Theorem 1.

Lemma 1. *It holds that, for any* $F = (v_1, \ldots, v_k) \in \mathcal{E}$,

$$\frac{1}{\gamma k^2} \cdot \sum_{i=1}^{k-1} \sum_{j=i+1}^{k} \text{opt-cost}(v_i, v_j) \leq \text{opt-cost}(F) \leq \sum_{i=1}^{k} \text{opt-cost}(v, v_i), \qquad (4)$$

where $v = \arg \max_{u \in F} \text{atime}(u)$.

For any perfect k-way matching $\mathcal{M}$, define an edge subset M such that $e \in M$ if and only if the pair e is contained in some set F of $\mathcal{M}$. Thus we represent each set in $\mathcal{M}$ with a complete graph of k vertices. We will show below that the characteristic vector for M is feasible to $(\mathcal{P}')$. Here, for a subset $X \subseteq E$, the characteristic vector $\mathbb{1}_X \in \{0, 1\}^E$ is defined to be

$$\mathbb{1}_X(x) = \begin{cases} 1 & (x \in X) \\ 0 & (x \notin X) \end{cases}.$$

Moreover, this implies that the optimal value of $(\mathcal{P}')$, denoted by $\mathcal{P}'(\sigma)$, is a lower bound of $\mathcal{OPT}(\sigma)$ for any instance σ.

Lemma 2. *Let $\mathcal{M}$ be a perfect k-way matching. Define an edge subset $M = \{(u, v) \in E \mid \exists F \in \mathcal{M} \text{ s.t. } u, v \in F\}$. Then $x = \mathbb{1}_M$ is a feasible solution to $\mathcal{P}'$. Furthermore, $\mathcal{P}'(\sigma) \leq \mathcal{OPT}(\sigma)$ holds.*

The dual linear programming problem of $(\mathcal{P}')$ is

$$(\mathcal{D}') \quad \begin{array}{rl} \text{max.} & \displaystyle\sum_{S \subseteq V} \text{sur}(S) \cdot (k - \text{sur}(S)) \cdot y_S \\[2ex] \text{s.t.} & \displaystyle\sum_{S: e \in \delta(S)} y_S \leq \frac{1}{\gamma k^2} \cdot \text{opt-cost}(e), \quad \forall e \in E \hspace{1cm} (5) \\[2ex] & y_S \geq 0, \hspace{3cm} \forall S \subseteq V \end{array}$$

The weak duality of LP implies that $\mathcal{D}'(\sigma) \leq \mathcal{P}'(\sigma)$, where $\mathcal{D}'(\sigma)$ is the dual optimal value.

3.2 Greedy Dual for k-MPMD (GD-k)

We present our proposed algorithm, called *Greedy Dual for k-MPMD*(GD-k). The proposed algorithm extends the one by Bienkowski et al. [9] for 2-MPMD using the LP $(\mathcal{P}')$.

In the algorithm GD-k, we maintain a family of subsets of requests, called *active sets*. At any time, any request v arrived so far belongs to exactly one active set, denoted by $A(v)$. We also maintain a k-way matching $\mathcal{M}$. A request not in $\bigcup_{F \in \mathcal{M}} F$ is called *free*, and, for a subset $S \subseteq V$ of requests, free(S) is the set of free requests in S.

When request v arrives, we initialize $A(v) = \{v\}$ and $y_S = 0$ for any subset $S \subseteq V$ such that $v \in S$. At any time, for an active set S such that free(S) is nonempty, we increase y_S with rate r, where r is set to be $1/(\gamma k^2)$. Then, at some point, there exists an edge $e = (u, v) \in E$ such that $\sum_{S: e \in \delta(S)} y_S = \frac{1}{\gamma k^2} \cdot \text{opt-cost}(e)$, which we call a *tight* edge. When it happens, we merge the active sets $A(u)$ and $A(v)$ to a large subset $S = A(u) \cup A(v)$, that is, we update

$A(w) = S$ for all $w \in S$. We also mark the tight edge e. If $|\text{free}(S)| \geq k$, we partition free(S) arbitrarily into subsets of size k with sur(S) free requests, and add these size-k subsets to $\mathcal{M}$.

The pseudo-code of the algorithm is given as in Algorithm 1.

Let T be the time when all requests are matched in the algorithm. For any subset S, we denote the value of y_S at time τ in the algorithm by $y_S(\tau)$.

Algorithm 1 Greedy Dual for k-MPMD

1: **procedure** GD-$k(\sigma)$
2:     $\mathcal{M} \leftarrow \emptyset$
3:     **for all** moments t **do**
4:         **if** a request v arrives **then**
5:             $A(v) \leftarrow \{v\}$
6:             **for all** subsets $S \ni v$ **do**
7:                 $y_S \leftarrow 0$
8:             **end for**
9:             **modify** constraints of $(\mathcal{D}')$.
10:        **end if**
11:        **if** there exists $e = (u, v) \in E$ such that $\sum_{S:e\in\delta(S)} y_S = \frac{1}{\gamma k^2} \cdot \text{opt-cost}(e)$ and $A(u) \neq A(v)$ **then**
12:            $S \leftarrow A(u) \sqcup A(v)$
13:            **for all** $v \in S$ **do**
14:                $A(v) \leftarrow S$
15:            **end for**
16:            **mark** e
17:            **while** $|\text{free}(S)| \geq k$ **do**
18:                choose arbitrarily a set F of k requests from S
19:                $\mathcal{M} \leftarrow \mathcal{M} \cup \{F\}$
20:            **end while**
21:        **end if**
22:        **for all** sets S which are active and free(S) $\neq \emptyset$ **do**
23:            increase continuously y_S at the rate of r per unit time
24:        **end for**
25:    **end for**
26: **end procedure**

We show that y_S's maintained in Algorithm 1 are always dual feasible.

Lemma 3. *For any request v, it holds that*

$$\sum_{S:v\in S} y_S(\tau) \leq r \cdot (\tau - \text{atime}(v)) \tag{6}$$

at any time $\tau \geq \text{atime}(v)$. This holds with equality while v is not matched.

Lemma 4. *Let $r = \frac{1}{\gamma k^2}$. Then, at any time τ, $y_S(\tau)$ maintained in Algorithm 1 is a feasible solution to $(\mathcal{D}')$.*

3.3   Competitive Ratio of GD-k

To bound the competitive ratio of GD-k, we evaluate the distance cost and the waiting cost separately. We will show that each cost is upper-bounded by the dual optimal value of $\mathcal{D}'(\sigma)$.

Waiting Cost. We can upper-bound the waiting cost of the output as follows.

Lemma 5. *Let* $\mathcal{M} = \{M_1, \ldots, M_p\}$ *be a perfect k-way matching returned by Algorithm 1, and let τ_ℓ be the time when we match M_ℓ. Then it holds that*

$$\sum_{\ell=1}^{p}\sum_{i=1}^{k}(\tau_\ell - \mathrm{atime}(v_{\ell,i})) = \frac{1}{r} \cdot \sum_{S \subseteq V} \mathrm{sur}(S) \cdot y_S(T) \leq \frac{1}{r} \cdot \mathcal{D}'(\sigma),$$

where we denote $M_\ell = \{v_{\ell,1}, \ldots, v_{\ell,k}\}$.

Distance Cost. We say that a set $S \subseteq V$ is *formerly-active at time τ* if S is not active at time τ, but has been active before time τ.

Lemma 6. *Let S be an active or formerly-active set at time τ. Then, marked edges both of whose endpoints are contained in S form a spanning tree in S.*

We now evaluate the distance cost.

Lemma 7. *Let* $\mathcal{M} = \{M_1, \ldots, M_p\}$ *be a perfect k-way matching returned by Algorithm 1. Then it holds that*

$$\sum_{\ell=1}^{p} d(\mathrm{pos}(v_{\ell,1}), \ldots, \mathrm{pos}(v_{\ell,k})) \leq 4\gamma mk \cdot \sum_{S} \mathrm{sur}(S) \cdot (k - \mathrm{sur}(S)) \cdot y_S(T) \leq 4\gamma mk \mathcal{D}'(\sigma),$$

where we denote $M_\ell = \{v_{\ell,1}, \ldots, v_{\ell,k}\}$.

Competitive Ratio. Summarizing the above discussion, we obtain Theorem 2.

Theorem 2. *Let d_H be an H-metric with parameter γ. Setting $r = 1/(\gamma k^2)$, Greedy Dual for k-MPMD achieves a competitive ratio $(4mk + k^2)\gamma$ for k-MPMD.*

Proof. Let σ be an instance of k-MPMD. It follows from Lemmas 5 and 7 that the cost of the returned perfect k-way matching is upper-bounded by $(4mk + k^2)\gamma \cdot \mathcal{D}'(\sigma)$. By the weak duality and Lemma 4, we observe that $\mathcal{D}'(\sigma) \leq \mathcal{P}'(\sigma) \leq \mathcal{OPT}(\sigma)$. Thus the theorem holds. □

Finally, we consider applying our algorithm to the problem with specific H-metrics such as $d_{\max}$ and d_{HC} given in Sect. 2.3. Since they have parameter $\gamma = 1$, it follows from Theorem 2 that GD-k achieves a competitive ratio $O(mk + k^2)$. In the case of $d_{\max}$, we can further improve the competitive ratio.

Theorem 3. *For the k-MPMD on a metric space $(\chi, d_{\max})$, GD-k achieves a competitive ratio $O(m + k^2)$.*

4   Lower Bound of GD-k for a Diameter on a Line

In this section, we show a lower bound on the competitive ratio for GD-k for the metric d_D. Recall that $d_D(p_1, \ldots, p_k) = \max_{i,j \in \{1,\ldots,k\}} |p_i - p_j|$ for $p_1, \ldots, p_k \in \mathbb{R}$.

We define an instance $\sigma_l = (V, \text{pos}, \text{atime})$ where $V = \{u_1, u_2, \ldots, u_m\}$ as follows. Suppose that the number m of requests is equal to $m = sk^2$ for some integer s. Let $p_1, \ldots, p_k$ be k points in $\mathbb{R}$ such that $d(p_i, p_{i+1}) = 2$ for any $i = 1, 2, \ldots, k-1$.

For $i = 1, 2, \ldots, sk$ and $j = 1, 2, \ldots, k$, define $\text{atime}(u_{k(i-1)+j}) = t_i$ and $\text{pos}(u_{k(i-1)+j}) = p_j$ for $j = 1, 2, \ldots, k$, where we define $t_1 = 0$ and $t_i = 1 + (2i - 3)\varepsilon$ for $i \geq 2$. Thus, at any time t_i ($i = 1, \ldots, sk$), the k requests $u_{k(i-1)+1}, \ldots, u_{k(i-1)+k}$ arrive at every point in $p_1, \ldots, p_k$, respectively.

Then it holds that $\mathcal{OPT}(\sigma_l) \leq k + k\varepsilon + k^3\varepsilon + mk\varepsilon$, while the output of GD-k has cost at least $m + k + (m - k)\varepsilon$.

Theorem 4. *For a metric space* $(\mathbb{R}, d_D)$, *there exists an instance* σ_l *of* m *requests such that GD-k admits a competitive ratio* $\Omega(\frac{m}{k})$.

References

1. Aggarwal, G., Goel, G., Karande, C., Mehta, A.: Online vertex-weighted bipartite matching and single-bid budgeted allocations. In: Proceedings of the 2011 Annual ACM-SIAM Symposium on Discrete Algorithms, pp. 1253–1264. SODA 2011, SIAM (2011)
2. Antoniadis, A., Barcelo, N., Nugent, M., Pruhs, K., Scquizzato, M.: A $o(n)$-competitive deterministic algorithm for online matching on a line. Algorithmica **81**, 2917–2933 (2019)
3. Ashlagi, I., Burq, M., Dutta, C., Jaillet, P., Saberi, A., Sholley, C.: Edge-weighted online windowed matching. Math. Oper. Res. **48**(2), 999–1016 (2023)
4. Assaf, S., Pal, K.: Partial n-metric spaces and fixed point theorems (2015). arXiv: 1502.05320
5. Azar, Y., Chiplunkar, A., Kaplan, H.: Polylogarithmic bounds on the competitiveness of min-cost perfect matching with delays. In: Proceedings of the Twenty-Eighth Annual ACM-SIAM Symposium on Discrete Algorithms (SODA 2017), pp. 1051–1061. SIAM (2017)
6. Azar, Y., Jacob Fanani, A.: Deterministic min-cost matching with delays. Theor. Comput. Syst. **64**(4), 572–592 (2020)
7. Azar, Y., Ren, R., Vainstein, D.: The min-cost matching with concave delays problem. In: Proceedings of the 2021 ACM-SIAM Symposium on Discrete Algorithms, pp. 301–320. SODA 2021, SIAM (2021)
8. Bansal, N., Buchbinder, N., Gupta, A., Naor, J.S.: An $O(\log^2 k)$-competitive algorithm for metric bipartite matching. In: Arge, L., Hoffmann, M., Welzl, E. (eds.) ESA 2007. LNCS, vol. 4698, pp. 522–533. Springer, Heidelberg (2007). https://doi.org/10.1007/978-3-540-75520-3_47
9. Bienkowski, M., Kraska, A., Liu, H.-H., Schmidt, P.: A primal-dual online deterministic algorithm for matching with delays. In: Epstein, L., Erlebach, T. (eds.) WAOA 2018. LNCS, vol. 11312, pp. 51–68. Springer, Cham (2018). https://doi.org/10.1007/978-3-030-04693-4_4

10. Deryckere, L., Umboh, S.W.: Online matching with set and concave delays (2023). arXiv: 2211.02394
11. Emek, Y., Kutten, S., Wattenhofer, R.: Online matching: haste makes waste! In: Proceedings of the Forty-Eighth Annual ACM Symposium on Theory of Computing, pp. 333–344. STOC 2016, ACM (2016)
12. Fuchs, B., Hochstättler, W., Kern, W.: Online matching on a line. Theoret. Comput. Sci. **332**(1), 251–264 (2005)
13. Goel, G., Mehta, A.: Online budgeted matching in random input models with applications to Adwords. In: Proceedings of the Nineteenth Annual ACM-SIAM Symposium on Discrete Algorithms, pp. 982–991. SODA 2008, SIAM (2008)
14. Gupta, A., Lewi, K.: The online metric matching problem for doubling metrics. In: Czumaj, A., Mehlhorn, K., Pitts, A., Wattenhofer, R. (eds.) ICALP 2012. LNCS, vol. 7391, pp. 424–435. Springer, Heidelberg (2012). https://doi.org/10.1007/978-3-642-31594-7_36
15. Karp, R.M., Vazirani, U.V., Vazirani, V.V.: An optimal algorithm for on-line bipartite matching. In: Proceedings of the Twenty-Second Annual ACM Symposium on Theory of Computing, pp. 352–358. STOC 1990, ACM (1990)
16. Khan, K.A.: On the possibitity of N-topological spaces. Int. J. Math. Arch. **3**(7), 2520–2523 (2012)
17. Koutsoupias, E., Nanavati, A.: The online matching problem on a line. In: Solis-Oba, R., Jansen, K. (eds.) WAOA 2003. LNCS, vol. 2909, pp. 179–191. Springer, Heidelberg (2004). https://doi.org/10.1007/978-3-540-24592-6_14
18. Liu, X., Pan, Z., Wang, Y., Wattenhofer, R.: Impatient online matching. In: 29th International Symposium on Algorithms and Computation, ISAAC 2018. LIPIcs, vol. 123, pp. 62:1–62:12. Schloss Dagstuhl - Leibniz-Zentrum für Informatik (2018)
19. Liu, X., Pan, Z., Wang, Y., Wattenhofer, R.: Online matching with convex delay costs (2022). arXiv:2203.03335
20. Mehta, A., Saberi, A., Vazirani, U., Vazirani, V.: AdWords and generalized on-line matching. In: 46th Annual IEEE Symposium on Foundations of Computer Science (FOCS 2005), pp. 264–273 (2005)
21. Mehta, A.: Online matching and ad allocation. Found. Trends Theor. Comput. Sci. **8**(4), 265–368 (2013)
22. Melnyk, D., Wang, Y., Wattenhofer, R.: Online k-Way Matching with Delays and the H-Metric (2021). arXiv:2109.06640
23. Nayyar, K., Raghvendra, S.: An input sensitive online algorithm for the metric bipartite matching problem. In: 2017 IEEE 58th Annual Symposium on Foundations of Computer Science (FOCS 2017), pp. 505–515 (2017)
24. Pavone, M., Saberi, A., Schiffer, M., Tsao, M.W.: Technical note-online hypergraph matching with delays. Oper. Res. **70**(4), 2194–2212 (2022)
25. Raghvendra, S.: A robust and optimal online algorithm for minimum metric bipartite matching. In: International Workshop on Approximation, Randomization, and Combinatorial Optimization. Algorithms and Techniques (APPROX/RANDOM 2016). LIPIcs, vol. 60, pp. 18:1–18:16. Schloss Dagstuhl-Leibniz-Zentrum fuer Informatik, Dagstuhl, Germany (2016)

Diversity and Freshness-Aware Regret Minimizing Set Queries

Hongjie Guo[1,2], Jianzhong Li[1,3], Fangyao Shen[2], and Hong Gao[1,2(✉)]

[1] Faculty of Computing, Harbin Institute of Technology, Harbin, Heilongjiang, China
lijzh@hit.edu.cn
[2] School of Computer Science and Technology, Zhejiang Normal University,
Jinhua, Zhejiang, China
{guohongjie,fyshen,honggao}@zjnu.edu.cn
[3] Faculty of Computer Science and Control Engineering, Shenzhen Institute
of Advanced Technology, Chinese Academy of Sciences, Shenzhen, China

Abstract. Multi-criteria decision-making often involves selecting a small representative set from a database. A recently proposed method is the regret minimization set (RMS) queries. It aims to rectify the shortcomings of needing a utility function in top-k queries and the overly large result size of skyline queries. However, the existing definition of RMS only ensures one result under any utility function, and do not consider the diversity and freshness of the returned results. In this paper, we define a strong regret set, which guarantees the utility value error of k data points under any utility function. Given this new definition, we propose two problems, namely the Minimum Size problem and the Max-sum Diversity and Freshness problem. Both proposed problems have been proven to be NP-hard. Correspondingly, we devise approximation algorithms for them, and analyze algorithms' time complexities and the approximation ratios of the solutions obtained.

Keywords: Regret minimizing set · Diversity · Freshness · NP-hard

1 Introduction

Finding a representative set of points from a database for multi-criteria decision-making is a critical issue. This problem is fundamental in numerous applications where users are primarily interested in selecting a few or even just one point within a large database. For example, Bob visits a large car database, in which each car point has two attributes: horsepower (HP) and miles per gallon (MPG). He is in search of a car with both high MPG and high HP. However, there is an inherent trade-off between these two objectives, as an increase in HP often results in decreased fuel efficiency. Consequently, it may be infeasible for Bob to examine every car in the database. A more feasible approach involves presenting a select

This work was supported by the National Natural Science Foundation of China under grants U22A2025 and U19A2059.

number of representative points based on specific criteria, such as those potentially favored by the majority of users. In existing literature, numerous queries have been proposed to address multi-criteria decision-making. Among these, the most extensively researched include top-k queries [1], skyline queries [2], and regret minimization set queries [3].

A top-k query [1] requires the user to specify a utility function that reflects his/her preference for each attribute. The database points are then ranked based on utility values computed from the utility function, and the k highest-ranked points are returned to the user. For instance, in a car database, a user's utility function may assign a weight of 0.8 (more important) to HP and 0.2 (less important) to MPG. A top-10 query would return the 10 cars with the highest utility values according to this function. However, it is challenging for non-expert users to provide specific utility functions for top-k queries. In contrast, a skyline query [2] does not necessitate any utility function from users. Instead, it relies on the concept of domination: point p is considered to dominate point q iff p is no worse than q in every attribute and is strictly superior to q in at least one attribute. In a car database, car A would be dominated by car B if B is both higher HP and higher MPG than B. A skyline query returns all points that are not dominated by any other point in the database. However, the result sizes of skyline queries are not controllable, particularly when the dimensionality (i.e., the number of attributes) of the database is high.

Recently, Nanongkai et al. [3] introduced the Regret Minimization Set (RMS) queries as a solution to overcome the drawbacks of top-k queries and skyline queries. RMS computes a smaller, more representative subset Q from a large dataset P, ensuring that for any utility function, the maximum utility value in Q has a small error compared to the maximum utility value in P. RMS incorporates features of both top-k and skyline queries, eliminating the need for users to supply utility functions and producing significantly smaller result sets compared to skyline queries. To further reduce the set size, Chester et al. [4] relaxed the RMS definition and introduced k-RMS. For k-RMS, the error in Q is defined as the gap between the highest utility value in Q and the kth highest utility value in P, with this gap referred to as the regret ratio. Studies [4–6] have demonstrated the complexity of the k-RMS problem, showing that it is NP-hard when the dimensionality $d \geq 3$. Subsequently, numerous algorithms have been proposed to improve the efficiency of computing k-RMS [7–9], as well as to consider its computation in dynamic environments [6,10,11].

A utility function represents a potential user preference, while both RMS and k-RMS only guarantee a result under any given utility function. However, users may require more results under this preference, and the more diverse these results, the better. Consider the following example: a car database has the following three products information: (1) 130 HP and 50 MPG, (2) 130 HP and 52 MPG, (3) 200 HP and 37 MPG. Suppose the utility values of these products, computed by a specific utility function, are 2.0, 1.9, and 1.8. If the desired result set size is 2, with a tolerance error of less than or equal to 0.2, providing products (1) and (3) offers more diverse choices for the user. A large body of research exists on query result diversity [12–15]. However, no work has considered diversity when computing RMS. Result diversity refers to the lower similarity between the returned results, the better, and the similarity between two points in the dataset is typically measured by the distance between them. The optimization objective of maximizing result diversity is to maximize the sum of the distances between any two points in the result set, which is called the Max-sum Diversification (MSD) problem [16]. In addition to diversity, data freshness is another data characteristic that needs to be considered when computing regret sets. Data points in a database often have temporal characteristics; even in read-oriented environments like data warehouses, data is usually updated in batch periodically. In particular, in sensor network, sensing devices continuously collect data and periodically transmit the data back to the database. During multi-criteria decision-making on these data, data freshness is an important indicator.

Based on the above discussion, this paper extends RMS and defines the Strong Regret Minimization Set (SRMS). For any utility function, the returned result set Q contains k data points whose the kth highest utility value have an error guarantee with the kth highest utility value of P. This error is called the maximum strong regret rate, denoted as $l_k(Q)$. Based on the SRMS definition, this paper studies the following two problems: (1) Min-size (MS) problem, which minimizes the size of Q with the premise that $l_k(Q) \leq \varepsilon$, and (2) Max-sum Diversity and Freshness (MSDF) problem, which jointly considers diversity and freshness in defining the scoring function of Q. Under the premise that $l_k(Q) \leq \varepsilon$, the score of Q is maximized. This paper discusses the computation of the MSDF problem under Manhattan distance (l_1), Euclidean distance (l_2), and Chebyshev distance (l_∞) for measuring the diversity between two points.

The rest of the paper is organized as follows: Firstly, Sect. 2 defines the problem to be solved in this paper and introduces the relevant background knowledge. Then, Sect. 3.1 proposes a method for solving the MS problem, and Sect. 3.2 provides the algorithms for the MSDF problem under l_1, l_2, and l_∞ distances, respectively. Finally, Sect. 4 concludes this paper.

2   Preliminaries

In this section, we first present several basic notions used throughout the paper in Sect. 2.1 and formally define the problems in Sect. 2.2. Then, we analyze the hardness of formalized problems in Sect. 2.3. Finally, In Sect. 2.4, we introduce the concepts of ε-coreset, δ-net and set multicover problem, on which our algorithms will be built.

2.1   Basic Notions

Let P be a set of n d-dimensional points. For each point $p \in P$, the value on the ith dimension is represented as $p[i]$. Before introducing the problem, we first present some related concept, following [4,6,10,11], we focus on the popular-in-practice linear utility functions.

Definition 1. *(Linear utility function) Given a class of linear utility functions* $\mathcal{U} = \{u \in \mathbb{R}^d : ||u|| = 1\}$, *a utility function* u *is a mapping* $u \colon \mathbb{R}^d \to \mathbb{R}$, *the utility value of a data point* p *is denoted as* $\omega(u, p)$, $\omega(u, p) = \langle u, p \rangle = \sum_{i=1}^{d} u_i p[i]$, *which shows how satisfied the user is with the data point.*

Given a utility function u and a integer $k \geq 1$, let $\varphi_k(u, P)$ represent the data point with the kth highest utility value in P, and $\omega_k(u, P)$ denotes its utility value, where $\omega_1(u, P)$ is abbreviated as $\omega(u, P)$. The definition of regret ratio of k-RMS problem is as follows [4,6,10,11]:

Definition 2. *(Regret ratio) Given a dataset* P, *a subset* Q *of* P *and a utility function* u. *The regret ratio of* Q, *represented as* $l(u, Q, P)$, *is defined as*

$$l(u, Q, P) = \frac{\max\{0, \omega_k(u, P) - \omega(u, Q)\}}{\omega_k(u, P)}.$$

Given a utility function u, the regret ratio of Q represents the loss of the highest utility value in Q relative to the kth highest utility value in P. For a class of utility functions $\mathcal{U}$, the definition of maximum regret ratio is as follows:

Definition 3. *(Maximum regret ratio) Given a dataset* P, *a subset* Q *of* P *and a class of utility functions* $\mathcal{U}$. *The maximum regret ratio of* Q, *represented as* $l(Q, P)$, *is defined as*

$$l(Q, P) = \max_{u \in \mathcal{U}} l(u, Q, P).$$

In this paper, when the context is clear, we will use the shorthand notation $l(u, Q)$ and $l(Q)$ for $l(u, Q, P)$ and $l(Q, P)$, respectively. The definition of $l(Q)$ indicates that it is monotonically decreasing with respect to k. That is, for any $k' \leq k$, $l(Q) \leq l(Q')$, where Q' is the regret set corresponding to k'. If the maximum regret rate of Q is $l(Q) \leq \varepsilon$, then Q is called a (k, ε)-RMS. Obviously, for any $k \geq 1$, a $(1, \varepsilon)$-RMS is also a (k, ε)-RMS.

The definition of (k, ε)-RMS only ensures the maximum utility value error for Q w.r.t. any $u \in \mathcal{U}$. Unlike this definition, we propose a more stringent definition of the regret rate, which guarantees the error of k utility values for Q w.r.t. u, is defined as

Definition 4. *(Strong regret ratio) Given a dataset* P, *a subset* Q *of* P *and a class of utility functions* $\mathcal{U}$. *The maximum regret ratio of* Q, *represented as* $l_k(u, Q)$, *is defined as*

$$l_k(u, Q) = \max_{1 \leq i \leq k} 1 - \frac{\omega_i(u, Q)}{\omega_k(u, P)}.$$

Correspondingly, this paper defines the maximum strong regret ratio as $l_k(Q) = \max_{u \in \mathcal{U}} l_k(u, Q)$. If the maximum strong regret ratio of Q, $l_k(Q) \leq \varepsilon$, then Q is called a (k, ε)-SRMS. Clearly, if Q is a (k, ε)-SRMS for P, then Q is also a (k, ε)-RMS for P.

Definition 5. *(Freshness) Given a data point p, its freshness represented as $F(p)$, which is computed by the following exponential decay function.*

$$F(p) = B^{(t_{cur} - t_p)}.$$

where $B \in (0,1)$ is the base number determining the decay rate, t_{cur} denotes the current time, and t_p represents the time when point p is inserted into the database.

For a dataset Q, its freshness is defined as the sum of the freshness decay of all the data in the set, i.e., $F(Q) = \sum_{p \in Q} F(p)$.

Definition 6. *(Diversity) Given data points p and q, the diversity between them is represented as $D(p,q)$, which is described by the distance $dist(p,q)$ between p and q.*

$$D(p,q) = dist(p,q).$$

For a dataset Q, its diversity is defined as the sum of the pairwise distances between all data points in the set, that is, $D(Q) = \sum_{p,q \in Q} D(p,q)$. In this paper, the distance function $dist(\cdot,\cdot)$ being studied includes the following three distance functions for numerical attributes: Manhattan distance (l_1), Euclidean distance (l_2), and Chebyshev distance (l_∞).

In this paper, we propose a scoring function that takes into account both freshness and diversity. For a given set Q. The scoring function is defined as follows: $f(Q) = \alpha D(Q) + (1-\alpha)F(Q)$. Here, the parameter $\alpha \in [0,1]$ is used to weigh the importance of diversity and freshness.

2.2 Problem Definition

In this paper, we study the following two problems.

Problem 1. (Minimization set problem, MS): Given a dataset P, an integer $k \geq 1$, a class of utility functions $\mathcal{U}$ and a parameter $\varepsilon > 0$, compute a subset $Q^* \subseteq P$ such that $l_k(Q^*) \leq \varepsilon$ and $|Q^*|$ is minimized, i.e., return a set

$$Q^* = \arg \min_{Q \subseteq P, l_k(Q) \leq \varepsilon} |Q|.$$

Problem 2. (Max-sum Diversity and Freshness, MSDF) Given a dataset P, an integer $k \geq 1$, a class of utility functions $\mathcal{U}$ and a parameter $\varepsilon > 0$, compute a subset $Q^* \subseteq P$ such that $l_k(Q^*) \leq \varepsilon$ and $f(Q^*)$ is maximized, i.e., return a set

$$Q^* = \arg \max_{Q \subseteq P, l_k(Q) \leq \varepsilon} f(Q).$$

We refer to the MSDF problem, which measures diversity using the l_1, l_2, and l_∞ distance functions, as the MSDF-l_1 problem, MSDF-l_2 problem, and MSDF-l_∞ problem, respectively.

2.3 Hardness

We prove the NP-hardness for MS and MSDF problems by reducing RMS problem to them.

Theorem 1. *The MC and MSDF problems are both NP-hard when $d \geq 3$.*

Proof. Given an instance I_1 of RMS problem with the dataset as P, a class of utility functions $\mathcal{U}$ and a parameter $\varepsilon > 0$. Construct an instance $I_2 = (P', \mathcal{U}', \varepsilon', k)$ of MS problem by the following function: $P' = P$, $\mathcal{U}' = \mathcal{U}$, $\varepsilon' = \varepsilon$ and $k = 1$. Let Q^* is the optimal solution of I_1, we can construct the solution as $Q' = Q^*$ for I_2. Apparently, Q' is the optimal solution for I_2.

Similarly, we can also reduce the RMS problem to the MSDF problem.

Clearly, the reduction could be accomplished in polynomial time. Since RMS is an NP-hard problem for any constant $d \geq 3$ [5], The MS and MSDF problem are both NP-hard problem for any constant $d \geq 3$. ∎

2.4 Background

δ-**net.** For a given parameter $\delta > 0$, a set $\mathcal{N} \subset \mathcal{U}$ is a δ-net iff for any $u \in \mathcal{U}$, there exists $v \in \mathcal{N}$ such that the angle between u and v is less than δ. The δ-net is widely used to transform the RMS problem into a hitting set problem or set coverage problem [6,10,11,18]. A δ-net of size $O(\frac{1}{\delta^{d-1}})$ can be obtained by extracting a uniform grid from $\mathcal{U}$. In practical applications, a δ-net can be obtained by randomly and uniformly sampling $O(\frac{1}{\delta^{d-1}} \log \frac{1}{\delta})$ points on $\mathcal{U}$ [19]. The δ-net in this paper is centrosymmetric, that is, if $u \in \mathcal{N}$, then $-u \in \mathcal{N}$.

Set Multicover Problem. The Set Multicover Problem (SMC) is a generalization of the set cover problem. Given a universe $U = a_1, \cdots, a_n$ of n elements, and a family of sets $\mathbb{S} = S_1, \cdots, S_m$, where for $i = 1, \cdots, m$, $S_i \subseteq U$. For each $a \in U$, there is a positive integer $\tau_a > 0$ that represents the number of times a needs to be covered. The goal of SMC is to select the smallest group of sets to cover the entire U according to the cover count requirement. The approximation ratio of the greedy algorithm for solving the SMC problem is $O(\log |U|)$ [20].

3 Algorithm and Algorithm Analysis

Due to the NP-hardness of MS and MSDF problems, this section first provides an approximate solution to the MS problem based on the computation of ε-coreset in Sect. 3.1, and analyzes the approximation ratio and time complexity of the proposed algorithm. Then, in Sect. 3.2, we discuss the computation of the MSDF-l_2 problem, MSDF-l_1 problem, and MSDF-l_∞ problem, respectively, and analyze the approximation ratio and time complexity of each algorithm.

3.1 Algorithm for MS

This section proposes an algorithm that approximately solves the MS problem based on computing ε-coreset. The core idea is to discretize the entire utility space $\mathcal{U}$ into a finite number of utility functions $\mathcal{W}$, and then use the smallest set Q to k-cover $\mathcal{W}$, thereby ensuring that $l_k(Q) \le \varepsilon$. Here, k-cover means that for any $u \in \mathcal{W}$, there are at least k points in Q whose utility values are not much less than $\omega_k(u, P)$.

Let $\mathcal{B}$ be the d standard basis vectors of $\mathbb{R}_+^d$: $v_1, \cdots, v_d$, $\mathcal{N}$ be the δ-net of $\mathcal{U}$, and $\epsilon = \varepsilon - \frac{2d\delta}{c}$, where $c = \min_{i \in [1,d]} \omega_k(v_i, P)$. The ϵ-approximate top-k results of P w.r.t. u are represented as $\Phi_{k,\epsilon}(u, P) = \{p \in P : \langle u, p \rangle \ge (1 - \epsilon) \cdot \omega_k(u, P)\}$. The pseudocode of the algorithm is shown in Algorithm 1. Specifically: construct the utility function set $\mathcal{W}$ composed of $\mathcal{B}$ and $\mathcal{N}$ (lines 1–3); for each $u \in \mathcal{W}$, $p \in P$, if $p \in \Phi_{k,\epsilon}(u, P)$, add u to the set S_p (lines 5–8); construct the set system $\Sigma = (\mathcal{W}, \mathcal{S})$, where $\mathcal{S} = \{S_p : p \in P\}$ (line 9); select the data set Q with the minimum size using a greedy method to make it k-cover $\mathcal{W}$ (lines 10–21); return Q (line 22).

Algorithm 1: MS

Input: Dataset P, parameter $\varepsilon \in (0,1)$
Output: Q
1 Construct $\mathcal{B}$ as d basis vectors
2 Construct $\mathcal{N}$ as a δ-net of $\mathcal{U}$
3 Let $\mathcal{W} = \mathcal{B} \cup \mathcal{N}$
4 Initialize $Q = \varnothing$ and $U' = \mathcal{W}$
5 For each $p \in P$, set $S_p = \varnothing$
6 **foreach** $u \in \mathcal{W}$ **do**
7 $\quad |\quad S_p \leftarrow S_p \cup \{u : p \in \Phi_{k,\epsilon}(u, P)\}$
8 **end**
9 Construct $\Sigma = (\mathcal{W}, \mathcal{S})$, where $\mathcal{S} = \{S_p : p \in P\}$
10 For each $u \in \mathcal{W}$, set $count(u) = 0$
11 **while** $U' \ne \varnothing$ **do**
12 $\quad |\quad p^* \leftarrow \arg\min_{p \in P \backslash Q} |S_p \cap U'|$
13 $\quad |\quad Q \leftarrow Q \cup \{p^*\}$
14 $\quad |\quad$ **foreach** $u \in S_{p^*}$ **do**
15 $\quad |\quad\quad count(u) + +$
16 $\quad |\quad\quad$ **if** $count(u) == k$ **then**
17 $\quad |\quad\quad\quad| U' \leftarrow U' \backslash u$
18 $\quad |\quad\quad$ **end**
19 $\quad |\quad$ **end**
20 $\quad |\quad \mathcal{S} \leftarrow \mathcal{S} \backslash S_{p^*}$
21 **end**
22 **return** Q

Theorem 2. *Q returned by Algorithm 1 is a (k, ε)-SRMS, i.e., $l_k(Q) \le \varepsilon$. The approximation ratio of Q is $O(d \log \frac{1}{\delta})$, and the time complexity of Algorithm 1 is $O(n|Q||W|)$, where $|W| = O(\frac{1}{\delta^{d-1}})$, $|Q| = O(\frac{k}{\varepsilon^{(d-1)/2}})$.*

Proof. Due to $\mathcal{N}$ is a δ-net of $\mathcal{U}$, $\forall u \in \mathcal{U}$, $\exists v \in \mathcal{N}$ such that $||v - u||_2 \le \delta$. Therefore, for each $p \in P$, the following formula holds [6,10].

$$|\langle v, p \rangle - \langle u, p \rangle| = |\langle v - u, p \rangle| \le ||v - u||_2 \cdot ||p|| \le \delta \cdot \sqrt{d}. \tag{1}$$

where we have used the Cauchy-Schwarz inequality for the first inequality, and the second inequality holds since $||v - u||_2 = \sqrt{2 - 2\cos(\angle vOu)} = 2 \sin \frac{\angle vOu}{2} \le \delta ||p|| \le d$.

Q is a k-cover solution of Σ, for each $v \in W$, $\exists Q_v \subseteq P$, $|Q_v| = k$, such that for each $q \in Q_v$, $\langle v, q \rangle \ge (1 - \epsilon) \cdot \omega_k(v, P)$. For a basis vector $v_i \in W, i \in [1, d]$, each $q \in Q_{v_i}$, $\langle v_i, q \rangle \ge (1 - \epsilon) \cdot \omega_k(v_i, P)$, hence, $\langle v_i, q \rangle \ge (1 - \epsilon) \cdot c$. Given that $||u|| = 1$, there exists an i for any $u \in \mathcal{U}$ such that $u[i] \ge \frac{1}{\sqrt{d}}$. Hence, for any $u \in \mathcal{U}$, there exists Q_{v_i} such that for each $q \in Q_{v_i}$, $\langle u, q \rangle \ge \langle v_i, q \rangle \cdot \frac{1}{\sqrt{d}} \ge (1 - \epsilon) \cdot \frac{c}{\sqrt{d}}$.

Next, we discuss two cases separately.

Case 1: $\omega_k(u, P) \le \frac{c}{\sqrt{d}}$: For each $u \in \mathcal{U}$, exists Q_{v_i} such that for each $q \in Q_{v_i}$, $\langle u, q \rangle \ge (1 - \epsilon) \cdot \omega_k(u, P) \ge (1 - \varepsilon) \cdot \omega_k(u, P)$. Since $|Q_{v_i}| = k$, therefore $l_k(u, Q) \le \varepsilon$.

Case 2: $\omega_k(u, P) > \frac{c}{\sqrt{d}}$: For each $u \in \mathcal{U}$, let $v \in W$ such that $||v - u||_2 \le \delta$. Let $p_1, \cdots, p_k$ represent the top-k results of P w.r.t. u. According to formula (1), for $i \in [1, k]$, we have $\langle v, p_i \rangle \ge \langle u, p_i \rangle - \delta \cdot \sqrt{d}$, which is $\langle v, p_i \rangle \ge \omega_k(u, P) - \delta \cdot \sqrt{d}$. In other words, there exist k points in P, the utility values of which w.r.t. v are at least $\omega_k(u, P) - \delta \cdot \sqrt{d}$, i.e., $\omega_k(v, P) \ge \omega_k(u, P) - \delta \cdot \sqrt{d}$. Hence, there exists a set Q_v, with $|Q_v| = k$, such that for each $q \in Q_v$,

$$\begin{aligned}
\langle u, q \rangle &\ge \langle v, q \rangle - \delta \cdot \sqrt{d} \ge (1 - \epsilon) \cdot \omega_k(v, P) - \delta \cdot \sqrt{d} \\
&\ge (1 - \epsilon) \cdot (\omega_k(u, P) - \delta \cdot \sqrt{d}) - \delta \cdot \sqrt{d} \\
&\ge (1 - \epsilon - \frac{(1 - \epsilon)d\delta}{c} - \frac{d\delta}{c}) \cdot \omega_k(u, P) \\
&\ge (1 - \epsilon - \frac{2d\delta}{c}) \cdot \omega_k(u, P) = (1 - \varepsilon) \cdot \omega_k(u, P).
\end{aligned}$$

Considering the above both cases, we have $l_k(Q) \le \varepsilon$.

The time complexity of the algorithm consists of the following parts: the time to construct W is $O(\frac{1}{\delta^{d-1}})$, the time to construct Σ is $O(n|W|)$. Due to the possible size of the top-k result set for an ϵ-approximate solution being $O(n)$, the time to greedily solve the set multicover problem is $O(n|Q||W|)$.

3.2 Algorithms for MSDF

This section first discusses the solution of the MSDF-l_2 problem, which measures diversity using the most common l_2 distance. It then separately solves the MSDF-l_1 and MSDF-l_∞ problems.

Theorem 2 proves that as long as Q is a k-cover of Σ, then for any $u \in \mathcal{U}$, there exists $Q_u \subseteq Q$, $|Q_u| = k$, $l_k(u, Q_u) \leq \varepsilon$. In fact, here $Q_u \subseteq \Phi_{k,\epsilon}(v, P)$, where $v \in \mathcal{W}$ such that $||v - u|| \leq \delta$, or v is a basis vector. Therefore, the MSDF-l_2 problem is transformed into finding the k subsets Q_v with the highest scores from $S_v = \Phi_{k,\epsilon}(v, P)$ for each $v \in \mathcal{W}$, that is, $Q_v^* = \arg\max_{Q_v \subseteq S_v, |Q_v| = k} f(Q_v)$. Then $Q^* = \cup_{v \in \mathcal{W}} Q_v^*$, $f(Q^*) = \sum_{v \in \mathcal{W}} f(Q_v^*)$. The pseudocode of the algorithm is shown in Algorithm 2.

Algorithm 2: MSDF-l_2

Input: Dataset P, parameter $\varepsilon \in (0, 1)$
Output: Solution Q to the MSDF-l_2 problem over P
1 Construct $\mathcal{B}$ as d basis vectors
2 Construct $\mathcal{N}$ as the δ-net of $\mathcal{U}$
3 Let $\mathcal{W} = \mathcal{B} \cup \mathcal{N}$
4 **foreach** $v \in \mathcal{W}$ **do**
5 $\quad$ Let $S_v = \Phi_{k,\epsilon}(v, P)$
6 $\quad$ Compute $Q_v^* = \text{MSD}(S_v, \mathcal{N}, k)$ by Algorithm 3
7 **end**
8 $Q^* = \cup_{v \in \mathcal{W}} Q_v^*$
9 **return** Q^*

Define a distance function f': for $p_i, p_j \in P$, if $p = q$, $f'(p, q) = 0$, otherwise $f'(p, q) = \alpha D(p, q) + \frac{1-\alpha}{k-1}(F(p) + F(q))$. For $Q_v \subseteq S_v$, $f(Q_v) = \sum_{p,q \in Q_v} f'(p, q)$, the MSDF-$l_2$ problem is to solve $Q_v^* = \arg\max_{Q_v \subseteq S_v, |Q_v| = k} \sum_{p,q \in Q_v} f'(p, q)$. Literature [21] presents an approximation algorithm to solve such problem, which selects the two furthest points to be included in the result set in each iteration, and obtains the approximate result with the maximum diversity through $k/2$ iterations. However, the time complexity of simply implementing this algorithm is $O(kn^2)$. This paper refers to the literature [14], using the relationship between the distance and the inner product of points in space, that is, $||p - q||_2 = \max_{u \in \mathcal{U}}\langle p - q, u \rangle$, define the following new distance function f'', and turn the search for the furthest point pair about f' into the search for the furthest point pair about f''. The new distance function f'': for $p_i, p_j \in P$, if $p = q$, $f''(p, q) = 0$, otherwise $f''(p, q) = \alpha \max_{u \in \mathcal{N}}\langle p - q, u \rangle + \frac{1-\alpha}{k-1}(F(p) + F(q))$. Next, this paper presents an approximate algorithm (Algorithm 3) for computing $Q_v = \arg\max_{Q_v \subseteq S_v, |Q_v| = k} \sum_{p,q \in Q_v} f''(p, q)$.

Define the score $s(p, u) = \alpha\langle p, u \rangle + \frac{1-\alpha}{k-1}F(p)$. Let $\varphi_k^s(u, S_v)$ represent the data point with the kth highest score under u in S_v, $\omega_k^s(u, S_v)$ is the corresponding score value, and $\Phi_k^s(u, S_v)$ represents the set of data points with the top-k scores under u in S_v. The pseudocode of the algorithm is as shown in Algorithm 3. Each round of the algorithm proceeds as follows: Firstly, for each $u, -u \in \mathcal{N}$, we consider the following 3 cases: Case 1: $\varphi_1^s(u, S_v) \neq \varphi_1^s(-u, S_v)$ (because when the freshness of a point dominates, $\varphi_1^s(u, S_v)$ may be equal to $\varphi_1^s(-u, S_v)$), we let $\xi_u = \varphi_1^s(u, S_v)$ and $\xi_{-u} = \varphi_1^s(-u, S_v)$ (lines 3-6). Case 2: $\varphi_1^s(u, S_v) = \varphi_1^s(-u, S_v)$, and $\omega_1^s(u, S_v) + \omega_2^s(-u, S_v) \geq \omega_1^s(-u, S_v) +$

$\omega_2^s(u, S_v)$, we let $\xi_u = \varphi_1^s(u, S_v)$ and $\xi_{-u} = \varphi_2^s(-u, S_v)$ (lines 7–10). Case 3: $\varphi_1^s(u, S_v) = \varphi_1^s(-u, S_v)$, and $\omega_1^s(u, S_v) + \omega_2^s(-u, S_v) < \omega_1^s(-u, S_v) + \omega_2^s(u, S_v)$, we let $\xi_u = \varphi_2^s(u, S_v)$ and $\xi_{-u} = \varphi_1^s(-u, S_v)$ (lines 11–13). Then, for all $u, -u \in \mathcal{N}$, add the ξ_{u^*}, ξ_{-u^*} that make $s(\xi_{u^*}, u^*) + s(\xi_{-u^*}, -u^*)$ the largest into Q_v (line 16). Finally, delete ξ_{u^*}, ξ_{-u^*} from S_v (lines 17–20). The algorithm iterates $k/2$ rounds in total, and when k is odd, the algorithm chooses the freshest point from the remaining points in the last round.

Algorithm 3: MSD(S_v, $\mathcal{N}$, k)

Input: S_v, $\mathcal{N}$, k
Output: Q_v^*

1 Initializes $round = 0$, $Q_v^* = \varnothing$
2 **while** $round \leq k/2, round + + $ **do**
3   **foreach** $u \in \mathcal{N}$ **do**
4   **if** $\varphi_1^s(u, S_v) \neq \varphi_1^s(-u, S_v)$ **then**
5             $\xi_u = \varphi_1^s(u, S_v)$, $\xi_{-u} = \varphi_1^s(-u, S_v)$
6   **end**
7   **if** $\varphi_1^s(u, S_v) == \varphi_1^s(-u, S_v)$ **then**
8   **if** $\omega_1^s(u, S_v) + \omega_2^s(-u, S_v) \geq \omega_1^s(-u, S_v) + \omega_2^s(u, S_v)$ **then**
9                 $\xi_u = \varphi_1^s(u, S_v)$, $\xi_{-u} = \varphi_2^s(-u, S_v)$
10   **end**
11   **else if** $\omega_1^s(u, S_v) + \omega_2^s(-u, S_v) < \omega_1^s(-u, S_v) + \omega_2^s(u, S_v)$ **then**
12                 $\xi_u = \varphi_2^s(u, S_v)$, $\xi_{-u} = \varphi_1^s(-u, S_v)$
13   **end**
14   **end**
15   **end**
16     $Q_v^* \leftarrow Q_v^* \cup \{\xi_{u^*}, \xi_{-u^*} : u^* = \arg\max_{u \in \mathcal{N}} s(\xi_u, u) + s(\xi_{-u}, -u)\}$
17   **foreach** $u \in \mathcal{N}$ **do**
18   If $\xi_v \in \Phi_k^s(u, S_v)$, then $\Phi_k^s(u, S_v) \leftarrow \Phi_k^s(u, S_v) \setminus \xi_v$
19   If $\xi_{-v} \in \Phi_k^s(u, S_v)$, then $\Phi_k^s(u, S_v) \leftarrow \Phi_k^s(u, S_v) \setminus \xi_{-v}$
20   **end**
21 **end**
22 **return** Q_v^*

Lemma 1. *The Algorithm 3 selects ξ_{u^*}, ξ_{-u^*} in each iteration, which are the farthest point pair concerning the distance function f''.*

Proof. Let the remaining data after the $r - 1$ round of the algorithm be S_{r-1}. For the point pairs ξ_{u^*}, ξ_{-u^*} added in the rth round of the algorithm, prove that $f''(\xi_{u^*}, \xi_{-u^*}) = \max_{p,q \in S_{r-1}} f''(p, q)$. First, we need to prove that for any u and any $p, q \in S_{r-1}$, $s(\xi_u, u) + s(\xi_{-u}, -u) \geq s(p, u) + s(q, -u)$. In the first case, according to the definition, $\omega_1^s(u, S_v) \geq s(p, u)$ and $\omega_1^s(-u, S_v) \geq s(q, -u)$, therefore $\omega_1^s(u, S_v) + \omega_1^s(-u, S_v) \geq s(p, u) + s(q, -u)$. Next, we discuss case 2 (case 3 is symmetric), when $p \neq \varphi_1^s(u, S_v) = \varphi_1^s(-u, S_v)$, according to the definition, $\omega_2^s(u, S_v) \geq s(p, u)$ and $\omega_1^s(-u, S_v) \geq s(q, -u)$, therefore $\omega_1^s(u, S_v) + \omega_2^s(-u, S_v) \geq \omega_2^s(u, S_v) + \omega_1^s(-u, S_v) \geq s(p, u) + s(q, -u)$. When $q \neq \varphi_1^s(u, S_v) =$

$\varphi_1^s(-u, S_v)$, according to the definition, $\omega_2^s(-u, S_v) \geq s(q, -u)$ and $\omega_1^s(u, S_v) \geq s(p, u)$, therefore $\omega_1^s(u, S_v) + \omega_2^s(-u, S_v) \geq s(p, u) + s(q, -u)$. In the rth round, ξ_{u^*} and ξ_{-u^*} are selected, where $u^* = \arg\max_{u \in \mathcal{N}} s(\xi_u, u) + s(\xi_{-u}, -u)$. Let $u_1 = \arg\max_{u \in \mathcal{N}} \langle p - q, u \rangle$, $s(\xi_{u^*}, u^*) + s(\xi_{-u^*}, -u^*) = f''(\xi_{u^*}, \xi_{-u^*}) \geq s(\xi_{u_1}, u_1) + s(\xi_{-u_1}, -u_1) \geq s(p, u_1) + s(q, -u_1) = f''(p, q)$. Since p, q are arbitrary, hence, ξ_{u^*}, ξ_{-u^*} are the farthest point pair under the current dataset w.r.t. the distance function f''.

Lemma 2. $f''(p, q) \geq (1 - \cos \delta)f'(p, q)$.

Proof. By the definition of the δ-net, for any $u \in \mathcal{U}$, there exists a $v \in \mathcal{N}$ such that $\angle vOu \leq \delta$. Let $u = \arg\max_{u \in \mathcal{U}} \langle p - q, u \rangle$, then $\frac{\alpha \max_{u \in \mathcal{N}} \langle p-q, u \rangle}{\alpha \langle p-q, u \rangle} \geq \frac{\alpha \langle p-q, v \rangle}{\alpha \langle p-q, u \rangle} \geq \cos \angle vOu \geq \cos \delta$. $f''(p, q) = \alpha \max_{u \in \mathcal{N}} \langle p - q, u \rangle + \frac{1-\alpha}{k-1}(F(p) + F(q)) \geq (1-\cos\delta)(\alpha\max_{u\in\mathcal{U}}\langle p-q, u\rangle + \frac{1-\alpha}{k-1}(F(p)+F(q))) = (1-\cos\delta)f'(p, q)$.

Lemma 3. *For a instance $(P, dist(\cdot, \cdot), k)$ of MSD problem, the algorithm iterates $k/2$ rounds, choosing the furthest pair of points under the distance function $dist(\cdot, \cdot)$ from P in each round. This algorithm obtains a solution with an asymptotically tight approximation ratio of 2.*

Proof. See Theorem 3.2 in [21].

Theorem 3. *The approximate ratio of Q_v^* computed by Algorithm 3 is $\frac{2}{\cos\delta}$, and the time complexity of Algorithm 3 is $O(|\mathcal{N}|k\log k)$.*

Proof. Let Q_1 and Q_2 be the solutions obtained under the distance functions f'' and f' by Algorithm 3, respectively, and let Q^* be the optimal solution of the problem. We have $\frac{f(Q^*)}{f(Q_2)} \leq 2$ and $\frac{f(Q_2)}{f(Q_1)} \leq \frac{1}{\cos\delta}$. Therefore, $\frac{f(Q_1)}{f(Q^*)} \geq \frac{2}{\cos\delta}$. The time complexity comes from the fact that for each $u \in \mathcal{N}$, the computation and maintenance of $\Phi_k^s(u, S_u)$ are respectively $O(n|\mathcal{N}|)$ and $O(\log k)$. The computation needs to be done only once, and the maintenance needs $O(k|\mathcal{N}|)$. Algorithm 3 is a subroutine of Algorithm 2. In fact, in practice, Algorithm 2 can compute both $S_u = \Phi_{k,\epsilon}(u, P)$ and $\Phi_k^s(u, S_u)$ (if $u \in \mathcal{N}$) for each $u \in \mathcal{W}$ at the same time. Therefore, the time complexity of Algorithm 3 is merely the cost of maintaining $\Phi_k^s(u, S_u)$.

Finally, we obtain the following corollary about the approximation ratio and time complexity for solving the MSDF-l_2 problem.

Corollary 1. *The Algorithm 2 solves the MSDF-l_2 problem with an approximation ratio of $\frac{2}{\cos\delta}$, and the time complexity is $O(|\mathcal{W}|(n + k\log k))$.*

When the diversity between two points is measured by the l_1 distance, since $\|p - q\|_1 \geq \|p - q\|_2 \geq \frac{\|p-q\|_1}{\sqrt{d}}$, using the solution obtained by Algorithm 2 as the solution to the MSDF-l_1 problem results in an approximation ratio of $\frac{2\sqrt{d}}{\cos\delta}$. In fact, there exist algorithms that solve the MSDF-l_1 problem with a smaller approximation ratio and lower complexity than Algorithm 2.

Let $S_a = a_1, \cdots, a_d$ be a d-dimensional point set of size d, where a_i represents a d-dimensional point with the ith dimension being 1 and the other dimensions being 0. For instance, when $d = 3$, $a_1 = (1,0,0)$, $a_2 = (0,1,0)$, and $a_3 = (0,0,1)$. Let S_b be a d-dimensional point set of size 2^d, where each point b in S_b has each dimension being either 1 or -1. Then, $||p - q||_1 = \sum_{i=1}^{d} |\langle p - q, a_i \rangle| = \max_{b \in S_b} \langle p, b \rangle + \langle q, -b \rangle$. Let $s(p, b) = \alpha \langle p, b \rangle + \frac{1-\alpha}{k-1} F(p)$. Analogous to the solution process for the MSDF-l_2 problem, we simply utilize (S_v, S_b, k) as the input to Algorithm 3. For the MSDF-l_∞ problem, we have $||p - q||_\infty = \max_{i \in [1,d]} |\langle p - q, a_i \rangle| = \max_{i \in [1,d]} |\langle p, a_i \rangle + \langle q, -a_i \rangle|$. Let $s(p, a) = \alpha \langle p, a \rangle + \frac{1-\alpha}{k-1} F(p)$, simply taking (S_v, S_a, k) as the input of Algorithm 3 will provide the solution to the problem. The pseudocode of the algorithm solving the MSDF-l_1 and MSDF-l_∞ problems is shown in Algorithm 4.

Algorithm 4: MSDF-l_1 and MSDF-l_∞

Input: Dataset P, parameter$\varepsilon \in (0, 1)$
Output: The solution Q of the MSDF-l_1 or MSDF-l_∞ problem on P
1 Construct $\mathcal{B}$ as d basis vectors
2 Construct $\mathcal{N}$ as the δ-net of $\mathcal{U}$
3 let $\mathcal{W} = \mathcal{B} \cup \mathcal{N}$
4 **foreach** $v \in \mathcal{W}$ **do**
5 $\quad$ $S_v = \Phi_{k,\epsilon}(v, P)$
6 $\quad$ **if** l_1 **then**
7 $\quad\quad |$ Compute $Q_v^* = \text{MSD}(S_v, S_b, k)$ by Algorithm 3
8 $\quad$ **end**
9 $\quad$ **if** l_∞ **then**
10 $\quad\quad |$ Compute $Q_v^* = \text{MSD}(S_v, S_a, k)$by Algorithm 3
11 $\quad$ **end**
12 **end**
13 $Q^* = \cup_{v \in \mathcal{W}} Q_v^*$
14 **return** Q^*

Theorem 4. *Algorithm 4 achieves an approximation ratio of 2 for both the MSDF-l_1 and MSDF-l_∞ problems. The time complexities for these problems are $O(n(|\mathcal{W}| + 2^d) + 2^d k \log k)$ and $O(n(|\mathcal{W}| + d) + dk \log k)$, respectively.*

Proof. The proof of this theorem follows the same process as Theorem 3. We omit the details here.

4   Conclusion

This paper addresses the shortcoming of the current definition of RMS, which can only ensure one result under a utility value and ignoring data diversity and freshness, by proposing the definition of SRMS. Based on this new definition, this paper studies the Minimum Set problem (MS) and the Max-sum Diversity and Freshness problem (MSDF). For the MS problem, an algorithm with an approximation ratio of $O(d \log \frac{1}{\delta})$ is proposed. For the MSDF problem, this paper transforms the MSDF problem into the MSD problem, and designs approximation algorithms with approximate ratios of $\frac{2}{\cos \delta}$, 2, and 2 respectively for diversity measurements under l_2, l_1, and l_∞ distances.

References

1. Ilyas, I.F., Beskales, G., Soliman, M.A.: A survey of top-k query processing techniques in relational database systems. ACM Comput. Surv. **40**(4), 1–58 (2008)
2. Borzsony, S., Kossmann, D., Stocker, K.: The skyline operator. In: ICDE, pp. 421–430. IEEE (2001)
3. Nanongkai, D., Sarma, A.D., Lall, A., Lipton, R.J., Xu, J.: Regret-minimizing representative databases. PVLDB **3**(1–2), 1114–1124 (2010)
4. Chester, S., Thomo, A., Venkatesh, S., Whitesides, S.: Computing k-regret minimizing sets. PVLDB **7**(5), 389–400 (2014)
5. Cao, W., et al.: k-regret minimizing set: efficient algorithms and hardness. In: ICDT, pp. 11:1–11:19. Schloss Dagstuhl-Leibniz-Zentrum fuer Informatik (2017)
6. Agarwal, P.K., Kumar, N., Sintos, S., Suri, S.: Efficient algorithms for k-regret minimizing sets. In: SEA, pp. 7:1–7:23. Schloss Dagstuhl-Leibniz-Zentrum fuer Informatik (2017)
7. Xie, M., Wong, R.C., Li, J., Long, C., Lall, A.: Efficient k-regret query algorithm with restriction-free bound for any dimensionality. In: SIGMOD, pp. 959–974. ACM (2018)
8. Asudeh, A., Nazi, A., Zhang, N., Das, G.: Efficient computation of regret-ratio minimizing set: a compact maxima representative. In: SIGMOD, pp. 821–834. ACM (2017)
9. Zheng, J., Dong, Q., Wang, X., Zhang, Y., Ma, W., Ma, Y.: Efficient processing of k-regret minimization queries with theoretical guarantees. Inf. Sci. **586**(1), 99–118 (2022)
10. Wang, Y., Li, Y., Wong, R.C.W., Tan, K.L.: A fully dynamic algorithm for k-regret minimizing sets. In: ICDE, pp. 1631–1642. IEEE (2021)
11. Zheng, J., Wang, Y., Wang, X., Ma, W.: Continuous k-regret minimization queries: a dynamic coreset approach. IEEE Trans. Knowl. Data Eng. **35**(6), 5680–5694 (2023)
12. Catallo, I., Ciceri, E., Fraternali, P., Martinenghi, D., Tagliasacchi, M.: Top-k diversity queries over bounded regions. ACM Trans. Database Syst. **38**(2), 1–44 (2013)
13. Moumoulidou, Z., McGregor, A., Meliou, A.: Diverse data selection under fairness constraints. In: ICDT 2021, Schloss Dagstuhl-Leibniz-Zentrum fuer Informatik (2021)
14. Agarwal, P.K., Sintos, S., Steiger, A.: Efficient indexes for diverse Top-k range queries. In: PODS, pp. 213–227. ACM (2020)
15. Zhang, M., Wang, H., Li, J., Gao, H.: Diversification on big data in query processing. Front. Comput. Sci. **14**(4), 1–20 (2020)
16. Borodin, A., Lee, H.C., Ye, Y.: Max-sum diversification, monotone submodular functions and dynamic updates. In: PODS, pp. 155–166. ACM (2012)
17. Wang, Y., Mathioudakis, M., Li, Y., Tan, K.L.: Minimum coresets for maxima representation of multidimensional data. In: PODS, pp. 138–152. ACM (2021)
18. Kumar, N., Sintos, S.: Faster approximation algorithm for the k-regret minimizing set and related problems. In: ALENEX, pp. 62–74. SIAM (2018)
19. Saff, E.B., Kuijlaars, A.B.J.: Distributing many points on a sphere. Math. Intelligencer **19**(1), 5–11 (1997)
20. Dobson, G.: Worst-case analysis of greedy heuristics for integer programming with nonnegative data. Math. Oper. Res. **7**(4), 515–531 (1982)
21. Hassin, R., Rubinstein, S., Tamir, A.: Approximation algorithms for maximum dispersion. Oper. Res. Lett. **21**(3), 133–137 (1997)

A Modified EXP3 in Adversarial Bandits with Multi-user Delayed Feedback

Yandi Li[2,3] and Jianxiong Guo[1,3(✉)]

[1] Advanced Institute of Natural Sciences, Beijing Normal University, Zhuhai 519087, China

[2] Hong Kong Baptist University, Hong Kong, China
liyandi@uic.edu.cn

[3] Guangdong Key Lab of AI and Multi-Modal Data Processing, Department of Computer Science, BNU-HKBU United International College, Zhuhai 519087, China
jianxiongguo@bnu.edu.cn

Abstract. For the adversarial multi-armed bandit problem with delayed feedback, we consider that the delayed feedback results are from multiple users and are unrestricted on internal distribution. As the player picks an arm, feedback from multiple users may not be received instantly yet after an arbitrary delay of time which is unknown to the player in advance. For different users in a round, the delays in feedback have no latent correlation. Thus, we formulate an adversarial multi-armed bandit problem with multi-user delayed feedback and design a modified EXP3 algorithm named MUD-EXP3, which makes a decision at each round by considering the importance-weighted estimator of the received feedback from different users. On the premise of known terminal round index T, the number of users M, the number of arms N, and upper bound of delay d_{max}, we prove a regret of $\mathcal{O}(\sqrt{TM^2 \ln N(Ne + 4d_{max})})$.

Keywords: Adversarial Bandit · Multi-user Delayed Feedback · EXP3 · Regret Analysis · Online Learning

1 Introduction

Multi-armed Bandit (MAB) problems are a collection of sequential decision-making problems that attract increasing attention for substantial application scenarios such as recommendation systems [10], online advertising [2], and clinical trials [13]. They refer to adopting an action at each round and collecting feedback information for subsequent action selection. In the conventional stochastic bandits, the feedback generated from actions is assumed to follow a fixed but

This work was supported in part by the National Natural Science Foundation of China (NSFC) under Grant No. 62202055, the Start-up Fund from Beijing Normal University under Grant No. 310432104, the Start-up Fund from BNU-HKBU United International College under Grant No. UICR0700018-22, and the Project of Young Innovative Talents of Guangdong Education Department under Grant No. 2022KQNCX102.

W. Wu and G. Tong (Eds.): COCOON 2023, LNCS 14423, pp. 263–278, 2024.
https://doi.org/10.1007/978-3-031-49193-1_20

unknown distribution, where the player can gradually estimate the expected feedback through continuous interaction with the environment. However, the potential feedback distribution tends to alter with time in the real world. For example, the audience of an advertisement change with time, leading to variation in the feedback distribution due to individuality. This induces the adversarial bandit problems [1], a.k.a. the non-stochastic bandits, of which the feedback of a certain action can arbitrarily change over time as if they are selected by an adversary. Furthermore, it experiences delays between conducting actions and receiving feedback. This triggers off the problem of adversarial bandits with delayed feedback [3,6,11,12,14].

However, the existing works only consider single-user feedback situations but not feedback from multiple users at a time. As an example, for online advertising, an advertisement is taken out for multiple users at a time and the delays in receiving those users' feedback are different. The new advertisement has to be put in before receiving all the user feedback of the last-round advertisement. In this situation, the user group can vary with time, corresponding to arbitrarily changed feedback in adversarial bandits. In this paper, we focus on oblivious adversary bandits with multi-user obliviously delayed feedback, where the feedback and delays for all arms, all users, and all rounds are arbitrarily chosen in advance. Specifically, the player executes an arm out of total N arms on M distinct users at round t. Then, the feedback from user j is observed at round $t + d_t^j$, where d_t^j is the delay of the feedback. To solve this problem, we propose a modified EXP3 algorithm [1] named MUD-EXP3 which effectively distinguishes the potential optimal arms. At each round, the player chooses an arm according to the importance-weighted estimator of the received feedback from different users. In addition, we conduct a detailed theoretical analysis and prove the upper bound of regret for the proposed algorithm.

The main contributions of the paper are summarized as the following points:

- We introduce a sequential decision-making problem with multi-user delayed feedback, which is ubiquitous in real-life situations. Then, we model it by using an adversarial bandit framework considering the trait of varying individual loss.
- We propose a modified EXP3 algorithm to adapt to our problem setting, which adopts an importance-weighted estimating method for the received feedback from different users in order to effectively balance exploration and exploitation.
- Sound and detailed theoretical analysis is presented to derive the regret upper bound of the proposed algorithm, achieving sublinear properties with regard to the terminal round index T.

2 Problem Formulation

Suppose the time is discretized into consecutive rounds. We consider an adversarial multi-armed bandit environment with N arms where the player selects an arm A_t at round t and the corresponding feedback is generated from M individual

---

Algorithm 1 MUD-EXP3

Input: $\mathcal{N}$, $\mathcal{M}$, $\mathcal{T}$, learning rate η, upper bound on the delays d_{max};

1: Truncate the learning rate: $\eta' = \min_{i \in \mathcal{N}} \left\{ \eta, \frac{1}{MNe(d_{max}+1)} \right\}$;

2: Initialize $\hat{L}_i(1) = 0$, $p_i(1) = \frac{1}{N}$ for any arm $i \in \mathcal{N}$;

3: **for** $t = 1, 2, \cdots, T$ **do**

4:    Draw an arm $A_t \in \mathcal{N}$ according to the distribution $\boldsymbol{p}(t)$;

5:    Observe a set of delayed losses $\{l_{A_s}^j(s)|(s,j) \in \Phi_t\}$;

6:    Update the cumulative estimated loss $\hat{L}_i(t)$ for all $i \in \mathcal{N}$:

7:    $\hat{l}_i^j(s) = \frac{\mathbb{I}\{A_s=i\} \cdot l_i^j(s)}{p_i(s)}$, $(s,j) \in \Phi_t$; $\ell_i(t) = \sum_{(s,j) \in \Phi_t} \hat{l}_i^j(s)$;

8:    $\hat{L}_i(t) = \hat{L}_i(t-1) + \ell_i^j(t)$

9:    Update the distribution $\boldsymbol{p}(t+1)$:

10:    $W_i(t) = \exp(-\eta' \hat{L}_i(t))$; $p_i(t+1) = \frac{W_i(t)}{\sum_{k=1}^N W_k(t)}$ for all $i \in \mathcal{N}$;

11:    $\boldsymbol{p}(t+1) = [p_1(t+1), \cdots, p_N(t+1)]$;

12: **end for**

---

users. We use the notation $[K] = \{1, 2, \cdots, K\}$ for brevity, then we define the set of arm indexes as $\mathcal{N} = \{i|i \in [N]\}$, the set of user indexes as $\mathcal{M} = \{j|j \in [M]\}$ and the set of round indexes as $\mathcal{T} = \{t|t \in [T]\}$. We use the *loss* rather than the *reward* to represent the feedback, denoted as $l_i^j(t)$ for loss from user j by selecting arm i at round t. The loss $l_{A_t}^j(t)$ is observed by the player after d_t^j rounds where the delay d_t^j is a non-negative integer and $d_{max} = \max\{d_t^j\}$. In other words, the player will observe a bunch of feedback losses $\{l_{A_s}^j(s)|s+d_s^j = t\}$ at round t. Without loss of generality, we assume $l_i^j(t) \in [0, 1]$. Note that there is no restriction on the distribution of d_t^j for generality. The losses of arms and the delays are arbitrarily chosen by an adversary prior to the start of the game, which is known as an oblivious adversary.

The objective is to find a policy of the player for the sequential arm selection in order to approximate the performance of the best fixed arm in hindsight. We use the expected *regret* as the measure of the approximation, which is defined as the difference between the expected cumulative loss induced by the player and the cumulative loss of the best arm in hindsight, as shown below:

$$\mathcal{R} = \mathbb{E}\left[\sum_{t=1}^T \sum_{j=1}^M l_{A_t}^j(t)\right] - \min_{i \in \mathcal{N}} \sum_{t=1}^T \sum_{j=1}^M l_i^j(t). \tag{1}$$

3   Algorithm

We propose an algorithm named MUD-EXP3 (**M**ulti-**U**ser **D**elayed **EXP3**) to solve this problem. MUD-EXP3 is devised based on the well-known EPX3 algorithm and takes into account the multi-user delayed feedback. The detailed algorithm is laid out in Algorithm 1. We assume the terminal round T and the upper bound on the delays d_{max} are known. The input learning rate η has to be truncated first if it is over $1/(MNe(d_{max}+1))$ for the guarantee of upper bound on regret.

At each round, MUD-EXP3 chooses an arm according to the softmax distribution $\boldsymbol{p}(t)$ derived from the cumulative estimated rewards, i.e. the minus cumulative expected losses, of each arm, where $\boldsymbol{p}(t) = [p_1(t), \cdots, p_N(t)]$ and the cumulative estimated loss of arm i is denoted by $\hat{L}_i(t)$. Note that the rounds and the users that contribute to the received losses at a round are arbitrary. For convenience, we first introduce the set of round-user pairs whose feedback losses are observed at round t as Φ_t, and represent the set of round-user pairs whose feedback losses are received out of the terminal round T as Ω, where $\Phi_t = \{(s,j)|s + d_s^j = t\}$ and $\Omega = \{(s,j) \in \Phi_t(t > T)\}$. The importance-weighted estimator $\hat{l}_i^j(t)$ is adopted to estimate the loss $l_i^j(t)$ for introducing the exploration into the algorithm, which is defined as:

$$\hat{l}_i^j(t) = \frac{\mathbb{I}\{A_t = i\} \cdot l_i^j(t)}{p_i(t)}. \tag{2}$$

We wrap up the sum of the estimated losses at round t as $\ell_i(t) = \sum_{(s,j) \in \Phi_t} \hat{l}_i^j(s)$. Then, the cumulative estimated loss is:

$$\hat{L}_i(t) = \hat{L}_i(t-1) + \ell_i(t). \tag{3}$$

Let $W_i(t) = \exp(-\eta' \hat{L}_i(t))$. The probability of choosing arm i at round t which is conditioned on the history observed after $t-1$ rounds is of the softmax function:

$$p_i(t) = \frac{W_i(t-1)}{\sum_{k=1}^{N} W_k(t-1)}. \tag{4}$$

Define the filtration $\mathcal{F}_t$ as $\mathcal{F}_t = \sigma\left(\{A_s|s + d_s^j \le t\}\right)$, where $\sigma\left(\{A_s|s + d_s^j \le t\}\right)$ denotes the σ-algebra generated by the random variables in $\{A_s|s + d_s^j \le t\}$. Note that the distribution $\boldsymbol{p}(t)$ is $\mathcal{F}_{t-1}$-measurable since $\boldsymbol{p}(t)$ is a function of all feedback received up to round $t-1$.

4   Regret Analysis

In this section, we establish the regret upper bound for MUD-EXP3. The regret is defined as Eq. (1) in terms of the optimal arm in hindsight. However, we can hardly acquire the optimal arm in practice. Thus, we transform the regret $\mathcal{R}$ into $\mathcal{R}_i$ as follows:

$$\mathcal{R}_i = \mathbb{E}\left[\sum_{t=1}^{T} \sum_{j=1}^{M} l_{A_t}^j(t)\right] - \sum_{t=1}^{T} \sum_{j=1}^{M} l_i^j(t), \tag{5}$$

which considers the difference between the proposed policy and any fixed arm i. Note that the bounding $\mathcal{R}_i$ is sufficient for bounding $\mathcal{R}$ since a regret bound that is applicable for any fixed arm i definitely fits the optimal arm. In the present, our target changes to bound $\mathcal{R}_i$ and then transfer the bound to $\mathcal{R}$.

We prove our regret upper bound by partially following some techniques of the existing work [3,6,9,12]. We will first introduce some auxiliary lemmas and intermediate theorems before reaching the analysis of eventual regret bound. For a brief explanation, Lemma 1 and Lemma 2 serve Lemma 3, Corollary 1 serves Lemma 4, and Lemma 3 and 4 support Theorem 1.

Lemma 1. *Under the setting of MUD-EXP3, for any round $t \geq 1$ and for any arm $i \in \mathcal{N}$, we have the following: $-\eta' p_i(t) \cdot \ell_i(t) \leq p_i(t + 1) - p_i(t) \leq \eta' p_i(t + 1) \sum_{k=1}^{N} p_k(t) \cdot \ell_k(t)$.*

Proof. We start the proof with upper bounding $p_i(t + 1) - p_i(t)$ and then lower bounding it by using the definition of $p_i(t)$ in terms of $W_i(t)$. According to the definition of $p_i(t)$, we can make the following transformation for the upper bound of $p_i(t + 1) - p_i(t)$:

$$p_i(t+1) - p_i(t) = p_i(t+1) - \frac{W_i(t-1)}{\sum_{k=1}^{N} W_k(t-1)}$$

$$= p_i(t+1) - p_i(t+1) \cdot \frac{\sum_{k=1}^{N} W_k(t)}{W_i(t-1)\exp\left(-\eta'\ell_i(t)\right)} \cdot \frac{W_i(t-1)}{\sum_{k=1}^{N} W_k(t-1)}$$

$$\underset{(a)}{\leq} p_i(t+1) - p_i(t+1) \cdot \frac{\sum_{k=1}^{N} W_k(t)}{\sum_{k=1}^{N} W_k(t-1)}$$

$$= p_i(t+1) - p_i(t+1) \cdot \frac{\sum_{k=1}^{N} W_k(t-1)\exp\left(-\eta'\ell_k(t)\right)}{\sum_{k=1}^{N} W_k(t-1)}$$

$$= p_i(t+1) - p_i(t+1) \cdot \sum_{l=1}^{N} \frac{W_l(t-1)}{\sum_{k=1}^{N} W_k(t-1)}\exp\left(-\eta'\ell_l(t)\right)$$

$$= p_i(t+1) - p_i(t+1) \cdot \sum_{l=1}^{N} p_l(t)\exp\left(-\eta'\ell_l(t)\right)$$

$$= p_i(t+1)\left(\sum_{l=1}^{N} p_l(t) - \sum_{l=1}^{N} p_l(t)\exp\left(-\eta'\ell_l(t)\right)\right)$$

$$= p_i(t+1)\sum_{l=1}^{N} p_l(t)\left(1 - \exp\left(-\eta'\ell_l(t)\right)\right)$$

$$\underset{(b)}{\leq} p_i(t+1)\sum_{l=1}^{N} \eta' p_l(t) \cdot \ell_l(t).$$

The Ineq. (a) results from $W_i(t-1)\exp\left(-\eta'\ell_i(t)\right) \leq W_i(t-1)$ since for $x \leq 0$ there exists $\exp(x) \leq 1$ and Ineq. (b) is obtained by using $1 - \exp(-x) \leq x$ for $x \in \mathbb{R}$ with $x = \eta'\ell_l(t)$ here.

We then infer the lower bound:

$$p_i(t+1) - p_i(t) = \frac{W_i(t-1)\exp(-\eta'\ell_i(t))}{\sum_{k=1}^{N} W_k(t)} - p_i(t)$$

$$= \frac{W_i(t-1)}{\sum_{k=1}^{N} W_k(t-1)} \cdot \exp(-\eta'\ell_i(t)) \cdot \frac{\sum_{k=1}^{N} W_k(t-1)}{\sum_{k=1}^{N} W_k(t)} - p_i(t)$$

$$= p_i(t)\exp(-\eta'\ell_i(t)) \cdot \sum_{l=1}^{N} \frac{W_l(t-1)}{\sum_{k=1}^{N} W_k(t)} - p_i(t)$$

$$= p_i(t)\exp(-\eta'\ell_i(t)) \cdot \sum_{l=1}^{N} \frac{W_l(t-1)\exp(-\eta'\ell_l(t))}{\sum_{k=1}^{N} W_k(t)} \frac{1}{\exp(-\eta'\ell_l(t))} - p_i(t)$$

$$= p_i(t)\exp(-\eta'\ell_i(t)) \cdot \sum_{l=1}^{N} p_l(t+1) \frac{1}{\exp(-\eta'\ell_l(t))} - p_i(t)$$

$$\geq p_i(t)\exp(-\eta'\ell_i(t)) \cdot \sum_{l=1}^{N} p_l(t+1) - p_i(t)$$

$$\underset{(c)}{=} p_i(t)\exp(-\eta'\ell_i(t)) - p_i(t)$$

$$= p_i(t)\left(\exp(-\eta'\ell_i(t)) - 1\right)$$

$$\underset{(d)}{\geq} -\eta' p_i(t) \cdot \ell_i(t),$$

where Eq. (c) results from $\sum_{l=1}^{N} p_l(t+1) = 1$ and Ineq. (d) results from the fact of $\exp(x) \geq 1 + x$ with $x = -\eta'\ell_l(t)$ here. ∎

Lemma 1 derives both the upper bound and lower bound of $p_i(t+1) - p_i(t)$, which is applied to the proof of Lemma 2 so that we can obtain the upper bound of $\frac{p_i(t+1)}{p_i(t)}$. Before moving to the next lemma, we need to infer a corollary from Lemma 1, upper bounding the sum of the absolute value of $p_i(t+1) - p_i(t)$ with respect to $p_i(t)$ and $\ell_i(t)$ for subsequent utilization.

Corollary 1. $\sum_{i=1}^{N}|p_i(t+1) - p_i(t)| \leq 2\eta' \sum_{k=1}^{N} p_k(t) \cdot \ell_k(t).$

Proof. Based on Lemma 1, we have

$$|p_i(t+1) - p_i(t)| \leq \max\left\{\eta' p_i(t) \cdot \ell_i(t),\ \eta' p_i(t+1) \sum_{k=1}^{N} p_k(t) \cdot \ell_k(t)\right\}$$

$$\leq \eta'\left(p_i(t) \cdot \ell_i(t) + p_i(t+1) \sum_{k=1}^{N} p_k(t) \cdot \ell_k(t)\right).$$

Then, the sum over all arms can be bounded as

$$\sum_{i=1}^{N} |p_i(t+1) - p_i(t)| \leq \eta' \sum_{i=1}^{N} \left(p_i(t) \cdot \ell_i(t) + p_i(t+1) \sum_{k=1}^{N} p_k(t) \cdot \ell_k(t) \right)$$

$$= \eta' \left(\sum_{i=1}^{N} p_i(t) \cdot \ell_i(t) + \sum_{i=1}^{N} p_i(t+1) \sum_{k=1}^{N} p_k(t) \cdot \ell_k(t) \right)$$

$$= \eta' \left(\sum_{i=1}^{N} p_i(t) \cdot \ell_i(t) + \sum_{k=1}^{N} p_k(t)\ell_k(t) \cdot \sum_{i=1}^{N} p_i(t+1) \right)$$

$$\underset{(a)}{=} \eta' \left(\sum_{i=1}^{N} p_i(t) \cdot \ell_i(t) + \sum_{k=1}^{N} p_k(t) \cdot \ell_k(t) \right)$$

$$= 2\eta' \sum_{k=1}^{N} p_k(t) \cdot \ell_k(t),$$

where Eq. (a) results from $\sum_{i=1}^{N} p_i(t+1) = 1$. ∎

Lemma 2. *Under the setting of MUD-EXP3, for any round $t \geq 1$ and for any arm $i \in \mathcal{N}$, if $\eta' \leq \frac{1}{MNe(d_{max}+1)}$, then we have $p_i(t+1) \leq \left(1 + \frac{1}{d_{max}} \right) p_i(t)$.*

Proof. We will use the strong mathematical induction method with the assistance of Lemma 1 to prove this result. First, we deal with the base case.

According to the initialization step of MUD-EXP3, $p_i(1) = 1/N$. When $t = 2$, the maximal increase of $p_i(2)$ compared with $p_i(1)$ should come if a different arm other than i has been chosen at round $t = 1$ and all feedback losses $l_i^j(1)$ are observed with no delays and are of value 1 for $j \in \mathcal{M}$ at round $t = 2$. Hence, we have the following:

$$p_i(2) \leq \frac{1}{N - 1 + e^{-\eta'N}} \underset{(a)}{\leq} \frac{1}{N - \eta'N}$$

$$= \frac{1}{N} \left(1 - \frac{1 - \eta'}{1 - \eta'} + \frac{1}{1 - \eta'} \right) = \frac{1}{N} \left(1 + \frac{\eta'}{1 - \eta'} \right) = p_i(1) \left(1 + \frac{1}{1/\eta' - 1} \right)$$

$$\underset{(b)}{\leq} p_i(1) \left(1 + \frac{1}{d_{max}} \right),$$

where Ineq. (a) follows since $\exp(x) \geq 1 + x$ for $x \in \mathbb{R}$ and Ineq. (b) holds for $\eta' \leq \frac{1}{MNe(d_{max}+1)}$.

Next, we assume $p_i(t) \leq \left(1 + \frac{1}{d_{max}}\right) p_i(t-1)$ holds for $p_i(2), \cdots, p_i(t)$. Before we prove the case for $p_i(t+1)$, an intermediate result need to be introduced:

$$\sum_{i=1}^{N} p_i(t) \cdot \hat{l}_i^j(s) = \sum_{i=1}^{N} p_i(t) \frac{\mathbb{I}\{A_s = i\} \cdot l_i^j(S)}{p_i(s)}$$

$$\leq \sum_{i=1}^{N} \frac{p_i(t)}{p_i(s)} = \sum_{i=1}^{N} \prod_{t=s}^{s+d_s} \frac{p_i(t)}{p_i(t-1)} \underset{(c)}{\leq} \sum_{i=1}^{N} \left(1 + \frac{1}{d_{max}}\right)^{d_s}$$

$$\leq \sum_{i=1}^{N} \left(1 + \frac{1}{d_{max}}\right)^{d_{max}} \leq Ne, \tag{6}$$

where Ineq. (c) adopts the inductive hypothesis. Then, we have

$$\sum_{i=1}^{N} p_i(t) \cdot \ell_i(t) = \sum_{i=1}^{N} p_i(t) \sum_{(s,j) \in \Phi_t} \hat{l}_i^j(s)$$

$$= \sum_{(s,j) \in \Phi_t} \sum_{i=1}^{N} p_i(t) \cdot \hat{l}_i^j(s) \underset{(d)}{\leq} \sum_{(s,j) \in \Phi_t} Ne \underset{(f)}{\leq} MNe, \tag{7}$$

where Eq. (6) brings about Ineq. (d) and $|\Phi_t| \leq M$ brings about Ineq. (f). According to this result and Lemma 1, we have

$$p_i(t+1) \underset{(g)}{\left(1 - \eta'MNe\right)} \leq p_i(t+1) \left(1 - \eta' \sum_{i=1}^{N} p_i(t) \cdot \ell_i(t)\right)$$

$$= p_i(t+1) - \eta' p_i(t+1) \sum_{i=1}^{N} p_i(t) \cdot \ell_i(t) \underset{(h)}{\leq} p_i(t+1) - (p_i(t+1) - p_i(t)) = p_i(t)$$

where Ineq. (g) follows Eq. (7) and Ineq. (h) follows Lemma 1. By taking into account the constraint $\eta' \leq \frac{1}{MNe(d_{max}+1)}$, we can prove the inductive case:

$$p_i(t+1) \leq \frac{1}{1 - \eta'MNe} p_i(t) \leq \frac{1}{1 - \frac{1}{d_{max}+1}} p_i(t) = \left(1 + \frac{1}{d_{max}}\right) p_i(t).$$

Finally, we finish the proof by combining the base case and the inductive case for a complete mathematical induction. ∎

The two expectation terms in Lemma 3 and Lemma 4 are key components that constitute the transformed regret expression. For clarity of expression, we analyze the upper bound for these two terms separately in advance, and then utilize their consequences in the analysis of Theorem 1 to obtain the final regret bound.

Lemma 3. *MUD-EXP3 satisfies the following inequality:*

$$\mathbb{E}\left[\sum_{t=1}^{T} \sum_{(s,j) \in \Phi_t} \sum_{k=1}^{N} p_k(t) \cdot l_k^j(s) - \sum_{t=1}^{T} \sum_{j=1}^{M} l_i^j(t)\right] \leq \frac{\ln N}{\eta'} + \frac{1}{2}\eta' M^2 TNe.$$

Proof. The proof will be shown in Appendix A. ∎

Lemma 4. *MUD-EXP3 satisfies the following inequality:*

$$\mathbb{E}\left[\sum_{t=1}^{T}\sum_{(s,j)\in\Phi_t}\sum_{k=1}^{N}p_k(s)\cdot l_k^j(s) - \sum_{t=1}^{T}\sum_{(s,j)\in\Phi_t}\sum_{k=1}^{N}p_k(t)\cdot l_k^j(s)\right] \le 2\eta' M^2 T d_{max}.$$

Proof. To prove this inequality, we first transform the expression into a form containing $\sum_{i=1}^{N}|p_i(t+1) - p_i(t)|$, then adopt Corollary 1 to upper bound it.

$$\mathbb{E}\left[\sum_{t=1}^{T}\sum_{(s,j)\in\Phi_t}\sum_{k=1}^{N}p_k(s)\cdot l_k^j(s) - \sum_{t=1}^{T}\sum_{(s,j)\in\Phi_t}\sum_{k=1}^{N}p_k(t)\cdot l_k^j(s)\right]$$

$$= \mathbb{E}\left[\sum_{t=1}^{T}\sum_{(s,j)\in\Phi_t}\sum_{k=1}^{N}l_k^j(s)\left(p_k(s) - p_k(t)\right)\right]$$

$$\le \mathbb{E}\left[\sum_{t=1}^{T}\sum_{(s,j)\in\Phi_t}\sum_{k=1}^{N}\left(p_k(s) - p_k(t)\right)\right]$$

$$= \mathbb{E}\left[\sum_{t=1}^{T}\sum_{(s,j)\in\Phi_t}\sum_{k=1}^{N}\sum_{r=s}^{t-1}\left(p_k(r) - p_k(r+1)\right)\right]$$

$$\le \mathbb{E}\left[\sum_{t=1}^{T}\sum_{(s,j)\in\Phi_t}\sum_{r=s}^{t-1}\sum_{k=1}^{N}|p_k(r) - p_k(r+1)|\right]$$

$$\underset{(a)}{\le} \mathbb{E}\left[\sum_{t=1}^{T}\sum_{(s,j)\in\Phi_t}\sum_{r=s}^{t-1}2\eta'\sum_{k=1}^{N}p_k(r)\cdot\ell_k(r)\right]$$

$$\underset{(b)}{=} 2\eta'\mathbb{E}\left[\sum_{t=1}^{T}\sum_{(s,j)\in\Phi_t}\sum_{r=s}^{t-1}\sum_{k=1}^{N}p_k(r)\cdot\mathbb{E}\left[\ell_k(r)|\mathcal{F}_{r-1}\right]\right]$$

$$= 2\eta'\mathbb{E}\left[\sum_{t=1}^{T}\sum_{(s,j)\in\Phi_t}\sum_{r=s}^{t-1}\sum_{k=1}^{N}p_k(r)\cdot\mathbb{E}\left[\sum_{(s',j')\in\Phi_r}\hat{l}_k^{j'}(s')|\mathcal{F}_{r-1}\right]\right]$$

$$\underset{(c)}{=} 2\eta'\mathbb{E}\left[\sum_{t=1}^{T}\sum_{(s,j)\in\Phi_t}\sum_{r=s}^{t-1}\sum_{k=1}^{N}p_k(r)\sum_{(s',j')\in\Phi_r}l_k^{j'}(s')\right]$$

$$\le 2\eta' M^2 T d_{max},$$

where Ineq. (a) holds by using Corollary 1, Eq. (b) uses $p_k(r) \in \mathcal{F}_{r-1}$, and Ineq. (c) follows the fact that $\hat{l}_k^{j'}(s')$ is $l_k^{j'}(s')/p_k(s')$ with probability $p_k(s')$ and zero otherwise. ∎

Theorem 1. *For any arm $i \in \mathcal{N}$, MUD-EXP3 guarantees the upper bound for $\mathcal{R}_i$ as shown below:*

$$\mathcal{R}_i \leq \frac{\ln N}{\eta'} + \frac{1}{2}\eta' M^2 TNe + 2\eta' M^2 Td_{max} + |\Omega|, \tag{8}$$

which implies the same regret upper bound as follows:

$$\mathcal{R} \leq \frac{\ln N}{\eta'} + \frac{1}{2}\eta' M^2 TNe + 2\eta' M^2 Td_{max} + |\Omega|. \tag{9}$$

Specially, for the known T and d_{max}, if $\eta = \sqrt{\frac{\ln N}{TM^2(Ne+4d_{max})}} \leq \frac{1}{MNe(d_{max}+1)}$, we have:

$$\mathcal{R} \leq \mathcal{O}\left(\sqrt{TM^2 \ln N(Ne + 4d_{max})}\right). \tag{10}$$

Proof. The expression of $\mathcal{R}_i$ can be transformed in order to approach Lemma 3 and Lemma 4 for upper bounding, as shown below:

$$\mathcal{R}_i = \mathbb{E}\left[\sum_{t=1}^{T}\sum_{j=1}^{M} l_{A_t}^{j}(t)\right] - \sum_{t=1}^{T}\sum_{j=1}^{M} l_i^{j}(t)$$

$$= \mathbb{E}\left[\sum_{t=1}^{T}\sum_{j=1}^{M}\mathbb{E}\left[l_{A_t}^{j}(t)|\mathcal{F}_t\right]\right] - \sum_{t=1}^{T}\sum_{j=1}^{M} l_i^{j}(t)$$

$$= \mathbb{E}\left[\sum_{t=1}^{T}\sum_{j=1}^{M}\sum_{k=1}^{N} p_k(t)\cdot l_k^{j}(t) - \sum_{t=1}^{T}\sum_{j=1}^{M} l_i^{j}(t)\right]$$

$$= \mathbb{E}\left[\sum_{t=1}^{T}\sum_{(s,j)\in\Phi_t}\sum_{k=1}^{N} p_k(s)\cdot l_k^{j}(s) + \sum_{(t,j)\in\Omega}\sum_{k=1}^{N} p_k(t)\cdot l_k^{j}(t) - \sum_{t=1}^{T}\sum_{j=1}^{M} l_i^{j}(t)\right]$$

$$\underset{(a)}{\leq} \mathbb{E}\left[\sum_{t=1}^{T}\sum_{(s,j)\in\Phi_t}\sum_{k=1}^{N} p_k(s)\cdot l_k^{j}(s) - \sum_{t=1}^{T}\sum_{j=1}^{M} l_i^{j}(t)\right] + |\Omega|$$

$$= \mathbb{E}\left[\sum_{t=1}^{T}\sum_{(s,j)\in\Phi_t}\sum_{k=1}^{N} p_k(t)\cdot l_k^{j}(s) - \sum_{t=1}^{T}\sum_{(s,j)\in\Phi_t}\sum_{k=1}^{N} p_k(t)\cdot l_k^{j}(s)\right]$$

$$+ \mathbb{E}\left[\sum_{t=1}^{T}\sum_{(s,j)\in\Phi_t}\sum_{k=1}^{N} p_k(s)\cdot l_k^{j}(s) - \sum_{t=1}^{T}\sum_{j=1}^{M} l_i^{j}(t)\right] + |\Omega|$$

$$= \mathbb{E}\left[\sum_{t=1}^{T}\sum_{(s,j)\in\Phi_t}\sum_{k=1}^{N} p_k(t)\cdot l_k^{j}(s) - \sum_{t=1}^{T}\sum_{j=1}^{M} l_i^{j}(t)\right]$$

$$+ \mathbb{E}\left[\sum_{t=1}^{T}\sum_{(s,j)\in\Phi_t}\sum_{k=1}^{N} p_k(s)\cdot l_k^{j}(s) - \sum_{t=1}^{T}\sum_{(s,j)\in\Phi_t}\sum_{k=1}^{N} p_k(t)\cdot l_k^{j}(s)\right] + |\Omega|$$

$$\underset{(b)}{\leq} \frac{\ln N}{\eta'} + \frac{1}{2}\eta' M^2 T N e + 2\eta' M^2 T d_{max} + |\Omega|,$$

where Ineq. (a) follows since $l_i^j(t) \leq 1$ and Ineq. (b) results from Lemma 3 and Lemma 4.   ∎

5   Related Work

Delayed Feedback. Joulani et al. [8] studied delayed feedback under full information setting rather than adversarial bandits, and proved regret bound of $\mathcal{O}(\sqrt{(T+D)\ln N})$ by reducing the problem from non-delayed feedback full information setting, where $D = \sum_{t=1}^{T} d_t$ is delay sum. Cesa-Bianchi et al. [6] proposed a cooperative version of the EXP3 for delayed feedback under the bandit setting. They achieved regret upper bound of $\mathcal{O}(\sqrt{(NT+D)\ln N})$. On the basis of [6], Thune et al. [12] proposed a wrapper algorithm to eliminate the restriction on T and D for adversarial bandit setting with delayed feedback, allowing regret bound of $\mathcal{O}(\sqrt{(NT+D)\ln N})$ for unknown T and D. Bistritz et al. [3] also proposed a modified EXP3 algorithm for the same problem as [12] and proved the Nash Equilibrium for a two-player zero-sum game with this algorithm. Zimmert and Seldin [17] presented a Follow the Regularized Leader algorithm for adversarial bandits with arbitrary delays and achieved the upper bound of $\mathcal{O}(\sqrt{NT} + \sqrt{DT\log N})$ on the regret, requiring no prior knowledge of D or T. In recent years there has been an increasing interest in algorithms that perform well in both regimes with no prior knowledge of stochastic or adversarial regimes [4,16], known as best-of-both-worlds problems. Masoudian et al. [11] followed Zimmert and Seldin [17] and introduce delayed feedback setting to the best-of-both-worlds problem. They proposed a slightly modified algorithm and achieved a best-of-both-worlds regret guarantee for both adversarial and stochastic bandits.

Composite Anonymous Feedback. On the basis of delayed feedback setting, Cesa-Bianchi et al. [5] raised a more complex problem, where the feedback is both anonymous and delayed, and can be partly observed at different rounds. It means the player can only observe the sum of partial feedback generated from different rounds. They proposed a general reduction technique that enables the conversion of a standard bandit algorithm to operate in this harder setting. Wang et al. [15] proposed a modified EXP3 algorithm for the problem in [5] with non-oblivious delays, requiring no knowledge of delays in advance and achieving an $\mathcal{O}(\sqrt{d+N}\log NT^{\frac{2}{3}})$ regret. Wan et al. [14] extended the problem into anonymous delayed feedback with non-oblivious loss and delays and prove both the lower and upper bounds.

6   Conclusion

In conclusion, this study addresses the adversarial multi-armed bandit problem with delayed feedback, where feedback results are obtained from multiple

274 Y. Li and J. Guo

users without any internal distribution restrictions. A modified EXP3 algorithm called MUD-EXP3 is proposed to solve this problem with the oblivious loss and oblivious delay adversary setting. MUD-EXP3 employs the importance-weighted estimator for the received feedback from different users, and selects an arm at each round stochastically according to the amount of the cumulative received loss. Under the assumptions of a known terminal round index, the number of users, the number of arms, and an upper bound on the delay, the study proves a regret bound of $\mathcal{O}(\sqrt{TM^2 \ln N}(Ne + 4d_{max}))$, demonstrating the algorithm's effectiveness. Overall, this research provides valuable insights for addressing complex real-life scenarios with multi-user delayed feedback.

A   Appendix: Proof of Lemma 3

According to the definition of $W_i(t)$, we can first calculate a lower bound of $\sum_{i=1}^{N} W_i(T)/\sum_{i=1}^{N} W_i(0)$ as follows:

$$
\frac{\sum_{i=1}^{N} W_i(T)}{\sum_{i=1}^{N} W_i(0)} = \frac{\sum_{i=1}^{N} \exp(-\eta' \hat{L}_i^T)}{\sum_{i=1}^{N} \exp(-\eta' \hat{L}_i^0)}
$$

$$
\geq \frac{\max_{i \in \mathcal{N}} \exp\left(-\eta' \sum_{t=1}^{T} \sum_{(s,j)\in\Phi_t} \hat{l}_i^j(s)\right)}{N}
$$

$$
\geq \frac{\exp\left(-\eta' \sum_{t=1}^{T} \sum_{(s,j)\in\Phi_t} \hat{l}_i^j(s)\right)}{N}.
$$

Before we derive the corresponding upper bound, we first analyze the upper bound of $\sum_{i=1}^{N} W_i(t)/\sum_{i=1}^{N} W_i(t-1)$ and then telescope this over T.

$$
\frac{\sum_{i=1}^{N} W_i(t)}{\sum_{i=1}^{N} W_i(t-1)} = \frac{\sum_{i=1}^{N} \exp(-\eta' \hat{L}_i^t)}{\sum_{i=1}^{N} \exp(-\eta' \hat{L}_i^{t-1})}
$$

$$
= \frac{\sum_{i=1}^{N} \exp(-\eta' \hat{L}_i^{t-1}) \exp(-\eta' \ell_i^t)}{\sum_{i=1}^{N} \exp(-\eta' \hat{L}_i^{t-1})}
$$

$$
= \sum_{i=1}^{N} p_i(t) \cdot \exp(-\eta' \ell_i^t) = \sum_{i=1}^{N} p_i(t) \exp\left(-\eta' \frac{1}{|\Phi_t|} \sum_{(s,j)\in\Phi_t} |\Phi_t| \cdot \hat{l}_i^j(s)\right)
$$

$$
\underset{(a)}{\leq} \sum_{i=1}^{N} p_i(t) \frac{1}{|\Phi_t|} \sum_{(s,j)\in\Phi_t} \exp\left(-\eta' |\Phi_t| \cdot \hat{l}_i^j(s)\right)
$$

$$
\underset{(b)}{\leq} \sum_{i=1}^{N} p_i(t) \frac{1}{|\Phi_t|} \sum_{(s,j)\in\Phi_t} \left(1 - \eta' |\Phi_t| \cdot \hat{l}_i^j(s) + \frac{1}{2} \eta'^2 |\Phi_t|^2 \cdot \hat{l}_i^j(s)^2\right)
$$

$$
= \sum_{i=1}^{N} p_i(t) \left(1 - \eta' \sum_{(s,j)\in\Phi_t} \hat{l}_i^j(s) + \frac{1}{2} \eta'^2 |\Phi_t| \sum_{(s,j)\in\Phi_t} \hat{l}_i^j(s)^2\right)
$$

$$= 1 - \eta' \sum_{(s,j)\in\Phi_t} \sum_{i=1}^{N} p_i(t) \cdot \hat{l}_i^j(s) + \frac{1}{2}\eta'^2 |\Phi_t| \sum_{(s,j)\in\Phi_t} \sum_{i=1}^{N} p_i(t) \cdot \hat{l}_i^j(s)^2$$

$$\underset{(c)}{\leq} \exp\left(-\eta' \sum_{(s,j)\in\Phi_t} \sum_{i=1}^{N} p_i(t) \cdot \hat{l}_i^j(s) + \frac{1}{2}\eta'^2 |\Phi_t| \sum_{(s,j)\in\Phi_t} \sum_{i=1}^{N} p_i(t) \cdot \hat{l}_i^j(s)^2\right),$$

where Ineq. (a) follows Jensen's inequality [7], Ineq. (b) follows $\exp(x) \leq 1 + x + (1/2)x^2$ for $x \in \mathbb{R}$, and Ineq. (c) follows $1 + x \leq \exp(x)$ for $x \in \mathbb{R}$. Then, we use it to upper bound $\sum_{i=1}^{N} W_i(T) / \sum_{i=1}^{N} W_i(0)$.

$$\frac{\sum_{i=1}^{N} W_i(T)}{\sum_{i=1}^{N} W_i(0)} = \prod_{t=1}^{T} \frac{\sum_{i=1}^{N} W_i(t)}{\sum_{i=1}^{N} W_i(t-1)}$$

$$\leq \prod_{t=1}^{T} \exp\left(-\eta' \sum_{(s,j)\in\Phi_t} \sum_{i=1}^{N} p_i(t) \cdot \hat{l}_i^j(s) + \frac{1}{2}\eta'^2 |\Phi_t| \sum_{(s,j)\in\Phi_t} \sum_{i=1}^{N} p_i(t) \cdot \hat{l}_i^j(s)^2\right)$$

$$= \exp\left(-\eta' \sum_{t=1}^{T} \sum_{(s,j)\in\Phi_t} \sum_{i=1}^{N} p_i(t) \cdot \hat{l}_i^j(s) + \frac{1}{2}\eta'^2 \sum_{t=1}^{T} |\Phi_t| \sum_{(s,j)\in\Phi_t} \sum_{i=1}^{N} p_i(t) \cdot \hat{l}_i^j(s)^2\right).$$

Combining the lower bound and upper bound of $\sum_{i=1}^{N} W_i(T) / \sum_{i=1}^{N} W_i(0)$ establishes the following expression:

$$\frac{\exp\left(-\eta' \sum_{t=1}^{T} \sum_{(s,j)\in\Phi_t} \hat{l}_i^j(s)\right)}{N}$$

$$\leq \exp\left(-\eta' \sum_{t=1}^{T} \sum_{(s,j)\in\Phi_t} \sum_{i=1}^{N} p_i(t) \cdot \hat{l}_i^j(s) + \frac{1}{2}\eta'^2 \sum_{t=1}^{T} |\Phi_t| \sum_{(s,j)\in\Phi_t} \sum_{i=1}^{N} p_i(t) \cdot \hat{l}_i^j(s)^2\right).$$

Take $\ln(\cdot)$ on both sides and do transposition:

$$\eta' \sum_{t=1}^{T} \sum_{(s,j)\in\Phi_t} \sum_{i=1}^{N} p_i(t) \cdot \hat{l}_i^j(s) - \eta' \sum_{t=1}^{T} \sum_{(s,j)\in\Phi_t} \hat{l}_i^j(s)$$

$$\leq \ln N + \frac{1}{2}\eta'^2 \sum_{t=1}^{T} |\Phi_t| \sum_{(s,j)\in\Phi_t} \sum_{i=1}^{N} p_i(t) \cdot \hat{l}_i^j(s)^2. \tag{11}$$

Before moving on, we conduct an analysis of the expectation on both sides of Ineq. (11), respectively. We first deal with the left part:

$$\mathbb{E}\left[\eta'\sum_{t=1}^{T}\sum_{(s,j)\in\Phi_t}\sum_{i=1}^{N}p_i(t)\cdot\hat{l}_i^j(s)-\eta'\sum_{t=1}^{T}\sum_{(s,j)\in\Phi_t}\hat{l}_i^j(s)\right]$$

$$\overset{=}{_{(d)}}\eta'\mathbb{E}\left[\sum_{t=1}^{T}\sum_{(s,j)\in\Phi_t}\sum_{i=1}^{N}p_i(t)\cdot\mathbb{E}\left[\hat{l}_i^j(s)|\mathcal{F}_{t-1}\right]-\sum_{t=1}^{T}\sum_{(s,j)\in\Phi_t}\mathbb{E}\left[\hat{l}_i^j(s)|\mathcal{F}_{t-1}\right]\right]$$

$$\overset{=}{_{(e)}}\eta'\mathbb{E}\left[\sum_{t=1}^{T}\sum_{(s,j)\in\Phi_t}\sum_{i=1}^{N}p_i(t)\cdot l_i^j(s)-\sum_{t=1}^{T}\sum_{(s,j)\in\Phi_t}l_i^j(s)\right], \tag{12}$$

where Eq. (d) uses $p_i(t)\in\mathcal{F}_{t-1}$ and Eq. (e) uses $p_i(s)\in\mathcal{F}_{t-1}$ together with the fact that $\hat{l}_i^j(s)$ is $l_i^j(s)/p_i(s)$ with probability $p_i(s)$ and zero otherwise. Then, the same operation is carried out for the right part:

$$\mathbb{E}\left[\ln N+\frac{1}{2}\eta'^2\sum_{t=1}^{T}|\Phi_t|\sum_{(s,j)\in\Phi_t}\sum_{i=1}^{N}p_i(t)\cdot\hat{l}_i^j(s)^2\right]$$

$$=\ln N+\frac{1}{2}\eta'^2\mathbb{E}\left[\sum_{t=1}^{T}|\Phi_t|\sum_{(s,j)\in\Phi_t}\sum_{i=1}^{N}p_i(t)\cdot\mathbb{E}\left[\hat{l}_i^j(s)^2|\mathcal{F}_{t-1}\right]\right]$$

$$=\ln N+\frac{1}{2}\eta'^2\mathbb{E}\left[\sum_{t=1}^{T}|\Phi_t|\sum_{(s,j)\in\Phi_t}\sum_{i=1}^{N}\frac{p_i(t)}{p_i(s)}l_i^j(s)^2\right]$$

$$\overset{\leq}{_{(f)}}\ln N+\frac{1}{2}\eta'^2M\mathbb{E}\left[\sum_{t=1}^{T}\sum_{(s,j)\in\Phi_t}\sum_{i=1}^{N}\frac{p_i(t)}{p_i(s)}\right]$$

$$\overset{\leq}{_{(g)}}\ln N+\frac{1}{2}\eta'^2M\mathbb{E}\left[\sum_{t=1}^{T}\sum_{(s,j)\in\Phi_t}\sum_{i=1}^{N}\left(1+\frac{1}{d_{max}}\right)^{d_s^j}\right]$$

$$\leq\ln N+\frac{1}{2}\eta'^2M\mathbb{E}\left[\sum_{t=1}^{T}\sum_{(s,j)\in\Phi_t}\sum_{i=1}^{N}\left(1+\frac{1}{d_{max}}\right)^{d_{max}}\right]$$

$$\leq\ln N+\frac{1}{2}\eta'^2M^2TNe, \tag{13}$$

where Ineq. (f) is due to $l_i^j(s)\leq 1$ and $|\Phi_t|\leq M$, and Ineq. (g) uses Lemma 2. By substituting Eq. (12) and Eq. (13) into Eq. (11), we have:

$$\eta'\mathbb{E}\left[\sum_{t=1}^{T}\sum_{(s,j)\in\Phi_t}\sum_{i=1}^{N}p_i(t)\cdot l_i^j(s)-\sum_{t=1}^{T}\sum_{(s,j)\in\Phi_t}l_i^j(s)\right]\leq\ln N+\frac{1}{2}\eta'^2M^2TNe.$$

Thus, the eventual result is presented as follows:

$$\mathbb{E}\left[\sum_{t=1}^{T}\sum_{(s,j)\in\Phi_t}\sum_{k=1}^{N}p_k(t)\cdot l_k^j(s) - \sum_{t=1}^{T}\sum_{j=1}^{M}l_i^j(t)\right]$$

$$\leq \mathbb{E}\left[\sum_{t=1}^{T}\sum_{(s,j)\in\Phi_t}\sum_{i=1}^{N}p_i(t)\cdot l_i^j(s) - \sum_{t=1}^{T}\sum_{(s,j)\in\Phi_t}l_i^j(s)\right] \leq \frac{\ln N}{\eta'} + \frac{1}{2}\eta' M^2 T N e.$$

The lemma has been proven.   ∎

References

1. Auer, P., Cesa-Bianchi, N., Freund, Y., Schapire, R.E.: Gambling in a rigged casino: the adversarial multi-armed bandit problem. In: Proceedings of IEEE 36th Annual Foundations of Computer Science, pp. 322–331. IEEE (1995)
2. Avadhanula, V., Colini Baldeschi, R., Leonardi, S., Sankararaman, K.A., Schrijvers, O.: Stochastic bandits for multi-platform budget optimization in online advertising. In: Proceedings of the Web Conference 2021, pp. 2805–2817 (2021)
3. Bistritz, I., Zhou, Z., Chen, X., Bambos, N., Blanchet, J.: Online EXP3 learning in adversarial bandits with delayed feedback. In: Advances in Neural Information Processing Systems, vol. 32 (2019)
4. Bubeck, S., Slivkins, A.: The best of both worlds: stochastic and adversarial bandits. In: Conference on Learning Theory. JMLR Workshop and Conference Proceedings, p. 42-1 (2012)
5. Cesa-Bianchi, N., Gentile, C., Mansour, Y.: Nonstochastic bandits with composite anonymous feedback. In: Conference on Learning Theory, pp. 750–773. PMLR (2018)
6. Cesa-Bianchi, N., Gentile, C., Mansour, Y., et al.: Delay and cooperation in nonstochastic bandits. J. Mach. Learn. Res. **20**(17), 1–38 (2019)
7. Jensen, J.L.W.V.: Sur les fonctions convexes et les inégalités entre les valeurs moyennes. Acta Math. **30**(1), 175–193 (1906). https://doi.org/10.1007/BF02418571
8. Joulani, P., Gyorgy, A., Szepesvári, C.: Delay-tolerant online convex optimization: unified analysis and adaptive-gradient algorithms. In: Proceedings of the AAAI Conference on Artificial Intelligence, vol. 30 (2016)
9. Lattimore, T., Szepesvári, C.: Bandit Algorithms. Cambridge University Press, Cambridge (2020)
10. Liu, W., Li, S., Zhang, S.: Contextual dependent click bandit algorithm for web recommendation. In: Wang, L., Zhu, D. (eds.) COCOON 2018. LNCS, vol. 10976, pp. 39–50. Springer, Cham (2018). https://doi.org/10.1007/978-3-319-94776-1_4
11. Masoudian, S., Zimmert, J., Seldin, Y.: A best-of-both-worlds algorithm for bandits with delayed feedback. In: Advances in Neural Information Processing Systems (2022)
12. Thune, T.S., Cesa-Bianchi, N., Seldin, Y.: Nonstochastic multiarmed bandits with unrestricted delays. In: Advances in Neural Information Processing Systems, vol. 32 (2019)

13. Villar, S.S., Bowden, J., Wason, J.: Multi-armed bandit models for the optimal design of clinical trials: benefits and challenges. Stat. Sci. A Rev. J. Inst. Math. Stat. **30**(2), 199 (2015)
14. Wan, Z., Sun, X., Zhang, J.: Bounded memory adversarial bandits with composite anonymous delayed feedback. In: Proceedings of the Thirty-First International Joint Conference on Artificial Intelligence, IJCAI 2022, Vienna, Austria, 23–29 July 2022, pp. 3501–3507 (2022)
15. Wang, S., Wang, H., Huang, L.: Adaptive algorithms for multi-armed bandit with composite and anonymous feedback. In: Proceedings of the AAAI Conference on Artificial Intelligence, vol. 35, pp. 10210–10217 (2021)
16. Wei, C.Y., Luo, H.: More adaptive algorithms for adversarial bandits. In: Conference on Learning Theory, pp. 1263–1291. PMLR (2018)
17. Zimmert, J., Seldin, Y.: An optimal algorithm for adversarial bandits with arbitrary delays. In: International Conference on Artificial Intelligence and Statistics, pp. 3285–3294. PMLR (2020)

Cabbage Can't Always Be Transformed into Turnip: Decision Algorithms for Sorting by Symmetric Reversals

Xin Tong[1], Yixiao Yu[1], Ziyi Fang[1], Haitao Jiang[1], Lusheng Wang[2]📵,
Binhai Zhu[3](✉)📵, and Daming Zhu[1]

[1] College of Computer Science and Technology, Shandong University, Qingdao, China
{xtong,yixiaoyu,fangziyi}@mail.sdu.edu.cn, {htjiang,dmzhu}@sdu.edu.cn
[2] Department of Computer Science, City University of Hong Kong,
Hong Kong, Kowloon, China
cswangl@cityu.edu.hk
[3] Gianforte School of Computing, Montana State University,
Bozeman, MT 59717, USA
bhz@montana.edu

Abstract. Sorting a permutation by reversals is a famous problem in genome rearrangements, and has been well studied over the past thirty years. But the involvement of repeated segments is sometimes inevitable during genome evolution, especially in reversal events. Since 1997, quite some biological evidence were found that in many genomes the reversed regions are usually flanked by a pair of inverted repeats. For example, a reversal will transform $+a + x - y - z - a$ into $+a + z + y - x - a$, where $+a$ and $-a$ form a pair of inverted repeats. This type of reversals are called symmetric reversals, which, unfortunately, were largely ignored in algorithm design. While sorting genomes with a mixture of reversals and symmetric reversals sees more practical in many scenarios, it is certainly a much harder problem (which is out the scope of this paper). In this paper, we investigate the decision problem of sorting by symmetric reversals ($SSR(A,B)$), which requires a series of symmetric reversals to transform one chromosome A into the another chromosome B. Given a pair of chromosomes A and B with n repeats, we present an $O(n^2)$ time algorithm to solve the decision problem $SSR(A,B)$. This result is achieved by converting the problem to the circle graph, which has been augmented significantly from the traditional circle graph and a list of combinatorial properties must be proved to successfully answer the decision question.

Keywords: Genome rearrangements · Sorting by symmetric reversals · Decision algorithms · NP-hardness

1 Introduction

In the 1980 s, quite some evidence were found that some species have essentially the same set of genes, but their gene order differs [12,17]. Since then, sorting

W. Wu and G. Tong (Eds.): COCOON 2023, LNCS 14423, pp. 279–294, 2024.
https://doi.org/10.1007/978-3-031-49193-1_21

permutations with rearrangement operations has gained a lot of interest in the area of computational biology in the last thirty years. Sankoff *et al.* formally defined the genome rearrangement events with some basic operations on genomes, e.g., reversals and transpositions, etc. [20], where the reversal operation is adopted the most frequently [9,14,30].

The complexity of the problem of sorting permutations by reversals is closely related to whether the genes are signed or not. Watterson *et al.* pioneered the research on sorting an unsigned permutation by reversals [29]. In 1999, Caprara established the NP-hardness of this problem [7]. Soon after, Berman *et al.* showed it to be APX-hard [4] and polynomial time approximations have been designed with factors from 2 down to 1.375 [3,8,14]. As for the more realistic problem of sorting signed permutations by reversals, Hannenhalli and Pevzner proposed an $O(n^4)$ time exact algorithm for this problem, where n is the number of genes in the given permutations (or singleton genomes) [11]. The running time has been improved to $O(n^2)$, $O(n^{1.5}\sqrt{\log n})$ and finally to $O(n^{1.5})$ [10,13,24].

On the other hand, some evidence has been found that the breakpoints where reversals occur could have some special property in the genomes [15,19]. As early as in 1997, some studies showed that the breakpoints are often associated with repetitive elements on mammals and drosophila genomes [1,2,23,25]. In fact, the well-known "site-specific recombination", which has an important application in "gene knock out" [16,21,22], is responsible for many important DNA rearrangements, including insertion, deletion or inversion of a segment of DNA, and an inversion occurs when the two recombination sites are related to each other in an inverted repeat manner [28].

Recently, Wang *et al.* conducted a systematic study on comparing different strains of various bacteria such as *Pseudomonas aeruginosa*, *Escherichia coli*, *Mycobacterium tuberculosis* and *Shewanella* [26,27]. Their study further illustrated that repeats are associated with the ends of rearrangement segments for various rearrangement events such as reversal, transposition, inverted block interchange, etc, so that the left and right neighborhoods of those repeats remain unchanged after the rearrangement events. Focusing on reversal events, the reversed regions are usually flanked by a pair of inverted repeats [23]. Such a phenomenon can also better explain why the famous "breakpoint reuse", which was an interesting finding and discussed in details when comparing human with mouse, could happen [18].

In this paper, we propose a new model called *sorting by symmetric reversals*, which requires each inverted region on the chromosomes being flanked by a pair of mutually inverted repeats. Admittedly, in real datasets, not all reversals are symmetric. For instance, in [27], among the 17 inversions reported 12 are symmetric. Hence, a practical mixed model might be *sorting by (mixed) reversals and symmetric reversals*, which seems to be a much harder problem to be investigated. (A reversal could change breakpoints while a symmetric reversal does not, as can be seen a bit later.) In this paper, we solely focus on the sorting by symmetric reversals problem, which is certainly a theoretical model and is a subproblem of the more general mixed version — to solve the general problem, we must be able to solve this restricted theoretical problem.

We investigate the decision problem of sorting by symmetric reversals (SSR for short), which asks whether a chromosome can be transformed into the other by a series of symmetric reversals. We devise an $O(n^2)$ time algorithm to solve this decision problem. (Additional results on the optimization version can be found in arXiv:abs/2302.03797.)

This paper is organized as follows. In Sect. 2, we give some necessary definitions. Then we present the details in Sects. 3–4, focusing only on the decision algorithms and leaving out most of the proofs. Finally, we conclude the paper in Sect. 5.

2 Preliminaries

In the literature of genome rearrangement, we always have a set of integers $\Sigma_1 = \{1, \cdots, g\}$, where each integer stands for a long DNA sequence (syntenic block or a gene). For simplicity, we use "gene" hereafter. Since we will study symmetric reversals, we define $\Sigma_2 = \{r_0, r_1, r_2, \cdots, r_t\}$ to be a set of symbols, each of them is referred to as a *repeat* and represents a relative shorter DNA sequence compared with genes. We then set $\Sigma = \Sigma_1 \cup \Sigma_2$ to be the alphabet for the whole chromosome.

Since reversal operations work on a chromosome internally, a genome can be considered as a chromosome for our purpose, i.e., each genome is a singleton and contains only one chromosome. Here we assume that each gene appears exactly once on a chromosome, on the other hand, by name, a repeat could appear multiple times. A gene/repeat x on a chromosome may appear in two different orientations, i.e., either as $+x$ or $-x$. Thus, each chromosome of interest is presented by a sequence of signed integers/symbols.

The number of occurrences of a gene/repeat x in both orientations is called the *duplication number* of x on the chromosome π, denoted by $dp[x, \pi]$. The duplication number of a chromosome π, denoted by $dp[\pi]$, is the maximum duplication number of the repeats on it. For example, chromosome $\pi = [+r_0, +1, -r, +2, +r, -r_0]$, $dp[1, \pi] = dp[2, \pi] = 1$, $dp[r_0, \pi] = dp[r, \pi] = 2$, and $dp[\pi] = 2$. Two chromosomes π_1 and π_2 are *related* if their duplication numbers for all genes and repeats are identical. Let $|x| \in \Sigma$ be an integer or symbol, and $+|x|$ and $-|x|$ be two occurrences of $|x|$, where the orientations of $+|x|$ and $-|x|$ are different. A chromosome of n genes/repeats is denoted as $\pi = [x_1, x_2, \ldots, x_{n-1}, x_n]$. A linear chromosome has two ends, and it can be read from either end to the other, so the chromosome $\pi = [x_1, x_2, \ldots, x_{n-1}, x_n]$ can also be described as $[-x_n, -x_{n-1}, \ldots, -x_2, -x_1]$, which is called the *reversed and negated* form of π. (Note that $-(-x_i) = +x_i$.)

A *reversal* is an operation that reverses a segment of continuous integers/symbols on the chromosome. A *symmetric reversal* is a reversal, where the reversed segment is flanked by pair of identical repeats with different orientations, i.e., either $(+r, \cdots, -r)$ or $(-r, \cdots, +r)$ for some $r \in \Sigma_2$. In other words, let $\pi = [x_1, x_2, \ldots, x_n]$ be a chromosome. The reversal $\rho(i, j)$ $(1 \leq i < j \leq n)$

reverses the segment $[x_i, x_{i+1}, \ldots, x_{j-1}, x_j]$, and yields $\pi' = [x_1, x_2, \ldots, x_{i-1},$ $-x_j, -x_{j-1}, \ldots, -x_{i+1}, -x_i, x_{j+1}, \ldots, x_n]$. If $x_i = -x_j$, we say that $\rho(i,j)$ is a *symmetric reversal* on $|x_i|$. Reversing a whole chromosome will not change the relative order of the integers but their signs, so we assume that each chromosome is flanked by $+r_0$ and $-r_0$, then a chromosome will turn into its reversed and negated form by performing a symmetric reversal between $+r_0$ and $-r_0$.

Again, as a simple example, let $\pi = [+r_0, +1, -r_1, +2, +r_2, +r_1, +r_2, -r_0]$, then a symmetric reversal on r_1 yields $\pi' = [+r_0, +1, -r_1, -r_2, -2, +r_1, +r_2, -r_0]$.

Now, we formally define the problems to be investigated in this paper.

Definition 1. *Sorting by Symmetric Reversals,* **SSR** *for short.*

   Instance: *Two related chromosomes π and τ, such that $dp[\pi] = dp[\tau] \geq 2$.*

   Question: *Is there a sequence of symmetric reversals that transform π into τ?.*

There is a standard way to make a signed gene/repeat unsigned. Let $\pi = [x_0, x_1, \ldots, x_{n+1}]$ be a chromosome, each occurrence of gene/repeat of π, say x_i ($0 \leq i \leq n+1$), is represented by a pair of ordered nodes, $l(x_i)$ and $r(x_i)$. If the sign of x_i is $+$, then $l(x_i) = |x_i|^h$ and $r(x_i) = |x_i|^t$; otherwise, $l(x_i) = |x_i|^t$ and $r(x_i) = |x_i|^h$. (Here, h represents 'head' and t represents 'tail'.) Note that, if x_i and x_j ($i \neq j$) are different occurrences of the same repeat, i.e., $|x_i| = |x_j|$, $l(x_i)$, $l(x_j)$, $r(x_i)$ and $r(x_j)$ correspond to two nodes $|x_i|^h$ and $|x_i|^t$ only. Consequently, π will also be described as $[l(x_0), r(x_0), l(x_1), r(x_1), \ldots, l(x_{n+1}), r(x_{n+1})]$. We say that $r(x_i)$ and $l(x_{i+1})$, for $0 \leq i \leq n$, form an *adjacency*, denoted by $\langle r(x_i), l(x_{i+1}) \rangle$. (*In the signed representation of a chromosome π, $\langle x_i, x_{i+1} \rangle$ forms an adjacency, noting that $\langle x_i, x_{i+1} \rangle = \langle -x_{i+1}, -x_i \rangle$.*) Also, we say that the adjacency $\langle r(x_i), l(x_{i+1}) \rangle$ is associated with x_i and x_{i+1}. Let $\mathcal{A}[\pi]$ represent the multiset of adjacencies of π. We take the chromosome $\pi = [+r_0, +1, -r_1, +2, +r_1, -r_0]$ as an example to explain the above notations. The multi-set of adjacencies is $\mathcal{A}[\pi] = \{\langle r_0^t, 1^h \rangle, \langle 1^t, r_1^t \rangle, \langle r_1^h, 2^h \rangle, \langle 2^t, r_1^h \rangle, \langle r_1^t, r_0^h \rangle\}$, π can also be viewed as $[r_0^h, r_0^t, 1^h, 1^t, r_1^t, r_1^h, 2^h, 2^t, r_1^h, r_1^t, r_0^t, r_0^h]$.

Lemma 1. *Let π be a chromosome and π' is obtained from π by performing a symmetric reversal. Then $\mathcal{A}[\pi] = \mathcal{A}[\pi']$.*

Lemma 1 implies a necessary condition for answering the decision question of **SSR**.

Theorem 1. *Chromosome π cannot be transformed into τ by a series of symmetric reversals if $\mathcal{A}[\pi] \neq \mathcal{A}[\tau]$.*

A simple negative example would be $\pi = [+r_0, +r_1, -2, +r_1, -1, -r_0]$ and $\tau = [+r_0, -r_1, +2, -r_1, +1, -r_0]$. One can easily check that $\mathcal{A}[\pi] \neq \mathcal{A}[\tau]$, which means that there is no way to convert π to τ using symmetric reversals. In the next section, as a warm-up, we first solve the case when each repeat appears at most twice in π and τ. Even though the method is not extremely hard, we hope the presentation and some of the concepts can help readers understand the details for the general case in Sect. 4 better.

3 An $O(n^2)$ Algorithm for SSR with Duplication Number 2

In this section, we consider a special case, where the duplication numbers for the two related chromosomes π and τ are both 2. That is, $\mathcal{A}[\pi] = \mathcal{A}[\tau]$ and $dp[\pi] = dp[\tau] = 2$. We will design an algorithm with running time $O(n^2)$ to determine if there is a sequence of symmetric reversals that transform π into τ.

Note that $\mathcal{A}[\pi]$ is a multi-set, where an adjacency may appear more than once. When the duplication number of each repeat in the chromosome is at most 2, the same adjacency can appear at most twice in $\mathcal{A}[\pi]$.

Let $\pi = [x_0, x_1, \ldots, x_{n+1}]$ be a chromosome. Let x_i and x_j be the two occurrences of a repeat x, and x_{i+1} and x_{j+1} the two occurrences of some other repeat x' in π. We say that $|x_i|$ and $|x_{i+1}|$ are *redundant*, if $r(x_i) = r(x_j)$ and $l(x_{i+1}) = l(x_{j+1})$ (or $r(x_i) = l(x_j)$ and $l(x_{i+1}) = r(x_{j-1})$). In this case, the adjacency $\langle r(x_i), l(x_{i+1}) \rangle$ appears twice. In fact, it is the only case that an adjacency can appear twice. An example is as follows: $\pi = [+r_0, +r_1, -r_2, +1, +r_2, -r_1, -r_0]$, where the adjacency $\langle +r_1, -r_2 \rangle$ appears twice (the second negatively), hence r_1 and r_2 are redundant. The following lemma tells us that if x_i and x_j are redundant, we only need to use one of them to do reversals and the other can be deleted from the chromosome so that each adjacency appears only once.

Lemma 2. *Given two chromosomes* $\pi = [x_0, x_1, \ldots, x_{n+1}]$ *and* τ, *such that* $\mathcal{A}[\pi] = \mathcal{A}[\tau]$. *Let* $|x_i|$ *and* $|x_{i+1}|$ *be two repeats in* π *that are redundant. Let* π' *and* τ' *be the chromosomes after deleting the two occurrences of* $|x_{i+1}|$ *from both* π *and* τ, *respectively. Then* π *can be transformed into* τ *by a series of symmetric reversals if and only if* π' *can be transformed into* τ' *by a series of symmetric reversals.*

Regarding the previous example, $\pi = [+r_0, +r_1, -r_2, +1, +r_2, -r_1, -r_0]$, where r_1 and r_2 are redundant, following the above lemma, one can obtain $\pi' = [+r_0, +r_1, +1, -r_1, -r_0]$. This is in fact equivalent to replacing the adjacency $\langle +r_1, -r_2 \rangle$ by $+r_1$, and $\langle +r_2, -r_1 \rangle$ by $-r_1$.

A chromosome π is *simple* if every adjacency in $\mathcal{A}[\pi]$ appears only once. Based on Lemma 2, we can remove the two occurrences of a redundant repeat from the chromosomes. Thus, if $dp[\pi] = dp[\tau] = 2$, we can always assume that both π and τ are simple. Consequently, there is a unique bijection between the identical adjacencies in $\mathcal{A}[\pi]$ and $\mathcal{A}[\tau]$ respectively. We say that any pair of adjacencies, determined by the bijection, are *matched* to each other.

For each repeat x with $dp[\pi, x] = dp[\tau, x] = 2$, let x_i, x_j be the two occurrences of x in π, and $y_{i'}$, $y_{j'}$ be the two occurrences of x in τ, there are four adjacencies associated with x_i and x_j in π: $\langle r(x_{i-1}), l(x_i) \rangle$, $\langle r(x_i), l(x_{i+1}) \rangle$, $\langle r(x_{j-1}), l(x_j) \rangle$, $\langle r(x_j), l(x_{j+1}) \rangle$. Similarly, there are four adjacencies associated with $y_{i'}$ and $y_{j'}$ in τ. We say that x is a *neighbor-consistent* repeat, if $\langle r(x_{i-1}), l(x_i) \rangle$ and $\langle r(x_i), l(x_{i+1}) \rangle$ are matched to two adjacencies both associated with $y_{i'}$ or both associated with $y_{j'}$. That is, the left and right neighbors of x_i are identical in both chromosomes. Note that $\mathcal{A}[\pi] = \mathcal{A}[\tau]$ also implies that the left and right neighbors of the other occurrences x_j are also identical in both two chromosomes if x is neighbor-consistent. If $\langle r(x_{i-1}), l(x_i) \rangle$ and $\langle r(x_i), l(x_{i+1}) \rangle$ are matched to two adjacencies, one of which is associated with $y_{i'}$ and the other is associated with $y_{j'}$, then x is a *neighbor-inconsistent* repeat. The genes and the repeats which appear once in π are also defined to be neighbor-consistent. (See Fig. 1 for an example.) By definition and the fact that $\mathcal{A}[\pi] = \mathcal{A}[\tau]$, we have

Proposition 1. *Performing a symmetric reversal on a repeat will turn the repeat from neighbor-consistent to neighbor-inconsistent or vice versa. (See Fig. 1.)*

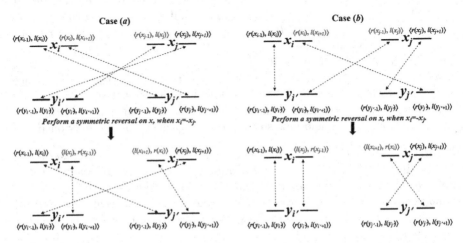

Fig. 1. x_i, x_j are the two occurrences of x in π with $x_i = -x_j$, and $y_{i'}$, $y_{j'}$ be the two occurrences of x in τ. Case (a): x is neighbor-consistent, and will turn to neighbor-inconsistent by a reversal on itself. Case (b): x is neighbor-inconsistent, and will turn to neighbor-consistent by a reversal on itself.

Theorem 2. *Given two simple related chromosomes π^* and τ with $dp[\pi^*] = dp[\tau] = 2$, $\pi^* = \tau$ if and only if $\mathcal{A}[\pi^*] = \mathcal{A}[\tau]$ and every repeat is neighbor-consistent.*

Based on Proposition 1 and Theorem 2, to transform π into τ, it is sufficient to perform an odd number (at least 1) of symmetric reversals on each neighbor-inconsistent repeat, and an even number (might be 0) of symmetric reversals on each neighbor-consistent repeat. Hereafter, we also refer a neighbor-consistent (resp. neighbor-inconsistent) repeat as an *even* (resp. *odd*) repeat.

The main difficulty to find a sequence of symmetric reversals between π and τ is to choose a "correct" symmetric reversal at a time. Note that, for a pair of occurrences (x_i, x_j) of a repeat x, the orientations may be the same at present but after some reversals, the orientations of x_i and x_j may differ. We can only perform a reversal on a pair of occurrences of a repeat with different orientations. Thus, it is crucial to choose a "correct" symmetric reversal at the right time. In the following, we will use "intersection" graph to handle this.

Suppose that we are given two simple related chromosomes π and τ with $dp[\pi] = dp[\tau] = 2$ and $\mathcal{A}[\pi] = \mathcal{A}[\tau]$. In this case, each repeat in the chromosomes represent an interval indicated by the two occurrences of the repeat. Thus, we can construct an *intersection graph* $IG(\pi, \tau) = (V[\pi], E[\pi])$. For each repeat x with $dp[\pi, x] = 2$, construct a vertex $x \in V_\pi$, and set its weight, $\omega(x) = 2$ if x is even, and $\omega(x) = 1$ if x is odd; set the color of x **black** if the signs of the two occurrences of x in π are different, and **white** otherwise. Construct an edge between two vertices x and y if and only if the occurrences of x and y appear alternatively in π, i.e., let x_i and x_j $(i < j)$ be the two occurrences of x, and x_k and x_l $(k < l)$ be the two occurrences of y in π, there will be an edge between the vertices x and y if and only if $i < k < j < l$ or $k < i < l < j$. There are three types of vertices in $V[\pi]$: black vertices (denoted as $V_b[\pi]$), white vertices of weight 1 (denoted as $V_w^1[\pi]$) and white vertices of weight 2 (denoted as $V_w^2[\pi]$). Thus, $V[\pi] = V_b[\pi] \cup V_w^1[\pi] \cup V_w^2[\pi]$. In fact, the intersection graph is a circle graph while ignoring the weight and color of all the vertices.

Lemma 3. *A single white vertex of weight 1 cannot be a connected component in $IG(\pi, \tau)$.*

For each vertex x in $IG(\pi, \tau)$, let $N(x)$ denote the set of vertices incident to x. For a black vertex, say x, in $IG(\pi, \tau)$, performing a symmetric reversal on x in π, yields π', where the intersection graph $IG(\pi') = (V[\pi'], E[\pi'])$ can be derived from $IG(\pi, \tau)$ following the three rules:

- rule-I: for each vertex $v \in N(x)$ in $IG(\pi, \tau)$, change its color from black to white, and vice versa.
- rule-II: for each pair of vertices $u, v \in N(x)$ of $IG(\pi, \tau)$, if $(u, v) \in E[\pi]$, then $E[\pi'] = E[\pi] - \{(u, v)\}$; and if $(u, v) \notin E[\pi]$, then $E[\pi'] = E[\pi] \cup \{(u, v)\}$.
- rule-III: decrease the weight of x by one, if $\omega(x) > 0$, then $V[\pi'] = V[\pi]$; and if $\omega(x) = 0$, then $V[\pi'] = V[\pi] - \{x\}$.

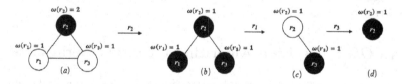

Fig. 2. $\pi = [+r_0, +r_1, +1, +r_2, +r_3, +r_1, -r_2, -2, +r_3, -r_0]$, and $\tau = [+r_0, +r_1, -r_2, -2, +r_3, +r_1, +1, +r_2, +r_3, -r_0]$. $\mathcal{A}[\pi] = \mathcal{A}[\tau] = \{\langle r_0^t, r_1^h \rangle, \langle r_1^t, 1^h \rangle, \langle 1^t, r_2^h \rangle, \langle r_2^t, r_3^h \rangle, \langle r_3^t, r_1^h \rangle, \langle r_1^t, r_2^t \rangle, \langle r_2^h, 2^t \rangle, \langle 2^h, r_3^h \rangle, \langle r_3^t, r_0^t \rangle\}$. The repeats r_1 and r_3 are odd, while the repeat r_2 is even. (a) The intersection graph $IG(\pi, \tau)$, performing the symmetric reversal on repeat r_2 will transform π into $\pi' = [+r_0, +r_1, +1, +r_2, -r_1, -r_3, -r_2, -2, +r_3, -r_0]$, (b) The intersection graph $IG(\pi', \tau)$. (c) The intersection graph after performing the symmetric reversal on the repeat r_1. (d) The intersection graph after performing the symmetric reversal on the repeat r_3.

If x is a black vertex in $IG(\pi, \tau)$ and $\omega(x) = 1$, then performing the symmetric reversal of x in π yields π'. Let $C_1, C_2, \ldots, C_m$ be the connected components introduced by the deletion of x in $IG(\pi', \tau)$, we go through some properties of performing this symmetric reversal.

Lemma 4. *In each C_i $(1 \leq i \leq m)$, there is at least one vertex z_i such that $z_i \in N(x)$ in $IG(\pi, \tau)$.*

Lemma 5. *Let x' be a black vertex, $\omega(x) = \omega(x') = 1$, and $x' \in N(x)$ in $IG(\pi, \tau)$. After performing the symmetric reversal on x in π, let y be a vertex in the connected component C_i, and x' is in the connected component C_j, $i \neq j$. Let π'' be the resulting chromosome after performing the symmetric reversal on x' in π, then the color of y is the same in $IG(\pi', \tau)$ and $IG(\pi'', \tau)$.*

Lemma 6. *Let x' be a black vertex, $\omega(x) = \omega(x') = 1$, and $x' \in N(x)$ in $IG(\pi, \tau)$. After performing the symmetric reversal of x in π, let y, z be two vertices in the connected component C_i, and x' is in the connected component C_j, $i \neq j$. Let π'' be the resulting chromosome after performing the symmetric reversal on x' in π. If $(y, z) \in E[\pi']$, then $(y, z) \in E[\pi'']$.*

Theorem 3. *If a connected component of $IG(\pi, \tau)$ contains at least one black vertex, then there exists a symmetric reversal, after performing it, any newly created connected component containing a white vertex of weight 1 also contains a black vertex.*

The main contribution of this section is the following theorem.

Theorem 4. *A chromosome π can be transformed into the other chromosome τ if and only if (I) $\mathcal{A}[\pi] = \mathcal{A}[\tau]$, and (II) each white vertex of weight 1 belongs to a connected component of $IG(\pi, \tau)$ containing a black vertex.*

The above theorem implies that a breadth-first search of $IG[\pi, \tau]$ will determine whether π can be transformed into τ, which takes $O(n^2)$ time, because $IG(\pi, \tau)$ contains at most n vertices and n^2 edges. We will show the details of the algorithm in Sect. 4, since it also serves as a decision algorithm for the general case.

4   An $O(n^2)$ Decision Algorithm for the General Case

For the general case, i.e., when the duplication number for the two related input genomes is arbitrary, the extension of the algorithm in Sect. 3 is non-trivial as it is impossible to make the genomes simple. Our overall idea is to fix any bijection f between the (identical) adjacencies of the input genomes, and build the corresponding alternative-cycle graph. This alternative-cycle graph is changing according to the corresponding symmetric reversals; and we show that, when the graph contains only 1-cycles, then the target τ is reached. Due to the changing nature of the alternative-cycle graph, we construct a blue edge intersection graph to capture these changes. However, this is not enough as the blue intersection graph built from the alternative-cycle graph could be disconnected and we need to make it connected by adding additional vertices such that the resulting sequence of symmetric reversals are consistent with the original input genomes, and can be found in the new intersection graph (called IG, which is based on the input genomes π and τ as well as f). We depict the details in the following.

Suppose that we are given two related chromosomes $\pi = [x_0, x_1, \ldots, x_{n+1}]$ and $\tau = [y_0, y_1, \ldots, y_{n+1}]$, such that $x_0 = y_0 = +r_0$ and $x_{n+1} = y_{n+1} = -r_0$. Theorem 1 shows that $\mathcal{A}[\pi] = \mathcal{A}[\tau]$ is a necessary condition, thus there is a bijection f between identical adjacencies in $\mathcal{A}[\pi]$ and $\mathcal{A}[\tau]$, as shown in Fig. 3. Based on the bijection f, we construct the alternative-cycle graph $ACG(\pi, \tau, f)$ as follows. For each x_i in π, construct an ordered pair of nodes, denoted by $l(x_i)$ and $r(x_i)$, which are connected by a red edge. For each y_k in τ, assume that $\langle r(y_{k-1}), l(y_k) \rangle$ is matched to $\langle r(x_{i-1}), l(x_i) \rangle$, and $\langle r(y_k), l(y_{k+1}) \rangle$ is matched to $\langle r(x_{j-1}), l(x_j) \rangle$, in the bijection f. There are four cases:

1. $l(y_k) = l(x_i)$ and $r(y_k) = r(x_{j-1})$, then connect $l(x_i)$ and $r(x_{j-1})$ with a blue edge,
2. $l(y_k) = r(x_{i-1})$ and $r(y_k) = r(x_{j-1})$, then connect $r(x_{i-1})$ and $r(x_{j-1})$ with a blue edge,
3. $l(y_k) = l(x_i)$ and $r(y_k) = l(x_j)$, then connect $l(x_i)$ and $l(x_j)$ with a blue edge,
4. $l(y_k) = r(x_{i-1})$ and $r(y_k) = l(x_j)$, then connect $r(x_{i-1})$ and $l(x_j)$ with a blue edge.

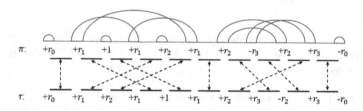

Fig. 3. The bijection between identical adjacencies in $\mathcal{A}[\pi]$ and $\mathcal{A}[\tau]$, and the corresponding alternative-cycle graph.

Actually, two nodes connected by a red edge implies they are from the same occurrence of some repeat/gene in π, so each occurrence of some repeat/gene in π corresponds to a red edge; and similarly, two nodes connected by a blue edge implies that they are from the same occurrence of some repeat/gene in τ, thus each occurrence of some repeat/gene in τ corresponds to a blue edge. Note that each node associates with one red edge and one blue edge, so $ACG(\pi, \tau, f)$ is composed of edge disjoint cycles, on which the red edges and blue edge appears alternatively. A cycle composed of c blue edges as well as c red edges is called a c-cycle, it is called a long cycle when $c \geq 2$.

Theorem 5. *Given two chromosomes π^* and τ, $\pi^* = \tau$ if and only if $\mathcal{A}[\pi^*] = \mathcal{A}[\tau]$, and there is a bijiection f between the identical adjacencies in $\mathcal{A}[\pi^*]$ and $\mathcal{A}[\tau]$, such that all the cycles in the resulting alternative-cycle graph $ACG(\pi^*, \tau, f)$ are 1-cycles.*

The above theorem gives us a terminating condition for our algorithm: let π and τ be the input chromosomes, and our algorithm keeps updating the alternative-cycle graph until all cycles in it become 1-cycles. Unfortunately, in the following, we observe that some cycles can not be performed by symmetric reversals directly, then we consider these cycles intersecting with each other as a connected component. But this is still not enough, since there could also be some connected components which do not admit any symmetric reversal, we managed to handle this case by joining all the cycles of the same repeat into a whole connected component.

Lemma 7. *In an alternative-cycle graph, each cycle corresponds to a unique repeat and every edge (both red and blue) in the cycle corresponds to an occurrence of the unique repeat.*

Proof. W.l.o.g, assume that $l(x_i)$ and $l(x_j)$ are connected with a blue edge, from the construction of the alternative-cycle graph, there must be an occurrence in τ, say y_k, such that $\{l(x_i), l(x_j)\} = \{l(y_k), r(y_k)\}$, thus, $|x_i| = |x_j| = |y_k|$, and the blue edge $(l(x_i), l(x_j))$ corresponds the occurrence y_k of the repeat $|y_k|$. □

Since each gene appears once in π, Lemma 7 implies that each gene has a 1-cycle in $ACG(\pi, \tau, f)$, these 1-cycles will be untouched throughout our algorithm.

Lemma 8. *In an alternative-cycle graph, if we add a green edge connecting each pairs of nodes $r(x_i)$ and $l(x_{i+1})$ (for all $0 \le i \le n$), then all the blue edges and green edges together form a (blue and green alternative) path.*

Proof. Actually, the green edge connecting $r(x_i)$ and $l(x_{i+1})$ $(0 \le i \le n)$ is the adjacency $\langle r(x_i), l(x_{i+1}) \rangle$ of $\mathcal{A}[\pi]$, which is identical to some adjacency $\langle r(y_j), l(y_{j+1}) \rangle$ of $\mathcal{A}[\tau]$ according to the bijection between identical adjacencies of $\mathcal{A}[\pi]$ and $\mathcal{A}[\tau]$. Therefore, y_j and y_{j+1} appears consecutively in τ, and following the construction of $ACG(\pi, \tau, f)$ and Lemma 7, they correspond to the two blue edges, one of which is associated with $r(x_i)$ and the other is associated with $l(x_{i+1})$ in $ACG(\pi, \tau, f)$, thus, the two blue edges are connected through the green edge $(r(x_i), l(x_{i+1}))$. The above argument holds for every green edge, therefore, all the blue edges and green edges constitute a path. We show an example in Fig. 3. □

Let $x \in \Sigma$ be a repeat. Let x_i and x_j be two occurrences of x in π, where $i \neq j$. A blue edge is *opposite* if it connects $l(x_i)$ and $l(x_j)$ or $r(x_i)$ and $r(x_j)$. A blue edge is *non-opposite* if it connects $l(x_i)$ and $r(x_j)$ or $r(x_i)$ and $l(x_j)$.

Specially, the blue edge on any 1-cycle (with a blue edge and a red edge) is non-opposite. A cycle is *opposite* if it contains at least one opposite blue edge.

Lemma 9. *Let x_i and x_j be two occurrences of repeat x in π. In the alternative-cycle graph $ACG(\pi,\tau,f)$, if $l(x_i)$ and $l(x_j)$ (or $r(x_i)$ and $r(x_j)$) are connected with an opposite edge, x_i and x_j has different orientations; and if $l(x_i)$ and $r(x_j)$ (or $r(x_i)$ and $l(x_j)$) are connected with a non-opposite edge, x_i and x_j has the same orientations.*

Proposition 2. *Given a k-cycle C of x, performing a symmetric reversal on two occurrences of x that are connected by an opposite blue edge, will break C into a $(k-1)$-cycle as well as a 1-cycle. Given a k_1-cycle C_1 and a k_2-cycle C_2 of x, performing a symmetric reversal on the two occurrences of $x_i \in C_1$ and $x_j \in C_2$, will join C_1 and C_2 into a (k_1+k_2)-cycle.*

Now, we construct the *blue edge intersection graph* $BG(\pi,\tau,f) = (BV_\pi, BE_\pi, f)$ according to $ACG(\pi,\tau,f)$, viewing each blue edge as an interval of the two nodes it connects. For each interval, construct an original vertex in BV_π, and set its weight to be 1, set its color to be black if the blue edge is opposite, and white otherwise. An edge in BE_π connects two vertices if and only if their corresponding intervals intersect but neither overlaps the other. An example of the blue edge intersection graph is shown in Fig. 4-(b).

Note that each connected component of $BG(\pi,\tau,f)$ forms an interval on π, for each connected component P in $BG(\pi,\tau,f)$, we use $\overline{P}$ to denote its corresponding interval on π.

Lemma 10. *Let P be some connected component of $BG(\pi,\tau,f)$, the leftmost endpoint of $\overline{P}$ must be a left node of some x_i, i.e., $l(x_i)$, and the rightmost endpoint of $\overline{P}$ must be a right node of some x_j, i.e., $r(x_j)$, where $i < j$.*

Lemma 11. *All the vertices in $BG(\pi,\tau,f)$ corresponding to the blue edges on the same long cycle in $ACG(\pi,\tau,f)$ are in the same connected component of $BG(\pi,\tau,f)$.*

As the two blue edges of a non-opposite 2-cycle do not intersect each other, we have,

Corollary 1. *A non-opposite 2-cycle can not form a connected component of $BG(\pi,\tau,f)$.*

For each repeat x, assume that it constitutes k cycles in $ACG(\pi,\tau,f)$. Let $x_{i_1}, x_{i_2}, \ldots, x_{i_k}$ be the k occurrences of x that are in distinct cycles in $ACG(\pi,\tau,f)$, where $1 \le i_1 < i_2 < \cdots < i_k \le n$. We construct $k-1$ *additional vertices* corresponding to the intervals $[r(x_{i_j}) - \epsilon, l(x_{i_{j+1}}) + \epsilon]$ to BV_π ($1 \le j \le k-1$), for each such vertex, set its weight to be 1, and set its color to be black if the signs of x_{i_j} and $x_{i_{j+1}}$ are distinct, and white otherwise. See the vertex marked with 10 in Fig. 4-(c) for an example. Also, there is an edge between two vertices of BV_π if and only if their corresponding intervals intersect, but none overlaps the other. The resulting graph is called the *intersection graph* of π, denoted as $IG(\pi,\tau,f) = (V[\pi], E[\pi])$. An example is shown in Fig. 4-(c). Let $V_\pi^w \subseteq V[\pi]$ be the subset of vertices which corresponding to non-opposite blue edges on long cycles in $ACG(\pi,\tau,f)$.

290   X. Tong et al.

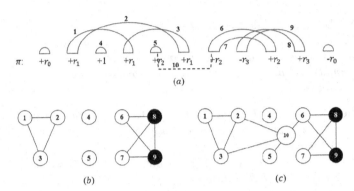

(a)

(b)   (c)

Fig. 4. (a) The alternative-cycle graph $ACG(\pi,\tau,f)$, where each blue edge is marked with a number. (b) The blue edge intersection graph $BG(\pi,\tau,f)$. (c) The intersection graph $IG(\pi,\tau,f)$ with additional vertices, where each number represents an interval. (Color figure online)

From Lemma 10 and the construction of the intersection graph of π, all the vertices corresponding to all the blue edges of the same repeat are in the same connected component. Note that the intersection graph of π may be distinct, when the bijection between identical adjacencies of $\mathcal{A}[\pi]$ and $\mathcal{A}[\tau]$ differs. Nevertheless, we have,

Lemma 12. *Let π and τ be two related chromosomes with $\mathcal{A}[\pi] = \mathcal{A}[\tau]$. Let x_i and x_j ($i < j$) be two occurrences of x in π, and $x_{i'}$ and $x_{j'}$ ($i' < j'$) be two occurrences of x' in π, if either $i < i' < j < j'$ or $i' < i < j' < j$ is satisfied, then, based on any bijection f between $\mathcal{A}[\pi] = \mathcal{A}[\tau]$, in the intersection graph $IG(\pi,\tau,f)$, the vertices corresponding to all the intervals of x and x' are in the same connected component.*

Actually, the connected components of the intersection graph partition the repeats on π into groups. From Lemma 10 and Lemma 12, the group partition of the repeats is independent of the bijection between identical adjacencies of $\mathcal{A}[\pi]$ and $\mathcal{A}[\tau]$. In other words, the group partition will be fixed once π and τ are given. Thus, each connected component corresponds to a group of sub-sequences in π and τ respectively. Then, we can check whether the sub-sequences in π are identical to these in τ. If so, we can check whether it is possible to convert all the cycles in this connected component into 1-cycles by changing f to another bijection f' according to the identical sub-sequences. Hence, in the following, we assume that, under the bijection f, the cycles in each connected component of $IG(\pi,\tau,f)$ cannot all be converted into 1-cycles by changing the bijection.

Similar to the intersection graph of chromosomes with a duplication number of 2, the intersection graph of chromosomes with unrestricted duplication number also admit the rule-I, rule-II, and rule-III, as in Sect. 3, while performing a symmetric reversal on π.

Algorithm 1. The decision algorithm for SSR

Input: Two related chromosomes π and τ.
Output: $YES\backslash NO$

1: **if** $\mathcal{A}[\pi] \neq \mathcal{A}[\tau]$ **then**
2:     return NO.
3: **end if**
4: Delete the redundant repeats from π and τ when $dp[\pi] = 2$ (by Lemma 2).
5: Build a bijection f between $\mathcal{A}[\pi]$ and $\mathcal{A}[\tau]$ when $dp[\pi] > 2$ by mapping identical adjacencies together and adopting any bijection among multiple occurrences of the same adjacency.
6: Construct the alternative-cycle graph $ACG(\pi, \tau, f)$ based on f when $dp[\pi] > 2$.
7: Construct the corresponding intersection graph $IG(\pi, \tau, f) = (V_\pi, E_\pi)$ (based on $ACG(\pi, \tau, f)$ when $dp[\pi] > 2$ and directly from π and τ if $dp[\pi] = 2$).
8: **if** $dp[\pi] = 2$ **then**
9:     set V_π^w to be the set of weight 1 white vertices
10: **end if**
11: **if** $dp[\pi] > 2$ **then**
12:     set V_π^w to be the set of white vertices corresponding to non-opposite blue edge on long cycle.
13: **end if**
14: Let V_π^b be set of black vertices in $IG(\pi, \tau, f)$.
15: Set queue $Q = V_\pi^b$.
16: Do a breadth first search using Q as the initial value to mark all vertices in the same component of every $x \in V_\pi^b$.
17: **if** there exists an $u \in V_\pi^w$, which is not marked **then return** NO.
18: **end if**
19: **return** YES.

Theorem 6. *If a connected component of $IG(\pi, \tau, f)$ contains a black vertex, then there exists a symmetric reversal, after performing it, we obtain π', any newly created connected component containing a white vertex, which corresponds to a blue edge on a non-opposite long cycle in $ACG(\pi', \tau, f)$, also contains a black vertex.*

Theorem 7. *A chromosome π can be transformed into the other chromosome τ by symmetric reversals if and only if (I) $\mathcal{A}[\pi] = \mathcal{A}[\tau]$, and (II) each white vertex in V_π^w belongs to a connected component of $IG(\pi, \tau, f)$ containing a black vertex.*

Now, we are ready to formally present the decision algorithm based on Theorem 7 for both the general case and the case, where the duplication number 2 in Algorithm 1. We just directly test conditions (I) and (II) in Theorem 7. Note that each connected component in $IG(\pi, \tau, f)$ may contain more than one black vertex. By setting $Q = V_\pi^b$ in line 11, we can guarantee that each connected component in $IG(\pi, \tau, f)$ is explored once during the breadth first search so that $O(n^2)$ running time can be kept.

Running time of Algorithm 1: Let us analyze the time complexity of Algorithm 1. Verifying whether $\mathcal{A}[\pi] = \mathcal{A}[\tau]$ can be done in $O(n^2)$ time. It takes

$O(n^2)$ time to build a bijection between $\mathcal{A}[\pi]$ and $\mathcal{A}[\tau]$, and construct the cycle graph $ACG(\pi, \tau, f)$, as well as the corresponding intersection graph $IG(\pi, \tau, f)$. It remains to analyze the size of $IG(\pi, \tau, f)$. For each repeat, say x, there are $dp[x, \pi]$ original vertices and $c[x] - 1$ additional vertices in $IG[\pi]$, where $c[x]$ is the number of cycles of x in $ACG(\pi, \tau, f)$. Note that $c[x] \leq dp[x, \pi]$ and $\sum_{x \in \Sigma} dp[x, \pi] = n$. Thus, the total number of vertices in $IG(\pi, \tau, f)$ is bounded by $\sum_{x \in \Sigma}(dp[x, \pi] + c[x] - 1) \leq 2\sum_{x \in \Sigma} dp[x, \pi] - 1 = 2n - 1$, then the number of edges in $IG[\pi]$ is at most $4n^2$. The whole breadth-search process takes $O(n^2)$ time, since there are at most $2n-1$ vertices and at most $4n^2$ edges in $IG(\pi, \tau, f)$. Therefore, Algorithm 1 runs in $O(n^2)$ time.

5 Concluding Remarks

This paper investigates a new theoretical model of genome rearrangements named sorting by symmetric reversals. We show that the decision problem, which asks whether a chromosome can be transformed into another by symmetric reversals, is polynomially solvable.

A key idea when the duplication number is 2, is that Lemma 2 shows two chromosomes can be converted into two new chromosomes which can be viewed as permutations of distinct adjacencies. At this point, some readers might think that the problem could then be solved by the famous HP-theory [6,11]. Actually, HP-theory does not always work in this symmetric reversal model. The reason is that, according to HP-theory, two "hurdles" can be mixed together into one component by performing a reversal between an element of one "hurdle" and some element of the other. Under the symmetric reversal model, this is not always possible — there might not be a pair of repeats with opposite signs in these two "hurdles" at all. In this paper, we manage to mix the hurdles with other oriented (good) components by using a series of "trivial" components, each composed of a single 1-cycle. These "trivial" components do not need to be considered by HP-theory while solving the classical sorting signed permutations by reversals problem.

References

1. Armengol, L., Pujana, M.A., Cheung, J., Scherer, S.W., Estivill, X.: Enrichment of segmental duplications in regions of breaks of synteny between the human and mouse genomes suggest their involvement in evolutionary rearrangements. Hum. Mol. Genet. **12**(17), 2201–2208 (2003)
2. Bailey, J.A., Baertsch, R., Kent, W.J., Haussler, D., Eichler, E.E.: Hotspots of mammalian chromosomal evolution. Genome Biol. **5**(4), R23 (2004)
3. Berman, P., Hannenhalli, S., Karpinski, M.: 1.375-approximation algorithm for sorting by reversals. In: Proceedings of the 10th European Symposium on Algorithms (ESA 2002), pp. 200–210 (2002)
4. Berman, P., Karpinski, M.: On some tighter inapproximability results (extended abstract). In: Proceedings of the 26th International Colloquium on Automata, Languages and Programming (ICALP 1099), pp. 200–209 (1999)

5. Bennetzen, J.L., Ma, J., Devos, K.M.: Mechanisms of recent genome size variation in flowering plants. Ann. Bot. **95**, 127–32 (2005)
6. Bergeron, A.: A very elementary presentation of the Hannenhalli-Pevzner theory. Discret. Appl. Math. **146**(2), 134–145 (2005)
7. Caprara, A.: Sorting permutations by reversals and eulerian cycle decompositions. SIAM J. Dis. Math. **12**(1), 91–110 (1999)
8. Christie, D.A.: A 3/2-Approximation algorithm for sorting by reversals. In: Proceedings of the 9th Annual ACM-SIAM Symposium on Discrete Algorithms (SODA 1998), pp. 244–252 (1998)
9. Fertin, G., Labarre, A., Rusu, I., Vialette, S., Tannier, E.: Combinatorics of genome rearrangements. MIT press (2009)
10. Han, Y.: Improving the efficiency of sorting by reversals. In: Proceedings of 2006 International Conference on Bioinformatics & Computational Biology (BIOCOMP 2006), pp. 406–409 (2006)
11. Hannenhalli, S., Pevzner, P.: Transforming cabbage into turnip: polynomial algorithm for sorting signed permutations by reversals. J. ACM **46**(1), 1–27 (1999)
12. Hoot, S.B., Palmer, J.D.: Structural rearrangements, including parallel inversions within the choroplast genome of anemone and related genera. J. Mol. Evol. **38**, 274–281 (1994)
13. Kaplan, H., Shamir, R., Tarjan, R.E.: A faster and simpler algorithm for sorting signed permutations by reversals. SIAM J. Comput. **29**(3), 880–892 (2000)
14. Kececioglu, J., Sankoff, D.: Exact and approximation algorithms for sorting by reversals, with application to genome rearrangement. Algorithmica **13**(1), 180–210 (1995)
15. Longo, M.S., Carone, D.M., Green, E.D., O'Neill, M.L., O'Neill, R.J.: Distinct retroelement classes define evolutionary breakpoints demarcating sites of evolutionary novelty. BMC Genomics **10**(1), 334 (2009)
16. Orban, P.C., Chui, D., Marth, J.D.: Tissue- and site-specific recombination in transgenic mice. Proc. Nat. Acad. Sci. USA **89**(15), 6861–6865 (1992)
17. Palmer, J.D., Herbon, L.A.: Tricicular mitochondrial genomes of brassica and raphanus: reversal of repeat configurations by inversion. Nucleic Acids Res. **14**, 9755–9764 (1986)
18. Pevzner, P., Tesler, G.: Human and mouse genomic sequences reveal extensive breakpoint reuse in mammalian evolution. Proc. Nat. Acad. Sci. USA **100**(13), 7672–7677 (2003)
19. Sankoff, D.: The where and wherefore of evolutionary breakpoints. J. Biology **8**, 66 (2009)
20. Sankoff, D., Leduc, G., Antoine, N., Paquin, B., Lang, B.F., Cedergran, R.: Gene order comparisons for phylogenetic interferce: Evolution of the mitochondrial genome. Proc. Nat. Acad. Sci. USA **89**, 6575–6579 (1992)
21. Sauer, B.: Functional expression of the Cre-Lox site-specific recombination system in the yeast Saccharomyces cerevisiae. Mol. Cell. Biol. **7**(6), 2087–2096 (1987)
22. Sauer, B., Henderson, N.: Site-specific DNA recombination in mammalian cells by the Cre recombinase of bacteriophage P1. Proc. Nat. Acad. Sci. USA **85**(14), 5166–5170 (1988)
23. Small, K., Iber, J., Warren, S.T.: Emerin deletion reveals a common X-chromosome inversion mediated by inverted repeats. Nat. Genet. **16**, 96–99 (1997)
24. Tannier, E., Bergeron, A., Sagot, M.-F.: Advances on sorting by reversals. Discret. Appl. Math. **155**(6–7), 881–888 (2007)
25. Thomas, A., Varr, J.-S., Ouangraoua, A.: Genome dedoubling by DCJ and reversal. BMC Bioinform. **12**(9), S20 (2011)

26. Wang, D., Wang, L.: GRSR: a tool for deriving genome rearrangement scenarios from multiple unichromosomal genome sequences. BMC Bioinform. **19**(9), 11–19 (2018)
27. Wang, D., Li, S., Guo, F., Wang, L.: Core genome scaffold comparison reveals the prevalence that inversion events are associated with pairs of inverted repeats. BMC Genomics **18**, 268 (2017)
28. Watson, J., Gann, A., Baker, T., Levine, M., Bell, S., Losick, R., Harrison, S.: Molecular Biology of the Gene. Cold Spring Harbor Laboratory Press, New York (2014)
29. Watterson, G.A., Ewens, W.J., Hall, T.E., Morgan, A.: The chromosome inversion problem. J. Theor. Biol. **99**(1), 1–7 (1982)
30. Wenger, A.M., Peluso, P., Rowell, W.J., Chang, P.C., Hunkapiller, M.W.: Accurate circular consensus long-read sequencing improves variant detection and assembly of a human genome. Nat. Biotechnol. **37**(11), 1155–1162 (2019)

k-Median/Means with Outliers Revisited: A Simple Fpt Approximation

Xianrun Chen[1], Lu Han[2], Dachuan Xu[3], Yicheng Xu[1(✉)], and Yong Zhang[1]

[1] Chinese Academy of Sciences, Shenzhen Institute of Advanced Technology,
Shenzhen, China
yc.xu@siat.ac.cn
[2] Beijing University of Posts and Telecommunications, Beijing, China
[3] Beijing University of Technology, Beijing, China

Abstract. We revisit the classical metric k-median/means with outliers in this paper, whose proposal dates back to (Charikar, Khuller, Mount, and Narasimhan SODA'01). Though good approximation algorithms have been proposed, referring to the state-of-the-art (6.994+ε)-approximation (Gupta, Moseley and Zhou ICALP'21) for k-median with outliers and (53.002+ε)-approximation (Krishnaswamy, Li, and Sandeep SODA'18) for k-means with outliers respectively, we are interested in finding efficient fpt (fixed-parameter tractable) approximations, following a recent research mainstream for constrained clusterings. As our main contribution, we propose a simple but efficient technical framework that yields a $(3 + \varepsilon)/(9 + \varepsilon)$-approximation for k-median/means with outliers, albeit in $((m + k)/\varepsilon)^{O(k)} \cdot n^{O(1)}$ time. It is notable that our results match with previous result (Goyal, Jaiswal, and Kumar IPEC'20) in terms of ratio and asymptotic running time. But as aforementioned, our technique is much more simplified and straightforward, where instead of considering the whole client set, we restrict ourselves to finding a good approximate facility set for coreset, which can be done easily in fpt time even with provably small loss. Similar idea can be applied to more constrained clustering problems whose coresets have been well-studied.

Keywords: Approximation algorithm · Fixed-parameter tractability · Clustering with outliers · Coreset

1 Introduction

The k-median/means problem is a classical optimization problem in which one must choose k facilities from a given set of candidate locations to serve a set of clients, so as to minimize the total distance cost between the clients and their closest facilities. This problem attracts research interests from various domains, such as computer science, data science, and operations research.

In general metric, both k-median and k-means are NP-hard, more precisely, APX-hard. The k-median is hard to approximate within a factor of $(1 + 2/e)$,

W. Wu and G. Tong (Eds.): COCOON 2023, LNCS 14423, pp. 295–302, 2024.
https://doi.org/10.1007/978-3-031-49193-1_22

and the k-means is hard to approximate within a factor of $(1 + 8/e)$ under P$\neq$ NP [15,23]. Both problems have been extensively studied in the literature from the perspective of approximation algorithms. The state-of-art approximation algorithm for k-median, to the best of our knowledge, is 2.67059 by Cohen-Addad et al. [8]. Kanungo et al. [24] give a $(9+\varepsilon)$-approximation algorithm for k-means, which is improved to 6.357 by Ahmadian et al. [3].

However, the presence of outliers can significantly affect the solution of the k-median/means, where some clients may not be served by any facility (i.e. outliers) due to various reasons such as geographic constraints or capacity limitations. Excluding outliers may greatly reduce the clustering cost and improve the quality of the clustering. Despite practical considerations, identifying outliers in clustering is a crucial and intriguing part of algorithm design, for example, what we study in this paper, k-median with outliers (k-MedO) and k-means with outliers (k-MeaO), which we will formally define later in definition 1.

Both k-MedO and k-MeaO are more difficult than the vanilla version. In a seminal work, Charikar et al. [5] first introduce the problem of k-MedO in the literature (also called robust k-median), and design a bi-criteria $(1 + \lambda, 4+4/\lambda)$-approximation algorithm that always returns a clustering at cost at most $4 + 4/\lambda$ times the optimum with violation of the outlier constraint by a factor of $1 + \lambda$. No constant approximation algorithm has been proposed for k-MedO until Chen [6] presents the first true constant approximation algorithm, whose approximation ratio is significantly improved to $7.081+\varepsilon$ by Krishnaswamy et al. [26] and $6.994+\varepsilon$ by Gupta et al. [17] via iterative LP rounding. For k-MeaO, Gupta et al. [18] give a bi-criteria approximation algorithm that outputs a solution with a ratio of 274 using at most $O(m\log n)$ outliers, where m stands for the desired number of outliers and n corresponds to the total count of clients. Krishnaswamy et al. [26] first propose a true constant approximation algorithm of 53.002, which is quite far from the lower bound.

While there is a lot of work focusing on the approximation algorithms for k-MedO and k-MeaO, there is another research mainstream aiming at developing *fixed-parameter tractable* (fpt) approximations, which enables an additional factor of $f(k)$ in the running time. Fpt algorithms have demonstrated their ability to overcome longstanding barriers in the field of approximation algorithms in recent years [11,25], improving the best-known approximation factors in polynomial time for many classic NP-hard problems, e.g., k-vertex separator [27], k-cut [17] and k-treewidth deletion [16].

Coresets turn out to be useful in fpt algorithm design for clustering recently. Coresets are small representative subsets of the original dataset that capture specific geometric structures of the data, which can help to develop existing algorithms. Agarwal et al. [1] first introduce the framework of coreset in computing diameter, width, and smallest bounding box, ball, and cylinder, initializing a research path of the coreset for many other combinatorial problems. In the classic k-median and k-means, Har-Peled and Mazumdar [19] first prove the existence of small coreset for k-median and k-means with size $O(k \log n\varepsilon^{-d})$ in Euclidean metrics and near-optimal size bounds have been obtained in more recent works by Cohen-Addad et al. [10,22]. In general metric, a seminar paper by Chen [7] obtains coresets for k-median and k-means with size $O(k \log n/\varepsilon)$ based on hierarchical sampling. However, coresets for k-MedO and k-MeaO seem to be less understood — previous results either suffer from exponential dependence on $(m + k)$ [12], or violate the constraint of k or m [20]. Recently, Huang et al. [21] present a near-optimal coreset for k-MedO and k-MeaO in Euclidean spaces with size $O((m + k/\varepsilon))$ based on uniform sampling.

Building on the previous work of coresets, there are several fpt results for the clustering problem. For the classic k-median and k-means, $(1 + 2/e + \varepsilon)$ and $(1 + 8/e + \varepsilon)$ fpt approximations are obtained by Cohen-Addad et al. [9], which are proven to be tight even in $f(k, \varepsilon) \cdot n^{O(1)}$ time, assuming Gap-ETH. For the k-MedO and k-MeaO, existing work [2,28] mostly overcomes the difficulty of identifying outliers by reducing the k-MedO/k-MeaO into a related $(k + m)$-median/means problem, which leads to an exponential time dependency on the outlier number m. Agrawal et al. [2] present $(1 + 2/e + \varepsilon)$ and $(1 + 8/e + \varepsilon)$ fpt approximations for k-MedO and k-MeaO respectively, in $((k + m)/\varepsilon)^{O(m)} \cdot (k/\varepsilon)^{O(k)} \cdot n^{O(1)}$ time. In addition to coresets, k-means++ [4] is also considered as a dataset reduction for k-MeaO in Euclidean spaces in the literature [13, 28]. Though it is not stated explicitly, Statman et al. [28] yields a $(1 + \varepsilon)$-approximation for Euclidean k-MeaO in fpt time.

Our Contribution. In this paper, we propose a coreset-based technical framework for k-MedO and k-MeaO in general metric. We employ the coreset for k-MedO and k-MeaO as a reduction of search space, which would help us to avoid the exponential time dependency on m. We restrict ourselves to finding a good facility set based on the constructed client coreset with size $O((k + m) \log n/\varepsilon)$. We propose a provably good approximate facility set by finding substitute facilities for *leaders* of clients in coreset, where *leaders* represent the clients with minimum cost in each optimal cluster. Moreover, *leaders* can be found easily in fpt(k) time by enumeration of k-sized subset from the coreset. Based on this idea, we derive a $(3 + \varepsilon)$-approximation for k-MedO and a $(9 + \varepsilon)$-approximation for the k-MeaO in $((k + m)/\varepsilon)^{O(k)} \cdot n^{O(1)}$ time. It is worth noting that our result improves upon Akanksha et al. [2] in terms of running time, as the outlier count m is consistently much larger than the facility count k in practical but with a constant loss of approximation ratio. Also note that this matches with Goyal et al. [14] in terms of approximation ratio and asymptotic running time, but with a much simplified and straightforward technical framework, which is promising to apply to more (constrained) clustering problems.

The rest of the paper is organized as follows. Section 2 describes the fpt algorithm as well as its analysis in detail. We conclude this paper and provide some interesting directions in Sect. 3.

2   A Coreset-Based Fpt Approximation

For ease of discussion, we provide formal definitions for k-MedO and k-MeaO.

Definition 1. *(k-MedO/k-MeaO) An instance I of the k-median/means prob-lem with outliers contains a tuple $((X \cup F, d), k, m)$, where $(X \cup F, d)$ is a metric space over a set of n points with a function $d(i, j)$ indicating the distance between two points i, j in $X \cup F$. Furthermore, X and F are two disjoint sets referred to as "clients" and "facilities locations", respectively, while k and m are two positive parameters. The objective is to identify a subset S of k facilities in F and simultaneously exclude a subset O of m clients from X to minimize*

$$cost_m(X, S) = \min_{O \subseteq X: |O| = m} \sum_{x \in X \setminus O} d^z(x, S).$$

Here, $z = 1$ corresponds to k-MedO, and $z = 2$ corresponds to k-MeaO.

This definition implies that, for a fixed set S of k open facilities, the set of m outliers can be naturally determined, namely the set of m points that are farthest from S. This reminds us that we can overcome the difficulty of identifying outliers by focusing on selecting a good approximation for the facilities.

As a pre-processing step, we construct a coreset for k-MedO/k-MeaO to reduce the search space, in order to find a good facility set in fpt(k) time.

Definition 2. *(Coreset) Let $\mathcal{S} = \{S_1, S_2, \dots\}$ be a set of candidate solutions. Recall that*

$$cost_m(X, S) = \min_{O \subseteq X: |O| = m} \sum_{x \in X \setminus O} d^z(x, S).$$

Then a weighted subset $C \subseteq X$ is an ε-coreset, if for all candidate solution $S \in \mathcal{S}$ we have

$$|cost_m(X, S) - cost_m(C, S)| \leq \varepsilon \cdot cost_m(X, S).$$

In other words, coreset is a weighted subset for the client set which preserves a good approximation for the whole client set. In this paper, we make use of the coreset construction technique [2], based on a hierarchical sampling idea [7], which is suitable for clustering problems with outliers.

Theorem 1. *(Agrawal et al. [2]) For $\varepsilon > 0$, there exists an algorithm that for each instance $I = ((X \cup F, d), k, m)$ of k-MedO or k-MeaO, outputs an ε-coreset $C \subseteq X$ with size $|C| = O(((k + m) \log n/\varepsilon)^2)$ with constant probability, running in $O(nk) + poly(n)$ time.*

For any instance $I = ((X \cup F, d), k, m)$ and $\varepsilon > 0$, we run the hierarchical sampling in [2] on I to obtain a coreset $C \subseteq X$ with size $O(((k + m) \log n/\varepsilon)^2)$. By the definition of coreset, an α-approximate solution on $I' = ((C \cup F, d), k, m)$ implies a $(1+\varepsilon)\alpha$-approximate solution on I. Therefore, the only thing we need to do is to find a facility subset $S \subseteq F$ of size k to minimize the objective function $\min_{L \subseteq C: |L| = m} \sum_{x \in C \setminus L} d^z(x, S)$ on instance I'. Towards this end, we present our algorithm in Algorithm 1.

Let $F^* = \{f_1^*, f_2^*, \dots f_k^*\}$ be facility set of optimal solution on instance $I' = ((C \cup F, d), k, m)$. For any $f_i^* \in F^*$, let C_i^* be the clients served by facility f_i^*,

---

Algorithm 1. Coreset-based fpt approximation

Require: Instance $I = ((X \cup F, d), k, m)$
    Construct the coreset C based on hierarchical sampling
    Find *leaders* of C by enumeration
    $S \leftarrow$ the nearest facilities in F for every *leader* client
    return S

---

and denote c_i^* as the *leader* of C_i^* representing the client in C_i^* who is closest to f_i^*. We define *leaders* as all such *leader* clients, thus is a k-sized set.

We will find a good substitute facility for each C_i^* via *leader*, inspired by [9] except that their *leader* is defined on the origin client set in order to deal with capacity constraints. By the definition, we can find *leaders* eventually by enumeration of a k-sized subset, which is allowed in fpt(k) time. We will prove that a good approximate facility set can be found around the *leaders* with a constant approximation guarantee.

Lemma 1. *Algorithm 1 yields a (3+ε)-approximation for k-MedO and a (9+ε)-approximation for k-MeaO.*

Proof. We define f_i for each *leader* client c_i^* as the closest facility in F. As c_i^* is the closest client of f_i^* in C_i^*, it must satisfy that $d(c_i^*, f_i) \leq d(c_i^*, f_i^*) \leq d(c, f_i^*)$ for any client c in C_i^*, which is shown in Fig. 1.

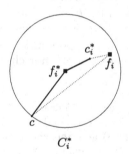

Fig. 1. Substitute facility in optimal cluster C_i^*

Thus, for each c in the coreset,

$$d(c, f_i) \leq d(c, f_i^*) + d(f_i^*, c_i^*) + d(c_i^*, f_i) \leq 3d(c, f_i^*),$$

where the first inequality follows from the triangle inequality. Combined with the definition of coreset (Theorem 1), it holds that

$$cost_m(X, S) \leq \frac{1}{1-\varepsilon} cost_m(C, S) \leq \frac{3^z}{1-\varepsilon} cost_m(C, F^*) \leq \frac{3^z(1+\varepsilon)}{1-\varepsilon} cost_m(X, F^*),$$

which implies a $(3 + \varepsilon)$-approximation/$(9 + \varepsilon)$-approximation for k-MedO/k-MeaO respectively. Though it is possible that some *leader* clients may share the same closest facility f_i, it will not affect the performance guarantee as it does not hurt to let f_i serve all C_i^* corresponding to these *leader* clients. This concludes the proof of Lemma 1.   □

By combining Theorem 1 and Lemma 1 together, we can establish that our algorithm yields a $(3+\varepsilon)\backslash(9+\varepsilon)$-approximation with a constant probability. To ensure a high probability of success, we can repeat the algorithm for a logarithmic number of rounds, which leads to the following theorem.

Theorem 2. *For any $\varepsilon > 0$, there exists a (3+ε)-approximation for k-MedO and a (9+ε)-approximation for k-MeaO with high probability, running in $((m + k)/\varepsilon)^{O(k)} \cdot n^{O(1)}$ time.*

3 Conclusion

To summarize, we propose a simple unified approach to obtain constant factor approximations for metric k-MedO/k-MeaO in fpt(k) time, more specifically, in $((m + k)/\varepsilon)^{O(k)} \cdot n^{O(1)}$ time. It is highlighted that the running time avoids exponential dependency on m, which partially answers (Agrawal et al. AAAI'23) who ask for faster fpt approximations for k-MedO/k-MeaO while obtaining the tight approximation ratios. The proposed approach leverages recent results on coresets for robust clustering, and presents a simple but novel idea to find a good substitute facility set for those *leaders* of coreset. We prove that the substitute facility set can be found easily in fpt(k) time and have provably small loss compared with the optimal facility set in terms of the k-MedO/k-MeaO objective for coreset. We believe similar idea has the potential to apply to a wide range of constrained clustering problems, for example, fair clustering, a recent mainstream in clustering field.

Acknowledgement. Xianrun Chen and Yicheng Xu are supported by Natural Science Foundation of China (No. 12371321), Fundamental Research Project of Shenzhen City (No. JCYJ20210324102012033) and Shenzhen Science and Technology Program (No. CJGJZD20210408092806017). Dachuan Xu is supported by Natural Science Foundation of China (No. 12371320). Yong Zhang is supported by National Key R&D Program of China (No. 2022YFE0196100) and Natural Science Foundation of China (No. 12071460).

References

1. Agarwal, P.K., Har-Peled, S., Varadarajan, K.R.: Approximating extent measures of points. J. ACM **51**(4), 606–635 (2004)
2. Agrawal, A., Inamdar, T., Saurabh, S., Xue, J.: Clustering what matters: optimal approximation for clustering with outliers. CoRR abs/ arXiv: 2212.00696 (2022)
3. Ahmadian, S., Norouzi-Fard, A., Svensson, O., Ward, J.: Better guarantees for k-means and euclidean k-median by primal-dual algorithms. SIAM J. Comput. **49**(4) (2020)

4. Arthur, D., Vassilvitskii, S.: k-means++: the advantages of careful seeding. In: SODA, pp. 1027–1035. SIAM (2007)
5. Charikar, M., Khuller, S., Mount, D.M., Narasimhan, G.: Algorithms for facility location problems with outliers. In: SODA, pp. 642–651. ACM/SIAM (2001)
6. Chen, K.: A constant factor approximation algorithm for k-median clustering with outliers. In: SODA, pp. 826–835. SIAM (2008)
7. Chen, K.: On coresets for k-median and k-means clustering in metric and euclidean spaces and their applications. SIAM J. Comput. **39**(3), 923–947 (2009)
8. Cohen-Addad, V., Grandoni, F., Lee, E., Schwiegelshohn, C.: Breaching the 2 LMP approximation barrier for facility location with applications to k-median. In: SODA, pp. 940–986. SIAM (2023)
9. Cohen-Addad, V., Gupta, A., Kumar, A., Lee, E., Li, J.: Tight FPT approximations for k-median and k-means. In: ICALP. LIPIcs, vol. 132, pp. 42:1–42:14. Schloss Dagstuhl - Leibniz-Zentrum für Informatik (2019)
10. Cohen-Addad, V., Larsen, K.G., Saulpic, D., Schwiegelshohn, C.: Towards optimal lower bounds for k-median and k-means coresets. In: STOC, pp. 1038–1051. ACM (2022)
11. Cohen-Addad, V., Li, J.: On the fixed-parameter tractability of capacitated clustering. In: ICALP. LIPIcs, vol. 132, pp. 41:1–41:14. Schloss Dagstuhl - Leibniz-Zentrum für Informatik (2019)
12. Feldman, D., Schulman, L.J.: Data reduction for weighted and outlier-resistant clustering. In: SODA, pp. 1343–1354. SIAM (2012)
13. Feng, Q., Zhang, Z., Huang, Z., Xu, J., Wang, J.: Improved algorithms for clustering with outliers. In: ISAAC. LIPIcs, vol. 149, pp. 61:1–61:12. Schloss Dagstuhl - Leibniz-Zentrum für Informatik (2019)
14. Goyal, D., Jaiswal, R., Kumar, A.: FPT approximation for constrained metric k-median/means. In: IPEC. LIPIcs, vol. 180, pp. 14:1–14:19. Schloss Dagstuhl - Leibniz-Zentrum für Informatik (2020)
15. Guha, S., Khuller, S.: Greedy strikes back: improved facility location algorithms. J. Algorithms **31**(1), 228–248 (1999)
16. Gupta, A., Lee, E., Li, J., Manurangsi, P., Wlodarczyk, M.: Losing treewidth by separating subsets. In: SODA, pp. 1731–1749. SIAM (2019)
17. Gupta, A., Moseley, B., Zhou, R.: Structural iterative rounding for generalized k-median problems. In: ICALP. LIPIcs, vol. 198, pp. 77:1–77:18. Schloss Dagstuhl - Leibniz-Zentrum für Informatik (2021)
18. Gupta, S., Kumar, R., Lu, K., Moseley, B., Vassilvitskii, S.: Local search methods for k-means with outliers. Proc. VLDB Endow. **10**(7), 757–768 (2017)
19. Har-Peled, S., Mazumdar, S.: On coresets for k-means and k-median clustering. In: STOC, pp. 291–300. ACM (2004)
20. Huang, L., Jiang, S.H., Li, J., Wu, X.: Epsilon-coresets for clustering (with outliers) in doubling metrics. In: FOCS, pp. 814–825. IEEE Computer Society (2018)
21. Huang, L., Jiang, S.H., Lou, J., Wu, X.: Near-optimal coresets for robust clustering. coRR abs/ arXiv: 2210.10394 (2022)
22. Huang, L., Vishnoi, N.K.: Coresets for clustering in euclidean spaces: importance sampling is nearly optimal. In: STOC, pp. 1416–1429. ACM (2020)
23. Jain, K., Mahdian, M., Saberi, A.: A new greedy approach for facility location problems. In: STOC, pp. 731–740. ACM (2002)
24. Kanungo, T., Mount, D.M., Netanyahu, N.S., Piatko, C.D., Silverman, R., Wu, A.Y.: A local search approximation algorithm for k-means clustering. Comput. Geom. **28**(2–3), 89–112 (2004)

25. Karthik C.S., Laekhanukit, B., Manurangsi, P.: On the parameterized complexity of approximating dominating set. J. ACM **66**(5), 33:1–33:38 (2019)
26. Krishnaswamy, R., Li, S., Sandeep, S.: Constant approximation for k-median and k-means with outliers via iterative rounding. In: STOC, pp. 646–659. ACM (2018)
27. Lee, E.: Partitioning a graph into small pieces with applications to path transversal. Math. Program. **177**(1–2), 1–19 (2019)
28. Statman, A., Rozenberg, L., Feldman, D.: k-means: Outliers-resistant clustering+++. Algorithms **13**(12), 311 (2020)

A Cost-Sharing Scheme for the k-Product Facility Location Game with Penalties

Xiaowei Li and Xiwen Lu$^{(\boxtimes)}$

School of mathematics, East China University of Science and Technology,
Shanghai 200237, China
xwlu@ecust.edu.cn

Abstract. In the k-product facility location game with penalties, each customer must be supplied with k different products or be rejected by paying the penalty cost. The game is considered in this paper. The cost-sharing scheme for this game refers to allocating the costs of the problem to all customers. We present a cross-monotonic and competitive scheme which satisfies 2 and $(\frac{3k}{2} - \frac{3}{2})$-approximate cost recovery when $k = 2$ and $k \geq 3$, respectively. Moreover, the lower bound of the cost-sharing scheme is obtained for the k-product facility location game with penalties.

Keywords: Facility location game · Cost-sharing scheme · Competitive · Cross-monotonic · Approximate cost recovery

1 Introduction

In the classic facility location problem (FLP), the inputs are a set of facilities F, a set of customers U, facility opening cost $f_i \geq 0$ for each facility $i \in F$, and service cost $c_{ij} \geq 0$ between each facility $i \in F$ and each customer $j \in U$. Each customer must be serviced by an opened facility. The objective is to open some facilities and service all customers such that the sum of the opening costs and service costs are minimized. The algorithm design of the facility location problem [3,8,9,12,16] and some variants of the facility location problem [1,2,4] has been extensively studied. The corresponding game of the facility location problem is called the facility location game. In this game, each customer is a player. For any given subset of customers $D \subseteq U$, let $c^*(D)$ denote the minimum costs required to serve all customers in D. The objective is to design a cost-sharing scheme that allocates the total costs to all customers. A cost-sharing scheme is an algorithm to compute the cost share $\alpha(D, j)$ of each customer $j \in D$. The cost-sharing scheme should be $fair$, $group\ strategy-proof$, $competitive$, $cross-monotonic$, and $exact\ or\ approximate\ cost\ recovery$ [15]. Pál and Tardos [15] showed that competitiveness and cross-monotonicity imply fairness. Moulin and Shenker [14] stated that cross-monotonicity implies group strategyproofness.

This research was supported by the National Natural Science Foundation of China under Grant 11871213.

Thus, researchers focus on developing a cost-sharing scheme that satisfies cross-monotonicity, competitiveness, and exact or approximate cost recovery.

A cost-sharing scheme is $cross-monotonic$ if the price charged to any individual in a group never goes up as the group expands, i.e., $\alpha(D,j) \leq \alpha(D',j)$ for all $D \subseteq D'$. Hence, customers have an economic incentive to cooperate. A cost-sharing scheme is $competitive$ if the customers are charged no more than the total costs, i.e., $\sum_{j \in D} \alpha(D,j) \leq c^*(D)$. A cost-sharing scheme satisfies $r-approximate\ cost\ recovery$ if the customers are required to recover $1/r$ of the total costs, i.e., $\sum_{j \in D} \alpha(D,j) \geq c^*(D)/r$, where $r \geq 1$. If $r = 1$, it is the $exact\ cost\ recovery$. However, Immorlica et al. [11] proved that no cross-monotonic and competitive cost-sharing scheme could recover more than one-third of the costs for the facility location game. Pál and Tardos [15] studied the facility location game and obtained a cost-sharing scheme that is cross-monotonic, competitive and 3-approximate cost recovery. Xu and Du [19] extended it to the k-level facility location game and obtained a 3-approximation cost recovery scheme. Wang and Xu [18] studied the facility location game with linear penalties. They presented a 3-approximate cost recovery scheme by adopting the ghost-process [15]; this process can be viewed as a smoothed version of a primal-dual algorithm [12]. Furthermore, Wang et al. [17] discussed the k-level facility location game with linear penalties. By applying the ghost-process outlined in [18] and [19], they obtained a 6-approximate cost recovery scheme.

In addition, researchers have also studied other forms of facility location games. Cheng et al. [7] discussed the mechanism design for an obnoxious facility location game where the customers want to be as far away from facilities as possible. Ye et al. [20] researched the problem where the objective is to minimize the sum of the squares of distances. Chen et al. [6] studied the optional preference model of the facility location game with two heterogeneous facilities on a line. Li et al. [13] studied the budgeted facility location games with strategic facilities. Chen et al. [5] proposed the facility location game with minimax envy.

The k-product facility location problem is an essential variant of the classic facility location problem. In the k-product facility location problem, there are k different kinds of products $m_1, m_2, ..., m_k$. Each facility can only produce one kind of product at most. The cost of opening facility i to produce product m_l is f_i^l. The customer is serviced if it is supplied with k kinds of products by a set of k different opened facilities. The goal is to open some facilities and service all customers while minimizing total costs, including facility opening costs and customer service costs. Huang and Li [10] obtained an approximation algorithm with an approximate ratio is $2k+1$. In particular, they considered the k-product facility location problem with no opening costs (k-PFLPN), in which the opening cost is zero for any facility. They proved that the 2-PFLPN is NP-hard. Then, they devised a $(2k-1)$-approximation algorithm for the k-PFLPN.

The k-product facility location problem with linear penalties is a variant of the k-product facility location problem. The study of this problem is driven by practical problems. On the one hand, the customers may have diverse product demands in actual situations, so it is necessary to consider the multi-product facility location problem. On the other hand, for many facility location applications, it may be economically essential to ignore distant customers by paying the penalty costs. The corresponding game of k-product facility location problem with linear penalties is the k-product facility location game with linear penalties. For this game, the objective is to design a cost-sharing scheme that allocates the total costs to all customers. The service costs between customers and facilities in this game have become more complex compared with the facility location game. This challenges algorithm design.

The main contribution of this paper is to develop a cost-sharing scheme for the k-product facility location game with penalties and no opening costs (k-PFLGPN), which can be applied in situations where the government bears the opening costs of some public facilities and customers only need to share other costs. Based on the linear-rounding technique and the greedy algorithm, we get a cross-monotonic and competitive cost-sharing scheme with 2-approximate cost recovery when $k = 2$. More generally, the scheme is $\frac{3k}{2} - \frac{3}{2}$-approximate cost recovery when $k \geq 3$. Moreover, we analyze the lower bound of the k-PFLGPN.

The rest of the paper is organized as follows. After describing the k-PFLPN and a preliminary algorithm in Sect. 2, we proceed with algorithm design and performance analysis for the k-PFLGPN in Sect. 3. Section 4 is devoted to conclusions and future works.

2   Preliminary Algorithm

Now, we present a model of the k-product facility location problem with no opening costs(k-PFLPN) and a preliminary algorithm. The algorithm will be used to design a cost-sharing scheme for the k-product facility location game with penalties in Sect. 3.1.

Given a set of facilities F and potential customers U. There are k kinds of products m_l, $l = 1, 2, ..., k$. Each facility can produce exactly one product. The service cost between any facility i and customer j is c_{ij}, which satisfies non-negativity, symmetry, and triangle inequality, where $i \in F, j \in U$. Given a subset $D \subseteq U$ of customers, each customer $j \in D$ demands k kinds of products. In the k-PFLPN, we need to open all facilities, specify the products they produce, and assign all customers to the opened facilities to meet their demands. Note that each customer must be served by a series of k facilities producing different products, which means each customer is assigned to a series of k facilities. The goal of the problem is to minimize the total service costs.

Let x_{ij}^l be equal to 1 if facility i supplies customer j with product m_l, for any $i \in F, j \in D$ and $l \in \{1, 2, \dots, k\}$. Otherwise, it is 0. Let y_i^l be 1 if facility i is opened and produce product m_l, and 0 otherwise. Thus, the k-PFLPN can be formulated as the following integer programming.

$$\text{P}_1 \quad \min \sum_{l=1}^{k} \sum_{i \in F} \sum_{j \in D} c_{ij} x_{ij}^l$$

$$\text{s.t.} \sum_{i \in F} x_{ij}^l = 1, \forall j \in D, l \in \{1, 2, \cdots, k\},$$

$$y_i^l \geq x_{ij}^l, \forall i \in F, j \in D, l \in \{1, 2, \cdots, k\}, \tag{1}$$

$$\sum_{l=1}^{k} y_i^l = 1, \forall i \in F,$$

$$x_{ij}^l \in \{0, 1\}, \forall i \in F, j \in D, l \in \{1, 2, \cdots, k\},$$

$$y_i^l \in \{0, 1\}, \forall i \in F, l \in \{1, 2, \cdots, k\}.$$

The first constraint guarantees that each customer j is supplied with k different products. The second constraint ensures that customer j is provided with product m_l by facility i only if facility i produces product m_l. The third constraint ensures that each facility produces exactly one product.

The LP-relaxation of the above programming is

$$\text{P}_2 \quad \min \sum_{l=1}^{k} \sum_{i \in F} \sum_{j \in D} c_{ij} x_{ij}^l$$

$$\text{s.t.} \sum_{i \in F} x_{ij}^l = 1, \forall j \in D, l \in \{1, 2, \cdots, k\},$$

$$y_i^l \geq x_{ij}^l, \forall i \in F, j \in D, l \in \{1, 2, \cdots, k\}, \tag{2}$$

$$\sum_{l=1}^{k} y_i^l = 1, \forall i \in F,$$

$$x_{ij}^l \geq 0, \forall i \in F, j \in D, l \in \{1, 2, \cdots, k\},$$

$$y_i^l \geq 0, \forall i \in F, l \in \{1, 2, \cdots, k\}.$$

The P_2 is analyzed below, and the results will help design the cost-sharing scheme for the k-product facility location game with linear penalties.

For the convenience of describing the algorithm, we regard the service cost between facility i and customer j as their distance for each $i \in F, j \in D$.

The following algorithm can obtain an optimal solution of P_2.

Algorithm 1

Step 1. Set $\bar{y}_i^l = \frac{1}{k}$, for any $l \in \{1, 2, \ldots, k\}, i \in F$.

Step 2. Any customer $j \in D$ is fractionally supplied by the k closest facilities to customer j. That is, if $i_1, i_2, \ldots, i_k$ are the k closest facilities to customer j, then we set

$$\bar{x}_{i_1 j}^l = \bar{x}_{i_2 j}^l = \cdots = \bar{x}_{i_k j}^l = \frac{1}{k}, \quad \forall l \in \{1, 2, \ldots, k\}.$$

$$\bar{x}_{ij}^l = 0, \quad \forall i \in F \backslash \{i_1, i_2, \ldots, i_k\}, l \in \{1, 2, \ldots, k\}.$$

For each customer $j \in D$, let $\bar{c}_j = \sum_{l=1}^{k} \sum_{i \in F} c_{ij} \bar{x}_{ij}^l$, where $(\bar{x}, \bar{y})$ is the solution obtained by Algorithm 1. The performance of Algorithm 1 is analyzed below.

Lemma 1. *For any customer $j \in D$ and any feasible solution (x,y) of P_2, $c_j \geq \bar{c}_j$, where $c_j = \sum_{l=1}^{k} \sum_{i \in F} c_{ij} x_{ij}^l$.*

Proof. For any customer $j \in D$, denote $F^{(j)}$ as the set of the k closest facilities to customer j. Without loss of generality, suppose that $F^{(j)} = \{i_1, i_2, \cdots, i_k\}$. According to Algorithm 1, we have $\bar{c}_j = \sum_{l=1}^{k} \sum_{i \in F} c_{ij} \bar{x}_{ij}^l = \sum_{i \in F^{(j)}} c_{ij}$. Since $|F^{(j)}| = k$, we can get that

$$\sum_{l=1}^{k} (1 - \sum_{i \in F^{(j)}} x_{ij}^l) = \sum_{i \in F^{(j)}} (1 - \sum_{l=1}^{k} x_{ij}^l).$$

Let $S^{(j)} = \{i \in F \mid \sum_{l=1}^{k} x_{ij}^l > 0\}$. We can know that, if $i \notin S^{(j)}, x_{ij}^l = 0$ for any $l \in \{1, 2, \cdots, k\}$. Combining with the first constraint of P_2, we have

$$\sum_{i \in S^{(j)}} x_{ij}^l = \sum_{i \in F} x_{ij}^l = \sum_{i \in F^{(j)}} x_{ij}^l + \sum_{i \in F \backslash F^{(j)}} x_{ij}^l = \sum_{i \in F^{(j)}} x_{ij}^l + \sum_{i \in S^{(j)} \backslash F^{(j)}} x_{ij}^l = 1$$

for any $l \in \{1, 2, \cdots, k\}$.

Let $c_{\max} = \max_{i \in F^{(j)}} c_{ij}$. According to the definition of $F^{(j)}$, it can be seen that $c_{ij} \geq c_{\max}$ for any facility $i \in S^{(j)} \backslash F^{(j)}$. This implies that

$$
\begin{aligned}
\sum_{l=1}^{k} \sum_{i \in S^{(j)}} c_{ij} x_{ij}^l &= \sum_{l=1}^{k} \sum_{i \in F^{(j)}} c_{ij} x_{ij}^l + \sum_{l=1}^{k} \sum_{i \in S^{(j)} \backslash F^{(j)}} c_{ij} x_{ij}^l \\
&\geq \sum_{l=1}^{k} \sum_{i \in F^{(j)}} c_{ij} x_{ij}^l + c_{\max} \sum_{l=1}^{k} \sum_{i \in S^{(j)} \backslash F^{(j)}} x_{ij}^l \\
&= \sum_{l=1}^{k} \sum_{i \in F^{(j)}} c_{ij} x_{ij}^l + c_{\max} \sum_{l=1}^{k} \left(1 - \sum_{i \in F^{(j)}} x_{ij}^l\right) \\
&= \sum_{l=1}^{k} \sum_{i \in F^{(j)}} c_{ij} x_{ij}^l + c_{\max} \sum_{i \in F^{(j)}} \left(1 - \sum_{l=1}^{k} x_{ij}^l\right) \\
&\geq \sum_{l=1}^{k} \sum_{i \in F^{(j)}} c_{ij} x_{ij}^l + \sum_{i \in F^{(j)}} c_{ij} \left(1 - \sum_{l=1}^{k} x_{ij}^l\right) \\
&= \bar{c}_j.
\end{aligned}
$$

Hence, we get that $c_j = \sum_{l=1}^{k} \sum_{i \in F} c_{ij} x_{ij}^l \geq \bar{c}_j$. The lemma is proved.   □

Lemma 2. *An optimal solution for P_2 can be obtained by Algorithm 1.*

Proof. We can find that the solution $(\bar{x}, \bar{y})$ obtained by Algorithm 1 satisfies all the constraints of P_2, so it is a feasible solution for P_2. By Lemma 1, for any feasible solution (x, y) of P_2, we have $\sum_{l=1}^{k} \sum_{i \in F} \sum_{j \in D} c_{ij} x_{ij}^l = \sum_{j \in D} c_j \geq \sum_{j \in D} \bar{c}_j$.

Therefore, the lemma is proved.   □

3   The Cost-Sharing Scheme for the k-PFLGPN

Now, we first consider the k-product facility location problem with penalties, in which the customer j can be rejected by paying the penalty cost p_j for any customer $j \in D$. The objective is to select some customers to reject their demands by paying penalty costs and assign the remaining customers to opened facilities to minimize the total costs, including service and penalty costs.

For any $i \in F, j \in D, l \in \{1, 2, ..., k\}$, let x_{ij}^l be 1 if facility i supplies customer j with product m_l, and 0 otherwise. Let y_i^l be 1 if facility i is opened to supply product m_l, and 0 otherwise. Let z_j be 1 if customer j is rejected, and 0 otherwise. The problem can be formulated as the following integer programming.

$$
\begin{aligned}
\mathrm{P}_3 \quad \min \quad & \sum_{l=1}^{k} \sum_{i \in F} \sum_{j \in D} c_{ij} x_{ij}^l + \sum_{j \in D} p_j z_j \\
\text{s.t.} \quad & \sum_{i \in F} x_{ij}^l + z_j = 1, \forall j \in D, l \in \{1, 2, \cdots, k\}, \\
& y_i^l \geq x_{ij}^l, \forall i \in F, j \in D, l \in \{1, 2, \cdots, k\}, \\
& \sum_{l=1}^{k} y_i^l = 1, \forall i \in F, \\
& x_{ij}^l \in \{0, 1\}, \forall i \in F, j \in D, l \in \{1, 2, \cdots, k\}, \\
& y_i^l \in \{0, 1\}, \forall i \in F, l \in \{1, 2, \cdots, k\}, \\
& z_j \in \{0, 1\}, \forall j \in D.
\end{aligned}
\tag{3}
$$

The first constraint guarantees that each customer is either served or rejected. If customer j is served, it is supplied with k different products. The second constraint ensures that customer j is provided with product m_l by facility i only if facility i produces product m_l. The third constraint ensures that each facility produces exactly one product.

The game corresponding to the k-product facility location problem with penalties is k-PFLGPN. In this game, each customer is a player. The objective is to construct a competitive and cross-monotonic cost-sharing scheme such that each customer pays a certain amount to recover the minimum costs of the

P_3. We can know that P_1 corresponds to P_3 when the penalty cost of each customer in P_3 is large enough. Huang and Li [10] proved that P_1 is an NP-hard problem, which means that P_3 is also an NP-hard problem. So, it is impossible to obtain the optimal solution of P_3 in polynomial time unless $P = NP$. Hence, this paper focuses on designing an approximation algorithm for the k-PFLGPN. The following algorithm will give a cost-sharing scheme that is cross-monotonic, competitive, and approximate cost recovery for the k-PFLGPN.

3.1   Algorithm for the k-PFLGPN

In the last section, we have known that the optimal solution of P_2 can be obtained by Algorithm 1. Then we use the method of greed to convert the optimal solution of P_2 into a feasible solution of P_3, and finally determine the cost shared by each customer. For the convenience of describing the algorithm, we regard the service cost between facility i and customer j as its distance for any $i \in F, j \in D$. Meanwhile, $\alpha(D, j)$ is abbreviated as α_j in this section.

Algorithm 2

Step 1. Let $\bar{D}$ denote the set of customers that have not been processed, $\widetilde{D}$ denote the set of customers to be rejected, and S_l denote the set of facilities to produce product m_l. For the convenience of algorithm analysis, subset $D^{'}$ is introduced. Initially, set $\bar{D} = D, \widetilde{D} = \emptyset, D^{'} = \emptyset, S_1 = S_2 = \cdots = S_k = \emptyset$.

Step 2. Use Algorithm 1 to solve P_2, and denote the obtained solution as $(\bar{x}, \bar{y})$. For each customer $j \in D$, let $\bar{c}_j = \sum\limits_{l=1}^{k} \sum\limits_{i \in F} c_{ij} \bar{x}_{ij}$, and denote $F^{(j)}$ as the set of the k closest facilities to customer j.

Step 3. Choose the customer in $\bar{D}$ with the minimum value of $\bar{c}_j$, i.e., choose $j^{'} = \arg\min\limits_{j \in \bar{D}} \{\bar{c}_j\}$.

Step 3.1. If $p_{j^{'}} < \bar{c}_{j^{'}}$, reject customer $j^{'}$ and set $z_{j^{'}} = 1$. Set $\bar{D} = \bar{D}\backslash\{j^{'}\}, \widetilde{D} = \widetilde{D} \cup \{j^{'}\}$.

Step 3.2. If $p_{j^{'}} \geq \bar{c}_{j^{'}}$, discuss the following two cases respectively.

- Case 1. If $\left| S_l \cap F^{(j^{'})} \right| \leq 1$ for each $l = 1, 2, ..., k$, put the facilities in $F^{(j^{'})}$ into S_l such that the updated S_l satisfies $\left| S_l \cap F^{(j^{'})} \right| = 1$ for each $l = 1, 2, ..., k$. Set $\bar{D} = \bar{D}\backslash\{j^{'}\}$.
- Case 2. If $\left| S_l \cap F^{(j^{'})} \right| \geq 2$ for some $l = 1, 2, ..., k$, set $\bar{D} = \bar{D}\backslash\{j^{'}\}, D^{'} = D^{'} \cup \{j^{'}\}$.

Then, repeat Step 3, until $\bar{D} = \emptyset$.

Step 4. For each facility $i \in F \backslash (\bigcup_{l=1}^{k} S_l)$, arbitrarily put it into any set of S_l, where $l \in \{1, 2, \cdots, k\}$. Use the facilities in S_l to produce products m_l, i.e., set $y_i^l = 1$ for any $i \in S_l$, $l \in \{1, 2, \cdots, k\}$. Otherwise, it is 0.

For each customer $j \in D \backslash \widetilde{D}$, select its closest facility in S_l to supply product m_l ($l \in \{1, 2, \cdots, k\}$), i.e., set $x_{ij}^l = 1$ if $i = i_l(j)$, where $i_l(j)$ is the closest facility in S_l to customer j. Otherwise, it is 0.

Step 5. Determine the costs allocated to all customers. For each customer $j \in D$, set $\alpha_j = \min(\bar{c}_j, p_j)$, i.e., $\alpha_j = p_j$, if $j \in \widetilde{D}$. Otherwise, it is $\bar{c}_j$.

3.2   Analysis of Algorithm 2

From now on, we prove that the cost-sharing scheme obtained in Sect. 3.1 is competitive, cross-monotonic, and 2-approximate cost recovery when $k = 2$. Moreover, we demonstrate that no competitive and cross-monotonic cost-sharing scheme can recover more than two-thirds of the costs.

Let the service and penalty costs be C^* and P^*, C and P in the optimal solution and the solution obtained by Algorithm 2, respectively.

Let the set of customers to be served and the set of customers to be rejected be D_1^* and D_2^*, D_1 and D_2 in the optimal solution and the solution obtained by Algorithm 2, respectively.

By Algorithm 2, we can find that $D_1 = D \backslash \widetilde{D}$, $D_2 = \widetilde{D}$. Let c_j^* denote the service cost of customer j in the optimal solution, where $j \in D_1^*$. By Lemma 1,

$$\bar{c}_j = \sum_{l=1}^{k} \sum_{i \in F} c_{ij} \bar{x}_{ij}^l = \sum_{i \in F(j)} c_{ij} \leq c_j^* \text{ for each customer } j \in D_1^*. \text{ We can get the}$$

following facts.

Fact 1. For each customer $j \in D_2$, we have $j \in D_2^*$.

Proof. By Algorithm 2, we can know that $\bar{c}_j > p_j$ for each customer $j \in D_2$. If there is a customer $j \in D_2$, and $j \notin D_2^*$, we can get that $j \in D_1^*$, which implies that $c_j^* \geq \bar{c}_j > p_j$. Therefore, we can obtain a better solution by transferring customer j from set D_1^* to set D_2^*. The fact is proved.   □

Since $D_1 \cup D_2 = D_1^* \cup D_2^* = D$, combining with Fact 1, we have $D_1^* \subseteq D_1$, $D_1 \backslash D_1^* = D_2^* \backslash D_2$. Now, we show that Algorithm 2 satisfies the following fact.

Fact 2. The customers are not charged more than $OPT(P_3)$, which is the optimal value of P_3. Specifically, $\sum_{j \in D} \alpha_j \leq OPT(P_3)$.

Proof. By Algorithm 2, we can get that $p_j \geq \bar{c}_j$ for each customer $j \in D_1$. Therefore, we have

$$\sum_{j \in D} \alpha_j = \sum_{j \in D_1} \bar{c}_j + \sum_{j \in D_2} p_j$$

$$= \sum_{j \in D_1^*} \bar{c}_j + \sum_{j \in D_1 \backslash D_1^*} \bar{c}_j + \sum_{j \in D_2} p_j$$

$$\leq \sum_{j \in D_1^*} c_j^* + \sum_{j \in D_1 \backslash D_1^*} p_j + \sum_{j \in D_2} p_j \qquad (4)$$

$$= \sum_{j \in D_1^*} c_j^* + \sum_{j \in D_2^*} p_j$$

$$= OPT(P_3).$$

The fact is proved.   □

We can get that Fact 2 implies competitiveness.

Furthermore, we can know that the cost $\alpha(D, j)$ shared by customer j will not go up as the set D expands, since $\alpha_j = \min(\bar{c}_j, p_j)$. Thus, we have the following fact.

Fact 3. The cost-sharing scheme obtained in Sect. 3.1 is cross-monotonic.

We analyze the approximation ratio of the cost-sharing scheme below. For each customer $j \in D_1$, let $\hat{c}_j$ denote the service cost of customer j obtained by Algorithm 2.

Lemma 3. *For any customer $j \in D_1$, we can get that $\hat{c}_j \leq 2\bar{c}_j$ when $k = 2$.*

Proof. According to Step 3 of Algorithm 2, we need to discuss two kinds of the customers in D_1.

For each customer $j \in D_1 \backslash D'$, $\hat{c}_j = \sum_{l=1}^{k} c_{i_l(j)j} = \bar{c}_j$.

For each customer $j \in D'$, we have $\bar{c}_j \geq p_j$, and $|S_l \cap F^{(j)}| > 1$ for $l = 1$ or $l = 2$. Without loss of generality, we assume that $|S_1 \cap F^{(j)}| = 2$, and $F^{(j)} = \{i_1, i_2\}$, $c_{i_1 j} \leq c_{i_2 j}$. By Case 2 in Step 3.2, we can know that there is a customer j' which satisfies the following three properties: (a) $i_1 \in F^{(j')}$. (b)$\bar{c}_{j'} \leq \bar{c}_j$. (c)$j' \in D_1 \backslash D'$. Without loss of generality, we assume that $F^{(j')} = \{i_1, i_3\}$. By Step 3.2 of Algorithm 2, we can know that $i_3 \in S_2$. Denote $i_l(j)$ as the closest facility in S_l to customer j. From the triangle inequality, we can get that

$$\hat{c}_j = \sum_{l=1}^{2} c_{i_l(j)j} \leq c_{i_1 j} + c_{i_3 j}$$

$$\leq c_{i_1 j} + c_{i_1 j} + c_{i_1 j'} + c_{i_3 j'}$$

$$\leq c_{i_1 j} + c_{i_2 j} + c_{i_1 j'} + c_{i_3 j'}$$

$$\leq \bar{c}_j + \bar{c}_{j'} \leq 2\bar{c}_j$$

The lemma is proved.   □

By Lemma 3, we can get the following theorem.

Theorem 1. *Algorithm 2 develops a 2-approximate cost recovery cost-sharing scheme when $k = 2$. More specifically, the cost of the solution obtained by Algorithm 2 is at most $2 \sum_{j \in D} \alpha_j$.*

Proof. Note that the solution obtained by Algorithm 2 is a feasible solution for P_3, which means that

$$OPT(P_3) \leq \sum_{j \in D_1} \hat{c}_j + \sum_{j \in D_2} p_j \leq \sum_{j \in D_1} 2\bar{c}_j + \sum_{j \in D_2} p_j \leq 2 \sum_{j \in D} \alpha_j$$

.

The theorem is proved.   □

We conclude that Algorithm 2 is a 2-approximate, cross-monotonic, and competitive cost-sharing scheme for 2-PFLGPN. In other words, the scheme (or algorithm) obtained in Sect. 3.1 can recover at least half of the costs when $k = 2$. Next, we analyze the lower bound of the 2-PFLGPN.

Theorem 2. *There doesn't exist a competitive and cross-monotonic cost-sharing scheme that can recover more than two-thirds of the costs when $k = 2$.*

Proof. We construct a complete graph $G = (V, E)$, where $V = \{v_1, v_2, ..., v_{2n}\}$, $E = \{e_1, e_2, ..., e_{(2n-1)n}\}$. Then, we consider the following instance I of P_3.

Let $F = \{i_1, i_2, ..., i_{2n}\}$, $D = \{j_1, j_2, ..., j_{(2n-1)n}\}$. For each customer $j \in D$, let $p_j = 6$. The service cost between $i \in F$ and $j \in D$ is as follows.

$$c_{ij} = \begin{cases} 1, & \text{vertex } i \text{ is an endpoint of edge } e_j, \\ 3, & \text{otherwise.} \end{cases}$$

For any feasible solution of the instance I, let the set of facilities producing product m_1 and those producing product m_2 be S_1 and S_2, respectively. Let $|S_1| = a$, then $|S_2| = 2n - a$. We can find that there are $a(2n - a)$ customers with a service cost of 2 and other customers with a service cost of 4. So the total costs are $2a(2n - a) + 4((2n - 1)n - a(2n - a))$ if no customer is penalized. If some customers are penalized, the total costs will be more since $p_j = 6$ for each customer $j \in D$.

Obviously, the optimal value of instance I is $6n^2 - 4n$ if and only if $a = n$.

We can know that $\bar{c}_j = 2$ for each customer $j \in D$. For any competitive and cross-monotonic cost-sharing scheme, we can get that $\alpha(D, j) \leq \alpha(\{j\}, j) \leq c^*(\{j\}) = \min(\bar{c}_j, p_j) = 2$ for any customer $j \in D$. Thus the recovery costs is no more than $\sum_{j \in D} \alpha(D, j) \leq 2 * (2n - 1) * n = 4n^2 - 2n$.

The ratio of $\frac{4n^2-2n}{6n^2-4n}$ tends to $\frac{2}{3}$ when n tends to infinity. Thus, we can know that there doesn't exist a competitive and cross-monotonic cost-sharing scheme that can recover more than two-thirds of the costs.   □

For the general cases, we analyze the approximate ratio of Algorithm 2 when $k \geq 3$.

Theorem 3. *Algorithm 2 develops a $\frac{3k}{2} - \frac{3}{2}$-approximate cost-sharing scheme when $k \geq 3$. More specifically, the cost of the solution obtained by Algorithm 2 is at most $(\frac{3k}{2} - \frac{3}{2}) \sum\limits_{j \in D} \alpha_j$.*

The proof of Theorem 3 is similar to Theorem 1, we only need to show that $\widehat{c}_j \leq (\frac{3k}{2} - \frac{3}{2})\bar{c}_j$ for each customer $j \in D_1$. At the same time, the key of the analysis is to discuss the closest facility to customer j among all facilities for each customer $j \in D_1$.

4 Conclusions

The k-product facility location game with penalties and no opening costs is proposed in this paper. We design a cross-monotonic and competitive cost-sharing scheme which is 2-approximate cost recovery and $(\frac{3k}{2} - \frac{3}{2})$-approximate cost recovery when $k = 2$ and $k \geq 3$, respectively. Moreover, it is proved that no competitive and cross-monotonic cost-sharing scheme can recover more than $\frac{2}{3}$ of the costs when $k = 2$. One direction of future research is to analyze the low bound for $k > 2$. Additionally, the next stage of work may be to design a cost-sharing scheme for the k-product facility location game with non-zero opening costs.

References

1. Aardal, K., Chudak, F.A., Shmoys, D.B.: A 3-approximation algorithm for the k-level uncapacitated facility location problem. Inf. Process. Lett. **72**, 161–167 (1999)
2. An, H.C., Singh, M., Svensson, O.: LP-based algorithms for capacitated facility location. SIAM J. Comput. **46**, 272–306 (2017)
3. Arya, V., Garg, N., Khandekar, R., Meyerson, A., Munagala, K., Pandit, V.: Local search heuristic for k-median and facility location problems. In: Proceedings of the Thirty-Third Annual ACM Symposium on Theory of Computing, pp. 21–29 (2001)
4. Charikar, M., Khuller, S., Mount, D.M., et al.: Algorithms for facility location problems with outliers. In: Proceedings of the Twelfth Annual Symposium on Discrete Algorithm, pp. 642–651 (2001)
5. Chen, X., Fang, Q., Liu, W., Ding, Y., Nong, Q.: Strategyproof mechanisms for 2-facility location games with minimax envy. J. Comb. Optim. **43**, 1628–1644 (2022)
6. Chen, Z., Fong, K.C., Li, M., Wang, K., Yuan, H., Zhang, Y.: Facility location games with optional preference. Theor. Comput. Sci. **847**, 185–197 (2020)
7. Cheng, Y., Yu, W., Zhang, G.: Strategy-proof approximation mechanisms for an obnoxious facility game on networks. Theor. Comput. Sci. **497**, 154–163 (2013)

8. Chudak, F.A., Shmoys, D.B.: Improved approximation algorithms for the unca-pacitated facility location problem. SIAM J. Comput. **33**, 1–25 (2003)
9. Guha, S., Khuller, S.: Greedy strikes back: improved facility location algorithms. J. Algor. **31**, 228–248 (1999)
10. Huang, H.C., Li, R.: A k-product uncapacitated facility location problem. Eur. J. Oper. Res. **185**, 552–562 (2008)
11. Immorlica, N., Mahdian, M., Mirrokni, V.S.: Limitations of cross-monotonic cost-sharing schemes. ACM Trans. Algor. **4**, 1–25 (2008)
12. Jain, K., Vazirani, V.V.: Approximation algorithms for metric facility location and k-median problems using the primal-dual schema and Lagrangian relaxation. J. ACM **48**, 274–296 (2001)
13. Li, M., Wang, C., Zhang, M.: Budgeted facility location games with strategic facili-ties. In: Proceedings of the Twenty-Ninth International Conference on International Joint Conferences on Artificial Intelligence, pp. 400–406 (2021)
14. Moulin, H., Shenker, S.: Strategyproof sharing of submodular costs: budget balance versus efficiency. Econ. Theory. **18**, 511–533 (2001)
15. Pál, M., Tardos, É.: Group strategy proof mechanisms via primal-dual algorithms. In: Proceedings of the 44th Annual IEEE Symposium on Foundations of Computer Science, pp. 584–593 (2003)
16. Shmoys, D.B., Tardos, É., Aardal, K.: Approximation algorithms for facility loca-tion problems. In: Proceedings of the Twenty-Ninth Annual ACM Symposium on Theory of Computing, pp. 265–274 (1997)
17. Wang, F.M., Wang, J.J., Li, N., Jiang, Y.J., Li, S.C.: A cost-sharing scheme for the k-level facility location game with penalties. J. Oper. Res. Soc. China. **10**, 173–182 (2022)
18. Wang, Z., Xu, D.: A cost-sharing method for an uncapacitated facility location game with penalties. J. Syst. Sci. Complex. **25**, 287–292 (2012)
19. Xu, D., Du, D.: The k-level facility location game. Oper. Res. Lett. **34**, 421–426 (2006)
20. Ye, D., Mei, L., Zhang, Y.: Strategy-proof mechanism for obnoxious facility location on a line. In: Proceedings of the 21st International Conference on Computing and Combinatorics, pp. 45–56 (2015)

Algorithm in Networks

Maximizing Diversity and Persuasiveness of Opinion Articles in Social Networks

Liman Du$^{(\boxtimes)}$ (iD), Wenguo Yang (iD), and Suixiang Gao

School of Mathematical Sciences,University of Chinese Academy of Sciences, Beijing 100049, China
duliman18@mails.ucas.edu.cn, {yangwg,sxg}@ucas.ac.cn

Abstract. The formation and evolution of opinions in social networks is a complex process affected by the interplay of different elements that incorporate peer interaction in social networks and the diversity of information to which each individual is exposed. Taking a step in this direction, we propose a model which captures the dynamic of both opinion and relationship in this paper. It not only considers the direct influence of friends but also highlights the indirect effect of group when individuals are exposed to new opinions. And it allows nodes which represent users of social networks to slightly adjust their own opinion and sometimes redefines friendships. A novel problem in social network whose purpose is simultaneously maximizing both the diversity and persuasiveness of new opinions that individuals have access to is formulated. This problem is proved to be NP-hard and its objective function is neither submodular nor supermodular. However, we devise an approximation algorithm based on the sandwich framework. And the influence of different seed selection strategies is experimentally demonstrated.

Keywords: Social Networks · Information Spread · Opinion Dynamics · Group Effect · Echo Chamber

1 Introduction

An opinion is usually defined as the degree of an individual's preference towards a particular phenomenon or thing [8]. Opinion articles can be shared through frequency interaction between friends [7]. During this process, humans' behaviors play an important role [3]. Toward a sensitive topic, people can be attracted, indifferent or repulsive. As part of their behaviors, they update their opinions and even change existing friendships, resulting in echo-chamber effect and polarization phenomenon. The echo-chamber effect is used to describe how people's opinion might be artificially enhanced as they are only exposed to information from like-minded individuals and opinions are reflected back at them through social interactions [2]. Polarization is a steady state where people do not change

---

Supported by the National Natural Science Foundation of China under grant numbers 12071459 and 11991022.

W. Wu and G. Tong (Eds.): COCOON 2023, LNCS 14423, pp. 317–328, 2024.
https://doi.org/10.1007/978-3-031-49193-1_24

their opinion anymore and form only two clusters opposing each other [14]. In social platforms, a controversial issue is often brought into public focus and people's initial opinions vary from person to person. Different from face-to-face debates, more related details can be persistently revealed and spread in social network during an online discussion. From the perspective of individuals participating in or watching the debate, they can be exposed to these external materials which incline toward a different view of the problem. Therefore, it is feasible to combat echo-chamber and polarization in social networks by selecting some initial publishers for those external materials.

How to select some individuals to share supplement materials for the purpose of maximizing their influence is studied in this paper. And as a prerequisite, it is necessary to propose a method to measure the influence of external materials shared by selected individuals. For the purpose of keeping an online discussion reasonable and rational, the external materials received by users in a social network should be diverse as much as possible. For everyone engaging in a debate and evolving their own opinions, reading widely, considering what they read critically and looking for holes in arguments can help them thinking more rationally. It not only makes a good atmosphere for discussion but also improve the quality of discussion. At the same time, one reasonable measure of the influence of an opinion expressed by a supplement material is the degree to which everyone's opinion changes in the discussion it triggers. Hence, we propose a novel problem which wants to find a wise information source selection strategy to achieve these two goals.

The remainder of this paper is organized as follows: Sect. 2 introduces some related work. The proposed model is introduced in Sect. 3. Then, Sect. 4 formulates the Opinions Diversity and Persuasiveness Maximization (ODPM) problem. Consequently, an algorithm designed to address ODPM problem is proposed and analyzed in Sect. 5. Section 6 shows some experimental results and we conclude in Sect. 7.

2   Related Work

As one of the domains which enjoy rapid theoretical growth, opinion dynamics has attracted the interest of researchers who are from different fields and study the evolution as well as formation of opinions in social systems. Individuals holds one of several possible opinions in discrete opinion models while continuous opinion models take value from a certain range of real numbers to represent people's opinions toward a given topic or issue [4,15]. And most of them confine values of opinions in the interval $[-1, 1]$. Social media can limit the exposure to diverse perspectives and favor the formation of groups of like-minded users framing and reinforcing a shared narrative. So, proposing mechanisms that expose online social media users to content that does not necessarily align with their prior beliefs could be a way to depolarize opinions.

The influence maximization (IM) problem has been one of the most attractive topics in the field of social networks and is well studied in the literature. The classical version of influence maximization problem which aims to find a set of seed nodes in the network to maximize the influence of information that spreads from the seed nodes is formalized in [9]. It is regarded as a discrete optimization problem and two basic propagation models are introduced to describe the process of information dissemination. As for complexity, it has been proven to be NP-hard [9]. And due to the stochasticity of information diffusion process, it is #P-hard to accurately compute the influence spread given a seed set under the commonly used diffusion models [1]. A great deal of research devoted to studying how to estimate the influence spread and using the design techniques to seek a balance between effectiveness, efficiency, and generalization ability [1,5,6,16].

3 The Model

Abstract a social network as directed graph $G = (V, E)$ where each element of V represents a social platform's user (individual). For each pair of users u and v who are friends or have direct social ties, $(u, v) \in E$ is used to reveal their friendship. $U = \{U_1, U_2, \dots\}$ is a group structure of given social network and $\bigcup_{U_k \in U} U_k \subset V$. For a given topic, denote M as the set of supplement materials (opinion articles) and the exposure order is given as $\langle m_1, m_2, \dots, m_{|M|} \rangle$. For each opinion article m_i, $i = 1, \dots, |M|$, its opinion $l(m_i) \in [-1, 1]$ is given. And for each node $v \in V$, only its initial opinion $l^0(v) = l(v) \in [-1, 1]$ is known.

Now, we propose our definition of several probability functions and an opinion update function based on the adopted configuration introduced in [3,18].

Denote $l_i(v)$ as the current opinion of node v after the dissemination of m_i and S_i^{t-1} as the set of nodes that have been exposed to m_i at time $t-1$. Denote $U(v) = \{k | U_k \in U, v \in U_k\}$ and $U(S_i^{t-1}) = \bigcup_{v \in S_i^{y-1}} U(v)$. Let $\Delta \in (0, 1)$. Then, we have the following definitions.

Definition 1. – For $v \in S_i^{t-1}$, the probability with which v shares the i-th opinion article m_i is defined as

$$p_s(v, m_i) = \cos^2 \left(\frac{\pi |l^{i-1}(v) - l(m_i)|}{2} \right). \tag{1}$$

– For each set, the probability with which $v \in V$ can successfully receive the opinion article m_i shared by S_i^{t-1} is defined as

$$p_r(S_i^{t-1}, v) = 1 - \prod_{u \in N(u) \cap S_i^{t-1}} (1 - p_r^i(u, v)) \prod_{U_k \in U(v)} (1 - p_r^i(U_j, v)). \tag{2}$$

Here, for each directed edge $(u, v) \in E$,

$$p_r^i(u, v) = p_s(u, m_i) \cos^2 \left(\frac{\pi |l^{i-1}(u) - l^{i-1}(v)|}{4} \right) \tag{3}$$

is the direct social influence probability with which $v \in V$ can successfully receive the opinion article m_i shared by $u \in V$. Denote $SU = S_i^{t-1} \cap U_k$. For each group U_k that v belongs to, the probability with which $v \in N(S_i^{t-1})$ can receive m_i resulting from U_k at time step t is defined as

$$p_r^i(U_j, v) = \begin{cases} \frac{\exp\{-|U_k|/|SU|\}}{\exp\{-|U_k|/|SU|\}+\exp\{-|U_k|/(|U_k|-|SU|)\}}, & SU \neq \emptyset, \\ 0, & otherwise. \end{cases} \quad (4)$$

- For $v \in V$, if v has not received m_i resulting from S_i, $l^{i-1}(v) = l^i(v)$, otherwise opinion update function $l^i(v)$ is defined as

$$l^i(v) = \begin{cases} l^{i-1}(v)(1+\Delta), & l^0(v) \cdot l(m_i) \geq 0, \\ l^{i-1}(v)(1-\Delta), & l^0(v) \cdot l(m_i) < 0. \end{cases} \quad (5)$$

- For each directed edge $(u,v) \in E$, the probability with which $(u,v) \in E$ is removed is defined as

$$p_b^i(u,v) = \begin{cases} \cos^2\left(\frac{\pi|l^i(u)-l^i(v)|}{2}\right), & |l^i(u)-l^i(v)| \geq 1, \\ 0, & otherwise. \end{cases} \quad (6)$$

Now, we introduce the Opinion influenced by Group - Independent Cascade (OG-IC) model which is proposed to model the opinion dynamic based on group effect in social network. Define a piece of writing which expresses a particular point of view as an opinion article. Let M be a set of opinion articles waiting to be shared and $\langle m_1, \ldots, m_{|M|} \rangle$ be a given sequence. We focus on m_1 at first and then check other elements in M following the given order. For m_1, the OG-IC model begins with choosing some individuals to form S_1^0 as initial publishers. And a selected individual $v \in S_1^0$ shares the given article with probability $p_s(v, m_1)$. When m_1 is successfully shared by publishers in S_1^0, both the direct and indirect social ties can play roles in the following spread. In other words, each individual v which has not received m_1 can receive the article with probability $p_r^1(S_1^0, v)$ which considers the influence of both neighbors and groups. Then, all new receivers can be persuaded and update their opinions according to the relationship between their initial opinion and $l(m_1)$. The update function is defined as Eq. 5. And the friendship between two individuals is influenced by opinion change. Even worse, the already established friendship will end, once the strong disagreements occur. The probability with which the relationship between two individuals ends is shown as Eq. 6. Then, these newly activated individuals have to decide whether to share the article. Such dissemination continues until there is no individual which can receive the given opinion article m_1. If there are more than one articles waiting to be shared, i.e. $|M| > 1$, successively repeat the above process for $m_2, m_3, \ldots, m_{|M|}$.

4 Problem Formulation

Given a series of articles with different opinions toward the same topic, every node can receive (be activated on) some of them in OG-IC model. Denote $\mathbb{N} =$

$\{(v,i)|v \in V, i \in \{1,\ldots,|M|\}\}$. Let $S = \{(v,i)|v \in S_i, i = 1,\ldots,|M|\} \subset \mathbb{N}$ be a seed set where $S_i \subset V$ consists of all initial publisher selected for $m_i \in M$. Then, denote $I_v(S) \subset M$ as a set of all opinion articles that v has received when dissemination of all the given opinion articles initially published by S terminates. Let $L(v, I_v(S))$ represent the set $\{l_v\} \cup \{l_{m_i} : m_i \in I_v(S)\} \cup \{-1, 1\}$ sorted by increasing values. As is mentioned above, l_v is the initial opinion of node v. $\{-1, 1\}$ includes two extreme opinions. And $\{l_{m_i} : m_i \in I_v(S)\}$ consists opinions of all the opinion articles received by node v. Now, we can define diversity score of node v resulting from S to quantify the diversity of opinions expressed by articles and received by node v. Taking a step further, a metric, defined as diversity function, can be proposed to quantity the diversity of opinions articles that are shared by S and spread in the social network.

Definition 2. *Given a seed set S, the diversity score function $div(v)$ of each node $v \in V$ can be calculated by*

$$div(v, S) = 1 - \sum_{j=1}^{|L(v,I_v(S))|-1} \frac{(l_{j+1} - l_j)^2}{4} \tag{7}$$

where $v \in V$ and $l_j \in L(v, I_v(S))$. Then, diversity score function $div(S)$ of S is defined as

$$div(S) = \sum_{v \in V} div(v, S). \tag{8}$$

Given a series of opinion articles with a determined order of arrival, there is another perspective from which we can measure the quality of individuals who are selected as the initial opinion articles publishers. When the spread of all the opinion articles terminates, the difference between v's initial opinion $l^0(v)$ and final opinion $l^{|M|}(v)$ can show the influence of opinion articles on v's opinion formation and evolution. We propose the following definition to formally define opinion articles' persuasiveness.

Definition 3. *Given a seed set S, the influence of all the opinion articles in M shared by S on v can be calculated by $per(v, S) = (l^0(v) - l^{|M|}(v))^2$ and $per(S)$ of S is defined as*

$$per(S) = \sum_{v \in V} per(v, S). \tag{9}$$

Denote c and o as the number of opinion articles m_i satisfying $l(m_i) \cdot l^0(v) \geq 0$ and $l(m_i) \cdot l(v) < 0$ among all the opinion articles received by v, respectively. Then, we have $|I_v(S)| = c + o \leq |M|$ and $per(v, S) = (l(v) - l^{|M|}(v))^2 = l(v)^2((1 + \Delta)^c(1 - \Delta)^o - 1)^2$.

On the foundation of these two definitions, opinion influence function which is a combination of exposure diversity and opinion articles' persuasiveness is defined as follows.

Definition 4. *Given $S = \{(v,i)|v \in S_i, i \in \{1,\ldots,|M|\}\}$, an opinion article set M, a given sequence $\langle m_1, m_2, \ldots \rangle$ where $m_i \in M$, $l(v)$ for $v \in V$, and $l(m_i)$*

for $m_i \in M$, opinion influence function of S is defined as a combination of diversity score function $div(S)$ and persuasiveness score function $per(S)$, i.e.,

$$\sigma(S) = div(S) + \alpha per(S) \tag{10}$$

where α is a parameter used to adjust the weight of persuasiveness score function $per(S)$.

What should be emphasized is that the order of opinion articles' publication plays an important role in the whole process of dissemination. In this paper, we assume that opinion articles are published by selected initial publishers one by one and the order of publication is given. Besides, due to the definition of opinion update function and persuasiveness score function, when the value of Δ is pretty small, the change of persuasiveness score function resulting from different seed sets may be too small to be ignored without a suitable setting of α.

Then, we introduce the definition of ODPM problem.

Definition 5 (ODPM Problem). *Given a directed social graph $G = (V, E)$ with group structure U, an opinion article set M, a sequence $\langle m_1, m_2, \dots \rangle$ where $m_i \in M$, individual's opinion $l(v)$ for $v \in V$, and article's opinion $l(m_i)$ for $m_i \in M$, positive integers a_v for $v \in V$, parameter α and a constant k, Opinions Diversity and Persuasiveness Maximization (ODPM) problem aims to find a seed set S, a union of seeds selected as initial publishers for each opinion article, such that the expectation value of opinion influence function $\sigma(S)$ is maximized under OG-IC model, i.e.,*

$$\max \quad \mathbb{E}[\sigma(S)]$$
$$s.t. \quad |S_{m_i}| \le k, m_i \in M$$
$$|S_v| \le a_v, v \in V$$

where $\sigma(S) = div(S) + \alpha per(S)$, $S_{m_i} \subset V$ is the set of initial publishers for opinion article $m_i \in M$ and $S_v \subset M$ is the set of opinion articles assigned to $v \in V$ to share.

Similar to [13], a bound constraint a_v for $v \in V$ is a positive integer. There are two main reasons for doing so. In social networks, if a user expresses too many different opinions on the same topic, there may be some mutually exclusive and contradictory opinions among them, leading to a decrease in the reliability and persuasiveness. Besides, human attention is a limited resource. Sharing too many similar opinion articles can make friends impatient and affect the probability with which these opinion articles are successfully spread.

The following theorem is proposed to describe properties of ODPM problem and its objective function.

Theorem 1. *ODPM problem is NP-hard. Given S, computing $\sigma(S)$ is #P-hard under the OG-IC model. $\sigma(S)$ is non-monotone, non-submodular and nonsupermodular.*

5 The Algorithm

Inspired by the concept of Sandwich Approximation [12], we bound $\sigma(\cdot)$ with two functions. The upper bound function of $\sigma(\cdot)$ is denoted as $\sigma_u(\cdot)$ while the lower bound function of $\sigma(\cdot)$ is denoted as $\sigma_l(\cdot)$. The first phase of ODPM algorithm aims to obtain optimal solutions for maximizing $\sigma(\cdot)$, $\sigma_u(\cdot)$ and $\sigma_l(\cdot)$ respectively. And the final solution S^* of ODPM problem is one of these three optimal solutions which maximizes the ODPM problem's objective function $\sigma(\cdot)$. Given that ODPM algorithm needs to select the best one from three candidates obtained by solving different optimization problems, we introduce all the methods used for addressing subproblems one by one.

Due to the fact that opinion influence function is neither submodular nor supermodular, algorithms designed for IM problem are not suitable for ODPM problem. Therefore, a sampling method based on the Monte Carlo simulation is used to estimate the value of opinion influence function for any given set S of initial publishers selected for each opinion article in M. Then, we propose an improved greedy algorithm to obtain a solution of maximizing the opinion influence function $\sigma(S)$. Following the order of publication which is given as

---

Algorithm 1. Activating

Require: $G = (V, E)$, U, M, l, S_i
Ensure: A
1: Initialize $A = S_i$ and $S_{new} = S_i$
2: **while** $S_{new} \neq \emptyset$ **do**
3:     $S_{temp} = \emptyset$
4:     **for all** $u \in S_{new}$ **do**
5:         $S_c = \{v | (u, v) \in E\} \cup \{v | v \in U_i, U_i \in U(u)\}$
6:     **for** $v \in S_c$ **do**
7:         Calculate $p_r^i(S_{new}, v)$
8:         Generate a random number $r \in [0, 1]$
9:         **if** $r \leq p_r^i(S_{new}, v)$ **then**
10:           $S_{temp} = S_{temp} \cup \{v\}$
11:     $S_{new} = S_{temp}$
12:     $A = A \cup S_{temp}$

$\langle m_1, m_2, \ldots \rangle$, a set of nodes that received m_i resulting from $S_i = \{v | (v, i) \in S\}$ is returned by Algorithm 1. Then, for activated nodes, update their opinion and the set of opinion articles assigned to them. Before finding nodes activated on m_{i+1} resulting from S_{i+1}, check each $(u, v) \in E$ and remove it with probability $p_b^i(u, v)$. After updating opinions of nodes activated on $m_{|M|}$, a sampling process finishes and the value of $\sigma(S)$ is returned by Algorithm 2. As is mentioned above, it is difficult to compute $\sigma(S)$ for any given S. Inspired by [18], Algorithm 3 is proposed to obtain an estimator $\hat{\sigma}(\cdot)$ of $\sigma(\cdot)$ with (ϵ, δ)-approximation where $\sigma_{max}(\cdot)$ is the maximum of $\sigma(\cdot)$. Now, on the basis of sub-algorithms introduced

---

Algorithm 2. Sampling
___
Require: : $G = (V, E)$, U, M, l, S,α
Ensure: : $\sigma(S)$
 1: **for** $i = 1, \ldots, |M|$ **do**
 2:     Initialize $A = \emptyset$
 3:     $S_i = \{v | (v, i) \in S\}$
 4:     Update A by Algorithm 1
 5:     **for all** $v \in A$ **do**
 6:         $I_v(S) = I_v(S) \cup \{m_i\}$
 7:         Update the opinion of v according to Eq. 5
 8:         Check $(u, v) \in E$ according to Eq. 6
 9: Calculate $\sigma(S)$

Algorithm 3. Estimating
___
Require: $G = (V, E)$, U, M, l, S,α
Ensure: $\hat{\sigma}(S)$
 1: $\Gamma = 1 + (1 + \varepsilon)(4(e - 2)\ln(2\backslash\delta)\backslash\varepsilon^2)$
 2: Calculate $\sigma(S)$
 3: $\gamma = \frac{\sigma(S)}{\sigma_{max}(\cdot)}$
 4: **while** $\gamma \leq \Gamma$ **do**
 5:     Calculate $\sigma(S)$
 6:     $\gamma = \gamma + \frac{\sigma(S)}{\sigma_{max}(\cdot)}$
 7: $\hat{\sigma}(S) = \gamma$

Algorithm 4. Selecting
___
Require: $G = (V, E)$, U, M, l, k,α
Ensure: S
 1: Initialize $S_i = \emptyset$ for $i = 1, \ldots, |M|$ and $S = \emptyset$
 2: **for** $i = 1, \ldots, |M|$ **do**
 3:     $V_c = V$, $A = \emptyset$
 4:     **while** $|S_i| < k$ **do**
 5:         $S_{cur} = S \cup \{(v, i) | v \in S_i\}$
 6:         $u = \arg\max_{v \in V_c \backslash V_i} \hat{\sigma}(S_{cur} \cup \{(v, i)\}) - \hat{\sigma}(S_{cur})$
 7:         **if** $S_u < a_u$ **then**
 8:             $S_i = S_i \cup \{u\}$
 9:         **else**
10:             $V_c = V_c \backslash \{u\}$
11:     Update A
12:     **for all** $w \in A$ **do**
13:         $I_w(S) = I_w(S) \cup \{m_i\}$
14:         Update the opinion of w according to Eq. 5
15:         Check $(u, v) \in E$ according to Eq. 6
16:     $S = S \cup \{(u, i) | u \in S_i\}$

above, we propose Algorithm 4. It describes the whole process of finding a set S such that the opinion influence function $\sigma(S)$ is maximized. Therefore, S consists of initial publishers for each opinion article $m_i \in M$ and is a feasible solution for ODPM problem.

To find a lower bound function $\sigma_l(\cdot)$ of $\sigma(\cdot)$, we change some settings of OG-IC model to obtain LOG-IC model, a modified version of the former. The new definition is shown as follows.

- $p_r^i(U_j, v) = 0.$

$$
p_b^i(u, v) = \begin{cases} \cos^2\left(\frac{\pi x}{4} + \frac{\pi}{2}\right) & x < 2 \\ 1, & \text{otherwise} \end{cases} \tag{11}
$$

where $x = (|l(u)| + |l(v)|)(1 + \Delta)^i$.

At the same time, define $per_l(S) = l^2(v)((1 - \Delta^2)^c - 1)^2$. Therefore, we have $\sigma_l(S) = div(S) + \alpha per_l(S)$.

Similarly, we start with proposing UOG-IC model to find $\sigma_u(\cdot)$. Compared with OG-IC model, the changed configuration is listed below.

- $p_r(S_i^{t-1}, v) = 1 - \prod_{u \in N(u) \cap S_i^{t-1}} (1 - p_r^i(u, v))$ if $U(v) \cap U(S_i^{t-1}) = \emptyset$, otherwise $p_r(S_i^{t-1}, v) = 1.$
- $l^i(v) = l^{i-1}(v)(1 + \Delta)$ only if $l(v) \cdot l(m_i) \geq 0$, otherwise $l^i(v) = l^{i-1}(v).$
- $p_b^i(u, v) = 0.$

Then, we have $per_u(v, S) = (l|M|(v) - l(v)))^2 = l^2(v)((1 + \Delta)^c - 1)^2$ and $\sigma_u(S) = div(S) + \alpha per_u(S)$.

For these two bound functions, we propose a theorem as follows.

Theorem 2. *Both $div_l(\cdot)$ and $per_l(\cdot)$ are submodular. $div_u(\cdot)$ is submodular while $per_u(\cdot)$ is supermodular. All of these four functions are monotone non-decreasing.*

Therefore, $\sigma_l(\cdot)$ is submodular and $\sigma_u(\cdot)$ can be represented as the sum of a submodular function and a supermodular function. Now, the maximization of $\sigma_l(\cdot)$ under LOG-IC model is the maximization of submodular set function subject to a matroid constraint while the maximization of $\sigma_u(\cdot)$ under UOG-IC model is the maximization of the sum of a submodular function and a supermodular function. Based on [13,17], Algorithm 4 can achieve approximation ratio $\frac{1 - k_{per_u}}{1 + (1 - k_{per_u})k_{div_u}}$ for $\sigma_u(\cdot)$ and $\frac{1}{2}$ for the maximization of $\sigma_l(\cdot)$, where k_{div_u} and k_{per_u} are the curvature of $div_u(\cdot)$ and $per_u(\cdot)$ respectively. However, according to [11], there is another algorithm that can produce a solution with better approximation ratio. On the foundation of these results, we propose the following theorem.

Theorem 3. *There is an algorithm such that the solution S^* of ODPM problem obtained by ODPM algorithm satisfies*

$$
\sigma(S^*) \geq \max\left\{ g_u \cdot \frac{\sigma(S_{\sigma_u})}{\sigma_u(S_{\sigma_u})}, \frac{1}{2} \cdot \frac{\sigma_l(S_{\sigma_l}^o)}{\sigma(S_{\sigma}^o)} \right\} \cdot \sigma(S_{\sigma}^o) \tag{12}
$$

where S^o is the optimal solution maximizing $\sigma(\cdot)$ subject to matroid constraint and $g_u = \min\{1 - k_{per_u}, 1 - k_{div_u}e^{-1}\}$.

6   Experiments Settings and Results

There are two data sets used in our experiments. One is denoted as SYN which is derived from a randomly generated directed graph containing 10 nodes and all the edges are directed. The other is MI (Moreno-Innovation data set introduced in [10]) which is a directed network capturing innovation shown as 25571 edges among physicians represented by 241 nodes. Set the maximum of $|M|$ is 10. For each $m_i \in M$, its opinion tendency $l(m_i)$ is randomly generated from $[-1, 1]$. And all the opinion articles are published following a given sequence. For each $v \in V$, the value of its opinion $l(v)$ is randomly generated from $[-1, 1]$. Generate a random integer number from the given range to represent the bound constraint a_v. Set $|U| = 2$ for SYN and $|U| = 65$ for MI. Then, each group randomly selects nodes from ground set V as its members and each node can be selected by many groups.

Now, we focus on the influence of $|M|$ on $\sigma(\cdot)$ with different settings of the number of initial publishers. In order to decrease time complexity, we use Algorithm 4 to find a solution for maximizing $\sigma_u(\cdot)$. According to [17], it also guarantees an approximation ratio.

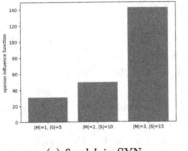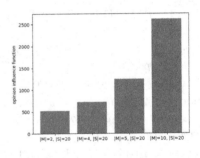

(a) fixed k in SYN   (b) fixed S in MI

Fig. 1. the influence of $|M|$ on $\sigma(\cdot)$

We focus on the influence of the number of external materials on the value of opinion influence function. And two series of experiments are conducted on two different data sets. The main difference between them are their setting about the number of initial publishers. Experiments conducted on SYN data set assume that the number of initial publishers selected for each opinion article is given whereas the total number of initial publishers is given for those experiments based on MI data set. Therefore, $|S|$ increases in the former and decreases in the later when M increases.

Based on SYN data set, set $|M| = 5$ and assume that $|S_i| = 5$ for $i = 1, \ldots, |M|$, i.e. $k = 5$. Both the order $\langle m_1, \ldots, m_5 \rangle$ and opinions $l(m_i)$ of article $m_i \in M$ are given. We conduct experiments in three different cases. In the first one, only the first opinion article m_1 spreads in the social network. And set $|M| =$

3 in the second case, that is, three opinion articles are published by individuals following the order $\langle m_1, m_2, m_3 \rangle$. And the last experiment is conducted under the assumption that all the opinion articles are considered, and the total number of initial publishers becomes 25. Experimental results are shown in Fig. 1a. And it is obvious that the value of opinion influence function increases with $|M|$ when the number of initial publishers selected for each opinion article is given.

Then, Fig. 1b shows results of another series of experiments conducted on MI data set. They also reflect the influence of the number of opinion articles spreading in the social network on the value of opinion influence function. Although their setting about constraint are different to those of experiments based on SYN data set, we can reach a similar conclusion that the $\sigma(\cdot)$ increases with $|M|$.

7   Conclusion

In this paper, we firstly introduce the OG-IC model reflecting the effect of group in opinion dynamic. Then, the novel Opinions Diversity and Persuasiveness Maximization (ODPM) problem which is NP-hard to approximate with any factor is proposed. Its objective function is neither submodular nor supermodular, which is a challenge in algorithm designed for address it. Hence, inspired by the method of sandwich framework, we propose the ODPM algorithm which can achieve an approximation guarantee with a high probability when finding a solution for ODPM problem. As a prerequisite, the upper and lower bound functions of opinion influence function are obtained by modifying the OG-IC model. And the theoretical analysis of their properties set the state for that of ODPM algorithm. And the experimental results on both synthetic and real-world data sets demonstrate the influence of different seed selection strategies.

References

1. Chen, W., Wang, C., Wang, Y.: Scalable influence maximization for prevalent viral marketing in large-scale social networks. In: Proceedings of the 16th ACM SIGKDD International Conference on Knowledge Discovery and Data Mining, pp. 1029–1038. KDD 2010, Association for Computing Machinery, New York, NY, USA (2010). https://doi.org/10.1145/1835804.1835934
2. Cinelli, M., Morales, G., Galeazzi, A., Quattrociocchi, W., Starnini, M.: The echo chamber effect on social media. Proc. Natl. Acad. Sci. **118**, e2023301118 (2021). https://doi.org/10.1073/pnas.2023301118
3. Ferraz de Arruda, H., et al.: Modelling how social network algorithms can influence opinion polarization. Inf. Sci. **588**, 265–278 (2022). https://doi.org/10.1016/j.ins.2021.12.069
4. Galam, S.: Real space renormalization group and totalitarian paradox of majority rule voting. Phys. A Statist. Mech. Appl. **285**(1), 66–76 (2000). https://doi.org/10.1016/S0378-4371(00)00272-7
5. Goyal, A., Lu, W., Lakshmanan, L.V.: Celf++: optimizing the greedy algorithm for influence maximization in social networks. In: Proceedings of the 20th International Conference Companion on World Wide Web, pp. 47–48. WWW 2011, Association for Computing Machinery, New York, NY, USA (2011). https://doi.org/10.1145/1963192.1963217

6. Guo, Q., Wang, S., Wei, Z., Lin, W., Tang, J.: Influence maximization revisited: efficient sampling with bound tightened. ACM Trans. Database Syst. **47**(3), 1–45 (2022). https://doi.org/10.1145/3533817

7. Noorazar, H.: Recent advances in opinion propagation dynamics: a 2020 survey. Eur. Phys. J. Plus **135**(6), 1–20 (2020). https://doi.org/10.1140/epjp/s13360-020-00541-2

8. Kaur, R., Kumar, R., Bhondekar, A., Kapur, P.: Human opinion dynamics: an inspiration to solve complex optimization problems. Sci. Rep. **3**, 1–7 (2013). https://doi.org/10.1038/srep03008

9. Kempe, D., Kleinberg, J., Tardos, E.: Maximizing the spread of influence through a social network. In: Proceedings of the Ninth ACM SIGKDD International Conference on Knowledge Discovery and Data Mining, pp. 137–146. KDD 2003, Association for Computing Machinery, New York, NY, USA (2003). https://doi.org/10.1145/956750.956769

10. Kunegis, J.: KONECT - The Koblenz network collection. In: Proceedings of International Conference on World Wide Web Companion, pp. 1343–1350 (2013). https://doi.org/10.1145/2487788.2488173

11. Lu, C., Yang, W., Yang, R., Gao, S.: Maximizing a non-decreasing non-submodular function subject to various types of constraints. J. Global Optim. **6**, 1–25 (2022). https://doi.org/10.1007/s10898-021-01123-x

12. Lu, W., Chen, W., Lakshmanan, L.V.S.: From competition to complementarity: comparative influence diffusion and maximization. Proc. VLDB Endow. **9**(2), 60–71 (2015). https://doi.org/10.14778/2850578.2850581

13. Matakos, A., Aslay, C., Galbrun, E., Gionis, A.: Maximizing the diversity of exposure in a social network. IEEE Trans. Knowl. Data Eng. **34**(9), 4357–4370 (2022). https://doi.org/10.1109/TKDE.2020.3038711

14. Sunstein, C.: The law of group polarization. J. Polit. Philos. **10**, 175–195 (2002). https://doi.org/10.1111/1467-9760.00148

15. Sznajd-Weron, K., Tabiszewski, M., Timpanaro, A.M.: Phase transition in the sznajd model with independence. EPL (Europhys. Lett.) **96**(4), 48002 (2011). https://doi.org/10.1209/0295-5075/96/48002

16. Tang, Y., Shi, Y., Xiao, X.: Influence maximization in near-linear time: a martingale approach. In: Proceedings of the 2015 ACM SIGMOD International Conference on Management of Data, pp. 1539–1554. SIGMOD 2015, Association for Computing Machinery, New York, NY, USA (2015). https://doi.org/10.1145/2723372.2723734

17. Wenruo, B., Jeffrey, A.B.: Greed is still good: maximizing monotone submodular+supermodular functions. CoRR abs/1801.07413 (2018). http://arxiv.org/abs/1801.07413

18. Zhu, J., Ni, P., Tong, G., Wang, G., Huang, J.: Influence maximization problem with echo chamber effect in social network. IEEE Trans. Comput. Soc. Syst. **8**(5), 1163–1171 (2021). https://doi.org/10.1109/TCSS.2021.3073064

Stochastic Model for Rumor Blocking Problem in Social Networks Under Rumor Source Uncertainty

Jianming Zhu[1]($\boxtimes$)(iD), Runzhi Li[1], Smita Ghosh[2], and Weili Wu[2]

[1] University of Chinese Academy of Sciences, Beijing, China
jmzhu@ucas.ac.cn
[2] The University of Texas at Dallas, Dallas, USA

Abstract. Either in real world social society or online social networks, rumor blocking is an important issue. Rumor sources spread negative information throughout the network, which may cause unbelievable results in real society, such as panic, unrest. Propagating positive information from several "protector" users is an effective method for rumor blocking once the rumor is detected. In this paper, we assume that user will not be influenced if they receive the positive information ahead of negative one. According to data analysis of user's activity, network manager may not know the exact positions of rumor but the probability of each user being a rumor, "protector" nodes need to be selected in order to prepare for rumor blocking. Given a social network $G = (V, E, P, Q)$, where P is the weight function on edge set E, $P_{(u,v)}$ is the probability that v is activated by u after u is activated, and Q is the weight function on node set V, Q_v is the probability that v will be a rumor source. Stochastic Rumor Blocking (SRB) problem is to select k nodes as "protectors" such that the expected influence of rumors on users is minimized eventually. SRB will be proved to be NP-hard and the objective function is supermodular. We present a Compound Reverse Influence Set (CRIS) sampling method for estimation of the objective value which can be represented as a compound set function. Based on CRIS, a randomized greedy algorithm with theoretical analysis will be presented.

Keywords: rumor blocking · stochastic · uncertainty · CRIS sampling · social networks

1 Introduction

Recently, smart phone plays an important role in every one's daily life, people are more convenient to access the world or each other through different kinds of online social networks by using the mobile smart phone. With the soaring

This work was supported in part by the National Natural Science Foundation of China under Grant No. 72074203 and the Fundamental Research Funds for the Central Universities.

popularity of online social networks, such as Twitter, Facebook, Wechat and Chinese Sina Weibo, etc., more and more people are able to become friends and share all kinds of information with each other. These information contain positive and negative. Sometimes the negative information may turn to a rumor. For example, in 2011, Japan Fukushima Daiichi nuclear-power was damaged by earthquake, and consumers in cities along the Chinese coastline, such as Shanghai and Guangzhou, and even in inland capital Beijing, began stockpiling table salt after problems while there were rumors that radiation would spread to China by air and sea and iodized table salt could protect body from radiation [5]. Salt was temporarily out of stock and salt price was 10 times over in some places. Another example is spread of misinformation on swine flu in Twitter [15]. Such negative information reached large scale panic in the population. Undoubtedly, rumors should be blocked as soon as possible once detected so that their negative influence can be minimized.

Lots of previous works studies the influence maximization problem aims to select k initially-influenced seed users to maximize the expected number of eventually-influenced users. Influence maximization finds its application in many domains, such as viral marketing [18]. In contrast, the negative influence minimization problem attempts to design effective strategies for blocking rumors and minimizing the negative influence. The first strategy is to block a limited number of links in a social network to minimize the propagation of rumor [12]. The second strategy is to block nodes in a social network [25]. Neither blocking links nor blocking nodes strategy do not take into account the issue of user experience in real social networks. The third strategy is to launch a positive information against the rumor from a set of nodes in the social network [4,23]. Positive information will spread in the network, and users will be immune to rumor if they receive the positive information ahead of rumor since first impressions are always strongest.

Unfortunately, the previous work on positive information spreading is based on rumor has been detected, while government sometimes needs to do preparation before rumors appear. In this paper, we will consider the uncertainty case that it is unknown who will be rumor sources in the social network. Fortunately, the probability of user being a rumor source may be analyzed from his activities in the Internet by data mining methods. For example, one will be a rumor source with high probability if he had spread a rumor or he always posted articles without information filtering on his own Twitter/Weibo. Government or network manager need to select nodes to spread positive information after rumor sources appear up to the probability of each node being a rumor source, which will be called "protector" in this paper. Sponsoring a "protector" always needs cost. Under the budget, k initially protectors are required to select such that rumor influence will be minimized. Then, robust decision of "protector" selection strategy should be made to deal with uncertainty of rumor sources. Models for uncertainty have widely application scenarios, such as pre-location of emergency resource, preparation of terrorist attack.

1.1   Related Works

In viral marketing, there may be multiple companies competing products. Bharathi et al. [2] modeled the innovation diffusion with multiple competing innovations as a game. They gave a $(1 - 1/e)$-approximate algorithm for computing the best response to an opponent's strategy. The rumor blocking was also studied from the game theory aspect [24], rumor blocking is viewed as a two-player game, in which one player, the rumor, will attempt to maximize the number of nodes accepting it while the second player, the protector, will attempt to minimize the rumor's influence. Both the rumor and the protector will choose their action sources (initial rumor sources and initial protector sources). In the zero-sum game context, the rumor's payoff is equal to the expected number of nodes infected, and the protector's payoff is the opposite of the rumor 's payoff. The authors propose a double oracle algorithm for this game. Budak et al. [4] considered the multi-campaign independent cascade model and investigated the problem of identifying a subset of individuals that need to adopt the "good" campaign in order to minimize the number of person that adopt the rumor. Li et al. [13] formulate the $\gamma - k$ rumor restriction problem and present $(1 - 1/e)$-approximation algorithm. And Fang et al. [9] propose an efficient random algorithm to solve the general rumor blocking problem. Several heuristic methods had been proposed by different works without performance guarantee [10, 20, 27].

When considering influence maximization for information propagation, Kempe et al. [11] were the first to formulate social Influence Maximization (IM) problem as an optimization problem under the IC model. They prove IM to be NP-hard under IC model and design a natural greedy algorithm that yields $(1 - 1/e - \epsilon)$-approximate solutions for any $\epsilon > 0$. Motivated by this celebrate work, a fruitful literature for IM [1, 8, 19, 26, 28, 30] has been developed. Zhu et al. [31] studied the influence maximization problem considering the crowd psychology influence. Also, several studies focus on uncertainty of the influence probability [29]. Chen et al. [6] and Lowalekar et al. [14] studied the robust influence maximization problem. However, most of the existing methods are either too slow for billion-scale networks such as Facebook, Twitter and World Wide Web or fail to retain the $(1 - 1/e - \epsilon)$-approximation guarantees.

TIM/TIM+ [22] and IMM [21] are two scalable methods with $(1 - 1/e - \epsilon)$-approximation guarantee for IM. Tang et al. utilize a novel Reverse Influence Set (RIS) sampling technique introduced by Borgs et al. [3]. TIM+ and IMM attempt to generate a $(1 - 1/e - \epsilon)$-approximate solution with minimal numbers of RIS samples. However, they may take days on billion-scale networks. Later, Nguyen et al. [17] make a breakthrough and proposed two novel sampling algorithms SSA and D-SSA. Unlike the previous heuristic algorithms, SSA and D-SSA are faster than TIM+ and IMM while providing the same $(1 - 1/e - \epsilon)$-approximate guarantee. SSA and D-SSA are the first approximation algorithms that use minimum numbers of samples, meeting strict theoretical thresholds characterized for IM.

The rest of this paper is organized as follows. In Sect. 2, we formulate the Stochastic Rumor Blocking (SRB) model. The statement of NP-hardness and properties of objective function will be given in Sect. 3. Algorithms for solving SRB are designed in Sect. 4 and we draw a conclusion in Sect. 5.

2 Problem Formulation

In this section, we present the rumor spreading model in this paper. Then Stochastic Rumor Blocking (SRB) model will be formulated.

2.1 Rumor Spreading Model

Since protectors and rumors are diffusing opposite information in the social network, the cascade arriving first will dominate the node and never change its state once activated. If a node is successfully activated by two or more neighbors with different cascades at the same time step, we assume the protector has a higher priority. Let S and R denote the seed sets of protectors and rumors, respectively. The information diffusion process is as follows. At time step $t = 0$, nodes in S and R are activated by protectors and rumors, respectively. At step $t > 0$, each node u which is activated at $t - 1$ will try to activate each of its inactive neighbors v with successful probability $p_{(u,v)}$. If v is successfully activated by rumors and protectors simultaneously, v will be activated by protectors. This means v will be protected against rumors. Finally, the diffusion process terminates when there is no node can be further activated.

2.2 Stochastic Rumor Blocking Problem

Firstly, we will introduce the concept of realization of rumors and random graph which helps to understand the rumor spreading model.

Let $G = (V, E, P, Q)$ be a social network with a node set V and a directed edge set with $|V| = n$ and $|E| = m$. Assume that each node $v \in V$ is associated with a weight $0 \le Q_v \le 1$ and Q_v is the probability that v will be a rumor source. For each directed edge e, definition of P_e has been stated in Sect. 2.1.

A realization rumor set R is a subset of V. The generation of R is as follows: (1) for each node $v \in V$, a random number r is generated from 0 to 1 in uniform; (2) add v into R if and only if $r \le Q_v$. Then, a rumor source set R is deterministic. Let $\mathcal{R}$ be the set of all possible realizations of rumor source set. Obviously, there are $2^{|V|} = 2^n$ realizations in $\mathcal{R}$. Let $P[R]$ be the probability that R can be generated. Then,

$$P[R] = \prod_{v \in R} Q_v \prod_{v \in V \setminus R} (1 - Q_v)$$

Further, a realization g of G is a graph where $V(g) = V(G)$ and $E(g)$ is a subset of $E(G)$. Each edge in $E(g)$ has the influence probability of 1 and is constructed in random. The construction process is as follows: (1) for each edge

$e \in E(G)$, a random number r is generated from 0 to 1 in uniform; (2) this edge e appears in g if and only if $r \leq P_e$. Then, g is a deterministic directed graph. Let $\mathcal{G}$ be the set of all possible realizations of G. Obviously, there are $2^{|E(G)|}$ realizations in $\mathcal{G}$. Let $P[g]$ be the probability that g can be generated. Then,

$$P[g] = \prod_{e \in E(g)} P_e \prod_{e \in E(G) \setminus E(g)} (1 - P_e)$$

Let $\sigma_r^g(S, R)$ be the number of nodes activated by rumor sources set R in g under a "protector" set S. According to the rumor spreading model, each node in g can be easily determined whether it is activated by R for given S and R by shortest path algorithm. Let $U_g(R, S)$ denote the node set that contains all nodes activated by rumor, where $U_g(R, S) \subset V$ and $R \subset V$. Then,

$$\sigma_r^g(R, S) = |U_g(R, S)| \tag{1}$$

Let $\sigma_r(R, S)$ be the expected number of nodes that are activated by rumor sources set R when S is selected as the seed set of protector. Therefore, $\sigma_r(R, S)$ can be expressed as

$$\sigma_r(R, S) = \mathbb{E}[\sigma_r^g(R, S)] = \sum_{g \in \mathcal{G}} P[g]\sigma_r^g(R, S) \tag{2}$$

While R is not exactly known before locating protectors, then we should consider the expected value for given S. Let $f(S)$ be expected number of nodes that are activated by all possible subset $R \in V$ with probability distribution Q, and $f(S) = \mathbb{E}[\sigma_r(R, S)]$ can be expressed as

$$f(S) = \mathbb{E}[\sigma_r(R, S)] = \sum_{R \in \mathcal{R}} P[R]\sigma_r(R, S) \tag{3}$$

Given a budget k, the Stochastic Rumor Blocking problem is given as follows.

Stochastic Rumor Blocking (SRB). Select k initially seed "protectors" S such that $f(S)$ is minimized.

2.3 Strategies for Choosing "Protector"

In this section, we will present two heuristic strategies for choosing "protector". One is called *Maximum Outdegree Seeds* (MOS) which picks up the first k nodes with maximum outdegree in G, while the other one is *Influence Maximization Seeds* (IMS) which picks up k nodes such that the expected eventually influenced nodes are maximized.

3 Properties of Stochastic Rumor Blocking

In this section, we first present statement of the hardness of the stochastic rumor blocking problem. Then discuss the properties of the objective function $f(S)$.

3.1   Hardness Results

First we review the well-known problem of *Influence Maximization* raised in [11]. Here, the information diffusion process is based on Independent Cascade (IC) model.

Influence Maximization (IM). Given a graph $G = (V, E, P)$ and a fixed budget k, we are required to select k initially seeds such that the expected eventually-influenced number of nodes is maximized.

It is known that any generalization of a NP-hard problem is also NP-hard. The influence maximization problem [11] has been proved NP-hard, which is a special case of our problem.

Theorem 1. *The Stochastic Rumor Blocking problem is NP-hard.*

Also, we can get the following complexity result of computing $\sigma_r(R, S)$ for given R and S since calculation for the objective function of IM was proved to be #P-hard under the IC model [11].

Theorem 2. *Given a protector node set S and a rumor node set R, computing $\sigma_r(R, S)$ is #P-hard under the IC model.*

3.2   Modularity of Objective Function

In this section, let $h(S) = \sigma_r(R, S)$ be set function of S for fixed R, then we will prove $h(S)$ is a supermodular function. Furthermore, we will discuss the modularity of $f(S)$. For given rumor source set R, $h(S) = \sigma_r(R, S)$ has the following property.

Theorem 3. *For given rumor source set R, $h(S) = \sigma_r(R, S)$ has the following properties:*

1. *$h(\emptyset) = \sigma_r(R, \emptyset)$.*
2. *$h(S) = 0$ for $|S| \geq |R|$.*
3. *$h(S)$ is monotone nonincreasing.*
4. *$h(S)$ is a supermodular function under IC model.*

Since $f(S)$ is the expected value of $h(S)$ for all possible rumor source sets R, we directly obtain the following corollary.

Corollary 1. *$f(S)$ has the following properties:*

1. *$f(\emptyset) = \mathbb{E}[h(\emptyset)] = \mathbb{E}[\sigma_r(R, \emptyset)] = \sum_{R \in \mathcal{R}} P[R]\sigma_r(R, \emptyset)$.*
2. *$f(S)$ is monotone nonincreasing.*
3. *$f(S)$ is a supermodular function under IC model.*

$f(S)$ is not a polymatroid function since $f(\emptyset) \neq 0$. Instead of analysis of $f(S)$, we define

$$y(S) = f(\emptyset) - f(S).$$

Then, SRB problem for minimizing $f(S)$ is equivalent to maximize $y(S)$. While $y(S)$ has the following properties.

Lemma 1. $y(S)$ *is a monotone nondecreasing function under IC model with* $y(\emptyset) = 0$.

Lemma 2. $y(S)$ *is a submodular function under IC model.*

Furthermore, the following theorem is directly obtained.

Theorem 4. $y(S)$ *is a polymatroid function.*

As shown in [16], greedy algorithm guarantees $(1 - 1/e)$-approximation for such a polymatroid maximization problem with cardinality constraints. Meanwhile, $y(S)$ is a compound function with two random variables R and g. Then, we will develop a new sampling method for approximating the objective value and design randomized greedy algorithm based on Compound Reverse Influence Set (CRIS) sampling method.

4   Algorithm

In this section, we will design a randomized greedy algorithm based on Compound Reverse Influence Set (CRIS) sampling method. Firstly, recall the (ϵ, δ)-approximation in [7]. Then, we will present Compound Reverse Influence Set (CRIS) sampling method to estimate $f(S)$ for given S and the sample complexity will be analyzed. Finally, a randomized greedy algorithm for solving SRB problem with theoretical results will proposed.

4.1   CRIS Sampling

Firstly, Our CRIS sampling method is based on Reverse Influence Set(RIS) sampling method [3]. Given a graph $G = (V, E, P)$, RIS captures the influence landscape of G through generating a set $\mathcal{C}$ of random *Reverse Reachable(RR) sets*. Different from RIS in [3], RR set in our method contains another parameter as shown in the following definition. Each RR set C_j is a subset of V and constructed as follows.

Definition 1. *(Reverse Reachable (RR) set). Given* $G = (V, E, P)$, *a random RR set* C_j *is generated from* G *by (1) selecting a random node* $v \in V$; *(2)generating a sample graph* g *from* G; *(3)returning* C_j *as the set of nodes that can reach* v *in* g; *(4)for each node* $u \in C_j$, *computing* $d(u)$ *as length of the directed shortest path from* u *to* v *and return* $d(u)$ *along with* C_j.

Then for given rumor sources set R and "protector" set S, an RR set C_j is *covered* by R if satisfies the following two constraints: (1) $C_j \cap R \neq \emptyset$; (2)$C_j \cap S = \emptyset$ or $\min\{d(u)|u \in R_j \cap R\} < \min\{d(u)|u \in C_j \cap S \neq \emptyset\}$. Assume $v \in C_j$ is the random node selected in construction process of RR set, then constraint (1) means v can be activated by R, while constraint (2) means either v can not be activated by S or the shortest distance from v to R is strictly smaller than that from v to S. Let $D(C_j, R) = 1$ if C_j is *covered* by R, otherwise $D(C_j, R) = 0$.

Let $\mathcal{C}$ be a collection of RR sets, and $Cov_{\mathcal{C}}(R, S) = \sum_{C_j \in \mathcal{C}} D(C_j, R)$ represents the number of RR sets in $\mathcal{C}$ covered by rumor source set R. $\sigma_r(R, S)$ can be estimated by computing $Cov_{\mathcal{C}}(R, S)$ of rumor source set R and "protector" set S. Then, the following lemma gives an estimation of $\sigma_r(R, S) = nPr[R$ covers $C_j]$.

Lemma 3. *Given $G = (V, E, P)$, a random RR set C_j generated from G. For each pair of sets $S, R \subseteq V$,*

$$\sigma_r(R, S) = nPr[R \ covers \ C_j]. \tag{4}$$

4.2 Randomized Greedy Algorithm

In this section, we will present a randomized greedy algorithm for solving SRB. Since $y(S)$ is a polymatroid function, we try to maximize $y(S)$ by applying greedy strategy and prove this Randomized Greedy Algorithm (RGA) guarantees $(1 - 1/e - \epsilon)$-approximation.

Let $\Delta_v y(S) = y(S \cup \{v\}) - y(S)$ be the increment of $y(S)$ by adding node v to S. We have $\Delta_v y(S) = y(S \cup \{v\}) - y(S) = (f(\emptyset) - f(S \cup \{v\})) - (f(\emptyset) - f(S)) = f(S) - f(S \cup \{v\})$. And $\Delta_v \hat{y}(S) = \hat{y}(S \cup \{v\}) - \hat{y}(S)$ represents the corresponding value in the case of sampling. We have

$$\Delta_v \hat{y}(S) = \hat{f}(S) - \hat{f}(S \cup \{v\})$$
$$= \frac{1}{N_1} \sum_R (\frac{n}{N_2} (Cov_{\mathcal{C}}(R, S) - Cov_{\mathcal{C}}(R, S \cup \{v\})))$$

The greedy algorithm is to pick node one by one according to increment maximization until all k nodes are selected. RGA is shown below.

The time complexity of Algorithm 1 is $O(knN^2)$ where $N = 4(e - 2)(1 + \epsilon')^2 \ln(2/\delta')(1/\epsilon'^2)$. And we also have the following theoretical result.

Theorem 5. *Given an instance of SRB G and $0 \le \epsilon, \delta \le 1$, Algorithm 1 returns a $(1 - 1/e - \epsilon)$-approximation solution S for objective function $y(S)$, i.e.*

$$\hat{y}(S) \ge (1 - \frac{1}{e} - \epsilon) y(S^*).$$

5   Conclusions

In this paper, we studied the rumor blocking problem which was an important research topic in social network. We proposed a Stochastic Rumor Blocking (SRB) model that aims to minimize the expected number of users that are activated by rumor sources under uncertainty scenario. We showed the SRB is NP-hard under IC model. Moreover, the objective function $f(S)$ of this problem was proved to be supermodular. A Compound Reverse Influence Set (CRIS) sampling method to estimate the objective value was developed. Further, we proved CRIS was near-optimal estimation method. We presented an randomized greedy algorithm for SRB problem with theoretical analysis on sampling complexity and performance. For future research, new rumor blocking models need to be developed such as considering crowd influence during rumor diffusion.

Algorithm 1: Randomized Greedy Algorithm (RGA)

Input: Given an instance of SRB $G = (V, E, P, Q)$, $0 \le \epsilon, \delta \le 1$, and the number of "protector" k.

Output: A $(1 - 1/e - \epsilon)$-approximation solution S for objective function $y(S)$ and $\hat{f}(S)$.

$S = \emptyset$ $\epsilon' = 1 - \sqrt{1 - \epsilon}$ $\delta' = 1 - \sqrt{1 - \delta}$

$\Upsilon \leftarrow 4(e - 2)(1 + \epsilon')^2 \ln(2/\delta')(1/\epsilon'^2)$ $\Upsilon_1 = 1 + (1 + \epsilon)\Upsilon$

$\mathcal{C} \leftarrow$ generate Υ_1 random RR sets by CRIS

$\mathcal{R} \leftarrow$ generate Υ_1 random rumor source sets

for $i = 1$ *to* k **do**

    for $v \in V$ **do**

        $f_{best} = 0$

        $L = \frac{1}{|\mathcal{C}|} \sum_{R \in \mathcal{R}} Cov_{\mathcal{C}}(R, S \cup \{v\})$

        while $L < \Upsilon_1$ **do**

            $R' \leftarrow$ generate a random rumor source set $\mathcal{R} = \mathcal{R} \cup \{R'\}$

            $L_{R'} = Cov_{\mathcal{C}}(R', S \cup \{v\})$

            while $L_{R'} < \Upsilon_1$ **do**

                $C' \leftarrow$ generate a random RR set by CRIS $\mathcal{C} = \mathcal{C} \cup \{C'\}$

                $L_{R'} = Cov_{\mathcal{C}}(R', S \cup \{v\})$;

            end

        end

        if $f_{best} \le \Delta_v y(\hat{S})$ **then**

            $f_{best} = \Delta_v y(\hat{S})$

            $v^* \leftarrow v$;

        end

    end

    Add v^* to S;

end

$\hat{f}(S) \leftarrow \frac{1}{|\mathcal{R}|} \sum_{R} (\frac{n}{|\mathcal{C}|} Cov_{\mathcal{C}}(R, S))$

return S and $\hat{f}(S)$

References

1. Aslay, C., Lakshmanan, L.V., Lu, W., Xiao, X.: Influence maximization in online social networks. In: Proceedings of the Eleventh ACM International Conference on Web Search and Data Mining, pp. 775–776. ACM (2018)

2. Bharathi, S., Kempe, D., Salek, M.: Competitive influence maximization in social networks. In: Deng, X., Graham, F.C. (eds.) WINE 2007. LNCS, vol. 4858, pp. 306–311. Springer, Heidelberg (2007). https://doi.org/10.1007/978-3-540-77105-0_31

3. Borgs, C., Brautbar, M., Chayes, J., Lucier, B.: Maximizing social influence in nearly optimal time. In: Proceedings of the Twenty-fifth Annual ACM-SIAM Symposium on Discrete Algorithms, pp. 946–957. SIAM (2014)

4. Budak, C., Agrawal, D., El Abbadi, A.: Limiting the spread of misinformation in social networks. In: Proceedings of the 20th International Conference on World Wide Web, pp. 665–674. ACM (2011)

5. Burkitt, L.: Fearing radiation, Chinese rush to buy table salt. The Wall Street Journal (2011)
6. Chen, W., Lin, T., Tan, Z., Zhao, M., Zhou, X.: Robust influence maximization. In: Proceedings of the 22nd ACM SIGKDD International Conference on Knowledge Discovery and Data Mining, pp. 795–804. ACM (2016)
7. Dagum, P., Karp, R., Luby, M., Ross, S.: An optimal algorithm for monte carlo estimation. SIAM J. Comput. **29**(5), 1484–1496 (2000)
8. Du, N., Liang, Y., Balcan, M.F., Gomez-Rodriguez, M., Zha, H., Song, L.: Scalable influence maximization for multiple products in continuous-time diffusion networks. J. Mach. Learn. Res. **18**(2), 1–45 (2017)
9. Fang, Q., et al.: General rumor blocking: an efficient random algorithm with martingale approach. Theoret. Comput. Sci. **803**, 82–93 (2020)
10. Garimella, K., Gionis, A., Parotsidis, N., Tatti, N.: Balancing information exposure in social networks (2017)
11. Kempe, D., Kleinberg, J., Tardos, É.: Maximizing the spread of influence through a social network. In: Proceedings of the ninth ACM SIGKDD International Conference on Knowledge Discovery and Data Mining, pp. 137–146. ACM (2003)
12. Kimura, M., Saito, K., Motoda, H.: Blocking links to minimize contamination spread in a social network. ACM Trans. Knowl. Dis. Data (TKDD) **3**(2), 9 (2009)
13. Li, S., Zhu, Y., Li, D., Kim, D., Huang, H.: Rumor restriction in online social networks. In: 2013 IEEE 32nd International Performance Computing and Communications Conference (IPCCC), pp. 1–10. IEEE (2013)
14. Lowalekar, M., Varakantham, P., Kumar, A.: Robust influence maximization. In: Proceedings of the 2016 International Conference on Autonomous Agents & Multiagent Systems, pp. 1395–1396. International Foundation for Autonomous Agents and Multiagent Systems (2016)
15. Morozov, E.: Swine flu: Twitters power to misinform. Foreign policy (2009)
16. Nemhauser, G.L., Wolsey, L.A., Fisher, M.L.: An analysis of approximations for maximizing submodular set functions-I. Math. Program. **14**(1), 265–294 (1978)
17. Nguyen, H.T., Thai, M.T., Dinh, T.N.: Stop-and-stare: optimal sampling algorithms for viral marketing in billion-scale networks. In: Proceedings of the 2016 International Conference on Management of Data, pp. 695–710. ACM (2016)
18. Nguyen, H., Zheng, R.: On budgeted influence maximization in social networks. IEEE J. Sel. Areas Commun. **31**(6), 1084–1094 (2013)
19. Ohsaka, N., Akiba, T., Yoshida, Y., Kawarabayashi, K.I.: Fast and accurate influence maximization on large networks with pruned monte-carlo simulations. In: AAAI, pp. 138–144 (2014)
20. Ping, Y., Cao, Z., Zhu, H.: Sybil-aware least cost rumor blocking in social networks. In: 2014 IEEE Global Communications Conference (GLOBECOM), pp. 692–697. IEEE (2014)
21. Tang, Y., Shi, Y., Xiao, X.: Influence maximization in near-linear time: a martingale approach. In: Proceedings of the 2015 ACM SIGMOD International Conference on Management of Data, pp. 1539–1554. ACM (2015)
22. Tang, Y., Xiao, X., Shi, Y.: Influence maximization: Near-optimal time complexity meets practical efficiency. In: Proceedings of the 2014 ACM SIGMOD International Conference on Management of Data, pp. 75–86. ACM (2014)
23. Tong, G., et al.: An efficient randomized algorithm for rumor blocking in online social networks. IEEE Trans. Netw. Sci. Eng. (2017)
24. Tsai, J., Nguyen, T.H., Tambe, M.: Security games for controlling contagion. In: AAAI (2012)

25. Wang, B., Chen, G., Fu, L., Song, L., Wang, X., Liu, X.: Drimux: dynamic rumor influence minimization with user experience in social networks. In: AAAI, vol. 16, pp. 791–797 (2016)
26. Yang, Y., Lu, Z., Li, V.O., Xu, K.: Noncooperative information diffusion in online social networks under the independent cascade model. IEEE Trans. Comput. Soc. Syst. 4(3), 150–162 (2017)
27. Zhang, H., Zhang, H., Li, X., Thai, M.T.: Limiting the spread of misinformation while effectively raising awareness in social networks. In: Thai, M.T., Nguyen, N.P., Shen, H. (eds.) CSoNet 2015. LNCS, vol. 9197, pp. 35–47. Springer, Cham (2015). https://doi.org/10.1007/978-3-319-21786-4_4
28. Zhu, J., Ghosh, S., Wu, W.: Group influence maximization problem in social networks. IEEE Trans. Comput. Soc. Syst. **PP**(99), 1–9 (2019)
29. Zhu, J., Ghosh, S., Wu, W.: Robust rumor blocking problem with uncertain rumor sources in social networks. World Wide Web (1), 24 (2021)
30. Zhu, J., Ni, P., Tong, G., Wang, G., Huang, J.: Influence maximization problem with echo chamber effect in social network. IEEE Trans. Comput. Soc. Syst. **PP**(99), 1–9 (2021)
31. Zhu, J., Zhu, J., Ghosh, S., Wu, W., Yuan, J.: Social influence maximization in hypergraph in social networks. IEEE Trans. Netw. Sci. Eng., 1–1 (2018). https://doi.org/10.1109/TNSE.2018.2873759

Algorithms for Shortest Path Tour Problem in Large-Scale Road Network

Yucen Gao[1], Mingqian Ma[1], Jiale Zhang[1], Songjian Zhang[2], Jun Fang[2], Xiaofeng Gao[1(✉)], and Guihai Chen[1]

[1] Shanghai Jiao Tong University, Shanghai, China
{guo_ke,mingqianma,zhangjiale100}@sjtu.edu.cn,
{gao-xf,gchen}@cs.sjtu.edu.cn
[2] Beijing Didi Co., Ltd., Beijing, China
{zhangsongjian,fangjun}@didiglobal.com

Abstract. Carpooling route planning becomes an important problem with the growth of low-carbon traffic systems. When each passenger has several potential locations to get on and off the car, the problem will be more challenging. In the paper, we discussed a simplified carpooling route planning problem, namely the Shortest Path Tour Problem (SPTP), whose aim is to find a single-origin single-destination shortest path through an ordered sequence of disjoint node subsets. We propose Stage Dijkstra and Global Dijkstra algorithms to find the optimal shortest path, with the time complexity of $O(l(n+m)\log n)$ and $O(l(n+m)\log(ln))$ respectively, where l represents the number of node subsets. To the best of our knowledge, $O(l(n+m)\log n)$ is the best time complexity of the exact algorithms for SPTP. Experiments conducted on large-scale road networks and synthetic datasets demonstrate the effectiveness and efficiency of our proposed algorithms.

Keywords: Carpooling route planning · Shortest path tour problem · Large-Scale Road Network · Stage & Global Dijkstra

1 Introduction

With the development of low-carbon transportation, carpooling becomes an increasingly important travel scenario [1]. Route planning is an essential problem in the carpooling scenario [2]. Compared with route planning for ride-hailing, route planning for carpooling that includes multiple pick-up and drop-off positions for passengers is more difficult. Given a many-to-one carpooling order that includes the driver's location and the locations of multiple passengers' pick-up and drop-off positions, the goal of route planning is to discover the shortest route from large-scale road networks. In addition, the problem becomes more challenging if each passenger has several possible candidate pick-up/drop-off locations, which is called the Multi-candidate Carpooling Route Planning (MCRP) problem. The MCRP problem is NP-hard, which can be proved by the reduction from the Travelling Salesman Problem (TSP), a famous NP-Complete problem [3].

W. Wu and G. Tong (Eds.): COCOON 2023, LNCS 14423, pp. 340–352, 2024.
https://doi.org/10.1007/978-3-031-49193-1_26

To solve the MCRP, we need to determine the pick-up and drop-off sequence and identify the most proper pick-up/drop-off position. In the paper, we focus on a simplified MCRP problem, namely the Shortest Path Tour Problem (SPTP), where the sequence has been determined. The aim of SPTP is to find the shortest path through at least one node in each node subset in a fixed order. It was first proposed in [4], as a variant of the Shortest Path Problem (SPP). The authors in [5] proved that SPTP belongs to the complexity class **P** by reducing it to a Single Source Shortest Path (SSSP) problem, which is a classical **P** problem. This means that the SPTP could be solved with the exact solution in polynomial time. However, SPTP is still worth studying because the time complexity of the algorithm needs to be as low as possible to meet the fast response requirements and to save computational resource overhead of the carpooling platform.

In addition to the carpooling scenario, applications of SPTP emerge in multiple other scenarios. For instance, automatic manufacture requires a robot to complete multiple stages of tasks [6]. Within each stage, the universal robot only needs to complete one of the selected tasks. This whole process can be seen as an SPTP in which tasks can be taken as nodes in a directed graph while the length of the arcs connecting every two nodes can be seen as the time needed for a robot to change tools to perform the following task.

Before reviewing the SOTA algorithms that will be described in Table 2 and summarizing our contributions, we first recall the definition of the SPTP.

Definition 1 (Shortest Path Tour Problem (SPTP)). *Given a directed graph $G = (V, E, W)$ with non-negative weights, the length $L(P)$ of a path P is defined as the sum of weights of the edges connecting consecutive nodes in P. Given an origin node $v_s \in V$, a destination node $v_d \in V$ and disjoint node subsets $V_1, V_2, \cdots, V_l$, ($|V_i| \leq r, \forall V_i$), where r represents the maximum size of node subsets, the aim is to find the shortest path P, which starts from v_s, sequentially passes through at least one node in each node subset, and ends at v_d.*

The notations are listed in Table 1.

Table 1. Symbols and Definitions

Symbol	Definition
$G = (V, E, W)$	directed graph with non-negative weights, $|V| = n$, $|E| = m$
v_s, v_d	origin and destination node
$V_1, V_2, \cdots V_l$	node subsets
l	number of node subsets
r	maximum size of the node subsets
d	average degree of nodes
$w(v_j, v_k)$	weight of edge from v_j to v_k
$dist(i, v_j)$	shortest distance from v_s to v_j that has passed through the first i node subsets in order

After the SPTP problem was proposed, a Dijkstra algorithm based on the SPTP reduction, some dynamic programming algorithms, shortest-path-based algorithms and a depth-first tour search algorithm were proposed [7,8]. Table 2 shows the comparison of the time complexity of existing algorithms.

Table 2. Comparison of Time Complexity of Existing Algorithms

Citation	Venue	Year	Algorithm	Time Complexity
[5]	Optim. Lett.	2012	Reduction + Dijkstra	$O(l^2 m^2 \log ln)$
			Dynamic Programming	$O(n^3 m)$
[7]	EJOR	2013	Modified Graph Method	$O(n^3)$
			Dynamic Programming	$O(l^2 rn^3)$
[8]	ICCCN	2017	DC-MPSP-1	$O(n^3) + O(lr^2)$
			DC-MPSP-2	$O(n^3) + O(lr^2)$
			DC-SSSP-1	$O(lr(n+m) \log n)$
			DC-SSSP-2	$O(l(n+m) \log n)$
			DFTS	$O(l^2(n+m) \log n)$

Assuming $V_0 \triangleq \{v_s\}, V_{l+1} \triangleq \{v_d\}$, SPTP reduction transforms an SPTP instance into an SSSP instance by calculating the shortest distances between nodes in adjacent stages, thus getting a trellis graph as shown in Fig. 1 [9]. Here, the original problem is transformed into finding the shortest path from v_s to v_d through all node subsets in order from the trellis graph. Hence, the problem can naturally be solved by applying the Dijkstra algorithm [10]. The first to sixth algorithms in Table 2 first use methods similar to SPTP reduction to get the trellis graph, and then use either the Viterbi algorithm [11], shortest-path algorithm or dynamic programming approach to achieve the exact solution.

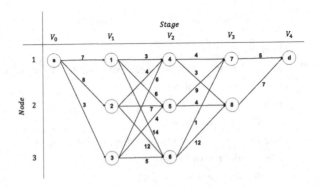

Fig. 1. A toy example of trellis graph

In order to reduce the time complexity of the algorithm, the latter three algorithms give up obtaining the fully connected trellis graph, and instead, compute only the edges that are possibly useful for the exact solution based on the SSSP or depth-first search methods. It is worth noting that although the DC-SSSP-2 algorithm has the lowest time complexity, the algorithm does not correctly derive the exact solution. We review the implementation of the DC-SSSP-2 algorithm.

DC-SSSP-2: When calculating distances between nodes in adjacent stages, it considers a virtual node $\hat{v}_i$ that connects to nodes in V_i with zero-weight edges, and applies Dijkstra to find the shortest paths from $\hat{v}_i$ to nodes in V_{i+1}. Hence, it has high efficiency as it only calls the Dijkstra algorithm once in each stage.

The problem with DC-SSSP-2 is that zero-weight edges are used to connect the virtual starting node with nodes in V_i. This can find the shortest paths from V_i to nodes in V_{i+1}, but it is not certain from which node in V_i the shortest path to the j_{th} node in V_{i+1} starts. This leads to the inability to find the eligible shortest path from v_s to v_d based on the idea of the shortest path algorithm.

Take the implementation of the DC-SSSP-2 in Fig. 1 as an example. We first calculate the shortest distance from V_0 to each node in V_1. Then it proceeds to calculate the shortest distance from V_1 to each node in V_2. However, it cannot correctly track the source of the shortest path from the previous stage to the nodes in the current stage when we want to form a complete path. For instance, we cannot actually add up the shortest distance 3 from V_0 to V_1 and 5 from V_1 to V_2 as they don't share the same node in V_1. This makes it impossible to concatenate the result from each stage to the eligible shortest path.

Therefore, the minimum time complexity of the existing algorithms that can correctly obtain the exact solution is $O(lr(n+m)\log n)$ or $O(l^2(n+m)\log n)$ [8]. In the paper, we propose an exact algorithm called Stage Dijkstra that can achieve a lower time complexity of $O(l(n+m)\log n)$ than all other existing methods. Besides, we propose a Global Dijkstra algorithm, though it has a higher time complexity, its running time is shorter in the large-scale road network.

2   Algorithm

2.1   Stage Dijkstra

We call the process of finding the shortest path from v_s to each node in V_i through the first i node subsets in order as stage i and propose the Stage Dijkstra algorithm. Algorithm 1 shows the pseudo-code of our proposed algorithm.

Stage Dijkstra first performs some initialization operations (Lines $1-2$), defining V_0, V_{l+1} and $dist[0, v_s]$ to facilitate the inclusion of subsequent per-stage computations. At stage i (Line 3), we first define the set of nodes whose shortest distance from v_s through the first $i-1$ node subsets in order are known (Line 4), and explore outward from this node set similar to the Dijkstra algorithm until the shortest distance from v_s through the first $i-1$ node subsets in order is found for all nodes in V_i (Lines $5-10$). Since these nodes belong to V_i, the shortest distance from v_s through the first $i-1$ node subsets is equal to the shortest distance from v_s through the first i node subsets in order. We perform the distance update operation (Lines $11 - 12$). When stage $l + 1$ is executed, we find the shortest path from v_s to v_d through all node subsets in order.

Algorithm 1: Stage Dijkstra

Input: Directed graph $G = (V, E, W)$, origin node v_s, destination node v_d, node subsets $V_1, V_2, \cdots V_l$.

Output: Shortest path P.

1 $V_0 \triangleq \{v_s\}, V_{l+1} \triangleq \{v_d\}$;

2 $dist[0, v_s] = 0$;

3 **for** *Stage* $i \leftarrow 1$ *to* $l + 1$ **do**

4      $S_i = V_{i-1}$;   //initialize the starting node set of Stage i

5   **while** $V_i \subset S_i \neq True$ **do** //calculate the shortest distance from v_s through the first $i - 1$ node subsets in order

6          $v_j = \arg\min_j dist[i - 1, v_j], v_j \in V/S_i$;

7          $S_i \leftarrow S_i \cup \{v_j\}$;

8   **for** $v_k \in G.Adj[v_j]$ **do**

9   **if** $dist[i - 1, v_j] + w(v_j, v_k) < dist[i - 1, v_k]$ **then**

10                 $dist[i - 1, v_k] = dist[i - 1, v_j] + w(v_j, v_k)$;

11   **for** $v_j \in V_i$ **do**   //update the shortest distance from v_s to each node in V_i through the first i node subsets

12         $dist[i, v_j] \leftarrow dist[i - 1, v_j]$;

Our algorithm implements an algorithm similar to the Dijkstra algorithm for single-source shortest path at each stage but differs in that we define a set of starting nodes and start exploring from them instead of exploring from a single node. We can also understand that at each stage i, we create a virtual starting node and connect the virtual starting node with the nodes in V_{i-1} using the shortest distance from v_s to each node in V_{i-1} obtained in the previous stage.

2.2 Global Dijkstra

In fact, we do not actually need to wait until the previous stage is completely finished before starting the next one. Instead, we can explore nodes in multiple stages simultaneously to increase computational efficiency. Algorithm 2 shows the pseudo-code of Global Dijkstra.

For the global algorithm, we perform the initialization operation first (Lines $1 - 6$). We construct a minimum heap to store distances to speed up the process of finding the nearest neighboring node (Line 5). We also set up a flag set to mark whether each stage is being executed. $f_i == 1$ means that stage i is being executed. We then perform the exploration process. When $V_{l+1} \subset S_{l+1} \neq True$, $dist[l+1, v_d]$ has not yet been found and the algorithm needs to continue running (Line 7). At each exploration, we take the nearest neighboring node from the top of the heap, thus determining the value of i and v_j (Line 8). If stage i is being executed and v_j has not been explored (Line 9), we update S_i and push the distances to neighboring nodes into the heap (Lines $10-12$). If $v_j \in V_i$, then $dist[i, v_j]$ can be determined (Lines $13-14$). If v_j is the first node in V_i that is added to S_i, stage $i+1$ can start execution (Lines $15-16$). If v_j is the last node in V_i added to S_i, all stages before stage i can be closed (Lines $17-19$).

Algorithm 2: Global Dijkstra

Input: Directed graph $G = (V, E, W)$, origin node v_s, destination node v_d, node subsets $V_1, V_2, \cdots V_l$.

Output: Shortest path P.

1 $V_0 \triangleq \{v_s\}, V_{l+1} \triangleq \{v_d\}$;
2 $dist[0, v_s] = 0$;
3 Construct a minimum heap $q = \emptyset$;
4 $q.push(dist[0, v_s])$;
5 For starting set $S = \{S_1, \cdots, S_{l+1}\}$, $S_i = \emptyset, \forall i \in \{1, 2, 3, \cdots, l+1\}$;
6 For flag set $F = \{f_1, \cdots f_{l+1}\}$, $f_1 = 1$, and $f_i = 0, \forall i \in \{2, 3, \cdots, l+1\}$;
   //initialize the flag set, which mark whether the stages are executing
7 **while** $V_{l+1} \subset S_{l+1} \neq True$ **do**   //not yet got $dist[l+1, v_d]$
8 $\quad dist[i-1, v_j] = q.pop()$;
9 $\quad$ **if** $v_j \notin S_i$ and $f_i == 1$ **then**   //search for the node with the minimum distance in the stages being executed
10 $\quad\quad S_i = S_i \cup \{v_j\}$;
11 $\quad\quad$ **for** $v_k \in G.Adj[v_j] \cap V/S_i$ **do**
12 $\quad\quad\quad q.push(dist[i-1, v_j] + w(v_j, v_k))$;
13 $\quad\quad$ **if** $v_j \in V_i$ **then**
14 $\quad\quad\quad dist[i, v_j] \leftarrow dist[i-1, v_j]$;
15 $\quad\quad\quad$ **if** $V_i \cap S_i = \{v_j\}$ **then**   //first node in V_i being added to S_i
16 $\quad\quad\quad\quad f_{i+1} = 1$;
17 $\quad\quad\quad$ **if** $V_i \subset S_i == True$ **then**   //last node in V_i being added to S_i
18 $\quad\quad\quad\quad$ **for** $k \leftarrow 1$ to i **do**   //not explore nodes in previous stages
19 $\quad\quad\quad\quad\quad f_k = 0$;

The advantage of the global algorithm is on the one hand that the use of minimum heap reduces the time to find the minimum distance value, and on the other hand that when stage i is finished, the previous stages before stage i can also be finished immediately, thus saving the unnecessary calculations. The essence of the global algorithm is to preferentially explore the nodes with smaller shortest distances starting from v_s. When arriving at v_d in such a way, the remaining unexplored paths will not yield eligible paths with shorter distances to v_d. We will give a more rigorous proof of correctness in Sect. 3.2.

Take the implementation of Global Dijkstra in Fig. 2 as an example. The exploration sequence for the nodes of the Global Dijkstra algorithm is $[v_s, v_1, v_2, v_3, v_4, v_6, v_d]$. Node involved are marked in red. Compared with Stage

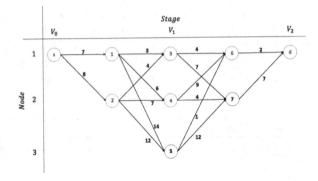

Fig. 2. A toy example of the implementation of Global Dijkstra

Dijkstra, the global algorithm does not need to calculated the shortest distance to v_5 and v_7, thus speeding up the execution time. This speedup comes from two aspects. One is that it is not necessary to explore all nodes in each stage, such as v_5, and the other is that some explorations of intermediate nodes between two adjacent stages can be saved, such as v_7. The effect of this acceleration will be even more pronounced in a larger scale real road network.

2.3   Complexity Analysis

Stage Dijkstra: Intuitively, Algorithm 1 has $l+1$ stages and the time complexity of each stage is equal to that of the Dijkstra algorithm for the single-source shortest path. Since the time complexity of Dijkstra is $O((n + m)\log n)$, the time complexity of Stage Dijkstra is $O(l(n + m)\log n)$.

From the perspective of formula calculation, the formula for calculating the time complexity of the Stage Dijkstra algorithm is

$$T^s_{stage} * T^s_{exploration} * (T^s_{extract-min} + T^s_{delete} + T^s_{add} * d) \tag{1}$$

where T^s_{stage} represents the number of stages (Line 3), $T^s_{exploration}$ represents the number of explorations in each stage (Line 5), $T^s_{extract_{min}}$ represents the number of operations required to find the nearest neighbor node (Line 6), T^s_{delete} represents the number of operations required to delete the explored node (Line 6), and T^s_{add} represents the number of operations required to add possible distances of neighboring nodes (Lines $8 - 10$). Since we use a minimum heap to store distances, the time complexity is $O((l+1) * (n-1) * (1 + \log n + \log n * d)) = O(ln(1 + d)\log n) = O(l(n + m)\log n)$.

Global Dijkstra: Compared with Stage Dijkstra, the time complexity of Global Dijkstra is higher even though the improved algorithm uses global computing to speed up the execution. The worst case is that all stage functions run simultaneously from stage 1 to stage $l+1$. In this case, we need to explore the neighboring node with the smallest shortest distance from the node set with the size of ln.

Intuitively, the algorithm is essentially equivalent to running the single-source shortest path algorithm on a graph with ln nodes and lm edges, which means that the time complexity of Global Dijkstra is $O(l(n+m)\log(ln))$.

From the perspective of formula calculation, the formula for calculating the time complexity of the Global Dijkstra algorithm is

$$T^g_{exploration} * (T^g_{extract-min} + T^g_{delete} + T^g_{add} * d) \tag{2}$$

Since the node size is ln, the time complexity is $O(ln-1)*(1+\log ln+\log ln* d)) = O(ln*(1+d)\log(ln)) = O(l(n+m)\log(ln))$. Though the time complexity is higher than the original Stage Dijkstra algorithm, the accurate execution time is faster, as demonstrated by the experimental results in Sect. 4.

3   Correctness Proof

3.1   Correctness Proof for Stage Dijkstra

We can prove that Algorithm 1 can find the shortest distance from v_s to v_d through all node subsets in order by the induction method and contradiction method. We first introduce Theorem 1 and give the proof.

Theorem 1. *Suppose $dist[i-1, v_j], \forall v_j \in V_{i-1}$ is the shortest distance from v_s to v_j through the first $i-1$ node subsets in order, and $v^{(1)}, v^{(2)}, \cdots, v^{(k)}$ are the nodes that are added to S_i in order at stage i. Then $dist[i-1, v^{(t)}], t \in \{1, 2, \cdots, k\}$ is the shortest distance from v_s to $v"(t)$ through the first $i-1$ node subsets in order.*

Proof. We use the induction method, which is divided into two steps.

Step 1: For stage i, $v^{(1)} = \arg\min_j dist[i-1, v_j], v_j \in V/V_{i-1}$, so $dist[i-1, v] \geq dist[i-1, v^{(1)}], \forall v \in V/(V_{i-1} \cup \{v^{(1)}\})$. This means that there is no node $v \in V/(V_{i-1} \cup \{v^{(1)}\})$ such that $dist[i-1, v] + w(v, v^{(1)}) < dist[i-1, v^{(1)}]$. That is, $dist[i-1, v^{(1)}]$ is the shortest distance from v_s to $v^{(1)}$ through the first $i-1$ node subsets in order.

Step 2: We need to prove that, assuming $dist[i-1, v^{(t-1)}]$ is the shortest distance from v_s to $v^{(t-1)}$ through the first $i-1$ node subsets in order, $dist[i-1, v^{(t)}]$ is also the shortest distance from v_s to $v^{(t)}$ through the first $i-1$ node subsets in order. After $v^{(t-1)}$ is added to S_i, $v^{(t)} = \arg\min_j dist[i-1, v_j], \forall v_j \in V/S_i$. That is to say, there is no node $v \in V/(S_i \cup \{v^{(t)}\})$ such that $dist[i-1, v] + w(v, v^{(t)}) < dist[i-1, v^{(t)}]$. Hence, $dist[i-1, v^{(t)}]$ is the shortest distance from v_s to $v^{(t)}$ through the first $i-1$ node subsets.

We then prove the correctness of Stage Dijkstra.

Theorem 2. *$dist[l+1, v_d]$ obtained by Stage Dijkstra is the shortest distance from v_s to v_d through all node subsets in order.*

Proof. We still use the induction method, which is divided into two steps.

Step 1: Since $dist[0, v_s] = 0$, $V_0 = \{v_s\}$, and the minimum distance in the non-negative weighted graph is 0, $dist[0, v_j], \forall v_j \in V_0$ is the shortest distance.

Step 2: We need to prove that, assuming $dist[i-1, v_j], \forall v_j \in V_{i-1}$ is the shortest distance through the first $i-1$ node subsets in order, then $dist[i, v_j], \forall v_j \in V_i$ is the shortest distance through the first i node subsets in order. According to Theorem 1, $dist[i-1, v_j], \forall v_j \in V_i$ is the shortest distance through the first $i-1$ node subsets. Since $dist[i, v_j] \geq dist[i-1, v_j], \forall v_j \in V$, and $dist[i, v_j] = dist[i-1, v_j], \forall v_j \in V_i$, $dist[i, v_j], \forall v_j \in V_i$ is the shortest distance through the first i node subsets.

Hence, $dist[l+1, v_j], \forall v_j \in V_{l+1}$ is the shortest distance through the first $l+1$ node subsets in order. Since $v_d \in V_{l+1}$, $dist[l+1, v_d]$ is the shortest distance through all node subsets in order.

3.2 Correctness Proof for Global Dijkstra

Though different stages are executed alternately, the order of nodes that are added to S_i in stage i in Algorithm 2 is constantly compared with that in Algorithm 1. Hence, $dist[i-1, v_j], \forall v_j \in V/V_{i-1}$ obtained by Algorithm 2 is still the shortest distance from v_s to v_j through the first $i-1$ node subsets in order, and $dist[i, v_j], \forall v_j \in V_i$ obtained by Algorithm 2 is still the shortest distance from v_s to v_j through the first i node subsets in order. That is to say, Theorem 1 and Theorem 2 still hold for Algorithm 2.

4   Experiments

4.1   Baseline Methods

Reduction Dijkstra [5]: The algorithm first uses the SPTP Reduction method to obtain the shortest distance between two nodes in transformed trellis graph G'. Then, it uses the Dijkstra algorithm to obtain the shortest path.

Depth-First Tour Search (DFTS). [8]: The algorithm constructs multiple sets with the distance as the object of comparison for all stages, and selects the node with the minimum distance in possible sets. Note that, the time-consuming operation of finding the minimum value may negatively affect the performance.

4.2   Dataset

Real-World Dataset: We use the Beijing, Shanghai and Qingdao road network from an anonymous carpooling service provider because they have different road network structures. We randomly select samples from the historical orders and use the popular pick-up and drop-off nodes based on historical data within 150m of the order's pick-up and drop-off positions as multiple candidate nodes. Table 3 shows the statistics corresponding to the different datasets.

Table 3. Statistics of the Datasets

Dataset	#Edges	#Nodes	Time Range
Beijing	2,575,216	1,119,143	11/07/2022-11/13/2022
Shanghai	2,840,477	1,225,903	08/03/2022
Qingdao	2,135,189	821,661	10/10/2022

(a) Square Grid Road Network   (b) Circular Radial Road Network   (c) Freestyle Road Network

Fig. 3. The Structure of Road Networks in Different Cities

Figure 3 shows the difference in the road network structure of different cities (labels on maps are marked in Chinese). Beijing has a relatively more regular road network structure as shown in Fig. 3(a), while road network structures of Shanghai and Qingdao are freer as shown in Fig. 3(b) and (c).

Synthetic Dataset: We use latitude and longitude to frame subplots from real datasets and use the Kosaraju algorithm to select the largest strongly connected directed graph. We construct instances by uniformly randomly selecting nodes.

4.3 Configurations and Metric

Table 4. Parameter Setting

Parameter	Value
No. of nodes	$[10, 100, 1000, \mathbf{10000}, 100000, 1000000]$
No. of stages	$[4, \mathbf{6}, 8]$
No. of nodes in each stages	$[3, \mathbf{5}, 7, 9]$

All the experiments are implemented on a Macbook Pro with 16 GB memory and a 2.9 GHz Intel Core i7 CPU. Our proposed algorithms and baselines are implemented in Python. Besides, we set the following parameters and change their values to test the algorithms' performance as shown in Table 4, where the default ones are marked in bold. Since the algorithms can find the exact shortest path, we use the running time as the metric to evaluate the performance.

4.4   Results

Results for Synthetic Dataset: Figure 4, 5 and 6 show the running time for different parameter values in Beijing, Shanghai and Qingdao.

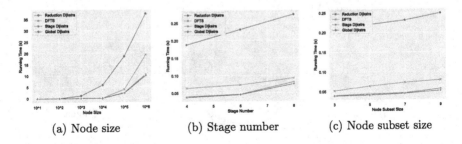

(a) Node size   (b) Stage number   (c) Node subset size

Fig. 4. Running time comparison for part of Beijing road network

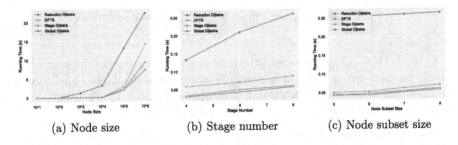

(a) Node size   (b) Stage number   (c) Node subset size

Fig. 5. Running time comparison for part of Shanghai road network

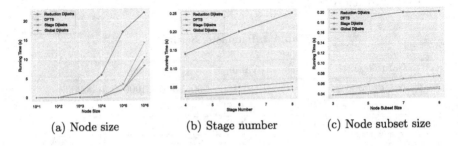

(a) Node size   (b) Stage number   (c) Node subset size

Fig. 6. Running time comparison for part of Qingdao road network

Regarding the node size, the algorithms have little variability in running time on a graph with less than a hundred nodes. However, as the node size increases, the gap in running time between the baseline algorithm and our proposed algorithms expands rapidly. Because the actual road network's node size is of the scale of millions or more, the baseline algorithms can hardly meet the demand for fast real-time response. Moreover, there are many solution requirements per day, thus using our proposed algorithm will save a lot of computational resources.

In terms of the number of node subsets and the node size in node subsets, the running time of our proposed algorithms grows linearly with either of them, which is consistent with the time complexity analysis in Sect. 2.3. The running time for different subset sizes doesn't change much for Reduction Dijkstra, as only the number of stages influences the size of the map it constructs.

Results for Real-World Dataset: We sample 100 orders uniformly randomly from real orders in three cities and took the average running time. Table 5 reports the corresponding results. In the actual application scenario, the Global Dijkstra achieves the best performance, while the Reduction Dijkstra algorithm that needs to calculate the shortest distance between nodes for adjacent stages is much slower than the other three algorithms. This indicates that although the time complexity of Global Dijkstra is higher than that of Stage Dijkstra, it can stop unnecessary exploration of nodes in intermediate stages in advance when the shortest path from v_s to v_d through all node subsets has been searched.

The running time of Reduction Dijkstra is positively correlated with the node size of the city. For the other three algorithms, in addition to the node size, the structure of road network can also considerably impact the running time. For instance, finding the shortest paths in Shanghai with more nodes takes less running time than Beijing except for Reduction Dijkstra. Observing the network structure, when exploring nodes in Beijing with a more regular road network structure, more intermediate nodes that may constitute the shortest path are explored, leading to a longer running time despite a smaller network size.

Table 5. Running Time of Different Algorithms for Different Cities (Seconds)

Algorithm	Dataset		
	Beijing	Shanghai	Qingdao
Reduction Dijkstra	33.370	46.426	20.477
DFTS	1.035	0.734	0.194
Stage Dijkstra	0.537	0.378	0.100
Global Dijkstra	**0.436**	**0.291**	**0.072**

5   Conclusion

In the paper, we propose a Stage Dijkstra algorithm for finding the shortest path with a single origin and a single destination passing through at least one node in multiple node subsets in order, which improves the time complexity of the problem from the previously known $O(lr(n+m)\log n)$ or $O(l^2(n+m)\log n)$ [8] to $O(l(n+m)\log n)$. We further propose an improved global algorithm to explore nodes in different stages simultaneously thus speeding up the execution time. We validate the effectiveness and efficiency of the Stage Dijkstra and Global Dijkstra algorithm through experiments based on a city-level road network. Although the worst time complexity of the Global Dijkstra is higher than that of Stage Dijkstra, its execution time is lower than that of Stage Dijkstra.

Acknowledgements. This work was supported by the National Key R&D Program of China [2020YFB1707900], the National Natural Science Foundation of China [62272302, 62172276], Shanghai Municipal Science and Technology Major Project [2021SHZDZX0102], and DiDi GAIA Research Collaboration Plan [202204].

References

1. Zhang, W.A., Yuan, C., Zhang, Y., Ye, J.: Taxi or hitchhiking: predicting passenger's preferred service on ride sharing platforms. In: ACM SIGIR Conference on Research & Development in Information Retrieval (SIGIR), pp. 1041–1044 (2018)
2. Zeng, Y., Tong, Y., Song, Y., Chen,L.: The simpler the better: An indexing approach for shared-route planning queries. In: Proceedings of the VLDB Endowment, vol. 13, pp. 3517–3530 (2020)
3. Dantzig, G.B., Ray Fulkerson, D., Johnson, S.M.: Solution of a large-scale traveling-salesman problem. Oper. Res. **2**(4), 393–410 (1954)
4. Bajajm, C.P.: Some constrained shortest-route problems. In: Unternehmensforschung, vol. 15(1), pp. 287–301 (1971)
5. Festa, P.: Complexity analysis and optimization of the shortest path tour problem. Optimiz. Lett. **6**(1), 163–175 (2012)
6. Osanlou, K., Bursuc, A., Guettier, C., Cazenave, T., Jacopin, E.: Optimal solving of constrained path-planning problems with graph convolutional networks and optimized tree search. In: International Conference on Intelligent Robots and Systems (IROS), pp. 3519–3525 (2019)
7. Festa, P., Guerriero, F., Laganà, D., Musmanno, R.: Solving the shortest path tour problem. Euro. J. Oper. Res. **230**(3), 464–474 (2013)
8. Bhat, S., Rouskas, G.N.: Service-concatenation routing with applications to network functions virtualization. In: International Conference on Computer Communication and Networks (ICCCN), pp. 1–9 (2017)
9. Prokopiak, R.: Using the Viterbi decoding trellis graph approach to find the most effective investment path. In: Smart Cities & Information and Communication Technology (CTTE-FITCE), pp. 1–6 (2019)
10. Dijkstra, E.W.: A note on two problems in connexion with graphs. Numer. Math. **1**, 269–271 (1959)
11. Viterbi, A.J.: Error bounds for convolutional codes and an asymptotically optimum decoding algorithm. IEEE Trans. Inform. Theory **IT-13**, 260–269 (1967)

Solving Systems of Linear Equations Through Zero Forcing Set

Jianbo Wang[ID], Chao Xu[✉][ID], and Siyun Zhou[ID]

University of Electronic Science and Technology of China, Chengdu, China
the.chao.xu@gmail.com, zhousiyun@std.uestc.edu.cn

Abstract. Let $\mathbb{F}$ be any field, we consider solving $Ax = b$ for a matrix $A \in \mathbb{F}^{n \times n}$ of m non-zero elements and $b \in \mathbb{F}^n$. If we are given a zero forcing set of A of size k, we can solve the linear equation in $O(mk + k^\omega)$ time, where ω is the matrix multiplication exponent. As an application, we show how the lights out game in an $n \times n$ grid is solved in $O(n^3)$ time, and then improve the running time to $O(n^\omega \log n)$ by exploiting the repeated structure in grids.

Keywords: Linear Algebra · Finite Field · Zero Forcing

1 Introduction

Solving the linear system $Ax = b$ has been a fundamental problem in mathematics. We consider the problem in its full generality. Given a field $\mathbb{F}$, a matrix $A \in \mathbb{F}^{n \times n}$ and a vector $b \in \mathbb{F}^n$, we are interested to find an $x \in \mathbb{F}^n$ such that $Ax = b$. Additionally, a zero forcing set associated with A is also given.

As arguably the most important algorithmic problem in linear algebra, there are many algorithms designed for solving systems of linear equations, we refer the reader to [13]. However, most works are for matrices over the real, complex or rational number fields. As this work is concerned with general fields, we will describe some known general algorithms below. The Gaussian elimination is a classic method for solving systems of linear equations, which requires $O(n^3)$ time. If the system is of full rank, one can reduce the problem to a single matrix multiplication, by taking $x = A^{-1}b$. In general cases, the more recent LSP decomposition, which is a generalization of the LUP decomposition, allows the running time to match the matrix multiplication [15,16]. If one views the non-zeros of the matrix as an adjacency matrix of a graph, the property of the graph can be then used to speed up the algorithm. For example, if the graph is planar, then the nested-dissection technique can generate an $O(n^{3/2})$ time algorithm when the field is real or complex [19]. Later, it was extended to any non-singular matrix over arbitrary fields, and the running time was improved to $O(n^{\omega/2})$ [3], where $\omega < 2.3728596$ is the matrix multiplication constant [2]. More recently, Fomin et al. made a major breakthrough in developing a fast algorithm that has a polynomial dependency on treewidth, pathwidth, and tree-partition

W. Wu and G. Tong (Eds.): COCOON 2023, LNCS 14423, pp. 353–365, 2024.
https://doi.org/10.1007/978-3-031-49193-1_27

width of the bipartite graph generated from the incidences of the row and the column [11,12]. In particular, when restricted to square matrices, there exist an $O(k^2 n)$ time algorithm for solving $Ax = b$ if a width k path decomposition or tree-partition is given, and an $O(k^3 n)$ time algorithm if a tree decomposition of width k is given. These algorithms are fairly complicated, but share one similarity with our work in taking the advantage of properties of the graph to improve the running time of the algorithm (Fig. 1).

Algorithm	Structural Requirement	Running Time	Algebraic Requirement
[19]	planar	$O(n^{3/2})$	positive definite
[3]	planar	$O(n^{\omega/2})$	square, non-singular
[12]	pathwidth/tree-partition width k	$O(k^2(n+p))$	−
[12]	treewidth k	$O(k^2(n+p))$	−
This Work	zero-forcing number k	$O(km + k^\omega)$	−

Fig. 1. Known algorithms using graph structure corresponding to the $n \times p$ matrix.

Zero forcing sets were first studied in [1] on graphs relating to the maximum nullity of a matrix, and the results were later expanded to the directed graphs [4]. The zero forcing set captures some "core" information of the linear system, or to be more specific, the system $Ax = b$ is uniquely determined by the values of x on a zero forcing set, which implies that one just needs to observe part of x to recover its entirety. This idea leads to the independent discovery of zero forcing set by physicists for control of quantum systems [8,20]. Later it was shown to be applicable to Phasor Measurement Units (PMU) for monitoring power networks [9], and also discovered as a graph searching technique [23]. Most studies in zero forcing set concentrate on its algebraic and combinatorial properties. On the computational front, finding the smallest zero forcing set of both the undirected and directed graphs is NP-hard [21,23], and some exact algorithms have been proposed [6,7].

This work was inspired by an algorithm for the lights out game. In the game, there is a light on each vertex of the graph, and each light is in a state either on (1) or off (0). There is also a button on each vertex. Pressing the button would flip the state of the light of itself and all its (out-)neighbors. The goal is to turn off all the lights. The lights out game is equivalent to solving a system of linear equations $Ax + b = 0$ in $\mathbb{F}_2$, where A is the adjacency matrix of the graph G, and b is the state of the lights and $\mathbb{F}_2$ is the finite field of order 2. The interpretation is that $x_v = 1$ means pressing the button at vertex v, and b_v is the initial state of the light at vertex v. Note that in $\mathbb{F}_2$, $-b = b$, and thus the game is equivalent to solving $Ax = b$. There is a large amount of research in the lights out game, see [10] for a survey. Finding a solution of the lights out game with the minimum number of button presses is NP-hard [5].

We focus on the case where G is an $n \times n$ grid, which corresponds to an $n^2 \times n^2$ matrix. Gaussian elimination would take $O(n^6)$ time. An alternative approach is the light-chasing algorithm [18], which is equivalent to zero forcing in the grid graph. Since the lights in the first row uniquely affect the states of the remaining rows, one can then look up which action on the first row should be taken according to the states of the last row. However, the literature does not provide the strategy for finding the lookup table. Wang claimed there is an $O(n^3)$ time algorithm that given the state of the last row, the state of the first row can be found through solving a linear equation of an n-square matrix [22]. Wang's result is the motivation behind this work. However, there is no proof of its correctness. Alternatively, the algorithm in [12] can be used to obtain an $O(n^4)$ time solver by observing that the pathwidth is n and the path decomposition is computed in linear time.

Our Contribution. We generalize Wang's method for the lights out game to solving an arbitrary system of linear equations $Ax = b$, and give a formal proof of its correctness. We define a structure, the core matrix B of A, such that one can solve the system of linear equations over B instead of A, and then lift it to a solution of A in linear time. As a consequence, we obtain the following algorithmic result.

Theorem 1 (main). *Given a matrix $A \in \mathbb{F}^{n \times n}$ of m non-zero elements, and a zero forcing set of A of size k where $k \leq n$. A data structure of size $O(k^2)$ can be computed in $O(mk + k^\omega)$ time, such that for each $b \in \mathbb{F}^n$, solving $Ax = b$ takes $O(k^2 + m)$ time.*

Note that the data structure is just the LSP decomposition of the core matrix.

We also show that the lights out game on an $n \times n$ grid can be solved in $O(n^\omega \log n)$ time by finding the core matrix using an alternative method. In addition, we prove some linear algebraic properties of the zero forcing set, which might be of independent interest.

Comparison with Previous Algorithms. If the pathwidth and zero-forcing number are within a constant of each other, the performance of our algorithm can be no worse than the existing algorithms. Actually, our algorithm can be much more efficient when the graph is sparse. In the extreme case where m, the number of edges, is $O(n)$, our algorithm is faster by a factor of k. For example, when the graph is an $n \times n$ grid graph, the pathwidth is n, and so is the zero-forcing number. Since there are n^2 nodes, the algorithm in [12] would then take $O(n^4)$ time, while our algorithm takes a faster $O(n^3)$ time.

On the other hand, zero forcing number can be much larger than the pathwidth. Indeed, the zero forcing number of a star on n vertices is $n - 1$, but the pathwidth is $O(1)$. Also, finding a small zero forcing set is difficult, and no good approximation is known so far. In this context, our algorithm may be very limited in practical applications.

2   Preliminaries

Define an index set $[n] = \{1, \ldots, n\}$. We consider an algebraic model, where every field operation takes $O(1)$ time, and each element in the field can be stored in $O(1)$ space.

It is useful to view a vector x indexed by elements X as a function $x : X \to \mathbb{F}$, and thus we can write $x \in \mathbb{F}^X$. Define $\mathrm{supp}(x)$ to be the set of non-zero coordinates of x. For a matrix A and a set of row indices R and a set of column indices C, we define $A_{R,C}$ to be the submatrix of elements indexed by the rows and columns in R and C, respectively. We use $*$ to represent the set of all row indices or column indices, depending on context. For example, $A_{R,*}$ stands for the submatrix of A that is composed of the rows indexed by R, $A_{i,*}$ is the ith row vector of A, and $A_{*,j}$ is the jth column vector. In addition, $A_{i,j}$ is the element in the ith row and jth column of A. For a vector x, x_i is the scalar at index i, and x_I is the subvector formed by the elements in the index set I.

Solving $Ax = b$ is finding a vector x such that $Ax = b$ holds. Given a directed graph $G = (V, E)$ with n vertices, and a field $\mathbb{F}$, we define $\mathcal{M}(G, \mathbb{F})$ to be the set of all matrices $A \in \mathbb{F}^{n \times n}$ such that for all $u \neq v$, $A_{u,v} \neq 0$ if and only if $(u, v) \in E$. We do not impose any restriction on $A_{v,v}$. Let $N^+(u) = \{v \mid (u, v) \in E\}$ be the set of all out-neighbors of u.

Then, we briefly review some basic concepts and results related to the zero forcing. Consider the process of coloring a graph. We start with a set of blue colored vertices, denoted by Z. If there exists a blue vertex v with exactly one non-blue out-neighbor, say $u \in V$, then color u blue. The operation of turning u blue is called *forcing*, and we say u is forced by v. If this process ends up with a situation where all vertices are colored blue, then we say the initial set Z is a *zero forcing set*. The *zero forcing number* $Z(G)$ of G is the size of the smallest zero forcing set. We also call Z a zero forcing set of A if $A \in \mathcal{M}(G, \mathbb{F})$ for some field $\mathbb{F}$. The following proposition gives the reason for the name "zero forcing".

Proposition 1 ([1,14]). *For a zero forcing set Z of A, if $x \in \ker(A)$, then $x_Z = 0$ implies $x = 0$. Namely, x vanishing at zero forcing set forces x to be 0.*

The converse is not true in general. As a counterexample, we can simply take A to be a 2×2 identity matrix. Then, $Z = \{1, 2\}$ is the unique zero forcing set of A, and $\ker(A) = \{0\}$. If $x \in \ker(A)$ and $x_1 = 0$, then we have $x = 0$. But $\{1\}$ is clearly not a zero forcing set.

The order of forces for a set Z is not unique, although the final coloring is [1,14]. For simplicity, we avoid this issue by considering a particular chronological list of forces π, which picks the smallest indexed vertex that can be forced. π is a total ordering of the vertices such that the elements in Z are ordered arbitrarily, and smaller than all elements in $V \setminus Z$. For each $v, u \in V \setminus Z$, $v \leq_\pi u$ if v is forced no later than u. The forcing graph is a graph where there is an edge between u and v if u forces v. It is well known that such graph is a set of node disjoint paths that begin in Z [14]. Hence, if v forces u, we can define $u^\uparrow = v$ to be the *forcing parent*, and correspondingly, u is called the *forcing child* of $u^\uparrow$. A vertex is a *terminal*, if it does not have a forcing child. Let T be the set of terminals, then $|T| = |Z|$. See Fig. 2 for sequence of forcing and terminal vertices.

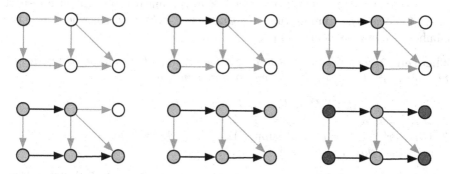

Fig. 2. The sequence of forcing starting from right most two vertices. In the final figure, the zero forcing set Z is the dark blue vertices, and the terminal set T is the red vertices. (Color figure online)

Consider the algorithm FORCING(A, Z, b) as given in Fig. 3, which takes a matrix A, a vector b, and a zero forcing set Z of A as inputs, and updates x_u for all $u \in V$ in each round. We set $x_u = 0$ during initialization. Then, for each $u \in V \backslash Z$, we update x_u according to $\left(b_{u\uparrow} - \sum_{v \in N^+(u\uparrow) \backslash \{u\}} A_{u\uparrow,v} x_v\right) / A_{u\uparrow,u}$, iteratively in the forcing order. Note that it is equivalent to setting $x_u \leftarrow \left(b_{u\uparrow} - A_{u\uparrow,*} x\right) / A_{u\uparrow,u}$, since x_u is previously 0.

FORCING(A, Z, b)
   $x \leftarrow 0$
   for $u \in V \setminus Z$ ordered by some forcing sequence π
      $x_u \leftarrow \left(b_{u\uparrow} - A_{u\uparrow,*} x\right) / A_{u\uparrow,u}$
   return x

Fig. 3. The forcing operation.

The following result formalizes the relation between the solution to a given linear system and the corresponding zero forcing set.

Proposition 2. *After the value of x_u is updated in* FORCING(A, Z, b), *we have* $A_{u\uparrow,*} x = b_{u\uparrow}$ *for all* $v \leq_\pi u$. *In particular, if* $Ax = b$ *has a solution where* $x_Z = 0$, *then* FORCING(A, Z, b) *finds such solution.*

Proof. The proof is by induction. The trivial case is when no x_u has been updated, then the conclusion is vacuously true. If some x_u is updated by FORCING(A, Z, b), we let x' be the vector before the update, and x'' be the one after the update. Then we get $x'_v = x''_v$ for all $v \neq u$, and $x'_u = 0$. And

$A_{u,*}x'' = A_{u,*}x' + A_{u\uparrow,u}x''_u = A_{u,*}x' + A_{u\uparrow,u}(b_{u\uparrow} - A_{u\uparrow,*}x')/A_{u\uparrow,u} = b_{u\uparrow}$. Also, for all $v <_\pi u$, $A_{v\uparrow,*}x'' = A_{v\uparrow,*}x' = b_{v\uparrow}$.

For some fixed A and Z, FORCING(A, Z, b) is a linear transform with respect to b. Moreover, the running time of FORCING(A, Z, b) is $O(m)$, where m is the number of non-zero elements in A.

Theorem 2. *Let Z be a zero forcing set of A, and L be the matrix such that $Lb =$ FORCING(A, Z, b). The following statements hold.*

1. *The columns in $V \backslash Z$ are linearly independent.*
2. *If $Ax = b$, $Ay = b$, and $x_Z = y_Z$, then $x = y$.*
3. *Given $x' \in \mathbb{F}^V$ such that $\mathrm{supp}(x') \subseteq Z$. If $Ax = b$ for some x such that $x_Z = x'_Z$, then $x = L(b - Ax') + x'$.*

Proof. The proofs of the first two statements can be found in [17]. Though for the second statement, only the $b = 0$ version was proven in [17], its proof still works for a general b. For the third statement, if $Ax = b$ where $x_Z = x'_Z$, then $A(x' + (x - x')) = b$, or in other words, $A(x - x') = b - Ax'$. Hence by Proposition 2, $L(b - Ax') = x - x'$, and therefore $(x - x') + x' = x$.

The *LSP decomposition* of $A \in \mathbb{F}^{m \times n}$ takes the form $A = LSP$, where $L \in \mathbb{F}^{m \times m}$ is a lower triangular matrix with value 1 in the diagonals, $S \in \mathbb{F}^{m \times n}$ can be reduced to a upper triangular matrix if all zero rows are deleted and elements in the main diagonal after the zero row deletion is non-zero, and $P \in \mathbb{F}^{n \times n}$ is a permutation matrix. The LSP decomposition can be found in $O(mnr^{\omega-2})$ time [16], where r is the rank of A. Given the LSP decomposition of an $n \times n$ matrix A, solving $Ax = b$ can be achieved by solving $Ly = b$ and $SPx = y$ based on the back substitution [15], which takes $O(n^2)$ time.

3   Solving $Ax = b$ Through Zero Forcing Set

In this section, we first introduce the *core matrix* that serves as the key part of our algorithm, followed by its theoretical guarantees. Based on the core matrix, we then present the detailed algorithm for solving $Ax = b$ with a given zero forcing set of size k, as well as the corresponding computational analysis.

For a clear exposition, we set up the instances we are working with. Throughout this section, we fix a directed graph $G = (V, E)$, a field $\mathbb{F}$, a matrix $A \in \mathcal{M}(G, \mathbb{F})$, a zero forcing set Z of G, a forcing order π, and the terminals T under the forcing order. Since the operations are performed on indices, without loss of generality, we assume that $V = [n]$ and $Z = [k]$.

Let $L(b) =$ FORCING(A, π, b). Given that L is a linear transform, we abuse the notation and let L be the matrix that induces the transform. Let $R = I - AL$. It can be observed that $\mathrm{supp}(Rb) \subset T$ and $\mathrm{supp}(Lb) \subset V \backslash Z$ for all b. Moreover, we have that $|V| = n$, $|Z| = k$, and the number of non-zero elements in A is m.

3.1   Core Matrix

As Theorem 2 suggests, the solution to $Ax = b$ can be obtained by knowing only x_Z. Hence, it is natural to think of finding the correct x_Z instead of acquiring the full information of x for solving $Ax = b$.

Let $a_v = A_{*,v}$ be the column of A indexed by v. Define a $k \times k$ matrix $B \in \mathbb{F}^{T \times Z}$ as $B = (RA)_{T,Z}$. In other words, the vth column equals $R(a_v)_T$. The matrix B is called the *core matrix* of A. We will show that for an arbitrary b, $Ax = b$ holds if and only if $Bx_Z = R(b)_T$. Hence, solving the equation $By = R(b)_T$ along with some post processing is sufficient to give a solution of $Ax = b$.

Lemma 1. $Ra_v = 0$ *for* $v \notin Z$.

Proof. The forcing algorithm given in Fig. 3 shows that if $v \neq u$, we will then have $x_v = 1$, and $x_u = 0$. Note that $Ax = a_v$, and therefore $Ra_v = a_v - Ax = 0$.

Before delving into the key theorem, we introduce some useful notations and definitions that will be repeatedly used in the remaining part of this section. Let $\text{rank}(A) = r \geq n - k$. Define $A' = A_{*,[k]} \in \mathbb{F}^{n \times k}$, and $A'' = A_{*,[n]\setminus[k]} \in \mathbb{F}^{n \times (n-k)}$. To facilitate the analysis, we further decompose A' into a 1×2 block matrix form $A' = [A'_1 \ A'_2]$ such that $A'_1 = [a_1 \ \cdots \ a_p]$, $A'_2 = [a_{p+1} \ \cdots \ a_k]$, $\text{rank}([A'_1 \ A'']) = r$, $\text{rank}(A'') = n - k$ and $\text{rank}(A'_1) = p = r - (n - k)$. The matrix A can be then rewritten as

$$A = [A'_1 \ A'_2 \ A'']. \tag{1}$$

The following result presents the rank-preserving property of A'_1 under the linear mapping R. Based on this property, together with the assumption of the existence of the solution to $Ax = b$.

Lemma 2. $\text{rank}(RA'_1) = \text{rank}(A'_1)$.

Proof. For any $s \in \mathbb{F}^n$, there exist $\gamma_{k+1}, \cdots, \gamma_n$, such that

$$s + \sum_{i=k+1}^{n} \gamma_i a_i = Rs,$$

which implies that $(R - I)s \in \text{span}(A'')$. Define another linear mapping Q as

$$Q = R - I, \tag{2}$$

and we have span $Q \subset$ span A''. Then, we can write RA'_1 as

$$RA'_1 = (Q + I)A'_1 = QA'_1 + A'_1,$$

where $QA'_1 \in$ span $Q \subset$ span A''. Hence, the columns of QA'_1 and the ones of A'_1 are linearly independent, which immediately gives $\text{rank}(RA'_1) = \text{rank}(A'_1)$.

The following theorem plays a pivotal role in the theoretical guarantees for our algorithm.

Theorem 3. *Given $A \in \mathbb{F}^{n \times n}$ of the form (1). Let $M \in \mathbb{F}^{k \times n}$ $(k \leq n)$ such that $A'' \in \ker(M)$, and $\mathrm{rank}(MA'_1) = \mathrm{rank}(A'_1)$. For $b \in \mathbb{F}^n$, suppose that $Ax = b$ has a solution. If $MA'y = Mb$ for some $y \in \mathbb{F}^k$, then there exists $x \in \mathbb{F}^n$ such that $Ax = b$ and $x_Z = y$.*

Proof. Since $A'_2 \in \mathrm{span}(A'_1, A'')$, we can rewrite A'_2 as $A'_2 = A'_1 C'_1 + A'' C''$ where $C'_1 \in \mathbb{F}^{p \times (k-p)}$ and $C'' \in \mathbb{F}^{(n-k) \times (k-p)}$ are coefficient matrices. The existence of the solution to $Ax = b$ indicates that $b \in \mathrm{span}(A'_1, A'')$. Similarly, we can write $b = A'_1 C'_b + A'' C''_b$ where $C'_b \in \mathbb{F}^p$ and $C''_b \in \mathbb{F}^{n-k}$ are coefficient vectors. Then, we have

$$M[A'_1 \quad A'_2]y = M[A'_1 \quad A'_1 C'_1 + A'' C'']y = M[A'_1 \quad A'_1 C'_1]y,$$
$$Mb = M(A'_1 C'_b + A'' C''_b) = MA'_1 C'_b,$$

where the last equalities of the above two formulas follow from $A'' \in \ker(M)$. Then, the equation $MA'y = Mb$ can be written as $[MA'_1 \quad MA'_1 C'_1]y = MA'_1 C'_b$. Decomposing $y \in \mathbb{F}^k$ into $y = \begin{bmatrix} y_1 \\ y_2 \end{bmatrix}$ with $y_1 \in \mathbb{F}^p, y_2 \in \mathbb{F}^{k-p}$ further leads to a new homogeneous linear system

$$MA'_1(y_1 + C'_1 y_2 - C'_b) = 0.$$

The condition $\mathrm{rank}(MA'_1) = \mathrm{rank}(A'_1)$ gives $\mathrm{rank}(MA'_1) = p$, or to say, MA'_1 is of full rank. Thus, we have $y_1 + C'_1 y_2 - C'_b = 0$.

In a similar manner, we decompose $x \in \mathbb{F}^n$ into three subvectors $x_1 \in \mathbb{F}^p$, $x_2 \in \mathbb{F}^{k-p}$, and $x_3 \in \mathbb{F}^{n-k}$. Then $Ax = b$ gives

$$[A'_1 \quad A'_1 C'_1 + A'' C'' \quad A''] \begin{bmatrix} x_1 \\ x_2 \\ x_3 \end{bmatrix} = A'_1 C'_b + A'' C''_b.$$

Taking $x_1 = y_1$, $x_2 = y_2$ yields

$$A'_1 y_1 + A'_1 C'_1 y_2 + A'' C'' y_2 + A'' x_3 = A'_1 C'_b + A'' C''_b,$$

which can be rearranged as

$$A'_1(y_1 + C'_1 y_2 - C'_b) + A''(C'' y_2 + x_3 - C''_b) = 0.$$

Since $C'' y_2 + x_3 - C''_b = 0$, we get

$$A''(C'' y_2 + x_3 - C''_b) = 0.$$

Now, we can set $x_3 = C''_b - C'' y_2$, which immediately gives a solution of $Ax = b$ of the form

$$x = \begin{bmatrix} y_1 \\ y_2 \\ C''_b - C'' y_2 \end{bmatrix}.$$

Theorem 4. *Let B be the core matrix of A. If there is a y such that $By = R_{T,*}b$, then there exists x such that $Ax = b$ and $x_Z = y$. Otherwise, $Ax = b$ has no solution.*

Proof. Assume there is a solution to $Ax = b$. Since $\operatorname{supp}(Rx), \operatorname{supp}(R_{T,*}x) \in T$ for all x, we have $\operatorname{rank}(RA_1') = \operatorname{rank}(R_{T,*}A_1')$. By Theorem 3 with $M = R_{T,*}$, we know there is a solution of $Ax = b$ such that $x_Z = y$.

Otherwise, suppose for contradiction that $Ax = b$ has no solution, but $By = R_{T,*}b$. From the definition of the core matrix B, we get $(RA)_{T,Z}y = (Rb)_T$. Then, we can construct an x^* as

$$x_Z^* = y, \quad x_{V/Z}^* = 0,$$

such that $(RA)_{T,*}x^* = (Rb)_T$ holds. Due to the fact that $\operatorname{supp}(RA) \subset T$ and $\operatorname{supp}(Rb) \subset T$, we obtain that $RAx^* = Rb$, and from (2) we further arrive at $QAx^* + Ax^* = Qb + b$. Since $QAx^*, Ax^*, Qb \in \operatorname{span}(A)$, we thus have $b \in \operatorname{span}(A)$, which gives a contradiction.

3.2   The Algorithm

We first provide the algorithm FINDCORE for effectively computing the core matrix in Fig. 4, with the computational cost given in Theorem 5.

FINDCORE(A, Z, b)
  $\pi, T \leftarrow$ the forcing ordering and terminal set
  for $v \in Z$
    $z \leftarrow$ FORCING$(A, Z, A_{*,v})$
    $B_{*,v} \leftarrow (A_{*,v})_T - A_{T,*}z$
  Compute the LSP decomposition of B
  return B

Fig. 4. Find the core matrix.

Theorem 5. *The algorithm* FINDCORE *takes $O(mk + k^\omega)$ time.*

Proof. Computing $L(A_{*,v})$ for all $v \in Z$ takes $O(m(1 + |Z|)) = O(mk)$ time. The computation of $B_{*,v}$ for a $v \in Z$ consists of a vector-vector difference and a matrix-vector product, which can be implemented in $O(m)$ linear time. Thus, computing the core matrix takes $O(mk)$ time in total. The time taken for LSP decomposition is $O(k^\omega)$.

Next, the algorithm SOLVELINEARSYSTEMGIVENCORE(A, Z, B, b) in Fig. 5 shows how the solution of $Ax = b$ is obtained using the computed core matrix and all the information about the zero forcing set. By Theorem 4, we know there exists a solution that matches y at the zero forcing set. By Theorem 2, we obtain the remaining part of the solution through forcing. The corresponding computational cost is provided in Theorem 6.

$\textsc{SolveLinearSystemGivenCore}(A, Z, B, b)$
$\pi, T \leftarrow$ forcing sequence and terminal set T
$z \leftarrow \textsc{Forcing}(A, Z, b)$
$b' \leftarrow b_T - A_{T,*}z$
$y \leftarrow$ solution to $By = b'$
if y does not exists
    return "NO SOLUTION"
$x' \leftarrow$ the vector where $x'_Z = y$ and 0 everywhere else
$x \leftarrow x' + \textsc{Forcing}(A, Z, b - Ax')$
return x

Fig. 5. Solve a linear system $Ax = b$ given a core matrix.

Theorem 6. *Given a matrix $A \in \mathbb{F}^{n \times n}$ of m non-zero elements, its zero forcing set of size k, and a core matrix B represented by its LSP decomposition. The system of linear equations $Ax = b$ can be solved in $O(k^2 + m)$ time.*

Proof. Following the algorithm in Fig. 5, the computation of $b' = R_{T,*}b$ by forcing takes $O(m)$ time. As mentioned earlier, if the LSP decomposition of the matrix is given, then solving a system of linear equations with k variables and k equations takes $O(k^2)$ time. Once the solution of the linear system for the core matrix is attained, we can find the solution to the original problem through forcing in $O(m)$ time. The total running time is thus $O(k^2 + m)$.

Combining Theorem 5 and Theorem 6, we finally arrive at our main theorem.

Theorem 1 (Main). *Given a matrix $A \in \mathbb{F}^{n \times n}$ of m non-zero elements, and a zero forcing set of A of size k where $k \leq n$. A data structure of size $O(k^2)$ can be computed in $O(mk + k^\omega)$ time, such that for each $b \in \mathbb{F}^n$, solving $Ax = b$ takes $O(k^2 + m)$ time.*

4   Lights Out Game on a Grid

In this section, we show that the lights out game in an $n \times n$ grid can be solved in $O(n^\omega \log n)$ time.

Consider the lights out game on an $n \times n$ grid graph. We number the vertices in position (i, j) with index $(i-1)n + j$, and hence, the vertices are in $[n^2]$. Let $Z = [n]$, which is obviously a zero forcing set. If we apply Theorem 1 directly to the adjacency matrix of the grid graph, then in this special case, we will obtain precisely the algorithm of [22]. Since $m = n^2$ and $k = n$, the algorithm takes $O(n^3)$ time. The computational bottleneck is the calculation of the core matrix, as forcing itself only takes $O(n^2)$ time. Fortunately, by exploiting the repeated structure in grids, we can significantly improve the running time of computing the core matrix to $O(n^\omega \log n)$.

The forcing operation for the lights out game is greatly simplified, because the field is $\mathbb{F}_2$. In this case, $x_u = 1$ or $x_u = 0$ can be interpreted as pressing the button at vertex u or not, respectively. The b_u can be understood as the state of light at vertex u, where the value 1 means on, and 0 means off. When operating on the uth vertex, the forcing operation sets $x_u = 1$ if and only if $b'_{u\uparrow} = 1$. Here b' is the state of the board after applying all previous button presses. In other words, the forcing operation is iteratively setting $x_u = b'_{u\uparrow}$.

We then encode the operation of forcing, where we are given the states of the first and second rows and the aim is to compute the force operations on the second row ensuring that the states of the first row can be all 0. The output is the states of the second and the third rows after all of the button presses. To this end, we define such matrix to be $N(n) \in \mathbb{F}_2^{(2n) \times (2n)}$. The vertices of the first row are indexed from 1 to n. The vertex of the second row below the vertex i is $n + i$. One can easily verify that N could be written in the following block form

$$N(n) = \begin{bmatrix} N'(n) & I_n \\ I_n & O_n \end{bmatrix},$$

where $N'(n)$ is the matrix satisfying that $N'(n)_{i,j} = 1$ if and only if $|i - j| \le 1$ (Fig. 6).

$$\begin{bmatrix} 1 & 1 & 0 & 1 & 0 & 0 \\ 1 & 1 & 1 & 0 & 1 & 0 \\ 0 & 1 & 1 & 0 & 0 & 1 \\ 1 & 0 & 0 & 0 & 0 & 0 \\ 0 & 1 & 0 & 0 & 0 & 0 \\ 0 & 0 & 1 & 0 & 0 & 0 \end{bmatrix}$$

Fig. 6. The matrix $N(3)$.

Now, we define $M = N^{n-1}(n)$, which requires $O(n^\omega \log n)$ computational time using the exponentiation by squaring. Let a_j be the jth column of A. Since $\mathrm{supp}(a_j)$ is a subset of the first two rows if $j \in [n]$, we first let $a'_j = (a_j)_{[2n]}$. Next, we compute $y = Ma'_j$. Let t be the last n coordinates of y, which then yields the desired $R_{T,*}a_j$. Note that we can batch all the multiplications together, that is to say, we can compute $M[a'_j | j \in [n]] = MA_{[n],[2n]}$ in one go, which takes $O(n^\omega)$ running time, and recover the desired core matrix from it. The procedure described above is summarized in Fig. 7.

FINDGRIDCORE(n)
  $A \leftarrow$ adjacency matrix of an $n \times n$ grid
  $M \leftarrow (N(n))^{n-1}$
  return $(MA_{[n],[2n]})_{[n],[n]}$

Fig. 7. Find core matrix for a grid graph.

At this point, we have provided an $O(n^\omega \log n)$ time algorithm for solving the lights out problem on an $n \times n$ grid. Moreover, the algorithm can also be applied to the $n \times m$ grids with $n \le m$, and the running time becomes $O(n^\omega \log m + nm)$ accordingly.

Acknowledgements. Chao like to thank Jephian C.-H. Lin for discussion on algorithmic application of zero forcing set.

References

1. AIM Minimum Rank – Special Graphs Work Group: Zero forcing sets and the minimum rank of graphs. Linear Algebra Appl. **428**(7), 1628–1648 (2008). https://doi.org/10.1016/j.laa.2007.10.009
2. Alman, J., Williams, V.V.: A refined laser method and faster matrix multiplication. In: Proceedings of the 2021 ACM-SIAM Symposium on Discrete Algorithms (SODA), pp. 522–539 (2021). https://doi.org/10.1137/1.9781611976465.32. https://epubs.siam.org/doi/abs/10.1137/1.9781611976465.32
3. Alon, N., Yuster, R.: Matrix sparsification and nested dissection over arbitrary fields. J. ACM **60**(4), 1–18 (2013). https://doi.org/10.1145/2508028.2505989
4. Barioli, F., et al.: On the minimum rank of not necessarily symmetric matrices: a preliminary study. Electron. J. Linear Algebra **18**, 126–145 (2009). https://doi.org/10.13001/1081-3810.1300
5. Berman, A., Borer, F., Hungerbühler, N.: Lights Out on graphs. Math. Semesterber. **68**(2), 237–255 (2021). https://doi.org/10.1007/s00591-021-00297-5
6. Brimkov, B., Fast, C.C., Hicks, I.V.: Computational approaches for zero forcing and related problems. Eur. J. Oper. Res. **273**(3), 889–903 (2019). https://doi.org/10.1016/j.ejor.2018.09.03. https://ideas.repec.org/a/eee/ejores/v273y2019i3p889-903.html
7. Brimkov, B., Mikesell, D., Hicks, I.V.: Improved computational approaches and heuristics for zero forcing. INFORMS J. Comput. **33**(4), 1384–1399 (2021). https://doi.org/10.1287/ijoc.2020.1032. https://pubsonline.informs.org/doi/abs/10.1287/ijoc.2020.1032
8. Burgarth, D., Giovannetti, V.: Full control by locally induced relaxation. Phys. Rev. Lett. **99**, 100501 (2007). https://doi.org/10.1103/PhysRevLett.99.100501. https://link.aps.org/doi/10.1103/PhysRevLett.99.100501
9. Dean, N., Ilic, A., Ramirez, I., Shen, J., Tian, K.: On the power dominating sets of hypercubes. In: 2011 14th IEEE International Conference on Computational Science and Engineering, pp. 488–491 (2011). https://doi.org/10.1109/CSE.2011.89
10. Fleischer, R., Yu, J.: A survey of the game "lights out!". In: Brodnik, A., López-Ortiz, A., Raman, V., Viola, A. (eds.) Space-Efficient Data Structures, Streams, and Algorithms. LNCS, vol. 8066, pp. 176–198. Springer, Heidelberg (2013). https://doi.org/10.1007/978-3-642-40273-9_13
11. Fomin, F.V., Lokshtanov, D., Pilipczuk, M., Saurabh, S., Wrochna, M.: Fully polynomial-time parameterized computations for graphs and matrices of low treewidth. In: Proceedings of the 2017 Annual ACM-SIAM Symposium on Discrete Algorithms (SODA), pp. 1419–1432 (2017). https://doi.org/10.1137/1.9781611974782.92. https://epubs.siam.org/doi/abs/10.1137/1.9781611974782.92

12. Fomin, F.V., Lokshtanov, D., Saurabh, S., Pilipczuk, M., Wrochna, M.: Fully polynomial-time parameterized computations for graphs and matrices of low treewidth. ACM Trans. Algorithms **14**(3), 1–45 (2018). https://doi.org/10.1145/3186898
13. Golub, G.H., Van Loan, C.F.: Matrix Computations. Johns Hopkins Studies in the Mathematical Sciences, 4th edn. Johns Hopkins University Press, Baltimore (2013)
14. Hogben, L.: Minimum rank problems. Linear Algebra Appl. **432**(8), 1961–1974 (2010). https://doi.org/10.1016/j.laa.2009.05.003
15. Ibarra, O.H., Moran, S., Hui, R.: A generalization of the fast LUP matrix decomposition algorithm and applications. J. Algorithms **3**(1), 45–56 (1982). https://doi.org/10.1016/0196-6774(82)90007-4
16. Jeannerod, C.P.: LSP matrix decomposition revisited. Technical report, 2006-28, École Normale Supérieure de Lyon (2006)
17. Kenter, F.H., Lin, J.C.H.: On the error of a priori sampling: zero forcing sets and propagation time. Linear Algebra Appl. **576**, 124–141 (2019). https://doi.org/10.1016/j.laa.2018.03.031. https://linkinghub.elsevier.com/retrieve/pii/S0024379518301411
18. Leach, C.D.: Chasing the lights in lights out. Math. Mag. **90**(2), 126–133 (2017). https://doi.org/10.4169/math.mag.90.2.126
19. Lipton, R.J., Rose, D.J., Tarjan, R.E.: Generalized nested dissection. SIAM J. Numer. Anal. **16**(2), 346–358 (1979). https://doi.org/10.1137/0716027. http://epubs.siam.org/doi/10.1137/0716027
20. Severini, S.: Nondiscriminatory propagation on trees. J. Phys. A: Math. Theor. **41**(48), 482002 (2008). https://doi.org/10.1088/1751-8113/41/48/482002
21. Trefois, M., Delvenne, J.C.: Zero forcing number, constrained matchings and strong structural controllability. Linear Algebra Appl. **484**, 199–218 (2015). https://doi.org/10.1016/j.laa.2015.06.025. https://www.sciencedirect.com/science/article/pii/S002437951500378X
22. Wang (axpokl), Z.: 点灯游戏 flip game 的 $O(n^3)$ 算法 (2018). https://zhuanlan.zhihu.com/p/53646257
23. Yang, B.: Fast-mixed searching and related problems on graphs. Theor. Comput. Sci. **507**, 100–113 (2013). https://doi.org/10.1016/j.tcs.2013.04.015. https://www.sciencedirect.com/science/article/pii/S0304397513002995. Combinatorial Optimization and Applications

Profit Maximization for Competitive Influence Spread in Social Networks

Qiufen Ni[1], Yun Bai[1], and Zhongzheng Tang[2]

[1] School of Computer Science and Technology, Guangdong University of Technology,
Guangzhou 510006, China
niqiufen@gdut.edu.cn
[2] School of Science, Beijing University of Posts and Telecommunications,
Beijing 100876, China
tangzhongzheng@amss.ac.cn

Abstract. Influence maximization is a classic problem in social networks and has been extensively studied in recent years. Viral marketing is an important application for influence maximization. Most of existing related research focus on influence maximization of a single product, but in reality, a marketer may promote multiple products in the social network at the same time. This paper studies the profit maximization problem for multiple kinds of products in viral marketing. We formulate it as the Profit Maximization Problem for Competitive Influence Spread (PMPCIS), which aims at selecting a set of seed users within the total budget B and the total number of seeds K to maximize the overall profit of k kinds of products. The objective problem is proved to be a monotone k-submodular maximization problem under the knapsack and cardinality constraint. We present a Singleton+Greedy-Local-Search Algorithm in four steps, and prove the approximation performance guarantee of the proposed algorithm.

Keywords: Social Network · Profit Maximization · k-submodular

1 Introduction

Online social networks, such as WeChat, Facebook, Twitter, have been important platforms for people to communicate and for business to advertise. People keep in touch with each other and make friends through social networks, they also like to share their innovations and ideas, etc., in the social networks [1,2]. According to statistics, there are 3.725 billion users active in the social networks by Dec. 2019. Companies make use of the advantage of large crowds and rapid information dissemination in social networks to promote their products. Motivated by the information propagation in social networks, Influence Maximization (IM) problem is put forward by Kempe *et al.* in [3]. They formulate the IM problem as: selecting a set of users as seeds to maximize the expected number of users who are influenced by seeds. Influence maximization finds applications in many domain, like viral marketing. Viral marketing makes good use of the word-of-mouth effect of the social networks, and it promotes products by giving discounts to a small set of customers to spread their product information. Two classic influence spread model: Independent Cascade (IC) model and Linear Threshold (LT) model, are proposed by Kempe et al. in [3]. They also prove that the influence propagation function is monotone and submodular, thus, the greedy algorithm can be applied to solve the IM problem and obtain an $1 - 1/e - \epsilon$ approximation ratio.

In real life, it is very common for multiple types of information propagate simultaneously in the social network. For example, in 2022 Apple Corporation issues two kinds of new iPhone, iPhone 14 is more cost-effective and iPhone 14 Pro is more powerful but more expensive. Although iPhone 14 and iPhone 14 Pro are the same kind of product, but they have different prices, appearances and performances, which can attract different groups of customers with different requirements. Based on the described background, consider the scenario where there are k kinds of products, a constraint K for the number of selected seeds, and a budget B for the activation cost of all selected seeds. Assuming that different seeds have different activation costs and different profits can be obtained when the product is purchased, which nodes to be selected as seeds and how to allocate the budget to seeds for k kinds of products such that the total profit is maximized? We study the profit maximization problem for competitive products in this paper, and assume that each seed can accept the discount of only one product for the fairness. We aim to allocate discounts to k sets of seed users for k kinds of products under two constraints K and B. The objective function that maximizes the total profit of k kinds of products can be formulated as a k-submodular function. Approximation algorithms with theoretical guarantee are proposed in our work. We summarize the main contributions in this paper as follows:

- We formulate the profit maximization for competitive influence spread problem as a k-submodular function with both a knapsack constraint and a cardinality constraint problem. To best of our knowledge, this is the first time to study a monotone k-submodular maximization problem under both the knapsack and cardinality constraints in social networks.
- We propose a Singleton+Greedy-Local-Search Algorithm in four steps, which obtains two approximation ratios: 0.216 and 0.158, in two different conditions.

2 Related Work

In this paper, we consider the profit maximization problem with multiple kinds of products in the network. We summarize the related studies on our work as follows.

Influence Maximization with Competitive Influence Spread: Most of the existing relevant studies on influence maximization consider the scenario of a single kind of information spreading in the social network. The competitive IM problem is firstly studied by Bharathi et al. [4]. They propose a game theory based method to solve it. Liang et al. [5] consider that multiple kinds of similar products promote in the network at the same time and the target of the promotion is a specific user group. They formulate such a problem as Targeted Influence Maximization in Competitive social networks (TIMC). A reverse reachable set based greedy method is proposed by them to solve the TIMC with approximation performance guarantee. Wu *et al.* [6] consider a scenario that multiple information propagate in a social network with different propagation probabilities. The problem is formulated as maximizing the total influence of all the different information under a constraint of seed budget k. They present a greedy algorithm with $\frac{1}{3}$ approximation ratio, and also propose a parallel algorithm which improves the efficiency of the algorithm.

Profit Maximization: Profit maximization problem is a transformation of the influence maximization problem, which aims at maximizing the profit of products by selecting a seed set with a limited budget. Zhang *et al.* [7] study the Profit Maximization with Multiple Adoptions (PM^2A) problem. Two different approximation algorithms are devised to maximize the total profit of multiple kinds of product by selecting a limited number of seeds. Chen *et al.* [8] propose a Randomized Modified Greedy (RMG) algorithm to solve the Profit Maximization with Multiple Adoptions (PM^2A) problem, which obtain an $(1 - 1/e - \epsilon)$ approximation ratio. Yuan *et al.* [9] design two discount allocation strategies under the non-adaptive setting and adaptive setting, respectively, which achieve the goal of maximizing the expected number of users who adopt the product finally.

k-Submodular Maximization: Huber *et al.* [10] firstly define the k-submodular function, which is a generalization of the submodular function. Ohsaka *et al.* [11] study the maximization for a monotone k-submodular function with two different size constraints, and propose greedy algorithms with constant approximation factor. Tang *et al.* [12] study the maximization for a non-negative monotone k-submodular function with a knapsack constraint, they present a greedy algorithm which can obtain an $\frac{1}{2} - \frac{1}{2e}$ approximation ratio with an $O(n^4k^3)$ time complexity. Wang *et al.* [13] propose a framework for relaxing a k-submodular function to continuous space with the technique of multilinear extension. They also improve the approximation ratio to $1/2 - \epsilon$ for maximizing a monotone k-submodular function with knapsack constraint. V. Pham *et al.* [14] explore the applications of maximizing the k-submodular function under the knapsack constraint in influence maximization of social networks and the sensor placement. However, for the monotone k-submodular maximization problem under the knapsack constraint and cardinality constraint, to best of our knowledge, there is no related conclusion about it in social networks. Thus, we try to fill this gap and apply it to the social networks.

3   Diffusion Model and Problem Definition

3.1   The Diffusion Model

A social network is constructed as a directed graph $G(V, E)$, where each node $v \in V$ represents a user, and each edge $(u, v) \in E$ represents that user v follows user u, u is an incoming neighbor of v, while v is an outgoing neighbor of u. The incoming neighbor set and the outgoing neighbor set of a node v are denoted as $N^-(v)$ and $N^+(v)$, respectively. IC model is used as the influence propagation model in our problem, its influence propagation process is described as follows.

Definition 1 (IC model). *Nodes in social networks have two different states: active and inactive, all nodes are initially inactive. There is a activation probability $p_{uv} \in (0,1]$ associated with each edge $e = (u,v) \in E$. When a node u is firstly activated at time t, then for each of his inactive outgoing neighbor $v \in N^+(u)$, he can activate them with a probability p_{uv} at time $t+1$. Finally, the influence propagation process terminates if there are no newly activated nodes in the future.*

3.2   Problem Formulation

Give a social network $G = (V, E)$. Assume that marketers wants to promote k kinds of products in the social network. k kinds of products information propagate under the IC model at the same time. We aim to choose k seed sets $S = \{S_1, S_2, \cdots, S_k\}$ and provide discounts to them, and assume that each seed user can only be used for propagating at most one kind of product information. As different user can bring different levels of influence, so we give different discounts to different seeds, and influential users get bigger discounts. Let $\sigma(S_i)$ be the expected influence spread of seed set S_i for product i, i.e., the expected number of users who adopt product i in the social network. Let $f(S_i)$ be the total profit that obtained by purchasing product i. Moreover, $\sigma(S|G)$ and $f(S|G)$ are the expected number of influenced people and the total profit obtained by adopting k kinds of products, respectively.

The profit maximization problem for competitive products marketing at the same time in the social network with a total activation cost B and a total number of seeds K constraints can be formulated as follows:

Problem 1 *(Profit Maximization Problem for Competitive Influence Spread (PMPCIS)).* *Given a social network graph $G = (V, E)$, k kinds of products, the IC model, the cost $c(a)$ that activating a node a to purchase a product, the profit p_i that a node can gain when he adopts product i, the seed set $S = \{S_1, S_2, \cdots, S_k\}$, where S_i is selected for propagating the information of the product i and $S_i \cap S_j = \varnothing$ for any $i,j \in [1,k]$ and $i \neq j$. The total number of selected seeds is represented by $|S| = \sum_{i=1}^{k} |S_i|$ and the upper bound is K. The total activation cost for the seed set S is denoted by $c(S) = \sum_{i=1}^{k} \sum_{a \in S_i} c(a)$ and the total budget is given as B. The expected influence spread for seed set S_i is expressed by $\sigma(S_i)$. Our target is to select an optimal seed set $S = \{S_1, S_2, \cdots, S_k\}$ such that the total profit $f(S)$ is maximized, i.e.,*

$$S^* = \arg\max_{S} \ f(S)$$

$$s.t. \quad c(S) \leq B \tag{1}$$

$$|S| \leq K.$$

From the definition of PMPCIS, we can intuitively get $f(S) = \sum_{i=1}^{k} p_i \sigma(S_i)$. In literature [3], Kempe *et al.* have proved that the classical influence maximization problem in IC model is a NP-hard problem. When the types of products are reduced to one, our PMPCIS is equivalent to the traditional IM problem for IC model. Thus, the PMPCIS is a NP-hard problem.

4   Solution for PMPCIS

We propose solution for PMPCIS in this section. At first, we analyze the properties of the objective function f for PMPCIS.

4.1   Properties for Objective Function f

Firstly, we introduce an important property for a set function: k-submodular. Let X be a finite non-empty set, and let $(k+1)^X := \{(U_1, \cdots, U_k) \mid U_i \subseteq X, \forall i \in \{1, 2, \cdots, k\}, U_i \cap U_j = \varnothing, \forall i \neq j\}$ be the family of k disjoint sets $(U_1, \cdots, U_k)$. A function $h: (k+1)^X \to \mathbb{R}$ is k-submodular if for any $U = \{U_1, \cdots, U_k\}$ and $W = \{W_1, \cdots, W_k\}$ in $(k+1)^X$, it satisfies,

$$h(U) + h(W) \geq h(U \sqcup W) + h(U \sqcap W),$$

where
$$U \sqcap W := (U_1 \cap W_1, \cdots, U_k \cap W_k),$$

$$U \sqcup W := \left(U_1 \cup W_1 \backslash \left(\bigcup_{i \neq 1} U_i \cup W_i \right), \cdots, U_k \cup W_k \backslash \left(\bigcup_{i \neq k} U_i \cup W_i \right) \right).$$

If a function satisfies the properties of orthant submodularity and pairwise monotonicity at the same time, which indicates that it is a k-submodular function. It is very intuitively to verify that our objective function is k-submodularity.

Then, we elaborate on additional characteristics and notations of k-submodular functions for our problem. Every k-tuple $\mathbf{x} = (X_1, \ldots, X_k) \in (k+1)^V$ uniquely corresponds to a set $A = \{(a, d) \mid a \in X_d, d \in [k]\}$ composed of *item-dimension pairs*. Hence, a user-product pair (a, d) is included in set A (termed as a *solution*) if and only if $a \in X_d$ in $\mathbf{x}$.

In our problem, for ease of presentation, we write $\mathbf{x}$ and its corresponding solution A interchangeably. For any solution $A \in (k+1)^V$, we define $U(A) := \{a \in V \mid \exists\, d \in [k],\ (a, d) \in A\}$ to be the set of seed included, and the *size* is $|A| = |U(A)|$.

The marginal gain of adding a user-product pair (a, d) to A is

$$\Delta_{a,d} f(A) = f(A \cup \{(a, d)\}) - f(A),$$

and the marginal density is $\frac{\Delta_{a,d} f(A)}{c(a)}$. As the profit maximization function f is monotone k-submodular, it satisfies the *pairwise monotonicity*

$$\Delta_{a,d} f(A) + \Delta_{a,l} f(A) \geq 0,$$

for any $d, l \in [k]$ and $d \neq l$. And it also satisfies the *orthant submodularity*

$$\Delta_{a,d} f(A) \geq \Delta_{a,d} f(C),$$

$$\forall A, C \in (k+1)^V \text{ with } A \subseteq C, a \notin U(C), d \in [k].$$

In this problem, each seed node $a \in V$ has a non-negative cost $c(a)$, and the cost of solution A is $c(A) = \sum_{a \in U(A)} c(a)$. The goal is to find a solution that maximizes the function value within the given budget $B \in \mathbb{Z}_+$ and size $K \in \mathbb{Z}_+$. Then we rewrite the PMPCIS in Eq. (1) as follows.

$$A^* = \arg \max_{A \in (k+1)^V} f(A)$$

$$s.t. \quad c(A) \leq B \tag{2}$$

$$|A| \leq K.$$

4.2 Proposed Algorithm

We present our solution for the proposed PMPCIS in four steps. In the first step, we consider unconstrained profit maximization problem and propose a simple greedy algorithm in Algorithm 1. Given a node set $V' = \{e_1, e_2, \ldots, e_m\}$ and $V' \subseteq V$. Firstly, we remove the two constraints of the problem in Eq. 2, then, Algorithm 1 is devised by greedily finding the maximum function value $\max_{A \in (k+1)^{V'}} f(A)$ without any constraint.

Let $T = \{(e_1, d_1^*), \ldots, (e_m, d_m^*)\}$ be an optimal solution that maximizes the function $f(A)$ over V'. Assume without loss of generality that the seed set are obtained by **Greedy Algorithm** in the order of $\{e_1, e_2, \ldots, e_m\}$, and denote the returned greedy solution by $A = \{(e_1, d_1), \ldots, (e_m, d_m)\}$. For $j = \{0, 1, \ldots, m\}$, define $A_j = \{(e_1, d_1), \ldots, (e_j, d_j)\}$. We have $A_0 = \varnothing$ and $A_m = A$. The following Lemma 1 is crucial to our subsequent proof.

Lemma 1 ([15, 16]). *If f is monotone, for $t = \{0, 1, \ldots, m\}$, we have $f(T) \leq 2f(A_t) + \sum_{e_i \in U(T) \setminus U(A_t)} \Delta_{e_i, d_i^*} f(A_t)$.*

Algorithm 1. A Simple Greedy Algorithm

Input: Social network subgraph $G = (V', E')$, objective function f.
Output: A solution $A \in (k+1)^{V'}$.
1: $A \leftarrow \varnothing$
2: **for** each item $a \in V'$ **do**
3:     $d_a \leftarrow \arg\max_{d \in [k]} \Delta_{a,d} f(A)$
4:     $A \leftarrow A \cup \{(a, d_a)\}$
5: **end for**
6: **return** A

Algorithm 2. Greedy-Knapsack Algorithm

Input: Social network graph $G = (V, E)$, objective function f, costs $c(a)$ for $a \in V$, budget B, seed size constraint K.

Output: Solution A_t in $(k+1)^V$.

1: $A_0 \leftarrow \varnothing$, $V_0 \leftarrow V$, card $\leftarrow 0$
2: **for** t from 1 to $|V|$ **do**
3:     Let $(a_t, d_t) = \arg \max\limits_{a \in V_{t-1}, d \in [k]} \frac{\Delta_{a,d} f(A_{t-1})}{c(a)}$ maximize the marginal density, and
       denote $\rho_t = \frac{\Delta_{a_t, d_t} f(A_{t-1})}{c(a_t)}$.
4:     **if** $c(A_{t-1}) + c(a_t) \leq B$ **then**
5:         $A_t \leftarrow A_{t-1} \cup \{(a_t, d_t)\}$
6:         card $\leftarrow$ card+1
7:         **if** card $= K$ **then**
8:             **break**
9:         **end if**
10:    **else**
11:        $A_t \leftarrow A_{t-1}$
12:    **end if**
13:    $V_t \leftarrow V_{t-1} \backslash \{a_t\}$
14: **end for**
15: **return** A_t.

Then we propose the following Algorithm 2. In Algorithm 2, we select the user who can bring the largest marginal density for objective function f in each iteration, and the selected users can not violate the budget and size constraint at the same time. Algorithm 2 returns the solution A_t.

Next, we devise the Algorithm 3, which aims at further optimizing the solution A_t. If we have solution with size K that is obtained with the Algorithm 2, we can execute the local search procedure utilizing Algorithm 3. We input the solution obtained by Algorithm 2, and denote it as a feasible solution A'. In each iteration, we try to swap a pair of selected and unselected items, aiming to augment the objective value while ensuring adherence to the knapsack constraint. In Algorithm 3, when we obtain the new seed set after the exchange in step 6, we invoke the Algorithm 1 **Greedy Algorithm** in step 7 to reassign the type of product that the seeds correspond to since the selected seed set may change.

Our main algorithm is as shown in Algorithm 4, which combines the singleton optimum, the greedy strategy in Algorithm 2 and the local search strategy in Algorithm 3. In step 1 of Algorithm 4, we find out the node that can maximize the objective function f among all the nodes, then A^* is the item pair with a single node set and the product it corresponds.

Algorithm 3. Local-Search Algorithm

Input: Social network graph $G = (V, E)$, objective function f, costs $c(a)$ for $a \in V$, budget B, seed size constraint K. a feasible solution $A' \in (k+1)^V$ with $|A'| = K$.
Output: A local optimal solution $A_t \in (k+1)^V$.

1: $A_K \leftarrow A'$, $t = K$, swap $\leftarrow$ true
2: **while** swap **do**
3:     swap $\leftarrow$ false
4:     Let $\rho_{(x,y)} = \max_{d \in [k]} \frac{f(A_t \setminus (y, d_y) \cup (x, d)) - f(A_t)}{c(x)}$ be the exchange marginal density. The swap (a_{t+1}, b_{t+1}) achieves the value $\max_{x \in V \setminus U(A_t), (y, d_y) \in A_t} \rho(x, y)$ and $\rho_{t+1} = \rho(a_{t+1}, b_{t+1})$
5:     **if** $\rho_{t+1} > 0$ and $c(a_{t+1}) - c(b_{t+1}) + c(A_t) \leq B$ **then**
6:        $V' \leftarrow U(A_t) \setminus \{b_{t+1}\} \cup \{a_{t+1}\}$
7:        $A_{t+1} \leftarrow$ **Greedy**(V', f)
8:        swap $\leftarrow$ true
9:        $t \leftarrow t + 1$
10:     **end if**
11: **end while**
12: **return** A_t

Algorithm 4. Singleton+Greedy-Local-Search Algorithm

Input: Social network graph $G = (V, E)$, objective function f, costs $c(a)$ for $a \in V$, budget B, seed size constraint K.
Output: The solution A^* in $(k+1)^V$.

1: Let $A^* \in \arg \max_{A: |A|=1, c(A) \leq B} f(A)$ be a size-1 solution giving the largest value.
2: $A \leftarrow$ **Greedy-Knapsack**(V, f, c, B, K)
3: **if** $f(A) > f(A^*)$ **then**
4:     $A^* \leftarrow A$
5: **end if**
6: **if** $|A| < K$ **then**
7:     **return** A^*
8: **end if**
9: $A \leftarrow$ **Local-Search**(V, f, c, B, K, A)
10: **if** $f(A) > f(A^*)$ **then**
11:     $A^* \leftarrow A$
12: **end if**
13: **return** A^*

If $f(A^*) \geq f(A)$, then A^* is returned as the final solution. We always find the node that maximize the marginal density as seeds in Algorithm 3, which may miss some nodes that can bring large profit and also need large costs at the same time, but have small marginal density. However, such nodes are actually a good solution. So the step 1 in Algorithm 4 is to find such kind of nodes. The returned solution A^* by the Algorithm 4 is a seed-product pair set, but in our original objective function in Eq. 1, we need to find an optimal seed set S^*, so the nodes in A^* is the returned solution for our PMPCIS in Eq. (1).

4.3   Approximation Performance Analysis

In this section, we analyze the approximation performance guarantee of the proposed Algorithm 4. We intend to establish the approximation ratio in Theorem 1 following the proof framework in [17]. The subsequent observation will be used twice in the proof of Theorem 1.

Lemma 2. *Let $A^1, A^2 \in (k+1)^V$ be two node sets with $|A^1| \leq K$ and $|A^2| = K$, then for any one-to-one function $y : U(A^1)\backslash U(A^2) \to U(A^2)\backslash U(A^1)$, we have $\sum_{x \in U(A^1)\backslash U(A^2)} f(A^2) - f(A^2\backslash\{(y(x), d_{y(x)})\}) \leq f(A^2)$ where $(y(x), d_{y(x)}) \in A^2$.*

Proof. Let $U(A^1)\backslash U(A^2) = \{x_1, x_2, \ldots, x_{K'}\}$. For any one-to-one function y, suppose $y_j = y(x_j)$ for each $j \in \{1, \ldots, K'\}$. Then we have

$$\sum_{x \in U(A^1)\backslash U(A^2)} f(A^2) - f(A^2\backslash\{(y(x), d_{y(x)})\})$$

$$\leq \sum_{j=1}^{K'} f(A^2\backslash\{(y_1, d_{y_1}), \ldots, (y_{j-1}, d_{y_{j-1}})\}) - f(A^2\backslash\{(y_1, d_{y_1}), \ldots, (y_j, d_{y_j})\})$$

$$\leq f(A^2),$$

where $j = 1$, $\{(y_1, d_{y_1}), \ldots, (y_{j-1}, d_{y_{j-1}})\}$ is an empty set.

Theorem 1. *Algorithm 4 has an approximation ratio at least $\frac{1}{6}(1 - e^{-3}) \approx 0.15836$.*

Proof. Let T be the optimal solution with $|T| \geq 2$ and A be the seed set output by Algorithm 4. If $|A| < K$, then A is a $\frac{1-e^{-2}}{4}$-approximation solution of maximum k-submodular with a knapsack constraint [16]. Thus, A is also a $\frac{1-e^{-2}}{4}$-approximation solution of maximum our k-submodular objective function f with a knapsack constraint and a cardinality constraint.

If $|A| = K$, we distinguish between two cases based on the last iteration of the algorithm.

Case 1: For any node $x \in U(T)\backslash U(A)$, node $y \in U(A)\backslash U(T)$, the swap (x, y) was rejected because $\rho_{(x,y)} \leq 0$. Since A is a greedy solution, by Lemma 1, we derive

$$f(T) \leq 2f(A) + \sum_{x \in U(T)\backslash U(A)} [f(A \cup \{(x, d_x^*)\}) - f(A)]$$

$$\leq 2f(A) + \sum_{x \in U(T)\backslash U(A)} [f(A \cup \{(x, d_x^*)\}\backslash\{(y(x), d_{y(x)})\}) - f(A\backslash\{(y(x), d_{y(x)})\})]$$

$$\leq 2f(A) + \sum_{x \in U(T)\backslash U(A)} [f(A) - f(A\backslash\{(y(x), d_{y(x)})\})]$$

$$\leq 3f(A),$$

where $(x, d_x^*) \in T$ and $y(x)$ is a one-to-one function with $(y(x), d_{y(x)}) \in A$. The second inequality holds because of submodularity. The third inequality holds as

$\rho_{(x,y)} \le 0$. The last inequality holds by Lemma 2. Thus, in this case, we have Algorithm 4 yields a $\frac{1}{3}$-approximation on the optimum.

Case 2: At least one swap for node pair (x, y) with $\rho_{(x,y)} > 0$, for $x \in U(T)\backslash U(A)$, $y \in U(A)\backslash U(T)$ was rejected because $c(x) - c(y) + c(A) > B$.

Let A_t be the partial greedy solution after t iterations as shown in Algorithm 2 and 3. Let $l + 1$ be the first iteration in which a swap (a_{l+1}, b_{l+1}) was rejected since it violates the knapsack constraint where $a_{l+1} \in U(T)\backslash U(A_l)$. We can further assume that $l + 1$ is the first iteration t for which $A_t = A_{t-1}$. Since A_t is a greedy solution for $t = 0, 1, \ldots, l$, by Lemma 1, we have

$$f(T) \le 2f(A_t) + \sum_{x \in U(T)\backslash U(A_t)} [f(A_t \cup \{(x, d_x^*)\}) - f(A_t)],$$

where $(x, d_x^*) \in T$.
For $t = 0, \ldots, K - 1$, $|A_t| = t$, we have

$$f(T) \le 2f(A_t) + \sum_{x \in U(T)\backslash U(A_t)} [f(A_t \cup \{(x, d_x^*)\}) - f(A_t)] \le 2f(A_t) + B\rho_{t+1}.$$

The second inequality holds since $c(T) \le B$.
For $t = K, \ldots, l$, $|A_t| = K$, we have

$$
\begin{aligned}
f(T) &\le 2f(A_t) + \sum_{x \in U(T)\backslash U(A_t)} [f(A_t \cup \{(x, d_x^*)\}) - f(A_t)] \\
&\le 2f(A_t) + \sum_{x \in U(T)\backslash U(A_t)} [f(A_t \cup \{(x, d_x^*)\}\backslash\{(y(x), d_{y(x)})\}) - f(A_t\backslash\{(y(x), d_{y(x)})\})] \\
&\le 2f(A_t) + \sum_{x \in U(T)\backslash U(A_t)} [f(A_t \cup \{(x, d_x^*)\}\backslash\{(y(x), d_{y(x)})\}) - f(A_t)] + \\
&\quad \sum_{x \in U(T)\backslash U(A_t)} [f(A_t) - f(A_t\backslash\{(y(x), d_{y(x)})\})] \\
&\le 3f(A_t) + \sum_{x \in U(T)\backslash U(A_t)} [f(A_t \cup \{(x, d_x^*)\}\backslash\{(y(x), d_{y(x)})\}) - f(A_t)] \\
&\le 3f(A_t) + B\rho_{t+1}.
\end{aligned}
$$

The second inequality holds because of submodularity. The fourth inequality holds by Lemma 2. The last inequality holds since $c(T) \le B$.
Therefore, for each $t = 0, 1, \ldots, l$, we have

$$f(T) \le 3f(A_t) + B\rho_{t+1}.$$

Take advantage of the techniques in [17, 18], we get

$$\frac{f(A_l\backslash\{(b_{l+1}, d_{b_{l+1}})\} \cup \{(a_{l+1}, d_{a_{l+1}})\})}{f(T)} \ge \frac{1}{3}(1 - e^{-3}),$$

where $\rho_{(a_{l+1},b_{l+1})} = \frac{f(A_l \setminus \{(b_{l+1},d_{b_{l+1}})\} \cup \{(a_{l+1},d_{a_{l+1}})\}) - f(A_l)}{c(a_{l+1})}$.

Therefore, **Singleton+Greedy-Local-Search** has a function value at least

$$\max\{f(A_l), f(\{(a_{l+1},d_{a_{l+1}})\})\} \geq \frac{1}{2}f(A_l \cup \{(a_{l+1},d_{a_{l+1}})\})$$

$$\geq \frac{1}{2}f(A_l \setminus \{(b_{l+1},d_{b_{l+1}})\} \cup \{(a_{l+1},d_{a_{l+1}})\}) \geq \frac{1}{6}(1-e^{-3})f(T).$$

The theorem is proved.

5   Conclusion

In this paper, we investigate the profit maximization problem for k kinds of competitive products information spreading at the same time in social networks. Considering that one seed user spreads multiple product information at the same time may disperse its followers' attention, one seed user can only propagate the influence of one kind of product. The goal of the proposed problem is to select k subsets of users as seeds with a budget B and a seed size K constraints such that the total profit for k kinds of products is maximized. Our optimal problem is formulated as maximizing a monotone k-submodular function under a knapsack constraint and a cardinality constraint. A Singleton+Greedy-Local-Search Algorithm is put forward in four steps to solve the profit maximization problem, which achieves a 0.216 and 0.158 approximation performance guarantees in two different cases, respectively.

Acknowledgment. This work is supported in part by the CCF-Huawei Populus Grove Fund under Grant No. CCF-HuaweiLK2022004, in part by the National Natural Science Foundation of China under Grant No. 62202109, No. 12101069 and in part by the Guangdong Basic and Applied Basic Research Foundation under Grant No. 2021A1515110321 and No. 2022A1515010611 and in part by Guangzhou Basic and Applied Basic Research Foundation under Grant No. 202201010676.

References

1. Ni, Q., Guo, J., Weili, W., Wang, H., Jigang, W.: Continuous influence-based community partition for social networks. IEEE Trans. Netw. Sci. Eng. **9**(3), 1187–1197 (2022)
2. Ni, Q., Guo, J., Weili, W., Wang, H.: Influence-based community partition with sandwich method for social networks. IEEE Trans. Comput. Soc. Syst. **10**(2), 819–830 (2023)
3. Kempe, D., Kleinberg, J., Tardos, É.: Maximizing the spread of influence through a social network. In: Proceedings of the Ninth ACM SIGKDD International Conference on Knowledge Discovery and Data Mining, pp. 137–146. ACM (2003)
4. Bharathi, S., Kempe, D., Salek, M.: Competitive influence maximization in social networks. In: Deng, X., Graham, F.C. (eds.) WINE 2007. LNCS, vol. 4858, pp. 306–311. Springer, Heidelberg (2007). https://doi.org/10.1007/978-3-540-77105-0_31

5. Liang, Z., He, Q., Hongwei, D., Wen, X.: Targeted influence maximization in competitive social networks. Inf. Sci. **619**, 390–405 (2023)
6. Guanhao, W., Gao, X., Yan, G., Chen, G.: Parallel greedy algorithm to multiple influence maximization in social network. ACM Trans. Knowl. Discov. Data (TKDD) **15**(3), 1–21 (2021)
7. Zhang, H., Zhang, H., Kuhnle, A., Thai, M.T.: Profit maximization for multiple products in online social networks, pp. 1–9 (2016)
8. Chen, T., Liu, B., Liu, W., Fang, Q., Yuan, J., Weili, W.: A random algorithm for profit maximization in online social networks. Theor. Comput. Sci. **803**, 36–47 (2020)
9. Yuan, J., Tang, S.-J.: Adaptive discount allocation in social networks. In: Proceedings of the 18th ACM International Symposium on Mobile Ad Hoc Networking and Computing, pp. 1–10 (2017)
10. Huber, A., Kolmogorov, V.: Towards minimizing k-submodular functions. In: Mahjoub, A.R., Markakis, V., Milis, I., Paschos, V.T. (eds.) ISCO 2012. LNCS, vol. 7422, pp. 451–462. Springer, Heidelberg (2012). https://doi.org/10.1007/978-3-642-32147-4_40
11. Ohsaka, N., Yoshida, Y.: Monotone k-submodular function maximization with size constraints. In: Advances in Neural Information Processing Systems, pp. 694–702 (2015)
12. Tang, Z., Wang, C., Chan, H.: On maximizing a monotone k-submodular function under a knapsack constraint. Oper. Res. Lett. A J. Oper. Res. Soc. Am. **50**(1), 28–31 (2022)
13. Wang, B., Zhou, H.: Multilinear extension of k-submodular functions. CoRR abs/2107.07103. https://arxiv.org/abs/2107.07103, p. eprint2107.07103 (2021)
14. Pham, C.V., Vu, Q.C., Ha, D.K.T., Nguyen, T.T., Le, N.D.: Maximizing k-submodular functions under budget constraint: applications and streaming algorithms. J. Comb. Optim. **44**, 723–751 (2022). https://doi.org/10.1007/s10878-022-00858-x
15. Ward, J., Živný, S.: Maximizing k-submodular functions and beyond. ACM Trans. Algorithms (TALG) **12**(4), 47 (2016)
16. Tang, Z., Chen, J., Wang, C.: An improved analysis of the greedy+singleton algorithm for k-submodular knapsack maximization. In: Li, M., Sun, X., Wu, X. (eds.) IJTCS-FAW 2023. LNCS, vol. 13933, pp. 15–28. Springer, Cham (2023). https://doi.org/10.1007/978-3-031-39344-0_2
17. Sarpatwar, K.K., Schieber, B., Shachnai, H.: Constrained submodular maximization via greedy local search. Oper. Res. Lett. **47**(1), 1–6 (2019)
18. Sviridenko, M.: A note on maximizing a submodular set function subject to a knapsack constraint. Oper. Res. Lett. **32**(1), 41–43 (2004)

Improved Approximation Algorithms for Multidepot Capacitated Vehicle Routing

Jingyang Zhao and Mingyu Xiao[✉]

University of Electronic Science and Technology of China, Chengdu, China
myxiao@gmail.com

Abstract. The Multidepot Capacitated Vehicle Routing Problem (MCVRP) is a well-known variant of the classic Capacitated Vehicle Routing Problem (CVRP), where we need to route capacitated vehicles located in multiple depots to serve customers' demand such that each vehicle must return to the depot it starts, and the total traveling distance is minimized. There are three variants of MCVRP according to the property of the demand: unit-demand, splittable and unsplittable. We study approximation algorithms for k-MCVRP in metric graphs where k is the capacity of each vehicle, and all three versions are APX-hard for any constant $k \geq 3$. Previously, Li and Simchi-Levi proposed a $(2\alpha + 1 - \alpha/k)$-approximation algorithm for splittable and unit-demand k-MCVRP and a $(2\alpha + 2 - 2\alpha/k)$-approximation algorithm for unsplittable k-MCVRP, where $\alpha = 3/2 - 10^{-36}$ is the current best approximation ratio for metric TSP. Harks et al. further improved the ratio to 4 for the unsplittable case. We give a $(4 - 1/1500)$-approximation algorithm for unit-demand and splittable k-MCVRP, and a $(4 - 1/50000)$-approximation algorithm for unsplittable k-MCVRP. Furthermore, we give a $(3 + \ln 2 - \max\{\Theta(1/\sqrt{k}), 1/9000\})$-approximation algorithm for splittable and unit-demand k-MCVRP, and a $(3 + \ln 2 - \Theta(1/\sqrt{k}))$-approximation algorithm for unsplittable k-MCVRP under the assumption that the capacity k is a fixed constant. Our results are based on recent progress in approximating CVRP.

Keywords: Capacitated Vehicle Routing · Multidepot · Approximation Algorithms

1 Introduction

In the Multidepot Capacitated Vehicle Routing Problem (MCVRP), we are given a complete undirected graph $G = (V \cup D, E)$ with an edge weight w satisfying the symmetric and triangle inequality properties. The n nodes in $V = \{v_1, \ldots, v_n\}$ represent n customers and each customer $v \in V$ has a demand $d(v) \in \mathbb{Z}_{\geq 1}$. The m nodes in $D = \{u_1, \ldots, u_m\}$ represent m depots, with each containing an infinite number of vehicles with a capacity of $k \in \mathbb{Z}_{\geq 1}$ (we can also think that each depot contains only one vehicle, which can be used many times). A tour

© The Author(s), under exclusive license to Springer Nature Switzerland AG 2024
W. Wu and G. Tong (Eds.): COCOON 2023, LNCS 14423, pp. 378–391, 2024.
https://doi.org/10.1007/978-3-031-49193-1_29

is a walk that begins and ends at the same depot and the sum of deliveries to all customers in it is at most k. The traveling distance of a tour is the sum of the weights of edges in the tour. In MCVRP, we wish to find a set of tours to satisfy every customer's demand with a minimum total distance of all the tours. In the *unsplittable* version of the problem, each customer's demand can only be delivered by a single tour. In the *splittable* version, each customer's demand can be delivered by more than one tour. Moreover, if each customer's demand is a unit, it is called the *unit-demand* version.

In logistics, MCVRP is an important model that has been studied extensively in the literature (see [19] for a survey). If there is only one depot, MCVRP is known as the famous Capacitated Vehicle Routing Problem (CVRP). Since k-CVRP is APX-hard for any fixed $k \geq 3$ [3], it also holds for k-MCVRP.

Consider approximation algorithms for k-CVRP. Haimovich and Kan proposed [11] a well-known algorithm based on a given Hamiltonian cycle, called *Iterated Tour Partitioning (ITP)*. Given an α-approximation algorithm for metric TSP, for splittable and unit-demand k-CVRP, ITP can achieve a ratio of $\alpha + 1 - \alpha/k$ [11]. For unsplittable k-CVRP, Altinkemer and Gavish [1] proposed a modification of ITP, called UITP, that can achieve a ratio of $\alpha + 2 - 2\alpha/k$ for even k. When k is arbitrarily large, k-CVRP becomes metric TSP. For metric TSP, there is a well-known 3/2-approximation Algorithm [6,21], and currently Karlin et al. [15,16] has slightly improved the ratio to $3/2 - 10^{-36}$. Recently, some progress has been made in approximating k-CVRP. Blauth et al. [4] improved the ratio to $\alpha + 1 - \varepsilon$ for splittable and unit-demand k-CVRP and to $\alpha + 2 - 2\varepsilon$ for unsplittable k-CVRP, where ε is a value related to α and satisfies $\varepsilon \approx 1/3000$ when $\alpha = 3/2$. Then, for unsplittable k-CVRP, Friggstad et al. [9] further improved the ratio to $\alpha + 1 + \ln 2 - \varepsilon$ based on an LP rounding method, where $\varepsilon \approx 1/3000$ is the improvement based on the method in [4]. There are other improvements for the case that the capacity k is a small fixed constant. Bompadre et al. [5] improved the classic ratios by a term of $\Omega(1/k^3)$ for all three versions. Zhao and Xiao [27] proposed a $(5/2 - \Theta(1/\sqrt{k}))$-approximation algorithm for splittable and unit-demand k-CVRP and a $(5/2 + \ln 2 - \Theta(1/\sqrt{k}))$-approximation algorithm for unsplittable k-CVRP, where the improvement $\Theta(1/\sqrt{k})$ is larger than $1/3000$ for any $k \leq 10^7$.

Consider approximation algorithms for k-MCVRP. Few results are available in the literature. Note that $\alpha \approx 3/2$. Based on a modification of ITP, Li and Simchi-Levi [18] proposed a cycle-partition algorithm, which achieves a ratio of $2\alpha + 1 - \alpha/k \approx 4 - \Theta(1/k)$ for splittable and unit-demand k-MCVRP and a ratio of $2\alpha + 2 - 2\alpha/k \approx 5 - \Theta(1/k)$ for unsplittable k-MCVRP. The only known improvement was made by Harks et al. [12], where they proposed a tree-partition algorithm with an improved 4-approximation ratio for unsplittable k-MCVRP. Note that their algorithm also implies a 4.38-approximation ratio for a more general problem, called *Capacitated Location Routing*, where we need to open some depots (with some cost) first and then satisfy customers using vehicles in the opened depots. When k is arbitrarily large, k-MCVRP becomes metric m-depot TSP. For metric m-depot TSP, Rathinam et al. [20] proposed a simple

2-approximation algorithm, and Xu et al. [25] proposed an improved $(2 - 1/m)$-approximation algorithm. Then, based on an edge exchange algorithm, Xu and Rodrigues [24] obtained an improved 3/2-approximation algorithm for any fixed m. Traub et al. [22] further improved the ratio to $\alpha + \varepsilon$ for any fixed m. Recently, Deppert et al. [8] obtained a randomized $(3/2 + \varepsilon)$-approximation algorithm with a running time of $(1/\varepsilon)^{O(d \log d)} \cdot n^{O(1)}$, and hence their algorithm even works with a *variable* number of depots.

If the capacity k is fixed, we will see that splittable k-MCVRP is equivalent to unit-demand k-MCVRP. Moreover, both unit-demand and unsplittable k-MCVRP can be reduced to the *minimum weight k-set cover problem*. In minimum weight k-set cover, we are given a set of elements (called *universe*), a set system with each set in it having a weight and at most k elements, and we need to find a collection of sets in the set system with a minimum total weight that covers the universe. In the reduction, customers can be seen as the elements. There are at most $mn^{O(k)}$ feasible tours, and each tour can be seen as a set containing all customers in the tour with a weight of the tour. When k is fixed, the reduction is polynomial. It is well-known [7] that the minimum weight k-set cover problem admits an approximation ratio of H_k, where $H_k := 1 + 1/2 + \cdots + 1/k$ is the k-th harmonic number. Hassin and Levin [13] improved the ratio to $H_k - \Theta(1/k)$. Recently, using a non-obvious local search method, Gupta et al. [10] improved the ratio to $H_k - \Theta(\ln^2 k/k)$, which is better than 4 for any fixed $k \leq 30$. So, for some $k \leq 30$, the best ratios of k-MCVRP are $H_k - \Theta(\ln^2 k/k)$.

Note that each vehicle must return to the depot it starts in our setting, which is also known as the *fixed-destination* property [18]. Li and Simchi-Levi [18] also considered a non-fixed-destination version where each vehicle may terminate at any depot. The non-fixed-destination MCVRP can be reduced to CVRP easily with the approximation ratio preserved since one can regard all depots as a single super-depot and let the distance between a customer and the super-depot be the minimum weight of the edges between the customer and the depots.

Recently, Lai et al. [17] studied a variant of MCVRP, called *m-Depot Split Delivery Vehicle Routing*, where the number of depots is still m, but the number of vehicles in each depot is limited and each vehicle can be used for at most one tour (one can also think that each depot contains only one vehicle, which can be used a limited number of times). When m is fixed, they obtained a $(6 - 4/m)$-approximation algorithm. Carrasco Heine et al. [14] considered a bifactor approximation algorithm for a variant of Capacitated Location Routing, where each depot has a capacity as well.

1.1   Our Contributions

Motivated by recent progress in approximating k-CVRP, we design improved approximation algorithms for k-MCVRP. For the sake of presentation, we assume that $\alpha = 3/2$. The contributions are shown as follows.

Firstly, we review the cycle-partition algorithm in [18] and then propose a refined tree-partition algorithm based on the idea in [12]. Note that our refined algorithm has a better approximation ratio for fixed k. For splittable and unit-demand k-MCVRP, both of them are 4-approximation algorithms. By making a trade-off between them and further using the result in [4], we obtain an improved $(4 - 1/1500)$-approximation ratio. The cycle-partition algorithm itself may only lead to a $(4 - 1/3000)$-approximation ratio.

Secondly, using the LP-rounding method in [9], we obtain an LP-based cycle-partition algorithm that can achieve a $(4 + \ln 2 + \delta)$-approximation ratio for unsplittable k-MCVRP with any constant $\delta > 0$. By making a trade-off between the LP-based cycle-partition algorithm and the tree-partition algorithm and further using the result in [4], we obtain an improved $(4 - 1/50000)$-approximation ratio.

At last, we propose an LP-based tree-partition algorithm, which works for fixed k. Using the lower bounds of k-CVRP in [27], we obtain an improved $(3 + \ln 2 - \Theta(1/\sqrt{k}))$-approximation algorithm for all three versions of k-MCVRP, which is better than the current-best ratios for any $k > 11$. By making a trade-off between the LP-based tree-partition algorithm and the cycle-partition algorithm and further using the result in [4], we show that the ratio can be improved to $3 + \ln 2 - \max\{\Theta(1/\sqrt{k}), 1/9000\}$ for splittable and unit-demand k-MCVRP.

Due to limited space, the proofs of lemmas and theorems marked with "*" were omitted and they can be found in the full version of this paper [28].

2 Preliminaries

2.1 Definitions

In MCVRP, we let $G = (V \cup D, E)$ denote the input complete graph, where vertices in V represent customers and vertices in D represent depots. There is a non-negative weight function $w : E \to \mathbb{R}_{\geq 0}$ on the edges in E. We often write $w(u, v)$ to mean the weight of edge uv, instead of $w(uv)$. Note that $w(u, v)$ would be the same as the distance between u and v. The weight function w is a semi-metric function, i.e., it is symmetric and satisfies the triangle inequality. For any weight function $w : X \to \mathbb{R}_{\geq 0}$, we extend it to subsets of X, i.e., we define $w(Y) = \sum_{x \in Y} w(x)$ for $Y \subseteq X$. There is a demand function $d : V \to \mathbb{N}_{\geq 1}$, where $d(v)$ is the demand required by $v \in V$. We let $\Delta = \sum_{v \in V} \min_{u \in D} d(v) w(u, v)$. For a component S, we simply use $v \in S$ (resp., $e \in S$) to denote a vertex (resp., an edge) of S, and let $w(S) := \sum_{e \in S} w(e)$ and $d(S) := \sum_{v \in S} d(v)$.

A *walk* in a graph is a succession of edges in the graph, where an edge may appear more than once. We will use a sequence of vertices to denote a walk. For example, $v_1 v_2 v_3 \ldots v_l$ means a walk with edges $v_1 v_2$, $v_2 v_3$, and so on. A *path* in a graph is a walk such that no vertex appears more than once in the sequence, and a *cycle* is a walk such that only the first and the last vertices are the same. A cycle containing l edges is called an *l-cycle* and the *length* of it is l. A *spanning forest* in a graph is a forest that spans all vertices. A *constrained spanning forest* in graph G is a spanning forest where each tree contains only one depot.

An *itinerary* I is a walk that starts and ends at the same depot and does not pass through any other depot. It is called an *empty itinerary* and denote it by

$I = \emptyset$ if there are no customer vertices on I, and a *non-empty itinerary* otherwise. A non-empty itinerary can be split into several minimal cycles containing only one depot, and each such cycle is called a *tour*. The Multidepot Capacitated Vehicle Routing Problem (k-MCVRP) can be described as follows.

Definition 1 (k-MCVRP). *An instance $(G = (V \cup D, E), w, d, k)$ consists of:*

- *a complete graph G, where $V = \{v_1, \ldots, v_n\}$ represents the n customers and $D = \{u_1, \ldots, u_m\}$ represents the m depots,*
- *a weight function $w \colon (V \cup D) \times (V \cup D) \to \mathbb{R}_{\geq 0}$, which represents the distances,*
- *a demand function $d \colon V \to \mathbb{N}_{\geq 1}$, where $d(v)$ is the demand required by customer $v \in V$,*
- *the capacity $k \in \mathbb{Z}_{\geq 1}$ of vehicles that initially stays at each depot.*

A feasible solution is a set of m itineraries, with each having one different depot:

- *each tour delivers at most k of the demand to customers on the tour,*
- *the union of tours over all itineraries meets every customer's demand.*

Specifically, the goal is to find a set of itineraries $\mathcal{I} = \{I_1, \ldots, I_m\}$ where I_i contains depot u_i, minimizing the total weight of the succession of edges in the walks in $\mathcal{I}$, i.e., $w(\mathcal{I}) := \sum_{I \in \mathcal{I}} w(I) = \sum_{I \in \mathcal{I}} \sum_{e \in I} w(e)$.

According to the property of the demand, we define three well-known versions. If each customer's demand must be delivered in one tour, we call it *unsplittable k-MCVRP*. If a customer's demand can be split into several tours, we call it *splittable k-MCVRP*. If each customer's demand is a unit, we call it *unit-demand k-MCVRP*.

In the following, we use CVRP to denote MCVRP with $m = 1$, i.e., only one depot. Unless otherwise specified, we think that k-MCVRP satisfies the fixed-destination property. Moreover, if it holds the non-fixed-destination property, we called it *non-fixed k-MCVRP*.

2.2   Assumptions

Note that in our problem the demand $d(v)$ may be very large since the capacity k may be arbitrarily larger than n. For the sake of analysis, we make several assumptions that can be guaranteed by some simple observations or polynomial-time reductions (see the full version).

Assumption 1. *For splittable and unsplittable k-MCVRP, each customer's demand is at most k.*

Assumption 2. *For splittable k-MCVRP with fixed k, each customer's demand is a unit.*

Assumption 3. *For unsplittable, splittable, and unit-demand k-MCVRP, there exists an optimal solution where each tour delivers an integer amount of demand to each customer in the tour.*

By Assumption 3, in the following, we may only consider a tour that delivers an integer amount of demand to each customer in the tour. Moreover, we know that for unit-demand k-MCVRP, there is an optimal solution consisting of a set of cycles, which intersect only at the depot.

3   Lower Bounds

To connect approximation algorithms for k-MCVRP with k-CVRP, we consider non-fixed k-MCVRP. The first reason is that non-fixed k-MCVRP is a relaxation of k-MCVRP, and then an optimal solution of the former provides a lower bound for the latter. Let OPT (resp., OPT$'$) denote the weight of an optimal solution for k-MCVRP (resp., k-CVRP). We have OPT$' \leq$ OPT. The second is that non-fixed k-MCVRP is equivalent to k-CVRP. The reduction is shown as follows.

Given $G = (V \cup D, E)$, we obtain a new undirected complete graph $H = (V \cup \{o\}, F)$ by replacing the m depots in D with a new single depot, denoted by o. There is a weight function $c : F \to \mathbb{R}_{\geq 0}$ on the edges in F. Moreover, it holds that $c(o, v) = \min_{u \in D} w(u, v)$ and $c(v, v') = \min\{c(o, v) + c(o, v'), w(v, v')\}$ for all $v, v' \in V$. We can verify that the weight function c is a semi-metric function. Note that $\Delta = \sum_{v \in V} d(v) c(o, v)$. Clearly, any feasible solution of k-CVRP in H corresponds to a feasible solution for non-fixed k-MCVRP in G with the same weight. Note that an edge vv' in E with $w(v, v') > c(o, v) + c(o, v')$ was also called a "dummy" edge in [18]. Any tour using a dummy edge vv' can be transformed into two tours with a smaller weight by replacing vv' with two edges uv and $u'v'$ incident to depots such that $c(o, v) = w(u, v)$ and $c(o, v') = w(u', v')$. So, any feasible solution of non-fixed k-MCVRP in G can also be modified into a feasible solution for k-CVRP in H with a non-increasing weight.

A *Hamiltonian cycle* in a graph is a cycle that contains all vertices in the graph exactly once. Let C^* be a minimum cost Hamiltonian cycle in graph H. We mention three lower bounds for k-CVRP, which also works for k-MCVRP.

Lemma 1 ([11]). *It holds that* $OPT \geq OPT' \geq c(C^*)$.

Lemma 2 ([11]). *It holds that* $OPT \geq OPT' \geq (2/k)\Delta$.

Let T^* denote an optimal spanning tree in graph H. Clearly, its cost is a lower bound of an optimal Hamiltonian cycle in H. By Lemma 1, we have

Lemma 3. *It holds that* $OPT \geq OPT' \geq c(T^*)$.

4   Review of the Previous Algorithms

4.1   The Cycle-Partition Algorithm

The main idea of the cycle-partition algorithm [18] is to construct a solution for non-fixed k-MCVRP based on the ITP or UITP algorithm for k-CVRP, and then modify the solution into a solution for k-MCVRP.

The ITP and UITP Algorithms. For splittable and unit-demand k-CVRP, given a Hamiltonian cycle C in graph H, the ITP algorithm is to split the cycle into segments in a good way, with each containing at most k of demand, and for each segment assign two edges between the new depot o and endpoints of the segment. The solution has a weight of at most $(2/k)\Delta + c(C)$ [2,11]. For unsplittable k-CVRP, the UITP algorithm is to use the ITP algorithm with a capacity of $k/2$ to obtain a solution for splittable and unit-demand k-CVRP, and then modify the solution into a feasible solution for unsplittable k-CVRP. Altinkemer and Gavish proved [1] that the modification does not take any additional cost.

Lemma 4 ([1,2,11]). *Given a Hamiltonian cycle C in graph H, for splittable and unit-demand k-CVRP, the ITP algorithm can use polynomial time to output a solution of cost at most $(2/k)\Delta + c(C)$; for unsplittable k-CVRP, the bound improves to $(4/k)\Delta + c(C)$.*

The Cycle-Partition Algorithm. For k-MCVRP, the cycle-partition algorithm uses the ITP or UITP algorithm to obtain a feasible solution for non-fixed k-MCVRP, and then modify the solution into a feasible solution for k-MCVRP using some additional cost. Li and Simchi-Levi [18] proved that the additional cost is exactly the cost of the Hamiltonian cycle used.

Lemma 5 ([18]). *Given a Hamiltonian cycle C in graph H, for splittable and unit-demand k-MCVRP, there is a polynomial-time algorithm to output a solution of cost at most $(2/k)\Delta + 2c(C)$; for unsplittable k-MCVRP, the bound improves to $(4/k)\Delta + 2c(C)$.*

Using the $3/2$-approximate Hamiltonian cycle [6,21], by Lemmas 1 and 2, the cycle-partition algorithm achieves a 4-approximation ratio for splittable and unit-demand k-MCVRP and a 5-approximation ratio for unsplittable k-MCVRP.

4.2   The Tree-Partition Algorithm

The tree-partition algorithm is based on an optimal spanning tree in graph H. Note that an optimal spanning tree in H corresponds to an optimal constrained spanning forest in G. The algorithm is to split the corresponding constrained spanning forest into small components in a good way such that each component has a demand of at most k, and moreover each that contains no depots in it has a demand of at least $k/2$. Note that each component that contains one depot can be transformed into a tour by doubling all edges in it and then shortcutting. For each component that contains no depots, the algorithm will add one edge with minimized weight connecting one depot to it. Then, it can be transformed into a tour by the same method: doubling and shortcutting.

Lemma 6 ([12]). *For all three versions of k-MCVRP, there is a polynomial-time algorithm to output a solution of cost at most $(4/k)\Delta + 2c(T^*)$.*

By Lemmas 2 and 3, the tree-partition algorithm achieves a 4-approximation ratio for all three versions of k-MCVRP.

5   An Improvement for Splittable MCVRP

In this section, we first propose a refined tree-partition algorithm based on the idea in [12]. Our algorithm is simpler due to the previous assumptions. Moreover, our algorithm has a better approximation ratio for the case that the capacity k is fixed. Then, based on recent progress in approximating k-CVRP [4], we obtain an improved $(4 - 1/1500)$-approximation algorithm for splittable and unit-demand k-MCVRP.

A Refined Tree-Partition Algorithm. In our algorithm, we first assign a single cheapest trivial tour for each $v \in V$ with $d(v) > \lfloor k/2 \rfloor$ since each customer's demand is at most k by Assumption 1. Let V' denote the rest customers. To satisfy customers in V', we find an optimal spanning tree T'^* in graph $H[V' \cup \{o\}]$. Note that T'^* corresponds to a constrained spanning forest in $G[V' \cup D]$, denoted by $\mathcal{F}$. Consider a tree $T_u \in \mathcal{F}$ that is rooted at the depot $u \in D$. Then, we will generate tours based on splitting T_u, like Hark et al. did in [12]. For each $v \in T_u$, we denote the sub-tree rooted at v and the children set of v by T_v and Q_v, and let $d(T_v) = \sum_{v' \in T_v} d(v')$.

- If $d(T_u) \leq k$, it can be transformed into a tour by doubling and shortcutting.
- Otherwise, we can do the following repeatedly until it satisfies that $d(T_u) \leq k$. We can find a customer $v \in T_u$ such that $d(T_v) > k$ and $d(T_{v'}) \leq k$ for every children $v' \in Q_v$. Consider the sub-trees $\mathcal{T}_v := \{T_{v'} \mid v' \in Q_v\}$. We can greedily partition them into l sets $\mathcal{T}_1, \ldots, \mathcal{T}_l$ such that $\lfloor k/2 \rfloor < d(\mathcal{T}_i) \leq k$ for each $i \in \{2, \ldots, l\}$. For each such set, saying $\mathcal{T}_2$, we can combine them into a component S by adding v with edges joining v and each tree in $\mathcal{T}_2$. Note that S is a sub-tree of T_v. Then, we find an edge e_S with minimized weight connecting one depot to one vertex in S. By doubling e_S with the edges in S and shortcutting (note that we also need to shortcut v), we obtain a tour satisfying all customers in trees of $\mathcal{T}_2$. After handling $\mathcal{T}_i$ for each $i \in \{2, \ldots, l\}$, we only have $\mathcal{T}_1$. If $d(\mathcal{T}_1) > \lfloor k/2 \rfloor$, we can handle it like $\mathcal{T}_i$ with $i > 1$. Otherwise, we have $d(\mathcal{T}_1) \leq \lfloor k/2 \rfloor$. The remaining tree is denoted by T_u'. Note that $d(\mathcal{T}_1) + d(v) \leq k$ since $d(v) \leq \lfloor k/2 \rfloor$. So, in T_u' the condition $d(T_v') > k$ and $d(T_{v'}') \leq k$ for every children $v' \in Q_v'$ will no longer hold which makes sure that the algorithm will terminate in polynomial time.

The algorithm is shown in Algorithm 1.

Theorem 1 (*). *For all three versions of k-MCVRP, the refined tree-partition algorithm can use polynomial time to output a solution of cost at most $\frac{2}{\lfloor k/2 \rfloor + 1} \Delta +$ $2c(T'^*)$.*

Lemma 7. *It holds that $c(C^*) \geq c(T'^*)$.*

Proof. Let C'^* denote an optimal Hamiltonian cycle in graph $H[V' \cup \{o\}]$. By the proof of Lemma 3, we have $c(C'^*) \geq c(T'^*)$. Note that we can obtain a Hamiltonian cycle in $H[V' \cup \{o\}]$ by shortcutting the optimal Hamiltonian cycle C^* in H. By the triangle inequality, we have $c(C^*) \geq c(C'^*)$.

Algorithm 1. A refined tree-partition algorithm for k-MCVRP

Input: Two undirected complete graphs: $G = (V \cup D, E)$ and $H = (V \cup \{o\}, F)$.
Output: A solution for k-MCVRP.
1: For each customer $v \in V$ with $d(v) > \lfloor k/2 \rfloor$, assign a trivial tour from v to its nearest depot.
2: Find an optimal spanning tree T'^* in graph $H[V' \cup \{o\}]$.
3: Obtain the constrained spanning forest $\mathcal{F}$ in $G[V' \cup D]$ with respect to T'^*.
4: **for** every tree $T_u \in \mathcal{F}$ **do**   ▷ T_u is rooted at the depot u
5:     **while** $d(T_u) > k$ **do**
6:         Find $v \in T_u$ such that $d(T_v) > k$ and $d(T_{v'}) \le k$ for each $v' \in Q_v$. ▷ Q_v is the children set of v
7:         Greedily partition trees in $\mathcal{T}_v := \{T_{v'} \mid v' \in Q_v\}$ into l sets $\mathcal{T}_1, \ldots, \mathcal{T}_l$ such that $\lfloor k/2 \rfloor < d(\mathcal{T}_i) \le k$ for each $i \in \{2, \ldots, l\}$.
8:         Initialize Index $:= \{2, \ldots, l\}$.
9:         **if** $d(\mathcal{T}_1) > \lfloor k/2 \rfloor$ **then**
10:             Index $:=$ Index $\cup \{1\}$.
11:         **end if**
12:         **for** $i \in$ Index **do**
13:             Combine trees in $\mathcal{T}_i$ into a component S by adding v with edges joining v and each tree in $\mathcal{T}_i$.
14:             Find an edge e_S with minimized weight connecting one depot to one vertex in S.
15:             Obtain a tour satisfying all customers in trees of $\mathcal{T}_i$ by doubling e_S with the edges in S and shortcutting.
16:             Update T_u by removing the component S except for v from T_u.
17:         **end for**
18:     **end while**
19:     Obtain a tour satisfying all customers in T_u by doubling and shortcutting.
20: **end for**

By Theorem 1 and Lemmas 1 and 7, the refined tree-partition algorithm has an approximation ratio of $\frac{k}{\lfloor k/2 \rfloor + 1} + 2 < 4$. Next, we consider the improvement for general k.

The Improvement. Blauth et al. [4] made a significant progress in approximating k-CVRP. We show that it can be applied to k-MCVRP to obtain an improved $(4 - 1/1500)$-approximation ratio for splittable and unit-demand k-MCVRP. The main idea is to make a trade-off between the cycle-partition algorithm and the refined tree-partition algorithm.

Lemma 8 ([4]). *If $(1 - \varepsilon) \cdot OPT' < (2/k)\Delta$, there is a function $f : \mathbb{R}_{>0} \to \mathbb{R}_{>0}$ with $\lim_{\varepsilon \to 0} f(\varepsilon) = 0$ and a polynomial-time algorithm to get a Hamiltonian cycle C in H with $c(C) \le (1 + f(\varepsilon)) \cdot OPT'$.*

Theorem 2 (*). *For splittable and unit-demand k-MCVRP, there is a polynomial-time $(4 - 1/1500)$-approximation algorithm.*

Note that if we only use the cycle-partition algorithm with a 3/2-approximate Hamiltonian cycle, we can merely get a $(4 - 1/3000)$-approximation algorithm.

6   An Improvement for Unsplittable MCVRP

In this section, we consider unsplittable k-MCVRP. Recently, Friggstad et al. [9] proposed an improved LP-based approximation algorithm for unsplittable k-CVRP. We show that it can be used to obtain an LP-based cycle-partition algorithm for unsplittable k-MCVRP with an improved $(4-1/50000)$-approximation ratio.

An LP-Based Cycle-Partition Algorithm. Recall that unsplittable k-CVRP can be reduced to the minimum weight k-set cover problem if k is fixed. The main idea of the LP-based approximation algorithm in [9] is that fixing a constant $0 < \delta < 1$ they build an LP (in the form of set cover) only for customers v with $d(v) \geq \delta k$. Then, the number of feasible tours is $n^{O(1/\delta)}$ which is polynomially bounded. Using the well-known randomized LP-rounding method (see [23]), they obtain a set of tours that forms a partial solution with a cost of $\ln 2 \cdot \text{OPT}$. Then, they design tours to satisfy the left customers based on a variant of UITP with an excepted cost of $\frac{1}{1-\delta} \cdot (2/k)\Delta + c(C)$, where C is a given Hamiltonian cycle in graph H.

Lemma 9 ([9]). *Given a Hamiltonian cycle C in graph H, for unsplittable k-CVRP with any constant $\delta > 0$, there is a polynomial-time algorithm to output a solution of cost at most $(\ln 2 + \delta) \cdot OPT' + (2/k)\Delta + c(C)$.*

For unsplittable k-MCVRP, we can use the same idea (see the full version). Fixing a constant $0 < \delta < 1$, we build an LP (in the form of set cover) only for customers v with $d(v) \geq \delta k$. A partial solution based on randomized LP-rounding has a weight of $\ln 2 \cdot \text{OPT}$. For left customers, we obtain a solution for k-CVRP in H with an excepted cost of $\frac{1}{1-\delta} \cdot (2/k)\Delta + c(C)$. Since the latter is based on the idea of the UITP algorithm, we can modify them into a set of feasible tours for k-MCVRP using an additional cost of $c(H)$ like Lemma 5. So, we can get the following theorem.

Theorem 3 (*). *Given a Hamiltonian cycle C in graph H, for unsplittable k-MCVRP with any constant $\delta > 0$, the LP-based cycle-partition algorithm can use polynomial time to output a solution of cost at most $(\ln 2 + \delta) \cdot OPT + (2/k)\Delta + 2c(C)$.*

The Improvement. By making a trade-off between the LP-based cycle-partition algorithm and the refined tree-partition algorithm, we can obtain an improved $(4 - 1/50000)$-approximation ratio for unsplittable k-MCVRP.

Theorem 4 (*). *For unsplittable k-MCVRP, there is a polynomial-time $(4 - 1/50000)$-approximation algorithm.*

Algorithm 2. An LP-based tree-partition algorithm for k-MCVRP

Input: Two undirected complete graphs: $G = (V \cup D, E)$ and $H = (V \cup \{o\}, F)$, and a constant $\gamma \geq 0$.
Output: A solution for k-MCVRP.
1: Solve the LP in $mn^{O(k)}$ time.
2: **for** $C \in \mathcal{C}$ **do** Put the tour C into solution with a probability of $\min\{\gamma \cdot x_C, 1\}$.
3: **end for**
4: For each customer contained in multiple tours, we shortcut it for all but one tour.
   ▷ Some customers may be contained in more than one tour due to the randomized rounding.
5: Let $\widetilde{V}$ be the customers that are still unsatisfied.
6: Obtain two new complete graphs: $\widetilde{G} = G[\widetilde{V} \cup D]$ and $\widetilde{H} = G[\widetilde{V} \cup \{o\}]$.
7: Call the refined tree-partition algorithm in Algorithm 1.

---

7 An Improvement for k-MCVRP with Fixed Capacity

In this section, we consider further improvements for the case that the capacity k is fixed. We propose an LP-based tree-partition algorithm based on the refined tree-partition algorithm with the LP-rounding method. The algorithm admits an approximation ratio of $3 + \ln 2 - \Theta(1/\sqrt{k})$. Then, by further using the result in Lemma 8, we also obtain a $(3 + \ln 2 - 1/9000)$-approximation algorithm for splittable and unit-demand k-MCVRP. Note that the former is better when k is a fixed constant less than 3×10^8.

An LP-Based Tree-Partition Algorithm. Due to Assumption 3 we can only consider a tour that delivers an integer amount of demand to each customer in the tour. Since k is fixed, there are at most $mn^{O(k)}$ feasible tours for k-MCVRP. Note that for splittable k-MCVRP each customer's demand is a unit by Assumption 2. Denote the set of feasible tours by $\mathcal{C}$, and define a variable x_C for each tour $C \in \mathcal{C}$. We have the following LP.

$$\text{minimize} \quad \sum_{C \in \mathcal{C}} w(C) \cdot x_C$$

$$\text{subject to} \quad \sum_{\substack{C \in \mathcal{C}: \\ v \in C}} x_C \geq 1, \quad \forall v \in V,$$

$$x_C \geq 0, \quad \forall C \in \mathcal{C}.$$

The LP-based tree-partition algorithm is shown in Algorithm 2.

Consider an optimal solution of k-CVRP in graph H. It consists of a set of simple cycles. Note that if we delete the longest edge from each cycle, we can obtain a spanning tree (by shortcutting if necessary since a customer may appear in more than one tour for the splittable case). Denote this spanning tree by T^{**}.

Theorem 5 (*). *For all three versions of k-MCVRP with any constant $\gamma \geq 0$, the LP-based tree-partition algorithm can use polynomial time to output a solution with an expected cost at most $\gamma \cdot OPT + e^{-\gamma} \cdot \frac{2}{\lfloor k/2 \rfloor + 1} \Delta + 2c(T^{**})$.*

The algorithm can be derandomized efficiently by conditional expectations [23].

The Analysis. Next, we show that the LP-based tree-partition algorithm achieves a ratio of $3 + \ln 2 - \Theta(1/\sqrt{k})$ for all three versions of k-MCVRP (see the full version). Note that the ratio $H_k - \Theta(\ln^2 k/k)$ for k-set cover in [10] is better than ours only for $k \leq 11$.

Theorem 6 (*). *For all three versions of k-MCVRP, the LP-based tree-partition algorithm achieves an approximation ratio of* $\max\{g(\lceil x_0 \rceil), g(\lfloor x_0 \rfloor)\}$, *where* $x_0 := \frac{\sqrt{4k+5}-1}{2}$ *and* $g(x) := 3 + \ln(\frac{k+1-x}{\lfloor k/2 \rfloor+1}) - \frac{1}{x}$.

A Further Improvement for Splittable k-MCVRP. By making a trade-off between the cycle-partition algorithm and the LP-based tree-partition algorithm, we can obtain an improved $(3 + \ln 2 - 1/9000)$-approximation ratio for splittable and unit-demand k-MCVRP.

Theorem 7 (*). *For splittable and unit-demand k-MCVRP, there is a polynomial-time $(3 + \ln 2 - 1/9000)$-approximation algorithm.*

The result $3 + \ln 2 - \Theta(1/\sqrt{k})$ in Theorem 6 is better than $3 + \ln 2 - 1/9000$ for any $k < 3 \times 10^8$. Note that for unsplittable k-MCVRP we cannot obtain further improvements using the same method. The reason is that even using an optimal Hamiltonian cycle the LP-based cycle-partition only achieves a ratio of about $3 + \ln 2$ by Theorem 3. So, there is no improvement compared with the LP-based tree-partition algorithm.

8 Conclusion

In this paper, we consider approximation algorithms for k-MCVRP. Previously, only a few results were available in the literature. Based on recent progress in approximating k-CVRP, we design improved approximation algorithms for k-MCVRP. When k is general, we improve the approximation ratio to $4 - 1/1500$ for splittable and unit-demand k-MCVRP and to $4 - 1/50000$ for unsplittable k-MCVRP; when k is fixed, we improve the approximation ratio to $3 + \ln 2 - \max\{\Theta(1/\sqrt{k}), 1/9000\}$ for splittable and unit-demand k-MCVRP and to $3 + \ln 2 - \Theta(1/\sqrt{k})$ for unsplittable k-MCVRP.

We remark that for unsplittable, splittable, and unit-demand k-MCVRP with fixed $3 \leq k \leq 11$ the current best approximation ratios are still $H_k - \Theta(\ln^2 k/k)$ [10]. In the future, one may study how to improve these results.

A more general problem than k-MCVRP is called *Multidepot Capacitated Arc Routing* (MCARP), where both vertices and arcs are allowed to require a demand. For MCARP, the current best-known approximation algorithms on general metric graphs are still based on the cycle-partition algorithm (see [26]). Some results in this paper could be applied to MCARP to obtain some similar improvements.

Acknowledgments. The work is supported by the National Natural Science Foundation of China, under the grants 62372095 and 61972070.

References

1. Altinkemer, K., Gavish, B.: Heuristics for unequal weight delivery problems with a fixed error guarantee. Oper. Res. Lett. **6**(4), 149–158 (1987)
2. Altinkemer, K., Gavish, B.: Heuristics for delivery problems with constant error guarantees. Transp. Sci. **24**(4), 294–297 (1990)
3. Asano, T., Katoh, N., Tamaki, H., Tokuyama, T.: Covering points in the plane by k-tours: towards a polynomial time approximation scheme for general k. In: STOC 1997, pp. 275–283. ACM (1997)
4. Blauth, J., Traub, V., Vygen, J.: Improving the approximation ratio for capacitated vehicle routing. Math. Program., 1–47 (2022)
5. Bompadre, A., Dror, M., Orlin, J.B.: Improved bounds for vehicle routing solutions. Discret. Optim. **3**(4), 299–316 (2006)
6. Christofides, N.: Worst-case analysis of a new heuristic for the travelling salesman problem. Carnegie-Mellon University, Tech. rep. (1976)
7. Chvatal, V.: A greedy heuristic for the set-covering problem. Math. Oper. Res. **4**(3), 233–235 (1979)
8. Deppert, M., Kaul, M., Mnich, M.: A $(3/2 + \varepsilon)$-approximation for multiple tsp with a variable number of depots. In: 31st Annual European Symposium on Algorithms (ESA 2023). Schloss Dagstuhl-Leibniz-Zentrum für Informatik (2023)
9. Friggstad, Z., Mousavi, R., Rahgoshay, M., Salavatipour, M.R.: Improved approximations for capacitated vehicle routing with unsplittable client demands. In: IPCO 2022. LNCS, vol. 13265, pp. 251–261. Springer (2022). https://doi.org/10.1007/978-3-031-06901-7_19
10. Gupta, A., Lee, E., Li, J.: A local search-based approach for set covering. In: SOSA 2023, pp. 1–11. SIAM (2023)
11. Haimovich, M., Kan, A.H.G.R.: Bounds and heuristics for capacitated routing problems. Math. Oper. Res. **10**(4), 527–542 (1985)
12. Harks, T., König, F.G., Matuschke, J.: Approximation algorithms for capacitated location routing. Transp. Sci. **47**(1), 3–22 (2013)
13. Hassin, R., Levin, A.: A better-than-greedy approximation algorithm for the minimum set cover problem. SIAM J. Comput. **35**(1), 189–200 (2005)
14. Heine, F.C., Demleitner, A., Matuschke, J.: Bifactor approximation for location routing with vehicle and facility capacities. Eur. J. Oper. Res. **304**(2), 429–442 (2023)
15. Karlin, A.R., Klein, N., Gharan, S.O.: A (slightly) improved approximation algorithm for metric TSP. In: STOC 2021, pp. 32–45. ACM (2021)
16. Karlin, A.R., Klein, N., Gharan, S.O.: A deterministic better-than-3/2 approximation algorithm for metric TSP. In: IPCO 2023. LNCS, vol. 13904, pp. 261–274. Springer (2023). https://doi.org/10.1007/978-3-031-32726-1_19
17. Lai, X., Xu, L., Xu, Z., Du, Y.: An approximation algorithm for k-depot split delivery vehicle routing problem. INFORMS J. Comput. (2023)
18. Li, C., Simchi-Levi, D.: Worst-case analysis of heuristics for multidepot capacitated vehicle routing problems. INFORMS J. Comput. **2**(1), 64–73 (1990)
19. Montoya-Torres, J.R., Franco, J.L., Isaza, S.N., Jiménez, H.F., Herazo-Padilla, N.: A literature review on the vehicle routing problem with multiple depots. Comput. Indust. Eng. **79**, 115–129 (2015)
20. Rathinam, S., Sengupta, R., Darbha, S.: A resource allocation algorithm for multivehicle systems with nonholonomic constraints. IEEE Trans. Autom. Sci. Eng. **4**(1), 98–104 (2007)

21. Serdyukov, A.I.: Some extremal bypasses in graphs. Upravlyaemye Sistemy **17**, 76–79 (1978)
22. Traub, V., Vygen, J., Zenklusen, R.: Reducing path TSP to TSP. SIAM J. Comput. **51**(3), 20–24 (2022)
23. Williamson, D.P., Shmoys, D.B.: The design of approximation algorithms. Cambridge University Press (2011)
24. Xu, Z., Rodrigues, B.: A 3/2-approximation algorithm for the multiple tsp with a fixed number of depots. INFORMS J. Comput. **27**(4), 636–645 (2015)
25. Xu, Z., Xu, L., Rodrigues, B.: An analysis of the extended christofides heuristic for the k-depot tsp. Oper. Res. Lett. **39**(3), 218–223 (2011)
26. Yu, W., Liao, Y.: Approximation and polynomial algorithms for multi-depot capacitated arc routing problems. In: Shen, H., et al. (eds.) PDCAT 2021. LNCS, vol. 13148, pp. 93–100. Springer, Cham (2022). https://doi.org/10.1007/978-3-030-96772-7_9
27. Zhao, J., Xiao, M.: Improved approximation algorithms for capacitated vehicle routing with fixed capacity. CoRR abs/ arXiv: 2210.16534 (2022)
28. Zhao, J., Xiao, M.: Improved approximation algorithms for multidepot capacitated vehicle routing. CoRR abs/ arXiv: 2308.14131 (2023)

On the Minimum Depth of Circuits with Linear Number of Wires Encoding Good Codes

Andrew Drucker[1] and Yuan Li[2(✉)]🆔

[1] Chicago, USA
andy.drucker@gmail.com
[2] Fudan University, Shanghai, China
yuan_li@fudan.edu.cn

Abstract. We determine, up to an additive constant 2, the minimum depth required to encode asymptotically good error-correcting codes using a linear number of wires. The inverse-Ackermann-type upper bound is guided by an encoding circuit construction due to Gál et al. [IEEE Trans. Inform. Theory 59(10), pp. 6611-6627, 2013] (which the authors showed asymptotically optimal for constant depths), but applies some new ideas in the construction and analysis to obtain shallower linear-size circuits. We also show our codes can obtain any constant rate and constant relative distance within the Gilbert-Varshamov bounds. The lower bound, which we credit to Gál *et al.*, since it directly follows their method (although not explicitly claimed or fully verified in that work), is obtained by making some constants explicit in a graph-theoretic lemma of Pudlák, extending it to super-constant depths.

We also study a subclass of MDS codes $C : \mathbb{F}^n \to \mathbb{F}^m$ characterized by the Hamming-distance relation $\text{dist}(C(x), C(y)) \geq m - \text{dist}(x, y) + 1$ for any distinct $x, y \in \mathbb{F}^n$. (For linear codes this is equivalent to the generator matrix being totally invertible.) We call these *superconcentrator-induced codes*, and we show their tight connection with superconcentrators. Specifically, we observe that any linear or nonlinear circuit encoding a superconcentrator-induced code must be a superconcentrator graph, and any superconcentrator graph can be converted to a linear circuit, over a sufficiently large field (exponential in the size of the graph), encoding a superconcentrator-induced code.

Keywords: Error-correcting codes · Circuit complexity · Superconcentrator

1 Introduction

Understanding the computational complexity of *encoding* error-correcting codes is an important task in theoretical computer science. Complexity measures of

An earlier version of this work appeared as Chapter 2 of the author's Ph.D. thesis [Li17]. Most of this work was done while the authors were affiliated with the University of Chicago Computer Science Dept. The second author is partly supported by the Shanghai Science and Technology Program under Project 21JC1400600.
A. Drucker—Independent.

W. Wu and G. Tong (Eds.): COCOON 2023, LNCS 14423, pp. 392–403, 2024.
https://doi.org/10.1007/978-3-031-49193-1_30

interest include time, space, and parallelism. Error-correcting codes are indispensable as a tool in computer science. Highly efficient encoding algorithms (or circuits) are desirable in settings studied by theorists including zero-knowledge proofs [Gol+21], circuit lower bounds [CT19], data structures for error-correcting codes [Vio19], pairwise-independent hashing [Ish+08], and secret sharing [DI14]. Besides that, the *existence* of error-correcting codes with efficient encoding circuits sheds light on the designing of practical error-correcting codes.

We consider codes with constant rate and constant relative distance, which are called asymptotically good error-correcting codes or *good codes* for short. The complexity of encoding good codes has been studied before. Bazzi and Mitter [BM05] proved that branching programs with linear time and sublinear space cannot encode good codes. By using the sensitivity bounds [Bop97], one can prove that AC^0 circuits cannot encode good codes; Lovett and Viola proved that AC^0 circuits cannot sample good codes [LV11]; Beck, Impagliazzo and Lovett [BIL12] strengthened the result.

Dobrushin, Gelfand and Pinsker [DGP73] proved that there exist linear-size circuits encoding good codes. Sipser and Spielman [Spi96,SS96] explicitly constructed good codes that are encodable by bounded fanin circuits of depth $O(\log n)$ and size $O(n)$, and decodable by circuits of size $O(n \log n)$. For bounded fan-in, the depth $O(\log n)$ is obviously optimal. Henceforth, unless otherwise stated, we consider circuits with unbounded fan-in, where the size is measured by the number of *wires* instead of gates.

Gál, Hansen, Koucký, Pudlák, and Viola [Gal+13] investigated the circuit complexity of encoding good codes. Gál *et al.* constructed circuits recursively and probabilistically, with clever recursive composition ideas, which resemble the construction of superconcentrators in [Dol+83]. They also proved size lower bounds for bounded depth, by showing that any circuit encoding good codes must satisfy some superconcentrator-like properties; the lower bound follows from the size bounds for a variant of bounded-depth superconcentrators studied by Pudlák [Pud94]. Their construction's wire upper bounds are of form $O_d(n \cdot \lambda_d(n))$ (in our notation[1]) and their lower bounds are of form $\Omega_d(n \cdot \lambda_d(n))$, matching up to a multiplicative constant c_d for constant values d. They also proved that there exist $O_d(n)$-size $O(\log \lambda_d(n))$-depth circuits encoding good codes. Here $\lambda_d(n)$ are slowly growing inverse Ackermann-type functions, e.g., $\lambda_2(n) = \Theta(\log n)$, $\lambda_3(n) = \Theta(\log \log n)$, $\lambda_4(n) = \Theta(\log^* n)$.

Druk and Ishai [DI14] proposed a randomized construction of good codes meeting the Gilbert-Varshamov bound, which can be encoded by linear-size logarithmic-depth circuits (with bounded fan-in). Their construction is based on linear-time computable pairwise independent hash functions [Ish+08].

Chen and Tell [CT19] constructed *explicit* circuits of depth d encoding linear code with constant relative distance and code rate $\Omega\left(\frac{1}{\log n}\right)$ using $n^{1+2^{-\Omega(d)}}$ wires, for every $d \geq 4$. They used these explicit circuits to prove bootstrapping results for threshold circuits.

---

[1] Our definition of $\lambda_d(n)$ follows Raz and Shpilka [RS03]. It is slightly different from Gál *et al.*'s. In [Gal+13], the function $\lambda_i(n)$ is actually $\lambda_{2i}(n)$ in our notation.

1.1   Background and Results

To encode good error-correcting codes, linear size is obviously required. It is natural to ask, what is the minimum depth required to encode goods using a linear number of wires? This question is addressed, but not fully answered, by the work of Gál et al. [Gal+13].

We show that one can encode error-correcting codes with constant rate and constant relative distance using $O(n)$ wires, with depth at most $\alpha(n)$, for sufficiently large n. Here, $\alpha(n)$ is a version of the inverse Ackermann function. This is nearly optimal, by a lower bound of $\alpha(n) - 2$ that we credit to Gál et al. [Gal+13] as discussed below. Our new upper bound states:

Theorem 1. *(Upper bound) Let $r \in (0,1)$ and $\delta \in (0, \frac{1}{2})$ such that $r < 1 - h(\delta)$. For sufficiently large n, there exists a linear circuit $C : \{0,1\}^n \to \{0,1\}^{\lfloor \frac{n}{r} \rfloor}$ of size $O_{r,\delta}(n)$ and depth $\alpha(n)$ that encodes an error-correcting code with relative distance $\geq \delta$.*

Our upper bound in Theorem 1 constructs certain binary codes encodable by linear-size circuits meeting any constant rate and relative distance within the Gilbert-Varshamov bound. For unbounded depth d, these circuits are smaller than the bounds provided by Gál et al., and this efficiency combined with a careful analysis allows us to achieve linear size in smaller and optimal (to within additive constant 2) depth. In comparison, Gál et al. constructed binary codes with rate 1/32 and relative distance 1/8, which are encodable by circuits of the size $O_d(n)$ and depth $O(\log \lambda_d(n))$, for any constant d. Their upper bound is strong, but suboptimal.

In terms of techniques, we follow Gál et al.'s upper-bound constructions and distill some clarifying concepts, with a few new ingredients. For example, in our recursive construction, the fanin of the gates on the last layer is bounded by an absolute constant (which makes the constant in the size bound $O(\lambda_d(n) \cdot n)$ an absolute constant, eliminating the dependence on d compared with Gál et al.'s); we use a property of the inverse Ackermann function to get rid of an $O(\log \alpha(n))$ factor in the analysis — this observation also improves the depth of the superconcentrator constructions [Dol+83]. Finally, by using a disperser graph at the bottom layer in the circuit and collapsing that layer afterward, we boost the rate and distance to any constants within the Gilbert-Varshamov bound. (This kind of rate or distance boosting technique is widely used, for instance, see [DI14], but it is not considered in [Gal+13]. Based on pairwise independent hash functions, Druk and Ishai [DI14] constructed codes meeting the Gilbert-Varshamov bound; in comparison, their circuits have bounded fanin and are of linear size and depth $O(\log n)$.) Our framework is also inspired by the superconcentrator construction by Dolev, Dwork, Pippenger, and Wigderson [Dol+83]. See the beginning of Section 3 for a detailed discussion.

Turning to the lower bounds: we credit the lower bound in the result below to Gál et al. (although it was not explicitly claimed or fully verified in that work), since it is directly obtainable by their size lower-bound method and the tool of

Pudlák [Pud94] on which it relies, when that tool is straightforwardly extended to super-constant depth.[2]

Theorem 2. *(Lower bound) [Gal+13] Let* $\rho \in (0,1)$ *and* $\delta \in (0, \frac{1}{2})$*, and let constant* $c > 0$*. Let* $C_n : \{0,1\}^n \to \{0,1\}^{\lfloor n/\rho \rfloor}$ *be a family of circuits of size at most* cn *that encode error-correcting codes with relative distance* $\geq \delta$*. Arbitrary Boolean-function gates of unrestricted fanin are allowed in* C_n*. If* n *is sufficiently large, i.e.,* $n \geq N(r, \delta, c)$*, the depth of the circuit* C_n *is at least* $\alpha(n) - 2$*.*

The proof for Theorem 2 closely follows [Gal+13] and is an application of a graph-theoretic argument in the spirit of [Val77, Dol+83, Pud94, RS03]. In detail, we use Pudlák's size lower bounds [Pud94] on "densely regular" graphs, and rely on the connection between good codes and densely regular graphs by Gál et al. [Gal+13]. Pudlák's bound was originally proved for bounded depth; in order to apply it to unbounded depth, we explicitly determine the hidden constants by directly following Pudlák's work, and verify that their decay at higher unbounded depths is moderate enough to allow the lower-bound method to give superlinear bounds up to depth $\alpha(n) - 3$.

Stepping back to a higher-level view, the strategy of the graph-theoretic lower-bound arguments in the cited and related works is as follows:

- Prove any circuit computing the target function must satisfy some superconcentrator-like connection properties;
- Prove any graph satisfying the above connection properties must have many edges;
- Therefore, the circuit must have many wires.

Valiant [Val75, Val76, Val77] first articulated this kind of argument, and proposed the definition of superconcentrators. Somewhat surprisingly, Valiant showed that linear-size superconcentrators exist. As a result, one cannot prove superlinear size bounds using this argument (when the depth is unbounded). Dolev, Dwork, Pippenger, and Wigderson [Dol+83] proved $\Omega(\lambda_d(n) \cdot n)$ lower bounds for bounded-depth (weak) superconcentrators, which implies circuit lower bounds for functions satisfying weak-superconcentrator properties. Pudlák [Pud94] generalized Dolev et al.'s lower bounds by proposing the definition of *densely regular graphs*, and proved lower bounds for bounded-depth densely regular graphs, which implies circuit lower bounds for functions satisfying densely regular property, including shifters, parity shifters, and Vandermonde matrices. Raz and Shpilka [RS03] strengthened the aforementioned superconcentrator lower bounds by proving a powerful graph-theoretic lemma, and applied it to

---

[2] The $\Omega(\cdot)$ notation in the circuit size lower bound of Gál *et al.*, for example, Theorem 1 in [Gal+13], involves an implicit constant which decays with the depth d, as can be suitable for constant depths; similarly for the tool of Pudlák, Theorem 3.(ii) in [Pud94], on which it relies. For general super-constant depths, more explicit work is required to verify the decay is not too rapid. In fact, even after our work, the precise asymptotic complexity of encoding good codes remains an open question for d in the range $[\omega(1), \alpha(n) - 3]$.

prove superlinear lower bounds for matrix multiplication. (This powerful lemma can reprove all the above lower bounds.) Gál et al. [Gal+13] proved that any circuits encoding good error-correcting codes must be densely regular. They combined this with Pudlák's lower bound on densely regular graphs [Pud94] to obtain $\Omega(\lambda_d(n) \cdot n)$ size bounds for depth-d circuits encoding good codes.

All the circuit lower bounds mentioned above apply even to the powerful model of *arbitrary-gate* circuits, that is,

- each gate has unbounded fanin,
- a gate with fanin s can compute any function from $\{0,1\}^s$ to $\{0,1\}$,
- circuit size is measured as the number of wires.

In this "arbitrary-gates" model, any function from $\{0,1\}^n$ to $\{0,1\}^m$ can be computed by a circuit of size mn.

It is known that any circuits encoding good codes must satisfy some superconcentrator-like connection properties [Spi96], [Gal+13]. Our other result is a theorem in the *reverse* direction in the algebraic setting over large finite fields. Motivated by this connection, we study *superconcentrator-induced codes* (Definition 6), a subclass of maximum distance separable (MDS) codes [LX04,GRS12], and observe its tight connection with superconcentrators.

Theorem 3. *(Informal) Given any (n, m)-superconcentrator, one can convert it to a linear arithmetic circuit encoding a code $C : \mathbb{F}^n \to \mathbb{F}^m$ such that*

$$\mathrm{dist}(C(x), C(y)) \geq m - \mathrm{dist}(x, y) + 1 \quad \forall x \neq y \in \mathbb{F}^n \tag{1}$$

by replacing each vertex with an addition gate and assigning the coefficient for each edge uniformly at random over a sufficiently large finite field (where $d2^{\Omega(n+m)}$ suffices, and d is the depth of the superconcentrator).

We also observe that any arithmetic circuit, linear or nonlinear, encoding a code $C : \mathbb{F}^n \to \mathbb{F}^m$ satisfying (1), viewed as a graph, must be a superconcentrator.

The proof of Theorem 3 relates the connectivity properties with the rank of a matrix, and uses Schwartz-Zippel lemma to estimate the rank of certain submatrices; these techniques are widely used, for example, in [CKL13,Lov18]. In addition, the idea of assigning uniform random coefficients (in a finite field) to edges, to form linear circuits, has appeared before in e.g. network coding [Ahl+00,LYC03]. The question we study is akin to a higher-depth version of the GM-MDS type questions about matrices with restricted support [Lov18].

Observe that any code satisfying the distance inequality (1) is a good code. The existence of depth-d size-$O(\lambda_d(n) \cdot n)$ superconcentrators [Dol+83,AP94], for any $d \geq 3$, immediately implies the existence of depth-d (linear) arithmetic circuits of size $O(\lambda_d(n) \cdot n)$ encoding good codes *over large finite field*.

In a subsequent work [Li23], inspired by this connection and using similar techniques, the second author proved that any (n, m)-superconcentrator can compute the shares of an (n, m) linear threshold secret sharing scheme. In other words, any (n, m)-superconcentrator-induced code induces an (n, m) linear threshold secret sharing scheme. Results in [Li23] can be viewed as an application of superconcentrator-induced codes.

2   Inverse Ackermann Functions

Definition 1. *(Definition 2.3 in [RS03]) For a function f, define $f^{(i)}$ to be the composition of f with itself i times. For a function $f : \mathbb{N} \to \mathbb{N}$ such that $f(n) < n$ for all $n > 0$, define*

$$f^*(n) := \min\{i : f^{(i)}(n) \le 1\}.$$

Let

$$\lambda_1(n) := \lfloor \sqrt{n} \rfloor \;,$$
$$\lambda_2(n) := \lceil \log n \rceil \;,$$
$$\lambda_d(n) := \lambda_{d-2}^*(n) \;.$$

As d gets larger, $\lambda_d(n)$ becomes extremely slowly growing, for example, $\lambda_3(n) = \Theta(\log\log n)$, $\lambda_4(n) = \Theta(\log^* n)$, $\lambda_5(n) = \Theta(\log^* n)$, etc.

We define the inverse Ackermann function as follows.

Definition 2 (Inverse Ackermann Function). *For any positive integer n, let*

$$\alpha(n) := \min\{even\ d : \lambda_d(n) \le 6\}.$$

There are different variants of the inverse Ackermann function; they differ by at most a multiplicative constant factor.

We need the definition of the Ackermann function.

Definition 3. *(Ackermann function [Tar75, Dol+83]) Define*

$$\begin{cases} A(0, j) = 2j, & for\ j \ge 1 \\ A(i, 1) = 2, & for\ i \ge 1 \\ A(i, j) = A(i-1, A(i, j-1)), & for\ i \ge 1, j \ge 2. \end{cases} \tag{2}$$

For notational convenience, we often write $A(i, j)$ as $A_i(j)$.

3   Upper Bound

In this section, we prove Theorem 1. That is, we non-explicitly construct, for any rate $r \in (0, 1)$ and relative distance $\delta \in (0, \frac{1}{2})$ satisfying $r < 1 - h(\delta)$, circuits encoding error-correcting codes $C : \{0, 1\}^n \to \{0, 1\}^{\lfloor n/r \rfloor}$ with relative distance δ, where the circuit is of size $O_{r,\delta}(n)$ and depth $\alpha(n)$.

First, we construct circuits encoding codes $C : \{0, 1\}^n \to \{0, 1\}^{32n}$ with relative distance $\frac{1}{8}$, where the constants 32 and $\frac{1}{8}$ are picked for convenience. Then, we use a simple trick to adjust the rate and boost the distance to achieve the Gilbert-Varshamov bound (without increasing the depth of the circuit).

Note that random linear codes achieve the Gilbert-Varshamov bound. However, circuits encoding random linear codes have size $O(n^2)$. In contrast, our circuits have size $O(n)$. Our circuits consist of XOR gates only; we call these *linear circuits* hereafter. We point out that the construction generalizes to any finite field, where XOR gates are replaced by addition gates (over that finite field).

Let $S_d(n)$ denote the minimum size of a depth-d linear circuit encoding a code $C : \{0,1\}^n \to \{0,1\}^{32n}$ with distance $4n$.

Our construction heavily relies our Gál et al. [Gal+13], who proved that $S_d(n) = O_d(\lambda_d(n) \cdot n)$, and proved that, for any fixed d, when the depth $d = O(\log(\lambda_d(n)))$, $S_d(n) = O_d(n)$. Our main technical contribution is to prove $S_d(n) = O(\lambda_d(n) \cdot n)$, where the hidden constant is an absolute constant. As a result, it implies that, $S_{\alpha(n)} = O(n)$, which is almost optimal. In terms of techniques,

- We distill a few clarifying concepts including partial good codes, composition lemmas, etc., which are implicit in [Gal+13]. Our framework of analysis is also inspired by [Dol+83].
- Step by step, we control the fanin of the output gates in the recursive construction. We make sure this property, bounded output fan-in, is preserved after composing a constant number of partial good codes and after boosting the rate. This is critical for the improvement of the upper bound, from $O_d(\lambda_d(n) \cdot n)$ to $O(\lambda_d(n) \cdot n)$, eliminating a dependence growing with d.
- Using a property of the inverse Ackermann function, we improve the analysis to get rid of an additive $O(\log \alpha(n))$ factor on the depth (compared with Corollary 32 in [Gal+13], or Corollary 1.1 in [Dol+83]). Specifically, in [Dol+83], a version of the inverse Ackermann function is (roughly) defined as the minimum d such that $\lambda_d(n) \le d$. We observe that $\lambda_{d+2}(n) = O(1)$, which implies that the depth-$(d+2)$ circuit has size $O(\lambda_{d+2}(n) \cdot n) = O(n)$. This observation improves the depth from $\alpha(n) + O(\log \alpha(n))$ to $\alpha(n)$.
- The above observation can also improve the depth bound on the linear-size superconcentrators in [Dol+83] from $\alpha(n) + O(\log \alpha(n))$ to $\alpha(n)$ (without change in the construction).

Definition 4. *(Partial good code)* $C : \{0,1\}^n \to \{0,1\}^{32n}$ *is called* (n,r,s)-partial good code *if for all* $x \in \{0,1\}^n$ *with* $\mathrm{wt}(x) \in [r,s]$, *we have* $\mathrm{wt}(C(x)) \ge 4n$.

Denote by $S_d(n,r,s)$ the minimum size of a linear circuit that encodes an (n,r,s)-partial good code, where r,s are real numbers.

The following theorem is our main construction, whose proof and the auxiliary lemmas are in the full version of our paper.

Theorem 4. *For any* $1 \le r \le n$, *and for any* $k \ge 3$, *we have*

$$S_{2k}\left(n, \frac{n}{A(k-1,r)}, \frac{n}{r}\right) = O(n). \tag{3}$$

Moreover, the output gates of the linear circuits encoding $\left(n, \frac{n}{A(k-1,r)}, \frac{n}{r}\right)$-*partial good code have bounded fanin.*

For any $1 \leq r \leq n$ and for any $k \geq 2$,

$$S_{2k}\left(n, \frac{n}{r}, n\right) = O(\lambda_{2k}(r) \cdot n). \tag{4}$$

By taking $r = n$ in Theorem 4, we immediately have

Corollary 1. *For any n,*

$$S_{\alpha(n)}(n) = O(n).$$

We have constructed a linear circuit of size $O(n)$ and depth $\alpha(n)$ encoding a code $C : \{0,1\}^n \to \{0,1\}^{32n}$ with relative distance $\frac{1}{8}$. By putting a rate booster at the bottom, we can achieve any constant rate and relative distance within the Gilbert-Varshamov bound. The proof of Theorem 1 is in the full version of our paper.

4 Depth Lower Bound

Definition 5. *(Densely regular graph [Pud94]) Let G be a directed acyclic graph with n inputs and n outputs. Let $0 < \epsilon, \delta$ and $0 \leq \mu \leq 1$. We say G is (ϵ, δ, μ)-densely regular if for every $k \in [\mu n, n]$, there are probability distributions $\mathcal{X}$ and $\mathcal{Y}$ on k-element subsets of inputs and outputs respectively, such that for every $i \in [n]$,*

$$\Pr_{X \in \mathcal{X}}[i \in X] \leq \frac{k}{\delta n}, \qquad \Pr_{Y \in \mathcal{Y}}[i \in Y] \leq \frac{k}{\delta n},$$

and the expected number of vertex-disjoint paths from X to Y is at least ϵk for randomly chosen $X \in \mathcal{X}$ and $Y \in \mathcal{Y}$.

Denote by $D(n, d, \epsilon, \delta, \eta)$ the minimal size of a (ϵ, δ, μ)-densely regular layered directed acyclic graph with n inputs and n outputs and depth d.

Theorem 5. *(Theorem 3 in [Pud94]) Let $\epsilon, \delta > 0$. For every $d \geq 3$, and every $r \leq n$,*

$$D(n, d, \epsilon, \delta, \frac{1}{r}) = \Omega_{d,\epsilon,\delta}(n\lambda_d(r)).$$

To apply the above lower bounds when d is not *fixed*, we need to figure out the hidden constant that depends on d, ϵ, δ.

Theorem 6. *Let $\epsilon, \delta > 0$. For every $d \geq 3$, and every $r \leq n$,*

$$D(n, d, \epsilon, \delta, \frac{1}{r}) \geq \Omega(2^{-d/2}\epsilon\delta^2\lambda_d(r)n).$$

The constant in $\Omega(2^{-d/2}\epsilon\delta^2\lambda_d(r)n)$ is an absolute constant. The proof of Theorem 6 is almost the same as Theorem 5. Due to the space constraints, the proof is deferred to the full version.

Corollary 2. *(Corollary 15 in [Gal+13]) Let $0 < \rho, \delta < 1$ be constants and C be a circuit encoding an error-correcting code $\{0,1\}^{\rho n} \to \{0,1\}^n$ with relative distance at least δ. If we extend the underlying graph with $(1-\rho)n$ dummy inputs, then its underlying graph is $(\rho\delta, \rho, \frac{1}{n})$-densely regular.*

The proof of Theorem 2 readily follows from Theorem 6 and Corollary 2; see the full version.

We would like to point out that, alternatively, one can use a powerful lemma by Raz and Shipilka [RS03], to prove the depth lower bound (i.e., Theorem 6).

5   Superconcentrator-induced Codes

It is known that circuits for encoding error-correcting codes must satisfy some superconcentrator-like connectivity properties. For example, Speilman [Spi96] observed that any circuits encoding codes from $\{0,1\}^n$ to $\{0,1\}^m$ with distance δm must have δn vertex-disjoint paths connecting any chosen δn inputs to any set of $(1-\delta)m$ outputs; using matroid theory, Gál et al. [Gal+13] proved that, for any $k \le n$, for any k-element subset of inputs X, taking a random k-element subset of outputs Y, the expected number of vertex-disjoint paths from X to Y is at least δk.

We observe a connection in the *reverse* direction by showing that *any* super-concentrator graph, converted to an arithmetic circuit over a sufficiently large field can encode a good code. (Recall that a directed acyclic graph $G = (V, E)$ with m inputs and n outputs is an (m, n)-*superconcentrator*, if for any equal-size inputs $X \subseteq [m]$ and outputs $Y \subseteq [n]$, the number of vertex-disjoint paths from X to Y is $|X|$.) Furthermore, the code $C : \mathbb{F}^n \to \mathbb{F}^m$ (encoded by the above circuits) satisfies a distance criterion, stronger than MDS (maximum distance separable) codes, captured by the following definition.

Definition 6. *(Superconcentrator-induced code) $C : \mathbb{F}^n \to \mathbb{F}^m$ is a superconcentrator-induced code if*

$$\text{dist}(C(x), C(y)) \ge m - \text{dist}(x, y) + 1$$

for any distinct $x, y \in \mathbb{F}^m$.

For a linear code $C : \mathbb{F}^n \to \mathbb{F}^m$, it is well known that C is an MDS code, i.e., C satisfies the Singleton bound

$$\text{dist}(C(x), C(y)) \ge m - n + 1 \quad \forall x \ne y \in \mathbb{F}^n,$$

if and only if any n columns of its generator matrix are linearly independent. Similarly, we show that a linear code is a superconcentrator-induced code if and only if every square submatrix of its generator matrix is nonsingular. (Such matrices are sometimes called *totally invertible matrices.*) See the full version for a proof.

We justify the naming "superconcentrator-induced codes" by proving the following two lemmas. The proofs are in the full version of our paper.

Lemma 1. *Any unrestricted arithmetic circuit encoding a superconcentrator-induced code must be a superconcentrator.*

Lemma 2. *Let G by an (n, m)-superconcentrator. Let $C_G : \mathbb{F}^n \to \mathbb{F}^m$ be an arithmetic circuit by replacing each vertex in G with an addition gate, and choosing the coefficient on each edge uniformly and random (over the finite field $\mathbb{F}$). With probability at least $1 - \sum_{i=1}^{n} \binom{n}{i} \binom{m}{i} \frac{di}{|\mathbb{F}|}$, C_G encodes a superconcentrator-induced code.*

6 Conclusion

In this work, we determine, up to an additive constant 2, the minimum depth required to encoding asymptotically good error-correcting codes, i.e., codes with constant rate and constant relative distance, using a linear number of wires. The minimum depth is between $\alpha(n) - 2$ and $\alpha(n)$, where $\alpha(n)$ is a version of the inverse Ackermann function. The upper bound is met by certain binary codes we construct (building on Gál *et al.* [Gal+13] with a few new ingredients) for any constant rate and constant relative distance within the Gilbert-Varshamov bounds. The lower bound applies to any constant rate and constant relative distance. We credit the lower bound to Gál *et al.* [Gal+13], although not explicitly claimed or fully verified in [Gal+13]; because our contribution is a routine checking of detail.

Valiant articulated graph-theoretic arguments for proving circuit lower bounds [Val75,Val76,Val77]. Since then, there have been fruitful results along this line. We show a result in the reverse direction, that is, we prove that any superconcentrator (after being converted to an arithmetic circuit) can encode a good code over a sufficiently large field (exponential in the size of the superconcentrator graph).

References

[Ahl+00] Ahlswede, R., Cai, N., Li, S.Y.R., Yeung, R.W.: Network information flow. IEEE Trans. Inform. Theory **46**(4), 1204–1216 (2000)

[AP94] Alon, N., Pudlák, P.: Superconcentrators of depths 2 and 3; odd levels help (rarely). J. Comput. Syst. Sci. **48**(1), 194–202 (1994)

[BM05] Bazzi, L.M.J., Mitter, S.K.: Endcoding complexity versus minimum distance. IEEE Trans. Inform. Theory **51**(6), 2103–2112 (2005)

[BIL12] Beck, C., Impagliazzo, R., Lovett, S.: Large deviation bounds for decision trees and sampling lower bounds for AC0 circuits. In: 2012 IEEE 53rd Annual Symposium on Foundations of Computer Science (FOCS). pp. 101–110. IEEE (2012)

[Bop97] Boppana, R.B.: The average sensitivity of bounded-depth circuits. Inform. Process. Lett. **63**(5), 257–261 (1997)

[CT19] Chen, L., Tell, R.: Bootstrapping results for threshold circuits "just beyond" known lower bounds. In: Proceedings of the 51st Annual ACM SIGACT Symposium on Theory of Computing, pp. 34–41 (2019)

[CKL13] Cheung, H.Y., Kwok, T.C., Lau, L.C.: Fast matrix rank algorithms and applications. J. ACM (JACM) **60**(5), 1–25 (2013)

[DGP73] Dobrushin, R.L., Gelfand, S.I., Pinsker, M.S.: On complexity of coding. In: Proceedings of 2nd International Symposium on Information Theory, pp. 174–184 (1973)

[Dol+83] Dolev, D., Dwork, C., Pippenger, N., Wigderson, A.: Superconcentrators, generalizers and generalized connectors with limited depth. In: Johnson, D.S., et al. (eds.) Proceedings of the 15th Annual ACM Symposium on Theory of Computing, 25–27 April 1983, Boston, Massachusetts, USA, pp. 42–51. ACM (1983)

[DI14] Druk, E., Ishai, Y.: Linear-time encodable codes meeting the gilbert-varshamov bound and their cryptographic applications. In: Proceedings of the 5th Conference on Innovations in Theoretical Computer Science, pp. 169–182 (2014)

[Gal+13] Gal, A., Hansen, K.A., Koucky, M., Pudlak, P., Viola, E.: Tight bounds on computing error-correcting codes by bounded-depth circuits with arbitrary gates. IEEE Trans. Inform. Theory **59**(10), 6611–6627 (2013)

[Gol+21] Golovnev, A., Lee, J., Setty, S., Thaler, J., SWahby, R.: Brakedown: linear-time and post-quantum SNARKs for R1CS. Cryptology ePrint Archive (2021)

[GRS12] Guruswami, V., Rudra, A., Sudan, M.: Essential coding theory. http://www.cse.buffalo.edu/atri/courses/coding-theory/book2.1 (2012)

[Ish+08] Ishai, Y., Kushilevitz, E., Ostrovsky, R., Sahai, A.: Cryptography with constant computational overhead. In: Proceedings of the Fortieth Annual ACM symposium on Theory of Computing, pp. 433–442 (2008)

[LYC03] Li, S.Y.R., Yeung, R.W., Cai, N.: Linear network coding. IEEE Trans. Inform. Theory **49**(2), 371–381 (2003)

[Li17] Li, Y.: Some Results in Low-Depth Circuit Complexity. The University of Chicago (2017)

[Li23] Li, Y.: Secret Sharing on Superconcentrator. arXiv preprint arXiv:2302.04482 (2023)

[LX04] Ling, S., Xing, C.: Coding theory: a first course. Cambridge University Press (2004)

[Lov18] Lovett, S.: MDS matrices over small fields: a proof of the GM-MDS conjecture. In: 2018 IEEE 59th Annual Symposium on Foundations of Computer Science (FOCS), pp. 194–199. IEEE (2018)

[LV11] Lovett, S., Viola, E.: Bounded-depth circuits cannot sample good codes. In: 2011 IEEE 26th Annual Conference on Computational Complexity (CCC), pp. 243–251. IEEE (2011)

[Pud94] Pudlak, P.: Communication in bounded depth circuits. Combinatorica **14**(2), 203–216 (1994)

[RS03] Raz, R., Shpilka, A.: Lower bounds for matrix product in bounded depth circuits with arbitrary gates. SIAM J. Comput. **32**(2), 488–513 (2003)

[SS96] Sipser, M., Spielman, D.A.: Expander codes. IEEE Trans. Inform. Theory **42**(6), 1710–1722 (1996)

[Spi96] Spielman, D.A.: Linear-time encodable and decodable errorcorrecting codes. IEEE Trans. Inform. Theory **42**(6), 1723–1731 (1996)

[Tar75] Tarjan, R.E.: Efficiency of a good but not linear set union algorithm. J. ACM (JACM) **22**(2), 215–225 (1975)

[Val75] Valiant, L.G.: On non-linear lower bounds in computational complexity. In: Proceedings of the Seventh Annual ACM Symposium on Theory of Computing, pp. 45–53 (1975)

[Val76] Valiant, L.G.: Graph-theoretic properties in computational complexity. J. Comput. Syst. Sci. **13**(3), 278–285 (1976)

[Val77] Valiant, L.G.: Graph-theoretic arguments in low-level complexity. In: 6th Symposium on Mathematical Foundations of Computer Science 1977, Tatranska Lomnica, Czechoslovakia, 5–9 September 1977, Proceedings, pp. 162–176 (1977)

[Vio19] Viola, E.: Lower bounds for data structures with space close to maximum imply circuit lower bounds". Theory Comput. **15**(1), 1–9 (2019)

Approval-Based Participatory Budgeting with Donations

Shiwen Wang[1], Chenhao Wang[1,2(✉)], Tian Wang[1,2], and Weijia Jia[1,2]

[1] BNU-HKBU United International College, Zhuhai, Guangdong, China
chenhwang@bnu.edu.cn
[2] Advanced Institute of Natural Sciences, Beijing Normal University,
Zhuhai, Guangdong, China

Abstract. Participatory budgeting (PB) is a democratic process that allows voters to directly participate in the decision-making process regarding budget spending. The process typically involves presenting voters with a range of proposed projects, and the goal is to select a subset of projects that will be funded. We explore the inclusion of donations in approval-based PB. By allowing voters to pledge donations to projects, the total available budget can increase, however, the concern arises that wealthier donors may wield disproportionate influence. Addressing this, we consider three broad classes of aggregation rules and examine whether they satisfy crucial desideratum in PB with donations.

Keywords: participatory budgeting · donation · aggregation rule

1 Introduction

Participatory budgeting (PB) is a democratic process that allows voters (or citizens, agents) to have a direct say in how public funds are allocated and spent within their communities [3,8,19]. It is a form of participatory democracy that involves community members in decision-making regarding the distribution of public resources. In a typical setting, the voters are presented a range of proposed projects, such as constructing a library or a park, and are asked to vote on these options. The goal is to select a subset of projects, known as a *bundle*, that will be funded. To ensure feasibility, the PB process considers the available budget as a constraint, and the total cost of the selected bundle should not exceed the budget.

Recently, Chen et al. [7] proposed an additional element for the PB process: the inclusion of *donations*. In this model, voters have the option to pledge financial contributions to the projects they support. If a project is selected, the donations pledged towards it are collected, and the remaining cost is covered by the public budget. This approach allows projects to be funded with a reduced impact on the public budget.

Incorporating donations into PB referenda offers several advantages. With an increased total available budget, it becomes possible to achieve a higher overall

W. Wu and G. Tong (Eds.): COCOON 2023, LNCS 14423, pp. 404–416, 2024.
https://doi.org/10.1007/978-3-031-49193-1_31

satisfaction by funding more projects. Additionally, voters who have a strong preference for a particular project can support it financially, thereby increasing its chances of being funded. However, allowing donations introduces a significant risk. Wealthier voters may be able to contribute more money, granting them a potentially disproportionate influence over the PB process. This raises concerns about fairness and equity. Thus, in the paper, the main goal is to explore whether it is possible to include donations in PB while mitigating this risk and still reaping the advantages mentioned earlier.

We consider approval preferences of agents in this paper, while Chen et al. [7] studied general preferences. In approval-based PB, voters specify subsets of the projects which they approve of, and the agents can only make donations to those projects they approve of.

We study three classes of normal PB aggregations rules (namely, $\mathcal{T}, \mathcal{G}, \mathcal{P}$) to tackle donations, by reducing the cost of a project by the amount of donations pledged to the project. $\mathcal{T}$ rules refer to global optimization, and select a feasible bundle with maximum score in total. $\mathcal{G}$ rules refer to greedy rules, which adds a project with the maximum marginal score in each step, until up to the budget. $\mathcal{P}$ rules refer to proportional greedy rules, which is similar to $\mathcal{G}$ except that it adds a project with the maximum marginal density (the score divided by cost). In addition, we consider two classes of aggregations rules proposed by Chen et al. [7] for the setting with donations, namely, Sequential-R and Pareto-R, where R can be any normal rule $\mathcal{T}, \mathcal{G}$ or $\mathcal{P}$. The first step of both approaches is to use a normal rule R to obtain a bundle without taking any donations into consideration. Then, Sequential-R uses the budget saved in the first round due to the donations and runs R in a second round, and so on. Pareto-R selects a bundle with maximum social welfare among the bundles that Pareto-dominate the one selected in the first step, taking donations into account.

Our Results. In this paper, following Chen et al. [7], we address the concern that wealthier voters may have additional power to influence the outcome, and consider four desiderata as crucial for aggregation rules in PB with donations: donation-no-harm (D1), donation-project-monotonicity (D2), donation-welfare-monotonicity (D3) and donation-voter-monotonicity (D4). D1 means that no voter may become less satisfied with the outcome than in a process without donations. D2 means that increasing the donations to a winning project will not make it lose. D3 means that increasing the donations to a project will not hurt the social welfare. D4 means that a voter will not be worse off if she donates money for a project than if she donates no money for that project.

In our axiomatic analysis, we find that in the generalized approach, D1 is not satisfied under any of the three rules $\mathcal{T}$, $\mathcal{G}$ and $\mathcal{P}$, but is always satisfied under Sequential and Pareto rules. D2 is always satisfied for all rules we consider. D3 is satisfied by $\mathcal{T}$ rules, Pareto-$\mathcal{T}$ rules and Pareto-$\mathcal{G}$ rules. Regarding D4, whether these rules satisfy it depends on the utility functions of voters and the scoring functions. We defer a summary of results to Tables 7, 8 and 9 in the last section of this work, because some necessary notations will be introduced later.

406 S. Wang et al.

We note that since Chen et al. [7] study a more general preferences than ours, their positive results straightforwardly apply to our setting. (For example, if a rule R is proven to satisfy property D2 in [7], then it also satisfies D2 in our setting.) But their negative results (e.g., a rule does not satisfy some property) may not hold for our setting. Moreover, we study a new class of rules $\mathcal{P}$ that is not considered in [7].

Organization. Section 2 presents the preliminaries. Section 3 considers the three classes of PB rules $\mathcal{T}, \mathcal{G}, \mathcal{P}$. Sections 4 and 5 consider the sequential rules and Pareto rules, respectively.

Related Works. Participatory budgeting has garnered significant attention within the realm of computational social choice. It is one of the most successful democratic innovations in recent years [21]. Following its initial implementation in Porto Alegre, Brazil, in 1989, PB has become a widely adopted practice around the world [8,9]. Researchers have explored various approaches in participatory budgeting, including ordinal-based budgeting methods where voters rank items [1,14,20], utility-based budgeting methods where voters assign numerical utilities to items [4,11], and approval-based budgeting methods where voters approve a set of items [2,13], considered in this paper. Other notable works in this area include [5,6,10,12,15–18].

Very recently, Rey and Maly [19] presented a comprehensive overview of the state of the research on PB, including both a general overview of the main research questions that are being investigated, and formal and unified definitions of the most important technical concepts from the literature.

The most related work is by Chen et al. [7], who initialized the study of PB with donations, where citizens may provide additional money to projects they want to see funded. They proposed two budgeting methods (Sequential-R and Pareto-R) in the donation setting under diversity constraints and analyze their axiomatic properties. They further investigated the computational complexity of determining the outcome of a PB process with donations and of finding a citizen's optimal donation strategy.

2 Preliminaries

An *instance* or a *budgeting scenario* of PB (without donations) is a tuple $E = (A, V, c, B)$, where $A = \{p_1, p_2, \ldots, p_m\}$ is a set of m projects, $V = \{1, 2, \ldots, n\}$ is a set of voters/agents, $c : A \to \mathbb{R}$ is a *cost function* so that the *cost* of a project $p \in A$ is $c(p)$, and $B \in \mathbb{R}$ is the budget limit. Each voter $i \in V$ has a dichotomous preference over the projects, i.e., i specifies an approval set $A_i \subseteq A$ that contains those projects she approves. If $p_j \in A_i$, then we say voter i has a satisfaction $sat_i(j) = 1$, otherwise $sat_i(j) = 0$. Given an instance E, a solution is a *bundle* of projects $S \subseteq A$ satisfying the budget constraint $c(S) := \sum_{p_j \in S} c(p_j) \le B$. Given a solution S, we say that a project p is a winner or is funded if $p \in S$. An aggregation rule (or a rule, for short) is a function R that maps an instance E to a feasible bundle $R(E) \subseteq A$.

In PB with donations, each voter $i \in V$ may additionally have a donation b_{ij} to each project $p_j \in A$, which indicates how much money she is willing to donate if project p_j is selected. The voter is only willing to donate those projects she approves, i.e., $b_{ij} > 0$ only if $p_j \subseteq A_i$. Denote the donation vector of voter i by $\mathbf{b}_i = (b_{i1}, \ldots, b_{im})$, and $\mathbf{b} = (b_i)_{i \in V}$. Such donations can decrease the cost of selected projects. Hence, the budget constraint is relaxed to

$$\sum_{p_j \in S} \max\{0, c(p_j) - \sum_{i \in V} b_{ij}\} \leq B.$$

We define two *utility functions* μ, which lift the approvals of voters to utilities over every bundle $S \subseteq A$. (1) $\mu_i^+(S) := |A_i \cap S|$ is the number of winners approved by voter i. (2) $\mu_i^{max}(S) = 0$ if $S \cap A_i = \varnothing$, and $\mu_i^{max}(S) = 1$ otherwise. Then, a scoring function *score* computes a number indicating the overall utilities of the voters towards S. We consider two types of scoring functions: the sum scoring functions and the min scoring functions. Precisely, for each $\star \in \{max, +\}$, define

$$score_{\Sigma}^{\star}(S) = \sum_{i \in V} \mu_i^{\star}(S), \quad score_{min}^{\star}(S) = \min_{i \in V} \mu_i^{\star}(S).$$

When two bundles S and S' have the same *score* value, then we compare their utility vectors lexicographically. We look at 12 rules based on global optimization, greedy, and proportional greedy, respectively. Let $\star \in \{max, +\}$ and $\diamond \in \{\Sigma, min\}$.

- Global optimization. Rule $R_{\diamond}^{\star}$ selects a feasible bundle S with maximum $score_{\diamond}^{\star}(S)$. Define $\mathcal{T} := \{\mathcal{T}_{\Sigma}^+, \mathcal{T}_{min}^+, \mathcal{T}_{\Sigma}^{max}, \mathcal{T}_{min}^{max}\}$.
- Greedy. Rule $G_{\diamond}^{\star}$ iteratively adds a feasible project p to the winning bundle C' that maximizes the *marginal gain* $score_{\diamond}^{\star}(C' \cup \{p\})$ at each step. Define $\mathcal{G} := \{\mathcal{G}_{\Sigma}^+, \mathcal{G}_{min}^+, \mathcal{G}_{\Sigma}^{max}, \mathcal{G}_{min}^{max}\}$
- Proportional greedy. Rule $P_{\diamond}^{\star}$ iteratively adds a feasible project p_j to the winning bundle C' that maximizes the *marginal density* $\frac{score_{\diamond}^{\star}(C' \cup \{p_j\})}{\max\{0, c(p_j) - \sum_{i \in V} b_{ij}\}}$ at each step. Define $\mathcal{P} := \{\mathcal{P}_{\Sigma}^+, \mathcal{P}_{min}^+, \mathcal{P}_{\Sigma}^{max}, \mathcal{P}_{min}^{max}\}$.

These rules can simply handle the setting with donations, in a way that allowing donation is equivalent to reducing the cost of the respective project.

Axioms. We consider four axioms that are crucial properties for rules in PB with donations [7]. Let $E = (A, V, c, B, \mathbf{b})$ be an instance with donations, and $E_0 = (A, V, c, B, \mathbf{0})$ be the instance derived from E where all donations are zero.

Definition 1 (Donation-no-harm, D1). *An aggregation rule R is donation-no-harm if for each PB instance E and each voter $i \in V$ it holds that $\mu_i(R(E)) \geq \mu_i(R(E_0))$.*

Definition 2 (Donation-project-monotonicity, D2). *An aggregation rule R is donation-project-monotone if for each PB instance E, each voter $i \in V$ and each donation $\mathbf{b}_i'$ with $b_{ij} < b_{ij}'$ and $b_{ik} = b_{ik}' \; \forall k \neq j$, it holds that if $j \in R(E)$ then $j \in R(E - \mathbf{b}_i + \mathbf{b}_i')$, where $E - \mathbf{b}_i + \mathbf{b}_i'$ is the instance that replaces voter i's donation vector $\mathbf{b}_i$ by $\mathbf{b}_i'$ in the instance E.*

Definition 3 (Donation-welfare-monotonicity, D3). *An aggregation rule R is donation-welfare-monotone if for each PB instance E, each voter $i \in V$ and each donation $\mathbf{b}'_i$ with $b_{ij} < b'_{ij}$ and $b_{ik} = b'_{ik}$ $\forall k \neq j$, it holds that $score(R(E)) \leq score(R(E - \mathbf{b}_i + \mathbf{b}'_i))$.*

Definition 4 (Donation-voter-monotonicity, D4). *An aggregation rule R is donation-voter-monotonicity if for each PB instance E, each voter $i \in V$ and each donation $\mathbf{b}'_i$ with $b'_{ij} = 0$ and $b_{ik} = b'_{ik}$ $\forall k \neq j$, it holds that $\mu_i(R(E)) \geq \mu_i(R(E - \mathbf{b}_i + \mathbf{b}'_i))$.*

3 Normal PB Rules

First, we look at the three classes of normal rules $\mathcal{T}, \mathcal{G}, \mathcal{P}$, and examine whether they satisfy the four desiderata.

Proposition 1. *All rules in $\mathcal{T}$, $\mathcal{G}$ and $\mathcal{P}$ do not satisfy D1.*

Proof. Recall that D1 means that no voter will receive a lower utility by allowing donations than that without donations. We can always construct the following instance. There are three projects $\{p_1, p_2, p_3\}$, where the cost of p_1 exceeds the total budget, and the sum of the costs of p_2 and p_3 is below budget. When donation is not allowed, the winning bundle can only be $\{p_2, p_3\}$. If the score of p_1 is greater than $\{p_2, p_3\}$, and the cost of p_1 reduced by the donations is equal to the budget, then p_1 will be the unique winning project after donations. Suppose that there is a voter who approves p_2 and p_3 but not p_1. In this case, the satisfaction score of this voter will decrease by allowing donations.

Precisely, for every rules $R \in \{\mathcal{T}_\Sigma^+, \mathcal{G}_\Sigma^+, \mathcal{P}_\Sigma^+\}$, consider an instance E with three projects $\{p_1, p_2, p_3\}$ and 10 voters. The budget is $B = 5$, and the project cost is $c(p_1) = 7$, $c(p_2) = 2$ and $c(p_3) = 3$, respectively. Voter 1 approves $A_1 = \{p_2, p_3\}$, and other 9 voters approve $\{p_1\}$. Voter 2 has a donation $b_{22} = 2$ to project p_1. The rule R without donations will return $\{p_2, p_3\}$, and the utility of voter 1 is 2. However, R with donations will return $\{p_1\}$, and the utility of voter 1 decreases to 0. It indicates that donation harms voter 1, and thus R does not satisfy D1. For those rules not in $\{\mathcal{T}_\Sigma^+, \mathcal{G}_\Sigma^+, \mathcal{P}_\Sigma^+\}$, we provide counterexamples in Appendix. □

Proposition 2. *All rules in $\mathcal{T}$, $\mathcal{G}$ and $\mathcal{P}$ satisfy D2.*

Proof. Recall that D2 means that increasing the donations to a winning project will not make it lose. By [7], $\mathcal{T}$ and $\mathcal{G}$ satisfies D2 even for general preferences. It remains to consider $\mathcal{P}$ rules.

Based on the definition of the proportional greedy rule, it selects a project with the largest marginal density at each time, which is defined as the marginal score divided by its cost deducted by the donations. Consider any winning project that is selected in time t. When increasing the donation to it, the marginal density increases, implying that this project would either be still selected in time t, or be selected even earlier. Therefore, the proportional greedy rules can always satisfy D2. □

Proposition 3. *All $\mathcal{T}$ rules satisfy D3, and $\mathcal{G}$ rules and $\mathcal{P}$ rules do not satisfy D3.*

Proof. Recall that D3 can be satisfied if increasing the donation to a project would not hurt the social welfare. By [7], $\mathcal{T}$ rules satisfy D3 even for general preferences, and thus we only need to consider $\mathcal{G}$ and $\mathcal{P}$.

For $\mathcal{G}$ and $\mathcal{P}$ rules, we can always construct an instance, where the winning bundle without donations is $\{p_1, p_2\}$. With donations, p_3's cost deducted by donations is slightly less than the budget and no other project fits, while its original cost exceeds the budget. As long as the marginal score (or marginal density) derived by p_3 is greater than p_1 and p_2, then p_3 would be selected as the unique winner. However, it may be the case that the social welfare of p_3 is less than $\{p_1, p_2\}$, and thus D3 is not satisfied.

For example, for rules $\mathcal{G}_{\Sigma}^{+}$ and $\mathcal{P}_{\Sigma}^{+}$, consider an instance with three projects p_1, p_2 and p_3. The costs are $c(p_1) = 2.5$, $c(p_2) = 2.5$ and $c(p_3) = 6$, and the budget is $B = 5$. Voter 1 approves $\{p_1, p_2, p_3\}$, voter 2 approves $\{p_1, p_2, p_3\}$ with a donation $b_{23} = 3$ to p_3, and voter 3 approves $\{p_3\}$ only. The rule R without donations will return $\{p_1, p_2\}$, and the social welfare is 4. However, $\mathcal{G}_{\Sigma}^{+}$ with donations will return $\{p_3\}$ only, and the social welfare decreases to 3. It indicates that increasing the donation to project p_3 harms the social welfare, and thus R does not satisfy D3. For other rules, counterexamples can be found in Appendix. □

Proposition 4. *For $\mathcal{T}$, $\mathcal{G}$ and $\mathcal{P}$ rules, the rules w.r.t. utility function u_i^{max} satisfy D4, and the rules w.r.t. u_i^{+} do not satisfy D4.*

Proof. Recall that D4 means that a voter will not be worse off if she donates money for a project than if she donates no money for that project.

For utility function u_i^{max}, suppose voter i's utility is 1 when she donates no money to project p which she approves, implying that there exists a winning project p' approved by i. If voter i denotes some money to p, then either p' is still a winner, or p becomes a winner. In both cases, the utility of voter i is 1, and thus she cannot be worse off by donating money to p. Thus, the rules w.r.t. utility function u_i^{max} satisfy D4.

For the rules w.r.t. utility function u_i^{+}, using the same analysis in the proof of Proposition 1, we can show that it does not satisfy the axiom D4. For example, consider an instance with three projects $\{p_1, p_2, p_3\}$ and a budget of $B = 5$. The costs are $c(p_1) = 6$, $c(p_2) = 2.5$ and $c(p_3) = 2.5$. Voter 1 approves $\{p_1\}$, voter 2 approves $\{p_1, p_2, p_3\}$ with a donation $b_{21} = 3$ to p_1, and voter 3 approves $\{p_1\}$. If voter 2 does not donate money to project p_1, then the winning bundle is $\{p_2, p_3\}$ and her utility is 2. However, if voter 2 donates $b_{21} = 3$ to p_1, the winning bundle becomes $\{p_1\}$, and her utility decreases to 1. It indicates that the donation to p_1 makes voter 2 less satisfied than not donating to p_1, and thus the rules do not satisfy D4. □

4   Sequential Rules

In this section, we consider the Sequential rules. For a normal rule $R \in \mathcal{T} \cup \mathcal{G} \cup \mathcal{P}$, Sequential-$R$, or simply S-R, is defined as follows [7]. Given instance E, it first applies R on E_0 (the instance without donations) to find out an allocation A_0. If afterwards some budget is left due to donations (that is, all the donations of the funded projects will be collected and included in the remaining budget), then R is applied again with the remaining budget but still without donations; repeat this step until no more project can be added. In the last step, R is applied directly with donations, thus guaranteeing an exhaustive bundle.

Next, we examined whether S-R rules satisfy the four desiderata. By [7], for any normal rule $R \in \mathcal{T} \cup \mathcal{G}$, the S-R rules satisfy D1 and D2 even for general utility-based setting. For the S-$\mathcal{P}$ rules, it is easy to see that in sequential selections rules, the projects selected in the first round (without any donations) will not be removed in the later rounds, and thus the winner bundle by allowing donations contains the winner bundle without donations. implying that D1 is satisfied. Moreover, by the definition of proportional greedy rules, a winning project remains a winner if some voter increases the donation to it, and thus D2 is also satisfied.

We only need to focus on desiderata D3 and D4.

Table 1. Both S-$\mathcal{T}_{\Sigma}^{+}$ and S-$\mathcal{G}_{\Sigma}^{+}$ do not satisfy D3 or D4.

	$c(\cdot)$	sat_1	sat_2	sat_3	sat_4	b_1	b_3
p_1	5	1	1	1	0	$0 \rightarrow 1$	3
p_2	4	0	1	1	1	0	0
p_3	3	1	0	1	0	0	3
p_4	3	1	1	0	0	0	3

Proposition 5. *Rule S-$\mathcal{T}_{\Sigma}^{+}$ and S-$\mathcal{G}_{\Sigma}^{+}$ do not satisfy D3, D4.*

Proof. Consider the instance in Table 1 with budget $B = 5$. When voter 1 donates $b_{11} = 0$ to project p_1, both S-$\mathcal{T}_{\Sigma}^{+}$ and S-$\mathcal{G}_{\Sigma}^{+}$ rules will return $\{p_1, p_3, p_4\}$. Then, the social welfare is 7 and the utility of voter 1 is 3. However, when voter 1 increases the donation to project p_1 from 0 to $b'_{11} = 1$, both rules will return $\{p_1, p_2\}$. Then the social welfare decreases to 6, and the utility of voter 1 decreases to 1. It indicates that S-$\mathcal{T}_{\Sigma}^{+}$ and S-$\mathcal{G}_{\Sigma}^{+}$ do not satisfy D3, D4.   □

Proposition 6. *Rule S-$\mathcal{P}_{\Sigma}^{+}$ does not satisfy D3 and D4.*

Proof. Consider the instance in Table 2 with budget $B = 5$. When voter 1 donates 2 money to project p_1, the rule S-$\mathcal{P}_{\Sigma}^{+}$ returns $\{p_1, p_3, p_4\}$. The social welfare is 10 and the utility of voter 1 is 3. If voter 1 increases the donation to

Table 2. $S\text{-}\mathcal{P}_\Sigma^+$ does not satisfy D3, D4

	$c(\cdot)$	sat_1	sat_2	sat_3	sat_4	sat_5	b_1
p_1	4	1	1	1	1	1	$2 \to 4$
p_2	3.5	1	1	1	1	0	0
p_3	3	1	1	1	0	0	2
p_4	2	1	1	0	0	0	2

project p_1 from 2 to 4, $S\text{-}\mathcal{P}_\Sigma^+$ would return a winning bundle $\{p_1, p_2\}$. Then the social welfare decreases to 9 and the utility of voter 1 decreases to 2. It indicates that increasing donations harms the social welfare and the voter's utility, $S\text{-}\mathcal{P}_\Sigma^+$ does not satisfy D3, D4. □

Table 3. Both $S\text{-}\mathcal{T}_{min}^+$ and $S\text{-}\mathcal{G}_{min}^+$ do not satisfy D3, D4

	$c(\cdot)$	sat_1	sat_2	sat_3	sat_4	sat_5	b_1	b_5
p_1	5	1	1	0	1	1	$0 \to 1$	3
p_2	4	1	1	0	0	1	0	0
p_3	3	1	0	1	0	0	0	3
p_4	3	1	0	0	0	1	0	3

Proposition 7. *Rules $S\text{-}\mathcal{T}_{min}^+$ and $S\text{-}\mathcal{G}_{min}^+$ do not satisfy D3 and D4.*

Proof. Consider the instance in Table 3 with budget $B = 5$. When voter 1 donates 0 money to project p_1, both rules $S\text{-}\mathcal{T}_{min}^+$ and $S\text{-}\mathcal{G}_{min}^+$ return $\{p_1, p_3, p_4\}$. The social welfare is 1 and the utility of voter 1 is 3. If voter 1 increases the donation to project 1 from 0 to 1, both rules would return $\{p_1, p_2\}$. Then the social welfare decreases to 0 and the utility of voter 1 decreases to 2. It indicates that increasing the donations harms both the social welfare and the voter's utility. Hence, $S\text{-}\mathcal{T}_{min}^+$ and $S\text{-}\mathcal{G}_{min}^+$ do not satisfy D3, D4. □

Table 4. $S\text{-}\mathcal{T}_\Sigma^{max}$, $S\text{-}\mathcal{T}_{min}^{max}$, $S\text{-}\mathcal{G}_\Sigma^{max}$ and $S\text{-}\mathcal{G}_{min}^{max}$ do not satisfy D3

	$c(\cdot)$	sat_1	sat_2	sat_3	sat_4	b_1
p_1	5	1	1	1	0	$3 \to 4$
p_2	4	1	1	0	0	0
p_3	3	0	0	0	1	0

Proposition 8. *Rules $S\text{-}\mathcal{T}_\Sigma^{max}$, $S\text{-}\mathcal{T}_{min}^{max}$, $S\text{-}\mathcal{G}_\Sigma^{max}$ and $S\text{-}\mathcal{G}_{min}^{max}$ satisfy D4, but do not satisfy D3.*

Proof. Regarding D3, consider the instance in Table 4 with budget $B = 5$. When voter 1 donates money 3 to project p_1, rules $S\text{-}\mathcal{T}_\Sigma^{max}$, $S\text{-}\mathcal{G}_\Sigma^{max}$, $S\text{-}\mathcal{T}_{min}^{max}$ and $S\text{-}\mathcal{G}_{min}^{max}$ return $\{p_1, p_3\}$. Then, the social welfare under rule $S\text{-}\mathcal{T}_\Sigma^{max}$ and $S\text{-}\mathcal{G}_\Sigma^{max}$ is 4 and the social welfare under rule $S\text{-}\mathcal{T}_{min}^{max}$ and $S\text{-}\mathcal{G}_{min}^{max}$ is 1. If voter 1 increases the donation to project p_1 from 3 to 4, all the above four rules will return $\{p_1, p_2\}$. Then, the social welfare under rule $S\text{-}\mathcal{T}_\Sigma^{max}$ and $S\text{-}\mathcal{G}_\Sigma^{max}$ is 3 and the social welfare under rule $S\text{-}\mathcal{T}_{min}^{max}$ and $S\text{-}\mathcal{G}_{min}^{max}$ is 0. It indicates that increasing donations harms the social welfare, and thus the rules do not satisfy D3.

Regarding D4, it is known that the utility for a certain voter in the above four rules can only be 1 or 0. Suppose that a voter is going to increase her donation to a project p that she approves. If the project p is in the winning bundle before increasing donation, then it will be still in the winning bundle after increasing donation, and the utility of this voter is 1. If p is not in the winning bundle before increasing donation, then the output will not change. Thus, increasing the donation to a project that she did not donate will not harm her utility, and thus D4 is satisfied. □

Proposition 9. *Rule $S\text{-}\mathcal{P}_{min}^+$ does not satisfy D3 and D4. $S\text{-}\mathcal{P}_{min}^{max}$ and $S\text{-}\mathcal{P}_\Sigma^{max}$ satisfy D4, and do not satisfy D3.*

5   Pareto Rules

In this section, we consider the Pareto rules. For a normal rule $R \in \mathcal{T} \cup \mathcal{G} \cup \mathcal{P}$, Pareto-$R$, or simply $P\text{-}R$, is defined as follows. Given instance E, the first step is to apply R to the instance E_0 (without considering all donations). Based on the winning bundle A_0, it returns a bundle with maximum social welfare among all bundles that Pareto-dominate A_0, taking donations into account.[1] Further, we consider a process of increasing a voter's donation to a project, and denote by A_1 the winning bundle before this process, and by A_2 the winning bundle after this process.

Next, we examine whether $P\text{-}R$ rules satisfy the four desiderata. By [7], for any normal rule $R \in \mathcal{T} \cup \mathcal{G}$, $P\text{-}R$ satisfies D1, D2 and D3 even for general utility-based setting. Thus, we turn to other cases.

Table 5. $P\text{-}\mathcal{T}_\Sigma^+$, $P\text{-}\mathcal{T}_{min}^+$, $P\text{-}\mathcal{G}_\Sigma^+$ and $P\text{-}\mathcal{G}_{min}^+$ do not satisfy D4

	$c(\cdot)$	sat_1	sat_2	sat_3	sat_4	b_1	b_3
p_1	5	1	1	1	0	$0 \to 1$	3
p_2	4	0	1	0	1	0	0
p_3	3	1	0	0	0	0	0

---

[1] We say that a bundle A Pareto-dominates another bundle A', if all voters have a utility under A at least as well as that under A', and at least one voter strictly prefers A to A'.

Proposition 10. *Rule $P\text{-}\mathcal{T}_{\Sigma}^{+}$, $P\text{-}\mathcal{T}_{min}^{+}$, $P\text{-}\mathcal{G}_{\Sigma}^{+}$ and $P\text{-}\mathcal{G}_{min}^{+}$ do not satisfy D4.*

Proof. Regarding D4, consider the instance in Table 5 with budget $B = 5$. When voter 1 does not donate to p_1, rules $P\text{-}\mathcal{T}_{\Sigma}^{+}$, $P\text{-}\mathcal{G}_{\Sigma}^{+}$, $P\text{-}\mathcal{T}_{min}^{+}$ and $P\text{-}\mathcal{G}_{min}^{+}$ return $\{p_1, p_3\}$, and the utility of voter 1 is 2. If voter 1 increases the donation to project p_1 from 0 to 1, then the rules would return $\{p_1, p_2\}$, and the utility of voter 1 decreases to 1. It indicates that increasing donations harms the utility of voter 1, and thus these rules do not satisfy D4. ☐

Proposition 11. *Rules $P\text{-}\mathcal{T}_{\Sigma}^{max}$, $P\text{-}\mathcal{T}_{min}^{max}$, $P\text{-}\mathcal{G}_{\Sigma}^{max}$ and $P\text{-}\mathcal{G}_{min}^{max}$ satisfy D4.*

Proof. If the increased donation is given to a project in A_0 or A_2, the utility of the voter must be 1, and thus the outcome will not be worse off. If the increased donation is given to A_1, the utility of this voter in A_1 is 1. Thus, only if the utility of this voter in A_2 is 1 will A_2 be chosen instead of A_1. Therefore, the utility of the voter who increases donation to a project would not be worse off. It indicates that D4 is satisfied. ☐

Proposition 12. *All P-$\mathcal{P}$ rules can satisfy D1 and D2.*

Proof. By the definition of Pareto rules, a bundle that Pareto-dominates A_0 is chosen with donations. Therefore, the winning bundle will not change if donation is not allowed. It indicates that donation will not make voters less satisfied and Pareto rules can satisfy D1 all the time.

Since the winning bundle under Pareto rules Pareto-dominates A_0, the utility of any voter must be no less than the original one. If the donation is given to the project in the bundle of A_0, then this project must be chosen as well. If the donation is given to A_1 (dominating A_0), then A_2 dominates A_1, and A_2 be selected. At this time, this project with donation must be selected as well. It indicates that project with donation will not lose any more, and Pareto rules always satisfy D2. ☐

Table 6. $P\text{-}\mathcal{P}_{\Sigma}^{+}$ does not satisfy D3, D4

	$c(\cdot)$	sat_1	sat_2	sat_3	sat_4	sat_5	b_1
p_1	4	1	1	1	1	1	$2 \to 4$
p_2	3.5	1	1	1	1	0	0
p_3	3	1	1	1	0	0	2
p_3	2	1	1	0	0	0	2

Proposition 13. *Rule $P\text{-}\mathcal{P}_{\Sigma}^{+}$ does not satisfy D3, D4.*

Proof. Consider the instance in Table 6 with budget $B = 5$. When voter 1 has a donation 2 to project p_1, rule $P\text{-}\mathcal{P}_{\Sigma}^{+}$ returns $\{p_1, p_3, p_4\}$. The social welfare is 10 and the utility of voter 1 is 3. If voter 1 increases the donation to project p_1

from 2 to 4, then $P\text{-}\mathcal{P}_\Sigma^+$ will return $\{p_1, p_2\}$. The social welfare decreases to 9 and the utility of voter 1 decreases to 2. It indicates that donation harms the social welfare and the utility of voter 1, and thus $P\text{-}\mathcal{P}_\Sigma^+$ does not satisfy D3, D4. □

Proposition 14. *Rule $P\text{-}\mathcal{P}_{min}^+$ does not satisfy D3 and D4, and rules $P\text{-}\mathcal{P}_{min}^{max}$ and $P\text{-}\mathcal{P}_\Sigma^{max}$ do not satisfy D3.*

6   Conclusion

We have conducted a comprehensive analysis of various PB aggregation rules for approval-based participatory budgeting with donations pledged by voters, including the global optimization approaches, greedy approaches, proportional greedy approaches, as well as Sequential rules and Pareto rules. All results are presented in Tables 7, 8 and 9. Basically, D1 and D2 can always be satisfied through Sequential and Pareto rules, while D1 can not be satisfied by any normal rule. Further, when the utility function is μ_i^{max}, D4 is satisfied by all rules.

Regarding the question whether it is possible to include donations in approved-based PB while maintaining the fairness and equity, our results show that it depends on the utility functions of voters and the scoring functions with respect to a bundle. For example, when the utility function is μ_i^{max}, the Pareto-$\mathcal{T}$ and the Pareto-$\mathcal{G}$ rules satisfy all of the four properties, while they are not compatible for other setting. For future work, it would be interesting to explore more aggregations rules and other properties for the PB with donations.

Table 7. A summary of the results on normal PB rules.

	$\mathcal{T}_\Sigma^+$	$\mathcal{T}_{min}^+$	$\mathcal{T}_\Sigma^{max}$	$\mathcal{T}_{min}^{max}$	$\mathcal{G}_\Sigma^+$	$\mathcal{G}_{min}^+$	$\mathcal{G}_\Sigma^{max}$	$\mathcal{G}_{min}^{max}$	$\mathcal{P}_\Sigma^+$	$\mathcal{P}_{min}^+$	$\mathcal{P}_\Sigma^{max}$	$\mathcal{P}_{min}^{max}$
D1	×	×	×	×	×	×	×	×	×	×	×	×
D2	✓	✓	✓	✓	✓	✓	✓	✓	✓	✓	✓	✓
D3	✓	✓	✓	✓	×	×	×	×	×	×	×	×
D4	×	×	✓	✓	×	×	✓	✓	×	×	✓	✓

Table 8. A summary of the results on Sequential rules.

	$\mathcal{T}_\Sigma^+$	$\mathcal{T}_{min}^+$	$\mathcal{T}_\Sigma^{max}$	$\mathcal{T}_{min}^{max}$	$\mathcal{G}_\Sigma^+$	$\mathcal{G}_{min}^+$	$\mathcal{G}_\Sigma^{max}$	$\mathcal{G}_{min}^{max}$	$\mathcal{P}_\Sigma^+$	$\mathcal{P}_{min}^+$	$\mathcal{P}_\Sigma^{max}$	$\mathcal{P}_{min}^{max}$
D1	✓	✓	✓	✓	✓	✓	✓	✓	✓	✓	✓	✓
D2	✓	✓	✓	✓	✓	✓	✓	✓	✓	✓	✓	✓
D3	×	×	×	×	×	×	×	×	×	×	×	×
D4	×	×	✓	✓	×	×	✓	✓	×	×	✓	✓

Table 9. A summary of the results on Pareto rules.

	$\mathcal{T}_\Sigma^+$	$\mathcal{T}_{min}^+$	$\mathcal{T}_\Sigma^{max}$	$\mathcal{T}_{min}^{max}$	$\mathcal{G}_\Sigma^+$	$\mathcal{G}_{min}^+$	$\mathcal{G}_\Sigma^{max}$	$\mathcal{G}_{min}^{max}$	$\mathcal{P}_\Sigma^+$	$\mathcal{P}_{min}^+$	$\mathcal{P}_\Sigma^{max}$	$\mathcal{P}_{min}^{max}$
D1	✓	✓	✓	✓	✓	✓	✓	✓	✓	✓	✓	✓
D2	✓	✓	✓	✓	✓	✓	✓	✓	✓	✓	✓	✓
D3	✓	✓	✓	✓	✓	✓	✓	✓	✗	✗	✗	✗
D4	✗	✗	✓	✓	✗	✗	✓	✓	✗	✗	✓	✓

References

1. Aziz, H., Lee, B.E.: Proportionally representative participatory budgeting with ordinal preferences. In: Proceedings of the AAAI Conference on Artificial Intelligence, vol. 35, pp. 5110–5118 (2021)
2. Aziz, H., Lee, B.E., Talmon, N.: Proportionally representative participatory budgeting: axioms and algorithms. In: 17th International Conference on Autonomous Agents and Multiagent Systems, AAMAS 2018, pp. 23–31 (2018)
3. Aziz, H., Shah, N.: Participatory budgeting: models and approaches. In: Pathways Between Social Science and Computational Social Science: Theories, Methods, and Interpretations, pp. 215–236 (2021)
4. Benade, G., Nath, S., Procaccia, A.D., Shah, N.: Preference elicitation for participatory budgeting. Manag. Sci. **67**(5), 2813–2827 (2021)
5. Cabannes, Y.: Participatory budgeting: a significant contribution to participatory democracy. Environ. Urban. **16**(1), 27–46 (2004)
6. Cabannes, Y., Lipietz, B.: Revisiting the democratic promise of participatory budgeting in light of competing political, good governance and technocratic logics. Environ. Urban. **30**(1), 67–84 (2018)
7. Chen, J., Lackner, M., Maly, J.: Participatory budgeting with donations and diversity constraints. In: Proceedings of the AAAI Conference on Artificial Intelligence, vol. 36, pp. 9323–9330 (2022)
8. Dias, N.: Hope for democracy: 30 years of participatory budgeting worldwide. Epopee Rec. Officinal Coordination **638** (2018)
9. Dias, N., Enríquez, S., Júlio, S.: The participatory budgeting world atlas. Epopeia and Oficina (2019)
10. Fain, B., Goel, A., Munagala, K.: The core of the participatory budgeting problem. In: Cai, Y., Vetta, A. (eds.) WINE 2016. LNCS, vol. 10123, pp. 384–399. Springer, Heidelberg (2016). https://doi.org/10.1007/978-3-662-54110-4_27
11. Fluschnik, T., Skowron, P., Triphaus, M., Wilker, K.: Fair knapsack. In: Proceedings of the AAAI Conference on Artificial Intelligence, vol. 33, pp. 1941–1948 (2019)
12. Freeman, R., Pennock, D.M., Peters, D., Vaughan, J.W.: Truthful aggregation of budget proposals. In: Proceedings of the 2019 ACM Conference on Economics and Computation, pp. 751–752 (2019)
13. Goel, A., Krishnaswamy, A.K., Sakshuwong, S.: Budget aggregation via knapsack voting: welfare-maximization and strategy-proofness. Collect. Intell. 783–809 (2016)
14. Goel, A., Krishnaswamy, A.K., Sakshuwong, S., Aitamurto, T.: Knapsack voting for participatory budgeting. ACM Trans. Econ. Comput. (TEAC) **7**(2), 1–27 (2019)

15. Jain, P., Talmon, N., Bulteau, L.: Partition aggregation for participatory budgeting. In: Proceedings of the 20th International Conference on Autonomous Agents and MultiAgent Systems, pp. 665–673 (2021)
16. Laruelle, A.: Voting to select projects in participatory budgeting. Eur. J. Oper. Res. **288**(2), 598–604 (2021)
17. Michorzewski, M., Peters, D., Skowron, P.: Price of fairness in budget division and probabilistic social choice. In: Proceedings of the AAAI Conference on Artificial Intelligence, vol. 34, pp. 2184–2191 (2020)
18. Munagala, K., Shen, Y., Wang, K.: Auditing for core stability in participatory budgeting. In: Hansen, K.A., Liu, T.X., Malekian, A. (eds.) WINE 2022. LNCS, vol. 13778, pp. 292–310. Springer, Heidelberg (2022). https://doi.org/10.1007/978-3-031-22832-2_17
19. Rey, S., Maly, J.: The (computational) social choice take on indivisible participatory budgeting. arXiv preprint arXiv:2303.00621 (2023)
20. Shapiro, E., Talmon, N.: A participatory democratic budgeting algorithm. arXiv preprint arXiv:1709.05839 (2017)
21. Wampler, B., McNulty, S., Touchton, M.: Participatory Budgeting in Global Perspective. Oxford University Press, Cambridge (2021)

Author Index

Printed in the United States
by Baker & Taylor Publisher Services